Lecture Notes in Mathematics

Edited by A. Dold and B. Eckmann

961

T0184771

Algebraic Geometry

Proceedings of the International Conference on Algebraic Geometry
Held at La Rábida, Spain, January 7 – 15, 1981

Edited by J. M. Aroca, R. Buchweitz, M. Giusti, and M. Merle

Springer-Verlag
Berlin Heidelberg New York 1982

Editors

José Manuel Aroca
Facultad de Ciencias, Prado de la Magdalena
Valladolid, Spain

Ragnar Buchweitz
Universität Hannover
Welfengarten 1, 3000 Hannover 1, Federal Republic of Germany

Marc Giusti
Michel Merle
Centre de Mathématiques, Ecole Polytechnique
91128 Palaiseau Cedex, France

ISBN 3-540-11969-8 Springer-Verlag Berlin Heidelberg New York
ISBN 0-387-11969-8 Springer-Verlag New York Heidelberg Berlin

This work is subject to copyright. All rights are reserved, whether the whole or part of the material is concerned, specifically those of translation, reprinting, re-use of illustrations, broadcasting, reproduction by photocopying machine or similar means, and storage in data banks. Under § 54 of the German Copyright Law where copies are made for other than private use, a fee is payable to "Verwertungsgesellschaft Wort", Munich.

© by Springer-Verlag Berlin Heidelberg 1982
Printed in Germany

Printing and binding: Beltz Offsetdruck, Hemsbach/Bergstr.
2146/3140-543210

ACKNOWLEDGEMENTS

The International Conference on Algebraic Geometry held at La Rábida University was sponsored by:

- The International Mathematical Union
- The Council for Scientific Research of Spain
- The Universities Complutense de Madrid, Sevilla and Valladolid
- The City of Palos de la Frontera.

Our gratitude toward all of them.

Also we are extremely grateful to all those persons who collaborated, either in the organization, or in the development of the Conference, mainly the Board of Governors of the University of La Rábida, and the staff of the Departments of Algebra of the Universities of Sevilla and Valladolid.

The Organizing Commitee

LIST OF PARTICIPANTS

ABELLANAS, Pedro
Facultad de Matemáticas
Ciudad Universitaria
Madrid-3, SPAIN

ABHYANKAR, S.S.
Division of Mathematical Sciences
Purdue University
West Lafayette, IN 47907, USA

ANGENIOL, Bernard
3, Avenue Jean Jaures
91400, Gometz le Chatel, FRANCE

AROCA, José Manuel
Facultad de Ciencias
Prado de la Magdalena
Valladolid, SPAIN

BRIALES, Emilio
Facultad de Matemáticas
Tarfia s/n
Sevilla-12, SPAIN

BRYLINSKI, Jean Luc
Centre de Mathématiques
Ecole Polytechnique
91128, Palaiseau, FRANCE

BUCHWEITZ, Ragnar
Universität Hannover
Welfengarten 1
D-3000 Hannover 1 GERMANY

CAMPILLO, Antonio
Facultad de Ciencias
Prado de la Magdalena
Valladolid, SPAIN

CANO, Felipe
Facultad de Ciencias
Prado de la Magdalena
Valladolid, SPAIN

CASAS,E.
Facultad de Matemáticas
Universidad Central de Barcelona
Barcelona. SPAIN.

CASTELLANOS, Julio
Facultad de Matemáticas
Ciudad Universitaria
Madrid-3, SPAIN

COSSART,Vincent
Bat 425- Mathématiques
Faculté des Sciences
91405 Orsay. FRANCE

FALTINGS, Kay
Universität Münster
MathematischesInstitut
Roxelerstrasse, 64
44, Munster, GERMANY

FINAT, Javier
Facultad de Ciencias
Prado de la Magdalena
Valladolid, SPAIN

FUERTES, Concepción
Facultad de Matemáticas
Ciudad Universitaria
Madrid-3, SPAIN

GAETA, Federico
Facultad de Matemáticas
Ciudad Universitaria
Madrid-3, SPAIN

GALLIGO, André
Departement de Mathématiques
Parc Valrose
06-Nice, FRANCE

GIUSTI, Marc
Centre de Mathématiques
Ecole Polytechnique
91128 Palaiseau, FRANCE

GONZALEZ, Gerardo
Centre de Mathématiques
Ecole Polytechnique
91128 Palaiseau, FRANCE

GRANGER, Jean-Michel
Departement de Mathématique
Parc Valrose
06034 Nice, FRANCE

GREUEL, Gert-Martin
Sonderforschungbereich Theoretische Mathematik
Universität Bonn
Beringstrasse 4
5300 Bonn 1 , GERMANY

HARTSHORNE, Robin
Department of Mathematics
University of California at Berkeley
Berkeley, CA, 94720, USA

HAUSER, H.
Inst. f. Math.
Univ. Innsbruck
A-6020 Innsbruck. AUSTRIA.

HENRY, Jean Pierre
Centre de Mathématiques
Ecole Polytechnique
91128 Palaiseau, FRANCE

HERMANN, Manfred
MathematischesInstitut
Universität Köln
5000 Köln 41, GERMANY

HERMIDA, José Angel
Facultad de Ciencias
Prado de la Magdalena
Valladolid, SPAIN

HERRERA, Javier
Facultad de Matemáticas
Tarfia s/n
Sevilla-12 , SPAIN

HERZOG, Jurgen
MathematischesInstitut
Universität Regensburg
Regensburg, GERMANY

HIRONAKA, Heisuke
Department of Mathematics
Harvard University
1, Oxford St.
Cambridge, Ma, 02138, USA

HIRSCHOWITZ, André
Departement de Mathématiques
Parc Valrose
06034 Nice-CEDEX. FRANCE

KHALED, A.
Institut de Mathématiques U.S.T.A.
B.P. nº 9 Dar El Beida. ALGERIA.

LAMARI, A.
Cité Anasser 11
B[t] 11A nº9 Kouba. ALGERIA.

LAUDAL, Olav
Matematisk Institutt
Universitelet Oslo.
P.O. BOX 1053 Blindern. Oslo-3. NORWAY

LEJEUNE-JALABERT, Monique
Departement de Mathématiques
Université de Grenoble
BP 116
38402, Saint Martin, FRANCE

LUENGO, Ignacio
Facultad de Matemáticas
Ciudad Universitaria
Madrid-3, SPAIN

MARUYAMA, Masaki
Faculty of Science
Kyoto University
Kyoto 606, JAPAN

MCPHERSON , Robert
I.H.E.S.
Route de Chartres
91440 Bures Sur Ivette, FRANCE

MERLE, Michel
Centre de Mathématiques
Ecole Polytechnique
91128 Palaiseau, FRANCE

ORBANZ, Ulrich
MathematischesInstitut
Universität Köln
5000 Köln 41, GERMANY

PIEDRA, Ramón
Facultad de Matemáticas
Tarfia s/n
Sevilla-12, SPAIN

SABBAH,C.
Centre de Mathématiques
Ecole Polytechnique
91128 Palaiseau, FRANCE

SANCHEZ, Tomás
Facultad de Ciencias
Prado de la Magdalena
Valladolid, SPAIN

SANCHO JR., Juan Bautista
Facultad de Matemáticas
Universidad de Salamanca
Salamanca, SPAIN

SLODOWY, Peter
Sonderforschungbereich Theoretische Mathematik
Universität Bonn
Beringstrasse 4
5300 Bonn 1 , GERMANY

STEENBRINK, Joseph
Mathematsch Institut der Rijksuniversiteit Leider
Wassenaarsweg 80
2333 AL Leiden, NETHERLANDS

TEISSIER, Bernard
Centre de Mathématiques
Ecole Polytechnique
91128 Palaiseau, FRANCE

TROTMAN, David
Boulevard Lavoisier
49045 Angers, FRANCE

VALEIRAS, Gerardo
Facultad de Matemáticas
Tarfia s/n
Sevilla-12, SPAIN

VICENTE, José Luis
Facultad de Matemáticas
Tarfia s/n
Sevilla-12, SPAIN

VILLANUEVA, Emilio
Facultad de Matemáticas
Universidad de Santiago
Santiago de Compostela. SPAIN

XAMBO,S.
Matemáticas, Universidad de Barcelona.
Plaza Universidad. Barcelona, SPAIN

CONTENTS

MODULES HOLONOMES A SINGULARITES REGULIERES

ET FILTRATION DE HODGE

J. L. BRYLINSKI

INTRODUCTION

Soit X une variété analytique complexe $\mathcal{O} = \mathcal{O}_X$ le faisceau des fonctions holomorphes sur X, $\mathcal{D} = \mathcal{D}_X$ le faisceau des opérateurs différentiels analytiques sur X. Ce sont des faisceaux d'anneaux cohérents ; $\mathcal{D}$ est un $\mathcal{O}$-Module quasi-cohérent, et $\mathcal{O}$ est un $\mathcal{D}$-Module cohérent, $\mathcal{D}$ agissant sur $\mathcal{O}$ de manière évidente. La variété caractéristique d'un $\mathcal{D}$-Module à gauche cohérent $\mathfrak{M}$ est une sous-variété $Ch(\mathfrak{M})$ du fibré cotangent T^*X, dont la dimension est partout au moins égale à la dimension de X. Lorsqu'on a $\dim Ch(\mathfrak{M}) = \dim(X)$, on dit que $\mathfrak{M}$ est holonome.

Un Module holonome $\mathfrak{M}$ est dit à singularités régulières si les opérateurs d'ordre m, dont le symbole principal s'annule sur $Ch(\mathfrak{M})$, accroissent lentement l'ordre dans une bonne filtration de $\mathfrak{M}$. La définition précise sera donnée dans le § 1.

Pour $\mathfrak{M}$ holonome, les faisceaux $\mathcal{E}xt^i_{\mathcal{D}}(\mathcal{O},\mathfrak{M})$ sont constructibles. Plus généralement, pour $\mathfrak{M}^\cdot$ un complexe de $\mathcal{D}$-Modules, borné, et tel que les faisceaux de cohomologie $\mathcal{H}^i(\mathfrak{M}^\cdot)$ sont holonomes, les faisceaux de cohomologie du complexe de faisceaux $\mathbb{R}\mathcal{H}om_{\mathcal{D}}(\mathcal{O},\mathfrak{M}^\cdot)$ sont constructibles.

Ce dernier complexe est seulement bien défini dans la catégorie dérivée $D(X)_c$ obtienne à partir de la catégorie des complexes de faisceaux bornés et à faisceaux de cohomologie constructibles, en inversant les quasi-isomorphismes.

On introduit une autre catégorie dérivée, obtenue à partir des complexes bornés de $\mathcal{D}$-Modules, à faisceaux de cohomologie holonomes à singularités régulières. On la note $D(\mathcal{D})_{h.r}$.

Mebkhout d'un côté [M1], [M2], [M3], Kashiwara et Kawai d'un autre [K-K], [K5], ont démontré que le foncteur

$$DR : D(\mathcal{D}_X)_{h.r} \longrightarrow D(X)_c$$

$$(\mathcal{M}^{\cdot}) \longmapsto \mathbb{R}\,\mathcal{H}om_{\mathcal{D}}(\mathcal{O},\mathcal{M}^{\cdot})$$

est une équivalence de catégories triangulées.

Deligne a récemment donné une caractérisation des objets de $D(X)_c$ qui sont de la forme $DR(\mathcal{M})$ avec $\mathcal{M}$ Module holonome à singularités régulières (vu comme complexe avec un seul terme non nul)♦. On donne au § 1 une démonstration de ce résultat. On en déduit que la catégorie $\mathcal{D}$- h.r des Modules holonomes à singularités régulières est équivalente à une catégorie Cons. Perv.(X).

On décrit au § 2 les objets simples de ces deux catégories.

- Du côté de $\mathcal{D}$- h.r, on sait associer un objet simple à la donnée d'un sous-espace analytique fermé Y de X, d'un sous-espace analytique fermé Z de Y contenant le lieu singulier de Y, et d'un faisceau localement constant $\mathcal{F}$ sur Y - Z.
- Du côté de Cons. Perv. (X), aux même données, correspond un complexe de faisceaux (bien défini à quasi-isomorphisme près) qui est le complexe d'homologie d'intersection de Y, tordu par $\mathcal{F}$.

A un objet de Cons. Perv. (X) correspond un objet $\mathcal{M}$ essentiellement unique de $\mathcal{D}$- h.r. $\mathbb{R}\,\mathcal{H}om_{\mathcal{D}}(\mathcal{O},\mathcal{M})$ a une réalisation naturelle comme complexe de de Rham. On rappelle au § 3 la construction d'une bonne filtration globale canonique de $\mathcal{M}$ due à Kashiwara et Kawai [K-K], et on s'en sert pour filtrer le complexe de de Rham de $\mathcal{M}$. On donne à cette filtration le nom de filtration de Hodge. On conjecture ensuite que pour X projective et $\mathcal{M}$ simple correspondant aux données $(Y, Y_{sing}, \mathbb{C}_{Y-Y_{sing}})$, on obtient grâce à cette filtration une structure de Hodge sur l'homologie de Goresky-MacPherson (ou d'intersection) de Y. On justifie cette conjecture par quelques exemples, dont le plus significatif concerne l'adhérence dans $\mathbb{P}_3(\mathbb{C})$ d'un cône cubique de $\mathbb{C}^3$ à singularité isolée à l'origine.

On discute brièvement quelques versions généralisées de cette conjecture et on tente de relier les Modules holonomes algébriques à la philosophie des motifs.

Je remercie Marie-Jo Lécuyer de ses efforts héroïques et finalement victorieux pour la frappe d'un épouvantable manuscrit.

♦ Kashiwara m'a informé que ce résultat lui était connu.

§ 1. LE PROBLEME DE RIEMANN-HILBERT.

Rappelons d'abord quelques définitions et résultats, pour lesquels on pourra consulter [Bj].

Si $\mathcal{M}$ est un $\mathcal{D}$-Module à gauche cohérent, localement sur X, on peut trouver une bonne filtration de $\mathcal{M}$, c'est-à-dire une filtration croissante $\{\mathcal{M}_k\}_{k\in\mathbb{Z}}$ par des sous-$\mathcal{O}$-Modules telle que

(i) $\mathcal{D}(m)\cdot\mathcal{M}_k \subset \mathcal{M}_{k+m}$, où $\mathcal{D}(m)$ dénote le faisceau des opérateurs différentiels d'ordre $\leq m$;

(ii) localement, on peut trouver un entier j_o tel que $\mathcal{M}_{j+m} = \mathcal{D}(m)\cdot\mathcal{M}_j$ pour $m \geq 0$, $j \geq j_o$.

Par exemple, $\mathcal{D}$ admet une bonne filtration par les $\mathcal{D}(m)$ (on a $\mathcal{D}(m) = 0$ pour $m < 0$). Le $\mathcal{D}$-Module $\mathcal{O}$ admet une bonne filtration telle que $\mathcal{O}_{-1} = \{0\}$, $\mathcal{O}_o = \mathcal{O}$.

La variété analytique Specan($\bigoplus_m \mathcal{D}(m)/\mathcal{D}(m-1)$) s'identifie au fibré cotangent T^*X ; $\mathcal{D}(m)/\mathcal{D}(m-1)$ s'identifie, par la construction "symbole principal" au faisceau sur X formé des fonctions sur T^*X, homogènes de degré m. On note $\sigma_m(P)$ la fonction homogène sur T^*X associée à une section locale P de $\mathcal{D}(m)$.

Avec les notations précédentes, $\bigoplus_{k\in\mathbb{Z}} (\mathcal{M}_k/\mathcal{M}_{k-1})$ est un Module sur $\bigoplus_m (\mathcal{D}(m)/\mathcal{D}(m-1))$. Il lui est associé un faisceau cohérent sur T^*X, dont le support est un sous-espace analytique homogène. Ce support, indépendant du choix de la bonne filtration, est donc globalement défini. C'est la variété caractéristique de $\mathcal{M}$, notée $\mathcal{C}h(\mathcal{M})$.

D'après un théorème de Kashiwara-Kawai-Sato ([K-K-S], [Ma], voir aussi [G]), on a partout :

$$\dim \mathcal{C}h(\mathcal{M}) \geq \dim (X) \quad .$$

Lorsqu'il y a partout égalité, on dit que $\mathcal{M}$ est holonome.

Exemples : Si Y est un sous-espace analytique fermé de X, $\mathcal{J}_Y$ son faisceau d'idéaux, pour tout entier k on introduit les faisceaux

$$\mathcal{H}^k_{[Y]}(\mathcal{O}_X) = \varinjlim_n \mathcal{E}xt^k(\mathcal{O}_X/\mathcal{J}^n_Y, \mathcal{O}_X)$$

$$\mathcal{H}^k_{[X/Y]}(\mathcal{O}_X) = \varinjlim_n \mathcal{E}xt^k(\mathcal{J}^n_Y, \mathcal{O}_X) \quad .$$

Ces faisceaux ont une structure naturelle de $\mathcal{D}$-Modules à gauche, et ils sont holonomes [K2].

Si Y est une hypersurface, on a une suite exacte

$$0 \longrightarrow \mathcal{O}_X \longrightarrow \mathcal{H}^o_{[X/Y]}(\mathcal{O}_X) \longrightarrow \mathcal{H}^1_{[Y]}(\mathcal{O}_X) \longrightarrow 0 \quad ,$$

$\mathcal{H}^{0}_{[X/Y]}(\mathcal{O}_X)$ est le faisceau des fonctions holomorphes sur X - Y, méromorphes le long de Y, et $\mathcal{H}^{1}_{[Y]}(\mathcal{O}_X)$ est le faisceau des "parties polaires" de ces fonctions.

Pour Y lisse dans X de codimension k, on pose :

$$\mathcal{B}_{Y/X} = \mathcal{H}^{k}_{[Y]}(\mathcal{O}_X) \quad .$$

Remarque 1 : Dans la suite exacte $0 \to \mathcal{M}' \to \mathcal{M} \to \mathcal{M}'' \to 0$ de $\mathcal{D}$-Modules cohérents, l'holonomie de $\mathcal{M}$ entraîne celle de $\mathcal{M}'$ et $\mathcal{M}''$ et réciproquement.

On dit que les Modules holonomes forment une sous-catégorie de Serre dans la catégorie des $\mathcal{D}$-Modules cohérents.

A un $\mathcal{D}$-Module $\mathcal{M}$ est associé un complexe de de Rham.

$$DR(\mathcal{M}) \; : \; 0 \longrightarrow \mathcal{M} \xrightarrow{d_1} \Omega^1 \otimes_{\mathcal{O}} \mathcal{M} \xrightarrow{d_2} \Omega^2 \otimes_{\mathcal{O}} \mathcal{M} \longrightarrow \cdots \xrightarrow{d_{n-1}} \Omega^n \otimes_{\mathcal{O}} \mathcal{M} \longrightarrow 0 \quad ,$$

$(n = \dim X)$, avec $d_i(\omega \otimes m) = (d\omega) \otimes m + \sum_{i=1}^{n} (dz_i \wedge \omega) \otimes (\frac{\partial}{\partial z_i} \cdot m)$ dans tout système de coordonnées locales $(z_1, \ldots, z_n)$.

Pour $\mathcal{M} = \mathcal{O}$, on retrouve le complexe de de Rham holomorphe de X. Pour Y une hypersurface de X, et $\mathcal{M} = \mathcal{H}^{0}_{[X/Y]}(\mathcal{O}_X)$ on a le complexe des formes différentielles holomorphes sur X - Y, méromorphes le long de Y.

Kashiwara [K1] a démontré que les faisceaux de cohomologie de $DR(\mathcal{M})$ sont constructibles pour $\mathcal{M}$ holonome.

Dans la catégorie dérivée des complexes de faisceaux sur X, on définit le complexe $Sol(\mathcal{M})$ des solutions de $\mathcal{M}$ par :

$$Sol(\mathcal{M}) = \mathbf{R}\,\mathcal{H}om_{\mathcal{D}}(\mathcal{M}, \mathcal{O}) \quad .$$

Pour $\mathcal{M}$ holonome, les faisceaux de cohomologie de $Sol(\mathcal{M})$ sont constructibles.

Dans la catégorie dérivée $D(X)_c$, on a :

$$DR(\mathcal{M}) = \mathbf{R}\,\mathcal{H}om_{\mathcal{D}}(\mathcal{O}, \mathcal{M}) \quad .$$

Cela résulte d'une résolution de $\mathcal{O}$ par des $\mathcal{D}$-Modules localement libres.

$$0 \to \mathcal{D}\otimes_{\mathcal{O}} \Lambda^n T \to \dots \mathcal{D}\otimes_{\mathcal{O}} \Lambda^i T \xrightarrow{d_i} \mathcal{D}\otimes_{\mathcal{O}} \Lambda^{i-1} T \to \dots \mathcal{D}\otimes_{\mathcal{O}} T \xrightarrow{d_1} \mathcal{D} \to \mathcal{O} \to 0 \quad ,$$

avec $d_i(P\otimes(v_1 \wedge \dots \wedge v_i)) = \sum_{k=1}^{i} (-1)^{k-1}(Pv_k)\otimes(v_1\wedge\dots\wedge\hat{v}_k\dots\wedge v_i)$

$$+ \sum_{1\le k<\ell\le i} (-1)^{k+\ell} P\otimes([v_k,v_\ell]\wedge v_1\wedge\dots\wedge\hat{v}_k\dots\wedge\hat{v}_\ell\dots\wedge v_i) .$$

Dans la catégorie $D(X)_c$,on dispose d'une involution contravariante de "dualité") D telle que

$$D(F^\cdot) = \mathbb{R}\mathcal{H}om(F^\cdot,\mathbb{C}_X) \quad .$$

Pour tout ouvert U de X, on a un accouplement :

$$\mathbb{H}^i(U,F^\cdot) \times \mathbb{H}_c^{2n-i}(U,D(F^\cdot)) \longrightarrow H_c^{2n}(U,\mathbb{C}_X) \simeq \mathbb{C} \quad .$$

C'est une dualité parfaite (dualité de Verdier, voir [V] par exemple).

<u>Théorème</u> (Mebkhout [1]) : Pour $\mathfrak{M}$ holonome, Sol($\mathfrak{M}$) est canoniquement isomorphe dans $D(X)_c$,à D(DR($\mathfrak{M}$)).

Il existe un foncteur contravariant $\mathfrak{M}\to\mathfrak{M}^*$ de la catégorie des $\mathcal{D}$-Modules à gauche dans elle-même, et un morphisme canonique

$$\mathrm{Bid} : \mathfrak{M} \longrightarrow (\mathfrak{M}^*)^*$$

tel que

(i) pour $\mathfrak{M}$ holonome, Bid est un isomorphisme ;

(ii) $\mathfrak{M}\mapsto\mathfrak{M}^*$ est un foncteur contravariant exact de la catégorie des $\mathcal{D}$-Modules holonomes dans elle-même ;

(iii) on a $\mathbb{R}\mathcal{H}om_{\mathcal{D}}(\mathfrak{M},\mathfrak{M}') \xrightarrow{\sim} \mathbb{R}\mathcal{H}om_{\mathcal{D}}(\mathfrak{M}'^*,\mathfrak{M}^*)$ pour $\mathfrak{M}$ et $\mathfrak{M}'$ holonomes ;

(iv) on a des isomorphismes naturels, pour $\mathfrak{M}$ holonome :

$$DR(\mathfrak{M}) \xrightarrow{\simeq} D(DR(\mathfrak{M}^*)) \simeq \mathrm{Sol}(\mathfrak{M}^*) \quad .$$

Si $D(\mathcal{D})_h$ est la catégorie dérivée de la catégorie des complexes bornés de $\mathcal{D}$-Modules à cohomologie holonome, on a des foncteurs DR et Sol de $D(\mathcal{D})_h$ vers $D(X)_c$ (il faut remarquer que DR($\mathfrak{M}^\cdot$) a seulement un nombre fini de faisceaux de cohomologie non nuls). Il est facile de voir que les assertions (i) et (iv) plus haut restent vraies dans la catégorie $D(\mathcal{D})_h$.

Le foncteur $\mathfrak{M}\mapsto DR(\mathfrak{M})$ n'étant pas pleinement fidèle, on restreint encore la catégorie des Modules holonomes en considérant ceux qui sont à singularités régulières (en abrégé R.S.).

$\mathfrak{M}$ est R.S. si, localement sur X, il existe une bonne filtration $\{\mathfrak{M}_k\}_{k\in\mathbb{Z}}$ de $\mathfrak{M}$ telle que, pour toute section locale P de $\mathcal{D}(m)$ avec $\sigma_m(P)$ nul sur $Ch(\mathfrak{M})$, on ait : $P\cdot\mathfrak{M}_k \subset \mathfrak{M}_{k+m-1}$. Kashiwara et Kawai ont démontré que si $\mathfrak{M}$ est R S., il admet globalement sur X une bonne filtration canonique ayant cette propriété. Nous rappellerons leur définition au § 2.

On trouvera dans [K-K] plusieurs conditions qui équivalent à R.S. On indiquera ici simplement le lien avec la théorie classique des équations aux dérivées partielles linéaires à une variation [D1], [W].

Soit $\mathfrak{M} = \mathcal{D}^N/\mathcal{D}^N \cdot P$ avec $P = z\cdot\frac{d}{dz}\cdot I_N - A(z)$, où A(z) est une matrice $N\times N$, dépendant holomorphiquement de z. La bonne filtration de $\mathfrak{M}$ par les $\mathfrak{M}_k$ = image dans $\mathfrak{M}$ de $\mathcal{D}(k)^N$, satisfait la condition précédente. En effet, $Ch(\mathfrak{M})$ est réunion de la section nulle de T^*X et de la fibre de T^*X en le point 0. On se convainc facilement qu'il suffit de montrer que l'opérateur $z\cdot\frac{d}{dz}$ d'ordre 1 (dont le symbole s'annule à l'ordre 1 sur $Ch(\mathfrak{M})$) satisfait $z\cdot\frac{d}{dz}(\mathfrak{M}_k)\subset\mathfrak{M}_k$. Mais pour $(P_1,\ldots,P_N)\in\mathcal{D}(k)^N$, on a :

$$z\cdot\frac{d}{dz}(P_1,\ldots,P_N) \equiv A(z)\cdot(P_1,\ldots,P_N) \text{ modulo } \mathcal{D}^N\cdot P \quad ,$$

d'où l'assertion.

Cette remarque se généralise aux équations à singularités régulières par rapport à un diviseur à croisements normaux [D1].

Soit maintenant $D(\mathcal{D})_{h.r}$ la catégorie dérivée de la catégorie des complexes bornés de $\mathcal{D}$-Modules, à cohomologie holonome R.S.

On a un foncteur $DR : D(\mathcal{D})_{h.r} \to D(X)_c$, $\mathfrak{M}^\bullet \to DR(\mathfrak{M}^\bullet)$.

<u>Théorème</u> (Mebkhout [M1], [M2], [M3] ; Kashiwara et Kawai [K-K], [K5]) : F est une équivalence de catégories.

Pour $\mathcal{F}^\bullet$ un objet de $D(X)_c$, on peut écrire $\mathcal{F}^\bullet \simeq DR(\mathfrak{M}^\bullet)$, pour $\mathfrak{M}^\bullet \in D(\mathcal{D})_{h.r}$ essentiellement unique. Deligne a donné un critère pour que l'on ait :

$$\mathcal{H}^i(\mathfrak{M}^\bullet) = 0 \quad \text{pour } i \neq 0 \quad .$$

<u>Théorème</u> : Pour $\mathcal{F}^\bullet$ objet de $D(X)_c$, l'existence d'un Module holonome R.S. $\mathfrak{M}$ tel que $\mathcal{F}^\bullet \simeq DR(\mathfrak{M})$ équivaut à la conjonction des deux conditions suivantes :

(i) on a $\mathcal{H}^i(\mathcal{F}^\bullet) = 0$ pour $i < 0$, et pour $i \geq 0$, le support de $\mathcal{H}^i(\mathcal{F}^\bullet)$ est partout de codimension $\geq i$;

(ii) on a $\mathcal{H}^i(D(\mathcal{F}^\bullet)) = 0$ pour $i < 0$, le support de $\mathcal{H}^i(D(\mathcal{F}^\bullet))$ est partout de codimension $\geq i$.

<u>Démonstration</u> : Prouvons d'abord que (i) et (ii) sont nécessaires. Tout d'abord, notons que comme $DR(\mathfrak{M}^*) = D(DR(\mathfrak{M}))$, il suffit de prouver (i) La condition $\mathcal{H}^i(DR(\mathfrak{M})) = 0$ pour $i < 0$ est claire. Il suffit ensuite de montrer que pour toute variété analytique localement fermée Y de condimension i-1, on a :

$$\mathcal{H}^i(DR(\mathfrak{M}))_{|U} = 0 \quad ,$$

pour U un ouvert dense de Y. Pour ce faire, on peut se localiser au voisinage d'un point lisse de Y, et donc supposer Y lisse et fermé dans X. On a

$$\mathcal{H}^i(DR(\mathfrak{M}))_{|Y} = \mathcal{H}^i(D_Y \mathbb{R}\Gamma_Y(D_X(DR(\mathfrak{M}))))[-2i+2] \quad .$$

Or $D_X(DR(\mathfrak{M})) \cong Sol(\mathfrak{M})$ et $\mathbb{R}\Gamma_Y(Sol(\mathfrak{M})) = \mathbb{R}\mathcal{H}om_{\mathcal{D}}(\mathfrak{M}, \mathbb{R}\Gamma_Y(\mathcal{O}))$. Comme $\mathcal{H}^k_Y(\mathcal{O}_X) = 0$ pour $k \neq i$, le complexe $\mathbb{R}\Gamma_Y(Sol(\mathfrak{M}))$ n'a de faisceaux de cohomologie non nuls qu'en degrés $\geq i-1$. Quitte à remplacer Y par un ouvert dense et à retirer le reste de X, on peut supposer ces faisceaux de cohomologie localement constants. Alors le dual D_Y de ce complexe n'a de cohomologie qu'en degrés $\leq 1-i$. Or $\mathcal{H}^i(DR(\mathfrak{M}))_{|Y}$ est le 2-ième faisceau de cohomologie de ce complexe. Il est donc nul.

Donnons-nous maintenant un objet $F^\cdot$ de $D(X)_c$ satisfaisant (i) et (ii). Si $\mathcal{D}^\infty$ est le faisceau des opérateurs différentiels d'ordre infini, et $\mathfrak{M}^\cdot$ le complexe de $D(\mathcal{D})_{h-r}$ tel que $F^\cdot \cong DR(\mathfrak{M}^\cdot)$, on a, d'après Mebkhout ([M1], [M]) :

$$\mathcal{D}^\infty \otimes_{\mathcal{D}} \mathfrak{M}^\cdot \cong \mathbb{R}\mathcal{H}om(D(F^\cdot), \mathcal{O}_X)$$

et il suffit clairement de prouver le

<u>Lemme</u> : La condition (i) entraîne : $\mathcal{E}xt^i(F^\cdot, \mathcal{O}_X) = 0$ pour $i < 0$.

Or on a la suite spectrale : $E_2^{p,q} = \mathcal{E}xt^p_X(\mathcal{H}^{-q}(F^\cdot), \mathcal{O}_X) \Rightarrow \mathcal{E}xt^{p+q}_X(F^\cdot, \mathcal{O}_X)$ et, vu l'hypothèse (i), tout se réduit au

<u>Lemme</u> : Pour G un faisceau constructible sur X dont le support est partout de codimension $\geq i$, on a :

$$\mathcal{E}xt^p_X(G, \mathcal{O}_X) = 0 \quad \text{pour } p < i \quad .$$

Comme G admet une filtration dont les sous-quotients sont des prolongements par zéro de faisceaux localement constants sur des sous-variétés localement fermées de codimension $\geq i$, on se ramène à supposer :

- qu'il existe un fermé rare Z de $Y = \mathrm{supp}(G)$, tel que la fibre de G soit nulle en dehors de $Y - Z$;
- que la restriction de G à $Y - Z$ est un faisceau localement constant.

D'après Łojasiewicz, il existe une triangulation $\mathcal{T}$ de Y de classe C^1 telle que Z soit réunion de simplexes de $\mathcal{T}$.
Pour un σ simplexe ouvert de dimension maximale, notons j_σ l'inclusion $\sigma \hookrightarrow X$.
On a une suite exacte de faisceaux :

$$0 \longrightarrow \bigoplus_{\substack{\sigma \\ \dim \sigma = \dim_{\mathbb{R}} Y}} (j_\sigma)_! \, (G_{|\sigma}) \longrightarrow G \longrightarrow H \longrightarrow 0 \quad ,$$

où H est concentré sur le squelette de dimension $\dim_{\mathbb{R}} Y - 1$ de Y, est nul en dehors de $Y - Z$ et localement constant sur $\mathrm{supp}(H) - Z$.
Il suffit de montrer la conclusion du lemme pour H et pour les faisceaux $(j_\sigma)_!(G_{|\sigma})$. Mais de la simple-connexité de σ, on déduit immédiatement :

$$(j_\sigma)_!(G_{|\sigma}) \cong \mathbb{C}_\sigma^{n_\sigma} \quad ,$$

où $\mathbb{C}$ est le faisceau constant sur σ, prolongé par zéro.
On veut montrer : $\mathcal{E}xt^p_X(\mathbb{C}_\sigma, \mathcal{O}_X) = 0$ pour $p < i$.
On a une suite exacte de faisceaux : $0 \to \mathbb{C}_\sigma \to \mathbb{C}_{\bar\sigma} \to \mathbb{C}_{\partial\sigma} \to 0$ et ce qu'on veut résultera des égalités :

$$\begin{cases} \mathcal{H}^p_{\bar\sigma}(\mathcal{O}_X) = 0 & \text{pour } p < i \\ \mathcal{H}^p_{\partial\sigma}(\mathcal{O}_X) = 0 & \text{pour } p \leq i \end{cases} .$$

Pour la première égalité, elle découle de la suite spectrale $E_2^{p,q} = \mathcal{H}^p_{\bar\sigma}(\mathcal{H}^q_Y(\mathcal{O}_X)) \Rightarrow \mathcal{H}^{p+q}_{\bar\sigma}(\mathcal{O}_X)$ et du fait que $\mathcal{H}^p_Y(\mathcal{O}_X) = 0$, pour $p < i$. De même on a on a $\mathcal{H}^p_{\partial\sigma}(\mathcal{O}_X)$ pour $p < i$, et $\mathcal{H}^i_{\partial\sigma}(\mathcal{O}_X) = \mathcal{H}^0_{\partial\sigma}(\mathcal{H}^i_Y(\mathcal{O}_X))$. Mais aucune section locale de $\mathcal{H}^i_Y(\mathcal{O}_X)$ n'est à support dans $\partial\sigma$. Sinon son support Z serait une sous-variété analytique W stricte de Y, et on aurait $\mathcal{H}^0_W(\mathcal{H}^i_Y(\mathcal{O}_X)) \neq 0$, d'où $\mathcal{H}^i_W(\mathcal{O}_X) \neq 0$, ce qui est impossible.

Il reste à montrer : $\mathcal{E}xt^p_X(H, \mathcal{O}_X) = 0$ pour $p < i$. Or on construit facilement une filtration

$$H = H_0 \supset H_1 \supset \dots \supset H_{2i-1} \supset H_{2i} = \{0\} \quad ;$$

telle que H_j / H_{j+1} est à support dans le squelette d'ordre j de $\mathcal{T}$, nul sur le squelette d'ordre j-1, et localement constant sur le complémentaire. Comme ce dernier ensemble est simplement connexe, on termine par le même argument que plus haut.

Corollaire : Il existe une catégorie abélienne dont les objets sont ceux de $D(X)_c$ satisfaisant (i) et (ii), le groupe des morphismes de $F^\cdot$ vers $G^\cdot$ étant $\mathrm{Ext}^o_X(F^\cdot, G^\cdot)$. Cette catégorie est artinienne pour X compact. On la note Cons. Perv. (X).

Remarque : Cons. Perv. (X) est équivalente à la catégorie $\mathcal{B}$(h-r) qui est artinienne lorsque X est compacte, car la longueur d'un objet $\mathfrak{M}$ est majorée par la somme des multiplicités de $\mathfrak{M}$ relativement aux composantes irréductibles de $\mathcal{C}h(\mathfrak{M})$ (voir [K3]).

Les objets de Cons. Perv. (X) sont des complexes pervers à cohomologie constructible (la perversité étant comprise au sens de Goresky-MacPherson [G-M])!

§ 2. MODULES HOLONOMES ET HOMOLOGIE D'INTERSECTION.

On rappelle qu'un objet d'une catégorie abélienne est simple s'il n'a pas de sous-objet propre.

Lemme : Soit $\mathfrak{M}$ un Module holonome simple, de support Y. Y est irréductible, et il existe un sous-espace analytique fermé Z de Y contenant le lieu singulier de Y tel que :

(a) $\mathfrak{M}_{|X-Z} \simeq \mathcal{F} \otimes_{\mathbb{C}} \mathcal{B}_{Y-Z|X-Z}$ avec $\mathcal{F}$ un faisceau localement constant sur Y - Z, prolongé par zéro,

(b) $\mathcal{H}^o_Z(\mathfrak{M}) = 0$,

(c) $\mathcal{H}^o_Z(\mathfrak{M}^*) = 0$.

Preuve : Il est d'abord clair que Y est irréductible. D'après [K1], $\mathcal{C}h(\mathfrak{M})$ est réunion des fibrés co-normaux aux strates d'une stratification de Whitney de Y. On choisit pour Z la réunion de Y_{sing} et des strates d'intérieur vide. D'après un théorème de Kashiwara [K1], on a alors (a). Le fait que $\mathfrak{M}$ est simple implique trivialement (b) et (c).

Il n'est pas nécessaire de modifier beaucoup la démonstration de la Proposition 8.5 de [B-K] pour obtenir le résultat suivant.

Proposition : Pour $Y \subset X$ espace analytique fermé, Z analytique fermé dans Y contenant Y_{sing} et $\mathcal{F}$ un faisceau localement constant sur $Y-Z$, il existe un Module holonome R.S. $\mathcal{L}$, unique à un isomorphisme unique près, tel que :

(a) $\mathcal{L}_{|X-Z} \cong \mathcal{F} \otimes_{\mathbb{C}} \mathcal{B}_{Y-Z|X-Z}$

(b) $\mathcal{H}^o_Z(\mathcal{L}) = 0$

(c) $\mathcal{H}^o_Z(\mathcal{L}^*) = 0$.

On note $\mathcal{L}(X,Y;\mathcal{F})$ ce Module. On a alors : $\mathcal{L}(X,Y;\mathcal{F})^* = \mathcal{L}(Y,X;\check{\mathcal{F}})$ où $\check{\mathcal{F}}$ est le faisceau localement constant dual de $\mathcal{F}$.

A ces données $(Y,Z,\mathcal{F})$ on peut attacher un complexe de faisceaux $\pi_Y(\mathcal{F})$ à cohomologie bornée et constructible, bien défini dans la catégorie dérivée $D(Y)_c$, caractérisé par les propriétés suivantes :

(1) $\mathcal{H}^k(\pi_Y(\mathcal{F})) = 0$ pour $k<0$;

(2) la restriction de $\pi_Y(\mathcal{F})$ à $Y-Z$ est quasi-isomorphe à $\mathcal{F}$ (placé en degré 0).

On se donne une stratification de Whitney $Y = \coprod Y_\alpha$ telle que $Y-Z$ soit la strate ouverte (donc Z est réunion de strates). Soit $d_\alpha = \mathrm{Codim}_Y(Y_\alpha)$, et soit Y_j la réunion des Y_α tels que $d_\alpha \geq j+1$. On note j_i l'inclusion de $X-Y_i$ dans $X-Y_{i+1}$. On doit alors avoir :

(3) les faisceaux $\mathcal{H}^k(\pi_Y(\mathcal{F}))$ sont localement constants sur $Y_i - Y_{i+1}$ (pour tous i et k).

(4) pour tout $i \geq 0$, on doit avoir :

$$\pi_Y(\mathcal{F})_{|X-Y_{i+1}} = \tau_{\leq i}(\mathbb{R}j_{i*}(\pi_Y(\mathcal{F})_{|X-Y_i}) ,$$

où $\tau_{\leq i}$ est l'opérateur "troncation d'un complexe en degrés $\leq i$", dû à Deligne [D2,]. Rappelons que pour un complexe $K^\cdot$, le complexe $\tau_{\leq i}(K^\cdot)$ est :

$$\longrightarrow \cdots \longrightarrow K^{i-2} \longrightarrow K^{i-1} \longrightarrow \ker d_i \longrightarrow 0 \longrightarrow 0 \ldots$$

En termes plus concrets, cette dernière condition signifie que pour tout point x de $Y_i - Y_{i+1}$ il existe un voisinage U de x dans $X - Y_{i+1}$ tel que

$$\mathbb{H}^j(U,\pi_Y(\mathcal{F})) = 0 \text{ pour } j \geq i+1$$

et que l'application de restriction :

$$\mathbb{H}^j(U,\pi_Y(\mathcal{F})) \longrightarrow \mathbb{H}^j(U-Y_i,\pi_Y(\mathcal{F}))$$

soit un isomorphisme pour $j \leq i$.

Cette construction est due à Deligne [D4]. Elle est exposée en détail

[G-M2]. Les résultats principaux sont les suivants :

<u>Théorème</u> [D4], [G-M2] :

(i) $D_Y(\pi_Y(\mathfrak{F}))$ est naturellement isomorphe à $\pi_Y(\check{\mathfrak{F}})$. En particulier $\pi_Y(\mathbb{C}_{Y-Y_{sing}})$ est "auto-dual".

(ii) Pour Y compacte, le groupe d'hypercohomologie $\mathbb{H}^k(Y,\pi_Y(\mathfrak{F}))$ s'identifie au groupe $IH_{2m-k}(Y,\mathfrak{F})$ d' "homologie d'intersection à coefficients tordus par $\mathfrak{F}$".

(iii) En conséquence $IH_\ell(Y,\mathfrak{F})$ et $IH_{2m-\ell}(Y,\check{\mathfrak{F}})$ sont en dualité parfaite, pour Y compacte.

Maintenant, le théorème 8.6 de [B-K], généralisé au cas des coefficients tordus, s'écrit

<u>Théorème</u> : Soit Y un espace analytique fermé, purement de codimension ℓ dans X. On a alors :

$$DR(\mathcal{L}(Y,X;\mathfrak{F})) = \pi_Y(\mathfrak{F})[-\ell]$$

(où on a prolongé $\pi_Y(\mathfrak{F})$ à X par zéro).

<u>Remarque</u> : Si Z' est un ouvert de Y inclus dans Z, et $\mathfrak{F}$ localement constant sur Z, alors les Modules holonomes $\mathcal{L}(Y,X;\mathfrak{F}_Z)$ et $\mathcal{L}(Y,X;\mathfrak{F}_{Z'})$ sont canoniquement isomorphes. C'est pourquoi on a omis Z de la notation. En pratique, on se donne un faisceau constructible $\mathfrak{F}$ sur $Y - Y_{sing}$, et pour Z on prend un ouvert où $\mathfrak{F}$ est localement constant.

§ 3. BONNES FILTRATIONS ET FILTRATIONS DE HODGE.

A un Module holonome R.S $\mathfrak{M}$, Kashiwara et Kawai associent dans [K-K] une bonne filtration canonique, qui est définie via une étude micro-locale de $\mathfrak{M}$. Nous utiliserons la formulation plus concrète de [K4]. Etant donné une section locale u de $\mathfrak{M}$, et Λ une composante irréductible de $\mathcal{C}h(\mathfrak{M})$, s'il existe localement sur T^*X, dans un voisinage du point générique de Λ, un opérateur micro-différentiel elliptique P tel que

$$u(x) = P(x,D) \cdot \delta(x_1,\dots,x_r)$$

dans des coordonnées locales $(x_1,\dots,x_n)$ telles que Λ soit le fibré conormal à la sous-variété V d'équation $x_1 = \dots = x_r = 0$, l'ordre de u est égal à $m+\frac{r}{2}$.

Ici l'élément $\delta(x_1,\dots,x_r)$, générateur du Module holonome $\mathcal{H}^r_{[Y]}(\mathcal{O}_X)$, est d'ordre $\frac{r}{2}$. Pour notre propos, cela est inadéquat, et nous dirons que l'<u>ordre modifié</u> de $\delta(x_1,\dots,x_r)$ est égal à 0, donc l'ordre modifié de $P.\delta(x_1,\dots,x_r)$ est égal à m. Cet <u>ordre modifié</u> est un entier.
Soit $\mathfrak{M}_r$ le sous-faisceau de $\mathfrak{M}$ formé des germes de sections qui sont d'ordre $\leq r$ par rapport à chaque composante irréductible de $\mathcal{C}h(\mathfrak{M})$. C'est un théorème profond de Kashiwara et Kawai que $\{\mathfrak{M}_r\}_{r\in\mathbb{Z}}$ est une bonne filtration (ce n'est vrai que si $\mathfrak{M}$ est R.S). Cette bonne filtration satisfait la condition suivante :

si P est une section locale de $\mathcal{D}(m)$ telle que $\sigma_m(P)$ s'annule sur $\mathcal{C}h(\mathfrak{M})$, on a : $P\,.\,\mathfrak{M}_j \subset \mathfrak{M}_{j+m-1}$.
Nous l'appellerons <u>la bonne filtration canonique</u> de $\mathfrak{M}$ (voir l' "Autocritique").
Il est facile de voir que cette construction est compatible au produit tensoriel externe $\mathfrak{M}\hat{\otimes}\mathfrak{N}$ (notations de Kashiwara) de deux Modules holonomes R.S sur deux variétés, en ce sens que

$$(\mathfrak{M}\hat{\otimes}\mathfrak{N})_k = \sum_{i+j=k} \mathfrak{M}_i \hat{\otimes} \mathfrak{N}_j$$

le point est que notre "ordre modifié" est toujours un entier.
Grâce à elle, nous définissons une filtration décroissante $F^{\cdot}$ du complexe de $DR(\mathfrak{M})$

$$F^p DR(\mathfrak{M}) \ :\ 0 \longrightarrow \mathfrak{M}_{-p} \longrightarrow \Omega^1 \otimes_{\mathcal{O}} \mathfrak{M}_{-p+1} \longrightarrow \dots \longrightarrow \Omega^i \otimes_{\mathcal{O}} \mathfrak{M}_{-p+i}$$
$$\longrightarrow \dots \longrightarrow \Omega^n \otimes_{\mathcal{O}} \mathfrak{M}_{-p+n} \longrightarrow 0 \quad .$$

Nous osons l'appeler filtration de Hodge. Avant de proposer une conjecture, considérons le cas d'une sous-variété lisse Y de X, et du Module holonome

$\mathcal{H}^{\mathrm{codim}\,Y}_{[Y]}(\mathcal{O}_X) = \mathcal{B}_{Y/X}$.

On a $\mathcal{B}_{Y/X} = \varinjlim_k \mathcal{E}xt^{\mathrm{codim}\,Y}_{\mathcal{O}_X}(\mathcal{O}_X/\mathcal{J}^k_Y, \mathcal{O}_X)$.

L'ordre modifié d'une section u de $\mathcal{B}_{Y/X}$ est clairement égal au plus petit entier k tel que u provienne d'une section locale de $\mathcal{E}xt^{\mathrm{codim}\,Y}_{\mathcal{O}_X}(\mathcal{O}_X/\mathcal{J}^k_Y, \mathcal{O}_X)$. On a donc, en posant $\mathfrak{M} = \mathcal{B}_{Y/X}$, et $\ell = \mathrm{codim}\,Y$, l'égalité : $\mathfrak{M}_{-1} = 0$, et par conséquent $F^p DR(\mathfrak{M})$ est nul en degrés inférieurs à p.

On a

$$\Omega^i_X \otimes_{\mathcal{O}_X} \mathfrak{M} = \Omega^i_X \otimes_{\mathcal{O}_X} \mathcal{H}^{\ell}_{[Y]}(\mathcal{O}_X) = \mathcal{H}^{\ell}_{[Y]}(\Omega^i_X) \quad .$$

On a une application résidu : res : $\mathcal{H}^{\ell}_{[Y]}(\Omega^i_X) \to \Omega^{i-\ell}_Y$. On obtient ainsi un morphismes de complexes de $DR(\mathfrak{M})$ vers $DR(Y)[\ell]$, le complexe de de Rham de Y décalé de ℓ vers la gauche.

On vérifie immédiatement que ce morphisme est compatible avec nôtre filtration $F^{\cdot}$ et la filtration bête de $DR(Y)[\ell]$. On vérifie par un calcul local que ce morphisme est un quasi-isomorphisme filtré [D2]. D'après loc. cit., la filtration induite par $F^{\cdot}$ sur

$$\mathbb{H}^k(X, DR(\mathfrak{M})) \cong \mathbb{H}^{k-\ell}(Y, DR(Y)) \cong H^{k-\ell}(Y, \mathbb{C})$$

est la filtration de Hodge décalée de ℓ sous l'hypothèse que Y est une variété algébrique projective.

Etudions maintenant le cas du Module holonome $\mathcal{H}^{o}_{[X/Y]}(\mathcal{O}_X) = \mathfrak{M}$, pour Y une hypersurface de X à croisements normaux.

Localement Y est défini par l'équation $z_1 \times \dots \times z_r = 0$, dans un système de coordonnées $(z_1, \dots, z_n)$. Pour I un sous-ensemble de $\{1, \dots, r\}$, soit Y_I la sous-variété lisse d'équation $z_i = 0$ $(i \in I)$. $\mathcal{C}h(\mathfrak{M})$ est la réunion, pour ces I, de Λ_I le fibré conormal à Y_I. $\mathcal{H}^{s}_{[X/Y]}(\mathcal{O}_X)$ est engendré par la fonction méromorphe $\frac{1}{z_1 \dots z_r}$. Pour tout I, au voisinage du point générique de Λ_I, $\frac{1}{z_1 \dots z_r}$ s'obtient à partir de $\frac{1}{\prod_{i \in I} z_i}$ par multiplication par une fonction inversible.

Cette section est donc d'ordre zéro. On en déduit que $\frac{1}{z_1 \dots z_n}$ est dans $\mathfrak{M}_o$ et que la bonne filtration $\{\mathfrak{M}_k\}$ n'est autre que la filtration par l'ordre du pôle de Deligne [D2]. Dans le cas où X est une variété projective, la filtration F induit donc sur $\mathbb{H}^k(X, DR(\mathcal{H}^{o}_{[X/Y]}(\mathcal{O}_X)) \cong \mathbb{H}^k(X, \mathbb{R}j_* \mathbb{C}_{X-Y}) \cong H^k(X - Y, \mathbb{C})$, la filtration de Hodge de [D2].

<u>Conjecture</u> : Soit X une variété algébrique projective, Y une sous-variété algébrique fermée de dimension m et de codimension ℓ, la filtration induite par

$F^{\cdot+\ell}$ sur

$$\mathbb{H}^k(Y,DR(\mathcal{L}(Y,X)) = \mathbb{H}^{k+\ell}(Y,\pi_{Y,X}) = (H_{2m-k-\ell}(Y,\mathbb{C})) \quad ,$$

ainsi que le réseau $(H_{2m-k-\ell}(Y,\mathbb{Z}))$, définissent une structure de Hodge de poids $k+\ell$.

Cette conjecture a été motivée par une conjecture présentée dans [C-G-M, § 4], à ceci près que le poids qui y est indiqué ne semble pas le bon.

Il y aurait lieu de généraliser cette conjecture au cas de $\mathcal{L}(Y,X;\mathbb{V})$ pour $\mathbb{V}$ une variation de structures de Hodge polarisées sur un ouvert Z de Y_{reg}. Mais pour cela, il faudrait définir sur le complexe de de Rham une filtration qui tienne compte à la fois de la filtration de Hodge sur $\mathbb{V}$ et, de la construction précédente. Le cas où $Y = X$ devrait être traité par une méthode s'inspirant de l'article de Zucker [Z], qui étudie le cas dim $X = 1$.

On remarquera que le cas où V est de type purement (0,0) correspond à une représentation unitaire du groupe fondamental de Z. Ce cas ne nécessite donc pas, en principe, de théorie de Hodge sur $\mathbb{V}$, et la conjecture pourrait s'y énoncer directement.

Nous passons maintenant au cas le plus significatif que nous ayons pu traiter à l'heure actuelle. Soit $X = \mathbb{C}^3$, $Y \hookrightarrow X$ un cône d'équation $F(X,Y,Z) = 0$ avec F homogène de degré 3. On suppose que Y a une singularité isolée en 0 (donc que la courbe projective E correspondante est lisse).

Déterminons d'abord le complexe π_Y. On a la stratification de Whitney $Y = Y - \{0\} \coprod \{0\}$. Par définition $\pi_{Y|Y-\{0\}} = \mathbb{C}_{Y-\{0\}}$, et la fibre de π_Y en 0 a pour groupes de cohomologie :

$$\begin{aligned} \mathcal{H}^i(\pi_Y)_0 &\cong H^i(B \cap Y - \{0\},\mathbb{C}) \quad \text{pour } i \le 1 \\ &= 0 \quad \text{pour } i \ge 2 \end{aligned}$$

(pour B une boule centrée en 0). On vérifie aisément : $\mathcal{H}^0(\pi_Y)_0 = \mathbb{C}$, $\mathcal{H}^1(\pi_Y)_0 = H^1(E)$. On a donc, dans la catégorie Cons. Perv(X), une suite exacte :

$$0 \longrightarrow H^1(E)^{[-3]}_{\{0\}} \longrightarrow \mathbb{C}_Y[-1] \longrightarrow \pi_Y[-1] \longrightarrow 0$$

qui s'obtient d'ailleurs en appliquant le foncteur Sol à la suite exacte suivante de Modules holonomes R.S

$$0 \longrightarrow \mathcal{L}(Y,X) \longrightarrow \mathcal{H}^1_{[Y]}(\mathcal{O}_X) \longrightarrow H^1(E) \otimes_{\mathbb{C}} \mathcal{B}_{\{0\}/X} \longrightarrow 0$$

telle que la filtration $F^{\cdot}$ de $DR(\mathcal{L}(Y,X)) = \pi_Y[-1]$ induise une filtration sur

$H^1(Y,\pi_Y) \cong H^o(Y,\mathcal{H}^1(\pi_Y)) \cong H^1(E)$. On s'attend à obtenir la filtration de Hodge de $H^1(E)$. (1). En effet, l'éclatement de l'origine définit une résolution des singularités $\widetilde{Y} \xrightarrow{P} Y$ pour laquelle on a :

$$\tau_{\leq 1}\mathbb{R}_p{}_*\mathbb{C}_{\widetilde{Y}} \cong \pi_Y \ .$$

Par conséquent, $\mathbb{H}^1(Y,\pi_Y) \cong \mathbb{H}^1(\widetilde{Y},\mathbb{C}) \cong H^1(E,\mathbb{C})$ et, d'après Cheeger-Goresky-Mac Pherson, ces isomorphismes doivent préserver les structures de Hodge [C-G-M].

Nous allons étudier la bonne filtration canonique sur $\mathcal{H}^1_{[Y]}(\mathcal{O}_X)$. Elle induit sur $\mathcal{L}(Y,X)$ la bonne filtration canonique. Nous trouverons aussi une base de l'espace vectoriel de dimension 2 : $\mathrm{Hom}_{\mathcal{D}}(\mathcal{H}^1_{[Y]}(\mathcal{O}_X),\mathcal{B}_{\{0\}/X})$. Tout d'abord, il résulte de [ ] que le polynôme de Bernstein de F est égal à $b(s) = s^2(s+\frac{1}{3})(s+\frac{2}{3})(s+1)$. On a une relation : $P(s,\underline{x},\frac{\partial}{\partial \underline{x}})F^s = b(s) \cdot F^{s-1}$, et donc F^{-2} est un générateur du $\mathcal{D}$-Module $\mathcal{H}^1_{[Y]}(\mathcal{O}_X)$. Il est clair que F^{-n} est d'ordre $n-1$ par rapport à la variété lagrangienne Λ_1 = adhérence du fibré conormal à $Y-\{0\}$, et d'ordre $3n-3$ par rapport à Λ_2 = fibré conormal à 0 dans X.

Plus généralement, pour R polynôme de degré r, non divisible par F, R/F^n est d'ordre $n-1$ pour Λ_1 et $3n-r-3$ pour Λ_2. Il en résulte que $\mathfrak{M}_k$ est le faisceau cohérent engendré par les éléments $R \cdot F^{-n}$ comme ci-dessus avec $n \leq k-1$ et $3n-3-r \leq k$, c'est-à-dire $r \geq 3n-3-k$.

En particulier $\mathfrak{M}_o$ est engendré par $\frac{1}{F}$, $\mathfrak{M}_1$ par les $\frac{P}{F^2}$, avec P homogène de degré 2, $\mathfrak{M}_2$ par les $\frac{Q}{F^3}$ avec Q homogène de degré 4, etc. $\frac{1}{F^n}$ est dans dans $\mathfrak{M}_{3n-3}$ mais pas $\mathfrak{M}_{3n-2}$.

Pour déterminer les morphismes de $\mathcal{H}^1_{[Y]}(\mathcal{O}_X)$ vers $\mathcal{B}_{\{0\}/X}$, il suffit de déterminer l'image de F^{-2}. Ce doit être un élément de $\mathcal{B}_{\{0\}/X}$, disons u, sur lequel le champ de vecteurs $\xi = X \cdot \frac{\partial}{\partial X} + Y \cdot \frac{\partial}{\partial Y}$ agit par $\xi \cdot u = -6u$. Donc u est homogène de degré -6, et on peut l'écrire :

$$n = \frac{a_1}{X^4YZ} + \frac{a_2}{XY^4Z} + \frac{a_3}{XYZ^4} + \frac{a_4}{X^3Y^2Z} + \frac{a_5}{X^3YZ^2} + \frac{a_6}{X^2Y^3Z} + \frac{a_7}{XY^3Z^2} + \frac{a_8}{X^2YZ^3} + \frac{a_9}{XY^2Z^3} + \frac{a_{10}}{X^2Y^2Z^2} \ .$$

D'autre part, F^{-2} est annulé par l'opérateur $(\frac{\partial F}{\partial Y}) \cdot \frac{\partial}{\partial X} - (\frac{\partial F}{\partial X}) \cdot \frac{\partial}{\partial Y}$, et deux opérateurs similaires. Il en est donc de même de u.

Pour faire le calcul, on supposera, comme il est loisible, les coordonnées choisies de manière à avoir : $F = X^3 + Y^3 + Z^3 + \lambda XYZ$. On a :

$$-\left(\frac{\partial F}{\partial Y}\right)\cdot\frac{\partial}{\partial X}(u) = 3\left(\frac{a_2}{X^2Y^2Z}+\frac{2a_6}{X^3YZ}+\frac{a_7}{X^2YZ^2}\right)$$

$$+\lambda\left(\frac{a_3}{XYZ^3}+\frac{2a_5}{X^3YZ}+\frac{a_7}{XY^3Z}+\frac{2a_8}{X^2YZ^2}+\frac{a_9}{XY^2Z^2}+\frac{2a_{10}}{X^2Y^2Z}\right)$$

$$-\left(\frac{\partial F}{\partial X}\right)\cdot\frac{\partial}{\partial Y}(u) = 3\left(\frac{a_1}{X^2Y^2Z}+\frac{2a_4}{XY^3Z}+\frac{a_5}{XY^2Z^2}\right)$$

$$+\lambda\left(\frac{a_3}{XYZ^3}+\frac{a_5}{X^3YZ}+\frac{2a_7}{XY^3Z}+\frac{a_8}{X^2YZ^2}+\frac{2a_9}{XY^2Z^2}+\frac{2a_{10}}{X^2Y^2Z}\right) \quad .$$

En égalant les coefficients, on trouve :

$$a_1 = a_3 \ , \quad 3a_6+\lambda a_5 = 0 \ , \quad 3a_4+\lambda a_7 = 0 \ , \quad 3a_7+\lambda a_8 = 0 \ , \quad 3a_5+\lambda a_9 = 0 \quad .$$

Les deux équations analogues pour u donnent $a_1 = a_2 = a_3$ et également $a_4 = a_5 = \ldots\ldots = a_9 = 0$.
En effet, on obtient :

$$a_4 = -\frac{\lambda}{3}\,a_7 = \left(-\frac{\lambda}{3}\right)^2 a_8 = -\left(\frac{\lambda}{3}\right)^3 a_4 \quad ,$$

d'où $a_4 = 0$ sauf si $\left(\frac{\lambda}{3}\right)^3 = -1$, cas exclu puisque la courbe E d'équation $F = 0$ dans $\mathbb{P}^2$ est lisse (cf. [D-R, page 148]). Les autres annulations s'obtiennent de même.

Ainsi donc u est dans l'espace de dimension 2 engendré par $\frac{1}{X^4YZ}+\frac{1}{XY^4Z}+\frac{1}{XYZ^4}$ et par $\frac{1}{X^2Y^2Z^2}$. Comme $\mathcal{H}om_{\mathcal{D}}(\mathcal{H}^1_{[Y]}(\mathcal{O}_X),\mathcal{B}_{\{0\}/X})$ est aussi de dimension 2, on en déduit sans coup férir l'existence de deux $\mathcal{D}$-morphismes φ, ψ avec

$$\varphi(F^{-2}) = \frac{1}{X^4YZ}+\frac{1}{XY^4Z}+\frac{1}{XYZ^4} \quad ,$$

et

$$\psi(F^{-2}) = \frac{1}{X^2Y^2Z^2} \quad .$$

On a :

$$\varphi(F^{-1}) = (X^3+Y^3+Z^3+\lambda XYZ)\cdot\left(\frac{1}{X^4YZ}+\frac{1}{XY^4Z}+\frac{1}{XYZ^4}\right) = \frac{3}{XYZ} \quad .$$

Alors que :

$$\psi(F^{-1}) = (X^3+Y^3+Z^3+\lambda XYZ)\cdot\left(\frac{1}{X^2Y^2Z^2}\right) = \frac{1}{XYZ} \quad .$$

Par conséquent $3\psi - \lambda\varphi$ s'annule sur F^{-1} et engendre l'espace des morphismes qui annulent F^{-1}.

La filtration de Hodge du complexe $DR(\mathcal{H}^1_{[Y]}(\mathcal{O}_X))$ en induit une sur $\mathcal{H}^2(DR(\mathcal{H}^1_{[Y]}(\mathcal{O}_X))_o \cong \mathcal{H}^2(DR(\mathcal{L}(Y,X))_o \cong \mathcal{H}^1(\pi_Y)_o$. On vérifie que cet espace de dimension 2 est engendré par les classes des éléments suivants de $\Omega^2_X \otimes_{\mathcal{O}_X} \mathcal{L}(Y,X)$:

$$\gamma = (dY \wedge dZ) \otimes (\frac{X}{F}) + (dZ \wedge dX) \otimes (\frac{Y}{F}) + (dX \wedge dY) \otimes (\frac{Z}{F})$$

$$\delta = (dY \wedge dZ) \otimes (\frac{X^2YZ}{F^2}) + (dZ \wedge dX) \otimes (\frac{XY^2Z}{F^2}) + (dX \wedge dY) \otimes (\frac{XYZ^2}{F^2}) \quad .$$

Le premier élément est dans $\Omega^2_X \otimes \mathcal{L}(Y,X)_o$, le deuxième dans $\Omega^2_X \otimes \mathcal{L}(X,Y)_1$.

Par conséquent, on a $0 = F^{+3} \subset F^{+2} = \mathbb{C} \cdot [\gamma] \subset F^{+1} = \mathcal{H}^1(\pi_Y)_o$.
Par ailleurs, on identifie cet espace à $H^1(E)$, la filtration de Hodge s'identifiant à la filtration habituelle sur $H^1(E)$, à un décalage de 1 près.

Il suffit en effet de restreindre nos 2-formes méromorphes au plan $Z = 1$, de voir qu'elles se prolongent à des formes sur $\mathbb{P}^2$, avec pôles seulement le long de la courbe E (d'équation $F(X,Y,Z) = 0$). On a $\tilde{\gamma} = \frac{dX \wedge dY}{F}$, $\tilde{\delta} = \frac{XY dX \wedge dY}{F^2}$.

D'après [D2, Proposition 9.2.6], les résidus de $\tilde{\gamma}$ et $\tilde{\delta}$ dans $H^1(E,\mathbb{C})$ sont respectivement dans $F^1 H^1(E,\mathbb{C})$ et $F^o H^1(E,\mathbb{C})$.

On a donc identifié $\mathcal{H}^1(\pi_Y)_0$ (avec sa filtration de Hodge en notre sens à $H^1(E)$, ce qui est compatible avec notre conjecture (le décalage de 1 dans notre conjecture étant bien sûr pris en compte).

Notons que par une méthode similaire on peut étudier la filtration de Hodge sur $\mathcal{H}^3(DR(\mathcal{H}^1_{[Y]}(\mathcal{O}_X))_0$, et montrer qu'elle coïncide à un décalage près avec la filtration de Hodge ordinaire sur $H^1(E)$.

On espère que ce genre de méthodes permettra de comprendre le cas d'un cône algébrique à singularité isolée à l'origine. On aurait pu aussi travailler avec l'adhérence $\overline{Y}$ de Y dans $\mathbb{P}_3(\mathbb{C})$, étant donné que $\mathbb{H}^1(\pi_{\overline{Y}}) \cong \mathbb{H}^1(Y,\pi_Y)$. Naturellement dans notre exemple, la non-compacité de Y n'était pas un problème sérieux, puisque tout se passait à l'origine.

On se donne maintenant une variété algébrique lisse, X. On notera $\mathcal{D}_X$

le faisceau de Zariski des opérateurs différentiels algébriques. On a alors : $\mathcal{D}_{X^{an}} \cong \mathcal{O}_{X^{an}} \otimes_{\mathcal{O}_X} \mathcal{D}_X$ (isomorphisme de faisceaux localement libres sur X^{an}). Il n'y a pas de difficulté à définir la notion de $\mathcal{D}_X$-Module holonome. Si $\mathfrak{M}$ est un tel $\mathcal{D}_X$-Module, il est quasi-cohérent comme $\mathcal{O}_X$-Module. On notera $\mathfrak{M}^{an}$ le $\mathcal{D}_{X^{an}}$-Module holonome correspondant.

<u>Proposition</u> : Si X est propre, le morphisme naturel, dans la catégorie des complexes bornés d'espaces vectoriels sur $\mathbb{C}$, de $\mathbb{R}\,\mathrm{Hom}_{\mathcal{D}_X}(\mathcal{O}_X,\mathfrak{M})$ vers $\mathbb{R}\,\mathcal{H}om_{\mathcal{D}_{X^{an}}}(\mathcal{O}_{X^{an}},\mathfrak{M}^{an})$, est un isomorphisme.

<u>Démonstration</u> : $\mathcal{O}_X$ admet comme $\mathcal{D}_X$-Module une résolution par le complexe

$$0 \longrightarrow \mathcal{D}_X \otimes_{\mathcal{O}_X} \Lambda^n(T_X) \longrightarrow \cdots\cdots \longrightarrow \mathcal{D}_X \otimes_{\mathcal{O}_X} T_X \longrightarrow \mathcal{D}_X$$

degré -n (sous $\mathcal{D}_X \otimes_{\mathcal{O}_X} \Lambda^n(T_X)$) ; degré 0 (sous le dernier $\mathcal{D}_X$)

La résolution similaire de $\mathcal{O}_{X^{an}}$ s'obtient par tensorisation avec $\mathcal{O}_{X^{an}}$ (à droite). Il suffit de montrer que pour $0 \le i \le n$, le morphisme naturel

$$\mathbb{R}\,\mathcal{H}om_{\mathcal{D}_X}(\mathcal{D}_X \otimes_{\mathcal{O}_X} \Lambda^i T_X,\mathfrak{M}) \longrightarrow \mathbb{R}\,\mathcal{H}om_{\mathcal{D}_{X^{an}}}(\mathcal{D}_{X^{an}} \otimes_{\mathcal{O}_{X^{an}}} \Lambda^i T_X,\mathfrak{M}^{an})$$

est un isomorphisme.

Or,cela revient à montrer que le morphisme naturel

$$\mathbb{R}\,\mathcal{H}om_{\mathcal{O}_X}(\Lambda^i T_X,\mathfrak{M}) \longrightarrow \mathbb{R}\,\mathcal{H}om_{\mathcal{O}_{X^{an}}}(\Lambda^i T_{X^{an}},\mathfrak{M}^{an})$$

est un isomorphisme, ou encore que $\mathrm{Ext}^j_{\mathcal{O}_X}(\Lambda^i T_X,\mathfrak{M})$ s'identifie à son avatar analytique. Or $\mathrm{Ext}^j_{\mathcal{O}_X}(\Lambda^i T_X , \mathfrak{M})$ est isomorphe à $\mathrm{Ext}^j_{\mathcal{O}_X}(\mathcal{O}_X,\Omega^i_X \otimes_{\mathcal{O}_X} \mathfrak{M}) \simeq H^j_X(\Omega^i_X \otimes_{\mathcal{O}_X} \mathfrak{M})$ et comme $\Omega^i_X \otimes_{\mathcal{O}_X} \mathfrak{M}$ est quasi-cohérent, il suffit d'appliquer GAGA.

<u>Remarque</u> : On peut se demander plus généralement si pour deux $\mathcal{D}_X$-Modules holonomes $\mathfrak{M}$ et $\mathfrak{N}$, on a :

$$\mathrm{Ext}^j_{\mathcal{D}_X}(\mathfrak{M},\mathfrak{N}) \xrightarrow{\approx} \mathrm{Ext}^j_{\mathcal{D}_{X^{an}}}(\mathfrak{M}^{an},\mathfrak{N}^{an}) \quad .$$

Je suppose que cela peut se déduire du cas particulier $\mathfrak{M} = \mathcal{O}_X$ par un procédé de "réduction à la diagonale", mais je ne l'ai pas vérifié.

Etant donné un $\mathcal{D}_X$-Module holonome, on se donne un sous-corps K de $\mathbb{C}$, de type fini sur $\mathbb{Q}$, sur lequel X et $\mathfrak{M}$ sont définis. On suppose donné un complexe borné $F^{\cdot}_{\mathbb{Z}}$ de faisceaux en groupes abéliens sur X, à cohomologie constructible (les fibres sont donc des groupes abéliens de type fini, et un isomorphisme dans $D(X)_c$:

$$F^{\cdot}_{\mathbb{C}} = \mathbb{C} \otimes_{\mathbb{Z}} F^{\cdot} \xrightarrow{\approx} \mathbb{R}\mathcal{H}om_{\mathcal{D}_{X^{an}}}(\mathcal{O}_{X^{an}}, \mathfrak{M}^{an}) \quad .$$

On suppose donné, pour chaque ℓ, un complexe borné $F^{\cdot}_{\ell}$ de $\mathbb{Z}_{\ell}$-faisceaux sur le site étale $X^{\text{ét}}$ de X_K, à cohomologie constructible (voir SGA IV, Exposé X). On a un morphisme de sites de X^{an} vers $X^{\text{ét}}$, noté $\varepsilon : X^{an} \to X^{\text{ét}}$ dans l'exposé XIII de SGA IV.

On suppose donné de plus un isomorphisme de $\varepsilon^{*}(F^{\cdot}_{\ell})$ vers $\mathbb{Z}_{\ell} \otimes_{\mathbb{Z}} F^{\cdot}$ (pour tout ℓ).

L'ensemble de ces données sera appelé <u>Module holonome</u> motivique.

Pour tout i, on dispose :

- d'un K-espace vectoriel avec une filtration de Hodge : $H^i_{DR} = \mathrm{Ext}^i_{\mathcal{D}_X}(\mathcal{O}_X, \mathfrak{M})$;
- d'un groupe abélien $H^i_{\mathbb{Z}} = \mathbb{H}^i(X, F^{\cdot}_{\mathbb{Z}})$ et d'un isomorphisme $\mathbb{C} \otimes_K H^i_{DR} \cong \mathbb{C} \otimes_{\mathbb{Z}} H^i_{\mathbb{Z}}$;
- d'un $\mathbb{Z}_{\ell}$-module $H^i_{\mathbb{Z}_{\ell}} = \mathbb{H}^i(\overline{X}^{et}_K, F^{\cdot}_{\mathbb{Z}_{\ell}})$ sur lequel $\mathrm{Gal}(\overline{K}/K)$ agit, et d'un isomorphisme $\mathbb{Z}_{\ell} \otimes_{\mathbb{Z}} H^i_{\mathbb{Z}} \cong H^i_{\mathbb{Z}_{\ell}}$.

Enfin, il y aurait lieu en général de définir sur chacun de ces objets une filtration par le poids au sens de Deligne. Dans le cas où $\mathfrak{M} = \mathcal{L}(Y,X)$ cela n'est pas nécessaire. On devrait alors obtenir un motif (en tout cas, on dispose de tout ce qui est nécessaire pour en obtenir un). J'espère que ce point de vue s'avèrera de quelque utilité. Dans le cas $X = G/B$ variété de drapeaux, on espère ainsi définir des motifs correspondant à $\mathrm{Ext}^i(M_w, M_w)$ ou $\mathrm{Ext}^i(M_w, L_w)$ (notations de [B-K]).

Dans le cas du Module holonome $\mathfrak{M}_w$, la filtration de Jantzen sur M_w (et donc sur $\mathfrak{M}_w$) devait correspondre à une filtration par le poids en cohomologie ℓ-adique, que Deligne s'applique à construire dans un cadre très général.

***

Autocritique

Bien que Kashiwara et Kawai définissent dans [K-K, chapitre V] la bonne filtration canonique d'un Module holonome R.S quelconque, la définition "naïve" que j'en ai donnée ne s'applique qu'à un Module qui a une "monodromie" unipotente relativement à toutes les composantes de sa variété caractéristique. Sous cette hypothèse, notre ordre modifié n'est pas toujours un entier. La compatibilité au produit tensoriel externe pose alors problème. Je ne sais pas si le Module holonome $\mathcal{L}(Y,X)$ satisfait cette hypothèse ; la conjecture demeure probablement raisonnable.

BIBLIOGRAPHIE

[B-B] Beilinson, A. et Bernstein, I. : Sur la localisation des $\mathcal{G}$-modules, Note aux C. R. Acad. Sc. Paris, à paraître.

[Bj] Björk, J.E. : Rings of Differential operators, North Holland 1980.

[B-K] Brylinski, J.L. et Kashiwara, M. : Kazhdan-Lusztig conjecture and holonomic systems, pré-publication du Centre de Mathématiques de l'Ecole Polytechnique, Novembre 1980.

[C-G-M] Cheeger, C. ; Goresky, M. et McPherson, R. : The L^2-cohomology and intersection homology for singular algebraic varieties, preprint.

[D1] Deligne, P. : Equations différentielles à points singuliers réguliers, Springer Lecture Notes in Mathematics No 163 (1970).

[D2] Deligne, P. : Théorie de Hodge II, Publ. Math. de l'I.H.E.S., vol. 40.

[D3] Deligne, P. : Théorie de Hodge III, Publ. Math. de l'I.H.E.S., vol. 44.

[D4] Deligne, P. : Lettre à D. Kazhdan et G. Lusztig, avril 1979.

[D-R] Deligne, P. et Rapoport, M. : Les schémas de modules de courbes elliptiques, in Modular Functions of the One Variable II, Springer Lecture Notes in Mathematics No 349.

[G] Gabber, O. : On the integrability of the characteristic variety, preprint, Tel-Aviv University, 1980.

[G-M1] Goresky, M. et MacPherson, R. : Intersection homology theory, Topology 19, (1980) p. 135-162.

[G-M2] Goresky, M. et MacPherson, R. : Intersection homology, II, à paraître.

[K1] Kashiwara, M. : On the maximally overdetermined systems of linear differential equations I, Publ. R.I.M.S., Kyoto Univ. 10 (1975) p. 563-579.

[K2] Kashiwara, M. : On the holonomic systems of linear differential equations II, Inventiones Math. 49 (1978) p. 121-135.

[K3] Kashiwara, M. : Systèmes d'équations microdifférentielles, cours rédigé par Teresa Monteiro Fernandes, Université de Paris-Nord (1978).

[K4] Kashiwara, M. : Micro-local calculus, preprint, Nagoya University.

[K5] Kashiwara, M. : Faisceaux constructibles et systèmes holonomes d'équations aux dérivées partielles, exposé au Séminaire Goulaouic-Schwartz 1979-1980, Centre de Mathématiques de l'Ecole Polytechnique.

[K-K] Kashiwara, M. et Kawai, T. : On holonomic systems of micro-differential equations III - Systems with regular singularities, R.I.M.S., Kyoto Univ. (1979).

[K-K-S] Kashiwara, M. ; Kawai, T. et Sato, M. Microfunctions and pseudo-differential equations, Springer Lecture Notes in Math. No 287 (1973) p. 264-529.

[Ma] Malgrange, B. : L'involutivité des caractéristiques des systèmes micro-différentiels, Séminaire Bourbaki, in Springer Lecture Notes in Math. No 710, p. 277-289.

[M1] Mebkhout, Z. : Thèse de doctorat d'Etat, Université de Paris VII (1979).

[M2] Mebkhout, Z. : Dualité de Poincaré, Exposé au Séminaire sur les Singularités, Publ. Math. de l'Univ. Paris VII, No 7.

[M3] Mebkhout, Z. : Sur le problème de Riemann-Hilbert, Note aux C. R. Acad. Sc. Paris, t. 290 (3 Mars 1980).

[V] Verdier, J.L. : Exposé VI au Séminaire de Géométrie Analytique de l'E.N.S., 1974-75, Astérisque No 36-37.

[W] Wasow, W. : Asymptotic expansions for ordinary differential equations, Wiley, 1963.

[Z] Zucker, S. : Hodge theory with degenerating coefficients : L^2 cohomology in the Poincaré metric, Annals of Math. 109 (1979), p. 415-476.

Centre de Mathématiques
Ecole Polytechnique
Plateau de Palaiseau
F-91128 PALAISEAU Cedex
(France)

ON PROJECTIONS OF SPACE ALGEBROID CURVES

by A. Campillo (*) and J. Castellanos (**)

(*)Departamento de Algebra y Fundamentos. Universidad de Valladolid.

(**)Departamento de Algebra y Fundamentos. Universidad Complutense de Madrid.

INTRODUCTION

Hamburger-Noether expansions are considered in [2] in order to describe the desingularization process of irreducible space algebroid curves by means of quadratic transformations. Moreover, a Hamburger-Noether expansion gives a complete information on the sequence of infinitely near points of the origin of the associated curve.

In this paper we use some of this information to determine some facts on space algebroid curves. The main result is the computation, in terms of a matrix, associated in a natural way to a Hamburger-Noether expansion, of the minimum integer d for which a given irreducible curve ϑ has a projection on a d-plane preserving the desingularization multiplicity sequence. Furthermore for such d the multiplicity sequence of ϑ agrees with that of a generic projection of ϑ on a d-plane, and, in fact, for any embedding of ϑ one of the coordinate d-planes provides a generic projection.

In the last section we determine, for a fixed multiplicity sequence, all the posible values of d, and for d also fixed, all the posible values of the embedding dimension.

1. Infinitely near points. Hamburger-Noether matrices.

Consider the N-dimensional algebroid space Spec A_N, $A_N = k[[X_1,\dots,X_N]]$, k an algebraically closed field, and the total blowing up $\pi : Bl_M(A_N) = \mathrm{Proj}(\bigoplus_{n=o}^{\infty} M^n) \longrightarrow \mathrm{Spec}\, A_N$ (M the maximal ideal of A_N). If O denotes the closed point of Spec A_N, the closed points of the k-subscheme $\pi^{-1}(O)$ (exceptional divisor of π) are called infinitely near points of O in the first order neigbourhood. $\pi^{-1}(O)$ is canonically isomorphic to the projective (N-1)-space $\mathrm{Proj}(\bigoplus_{n=o}^{\infty} M^n/M^{n+1})$, thus infinitely near points correspond bijectively to directions through the origin in Spec A_N. For a closed point $O_1 \in \pi^{-1}(O)$ one has $\hat{\mathcal{O}}_{Bl_M(A_N),O_1} \approx A_N$, and the closed points in the exceptional divisor of the global blowing up of $\hat{\mathcal{O}}_{Bl_M(A_N),O_1}$ are called infinitely near points of O in the second order neigbourhood. Points in higher order neigbourhoods are defined inductively.

An embedding of an irreducible algebroid curve Spec$\mathcal{O}$ Spec A_N determines a sequence $O, O_1, \dots, O_i, \dots$ of infinitely near points, where O is the closed point of SpecA_N and for all $i \geqslant 1$, O_i is a point in the first order neigbourhood of O_{i-1}. If $x_1 = x_1(t), \dots, x_N = x_N(t)$ are parametric equations for the curve, the sequence $O, O_1, \dots, O_i, \dots$ is represented by a Hamburger-Noether expansion corresponding to this parametrization, (see [2],2.4.):

$$
(1)\qquad
\begin{aligned}
Y &= A_{01} z_0 + \dots + A_{0h} z_0^h + Z_1 z_0^h \\
\bar{Z}_0 &= A_{11} z_1 + \dots + A_{1h_1} z_1^{h_1} + Z_2 z_1^{h_1} \\
&\dots\dots\dots\dots\dots\dots\dots\dots \\
&\dots\dots\dots\dots\dots\dots\dots\dots \\
\bar{Z}_{r-1} &= A_{r1} z_r + A_{r2} z_r^2 + \dots
\end{aligned}
$$

where A_{ji} are (N-1)x1-matrices with entries in k; Z_j and $\bar{Z}_j$ (N-1)x1-matrices with entries in k[[t]]; $z_j \in k[[t]]$ is a entry of Z_j and $\bar{Z}_j$ of minimum order among the entries in Z_j; $1 = \mathrm{ord}(z_r) < \dots < \mathrm{ord}(z_1) < \mathrm{ord}(z_0)$; z_0 is one of the power series $x_1(t), \dots, x_N(t)$ of minimum order, and Y is the (N-1)x1-matrix consisting of the other power series. Note that A_{j1}, $1 \leq j \leq r$, has "0" as entry exactly in the place in which z_{j-1} is placed in $\bar{Z}_{j-1}$.

A Hamburger-Noether expansion determines the multiplicity sequence of the desingularization process for Spec $\mathfrak{O}$, in fact this sequence is

$$\underbrace{n,\ldots,n}_{h \text{ times}},\ \underbrace{n_1,\ldots,n_1}_{h_1 \text{ times}},\ldots,\underbrace{n_{r-1},\ldots,n_{r-1}}_{h_{r-1} \text{ times}},\ n_r=1,\ldots$$

where $n_j=\text{ord}(z_j)$, $(n_o=n)$.

To see (1) in another way we will consider the matrix of dimension $N\times\infty$, called Hamburger-Noether matrix,

$$(2) \qquad (A'_{01},\ldots,A'_{0h} \mid A'_{11},\ldots,A'_{1h_1} \mid A'_{21},\ldots)$$

A'_{j1} being the (Nx1)-matrix obtained from A_{ji} by introducing a new entry 1 or 0 according as i=1 or i>1 in the k-th row if Z_j proceeds from x_k by the successive divisions indicated by the expansion. (For j=0 and $z_o=x_1$ the new entries are introduced in the 1-th row). Thus for each j, $0 \leq j \leq r$, there is in (2) a marked row of type

$$(3) \qquad (a^k_{01},\ldots,a^k_{0h} \mid a^k_{11},\ldots,a^k_{1h_1} \mid \ldots \mid 1,0,\ldots,0 \mid 0,a^k_{j+1,2},\ldots)$$

in which $a^k_{j+1,1}$ must be neccessarily zero and for $j<r$ some entry at the right hand side of $a^k_{j+1,1}$ is nonzero.

The expansion (1), and so the matrix (2), are not uniquely determined by the parametrization because they depend on the choice of the power series z_j. In this paper we will assume that the power series z_j are chosen in such a way that they verify the following property:

> (*) If for the index j, one row which has been marked for some $j'<j$ can be marked again, and if j' is the greatest integer with this property, then the j-th marked row is the j'-th one.

Note that such a selection is always possible and that (*) is preserved under quadratic transformation.

1.1. Proposition.- To each embedded curve is associated a matrix (2) composed by boxes of lengths $h, h_1, \ldots, h_r = \infty$ and having marked rows

$$(a^k_{01}, \ldots, a^k_{0h} 1 \ldots 1 1, \ldots 1 0, a^k_{j+1,2}, \ldots) , \quad 0 \leq j \leq r,$$

with some nonzero entry at the right hand side of $a^k_{j+1,2}$ for $j<r$ and verifying condition ($*$). Conversely, each matrix composed by boxes with marked rows as above verifying condition ($*$) corresponds to an algebroid curve.

Proof: It follows from above discussion and from the fact that any Hamburger-Noether expansion (and hence any Hamburger-Noether matrix) defines a parametrization and so an algebroid curve (see [2]).

One of the main features of the Hamburger-Noether matrix (2) is that it makes evident the sequence of infinitely near points associated to the curve, in such a way that A'_{ji} is the matrix of homogeneous coordintates of the infinitely near point $O_{h+\ldots+h_{j-1}+i}$ in an appropriate coordinate system of the corresponding exceptional divisor. Thus this property allows us to make a bijective correspondence between points in the sequence $O_1, \ldots, O_m, \ldots$ and matrices A'_{ji}; and therefore, geometric properties of the infinitely near points sequence can be obtained from (2) in an algebraic way. For instance, the satellitisme order of O_m, defined as the multiplicity of the total expceptional divisor at O_m, can be easely interpreted in terms of (2) according to the following

1.2. Proposition.- The satellitisme order of O_m is the number of different marked rows having zeroes from their marked 1's to the m-th column.

1.3. Remark.- In fact, satellitisme orders depends only on the multiplicity sequence as one may check directly from the Hamburger-Noether expansion. Furthermore, if $\mu_1, \ldots, \mu_m, \ldots$ is the multiplicity sequence, the satellitisme order of O_m exceeds exactly in an unit the number of integers $s<m$ satisfying

$\mu_{s-1} > \mu_s + \mu_{s+1} + \ldots + \mu_{m-1}$.

In the next section we will handle another set of invariants of ϑ obtained from (2) which do not depend only on the multiplicity sequence.

2. Projections preserving multiplicity sequences.

2.1. Definition.- Let Spec ϑ be an irreducible algebroid curve over k. The number of dimension of Spec ϑ is defined to be the least integer d for which there exists a projection of Spec ϑ on a d-plane preserving the multiplicity sequence.

2.2. Remark.- The number of dimension was introduced by Arf and Du Val in [1] and [4] respectively.

2.3. Theorem.- The number of dimension is equal to the number of total marked rows, (i.e., the number of rows which have been marked for at least one j) in any Hamburger-Noether matrix representing the curve. In particular this number of rows does not depend on the embedding.

Proof: We will give a geometric interpretation of the total number of marked rows in terms of infinitely near points. For it, take for each point O_m the set of projective linear varieties $\{L^m_s\}_{s<}$ defined inductively as follows: L^m_s is the projective linear variety at the s-th order neigbourhood consisting of O_s and all points P such that the line $\overline{O_sP}$ has at O_s a direction corresponding to a point in L^n_{s+1} . For s=m-1 the above definition applies setting $L^m_m = \{O_m\}$.

The set of numbers $d_{ms} = \dim L^m_s$, $s<m$, are obviously invariants of the local ring ϑ . If $m=h+\ldots+h_{r-1}+1$, the total number d' of marked rows is related to d_{m1} by the formula $d'=d_{m1}+1$. In fact, choosing in the first order neighbourhood the natural coordintates, L^m_1 is the projective linear variety generated by the directions $A'_{j_1,1},\ldots,A'_{j_{d'},1}$ at O_1 where $j_1,\ldots,j_{d'}$ are the indices for which the d' different marked rows appear marked the first time.

Now the relation d=d' is a consequence of the above observations. In fact, if $\operatorname{Spec}\vartheta \longrightarrow \operatorname{Spec}\vartheta_o$ is a projection of $\operatorname{Spec}\vartheta$ on a d-plane preserving the multiplicity sequence and if $\vartheta_o \hookrightarrow \vartheta$ is the corresponding injective ring homomorphism, we can choose a basis in the maximal ideal of ϑ, $\{x_1,...,x_N\}$ such that $\{x_1,...,x_d\}$ is a basis of the maximal ideal of ϑ_o. It is evident that for the embedding given by $\{x_1,...,x_N\}$ the matrix (2) can be taken in such a way that all the marked rows lie among the first ones. Hence $d' \leqslant d$. The other inequality is obvious.

<u>2.4. Corollary</u>.- If d is the number of dimension of $\operatorname{Spec}\vartheta$, then for any embedding of $\operatorname{Spec}\vartheta$ the multiplicity sequence of the generic projection on a d-plane is the multiplicity sequence of $\operatorname{Spec}\vartheta$. Moreover, one of the coordinate d-planes provides a generic projection.

<u>Proof:</u> For any embedding of $\operatorname{Spec}\vartheta$, as d is an invariant of ϑ there exists a coordinate d-plane preserving the multiplicity sequence by projection on it. Now, order the sequences of natural numbers by the lexicographic order and let $\underline{e}$, $\underline{e}_o$, $\underline{e}'$ be the multiplicity sequences respectively for $\operatorname{Spec}\vartheta$, for the projection on the selected coordinate d-plane and for the generic projection. One has $\underline{e} \leqslant \underline{e}' \leqslant \underline{e}_o$, and since $\underline{e} = \underline{e}_o$ we obtain $\underline{e} = \underline{e}'$ which completes the proof of the corollary.

<u>2.5. Remark.-</u> In [3] E. Casas gives ,in terms of infinitely near points, a neccessary and sufficient condition for an irreducible space curve to have d=2. Our last stated result includes this condition and puts it in an algebraic fashion. Moreover the used method work for any d and for any algebraically closed field of arbitrary characteristic.

<u>2.6. Remark</u>.- The number d (and in fact all d_s^m) are invariants of the Arf closure ϑ' of ϑ, i.e., of the maximum ring contained in the integral closure $\overline{\vartheta}$ having the same multiplicity sequence as ϑ (see [1] and [5]).

Rings between $\mathcal{O}$ and $\mathcal{O}'$ correspond to Hamburger-Noether matrices (2) obtained from one of $\mathcal{O}$ by introducing any arbitrary rows. In particular, it is obvious that the semigroup of values S' of $\mathcal{O}'$ consists of the integers $hn+h_1n_1+\ldots+h_{s-1}n_{s-1}+Kn_s$, $0\leqslant s\leqslant r$ and $K\leqslant h_s$. From this we can derive a method to compute a Hamburger-Noether matrix for $\mathcal{O}'$ from one of $\mathcal{O}$ in the following way: Consider the Apery basis of S' relative to n, i.e., the set of integers $\{a_o, a_1, \ldots, a_{n-1}\}$ defined by $a_o=n$ and $a_i=\min\{\gamma\in S \mid \gamma\equiv i \pmod{n}\}$, $0<i<n$. In the Hamburger-Noether matrix of $\mathcal{O}$ we drop the nonmarked rows and keep the remainder d ones. According to above observations, the first marked "1" in the ν-th of this rows corresponds to a number a_{j_ν} in the Apery basis, and it is obvious that $a_{j_1}, \ldots, a_{j_d}$ are pairwise different. Now for each $a_j\in\{a_o, \ldots, a_{n-1}\}-\{a_{j_1}, \ldots, a_{j_d}\}$ introduce the new row

$$(0,\ldots,0,1,0,\ldots)$$

where the place of the "1" is the right one to provide to S' of the value a_j. The obtained $n\times\infty$ matrix corresponds to $\mathcal{O}'$ due to the fact that the embedding dimension of $\mathcal{O}'$ is n (see [5]).

3. Other applications of the Hamburger-Noether matrix.

In this section we consider a fixed multiplicity sequence $\underline{e}$ given by parameters $n, h, n_1, h_1, \ldots, n_{r-1}, h_{r-1}, n_r=1$ and we will determine all the possible values for the number of dimension and for the embedding dimension of curves having $\underline{e}$ as their multiplicity sequence. Neccessary and sufficient conditions on the set $\{n, h, n_1, h_1, \ldots, n_{r-1}, h_{r-1}, n_r\}$ to be a multiplicity sequence can be found in [2], 5.3.

From proposition 1.1. the problem of locating curves with multiplicity sequence $\underline{e}$ is analogous to the problem of writting matrices composed by boxes of lengths $h, h_1, \ldots, h_r=\infty$ and marked rows verifying the following compatibility condition relative to $\underline{e}$: If $m_i=h_{i+1}n_{i+1}+\ldots+h_{s-1}n_{s-1}+Kn_s$, $i<s$, $K\leqslant h_s$, then the marked "1'

in the i-th box is followed in its row by exactly $(h_i-1)+h_{i+1}+\dots+(K-1)$ zeroes.

In order to construct matrices with this compatibility condition inductively we will need the following definition and technical lemma.

3.1. Definition.- A matrix with d rows and $h_1+\dots+h_{j-1}$ columns composed by boxes of consecutive columns of lengths $h, h_1, \dots, h_{j-1}$ and marked as in § 2 is said to be compatible with $\underline{e}$ if the above compatibility condition holds for indices $i<s$ with $s \leqslant j$.

In the sequel we will use the following notations:

$$d_o = 1 + \max_{1\leqslant e\leqslant r}\left\{\# \{k<e \mid n_k > h_{k+1}n_{k+1}+\dots+h_{e-1}n_{e-1}+n_e\}\right\} \quad (n_o=n)$$

$$d_1 = \# \left\{j \geqslant 0 \mid n_k \neq h_{k-1}n_{k-1}+\dots+h_{j-1}n_{j-1}+n_j \text{ for all } k<j\right\}$$

3.2. Lemma.- a) If $(B_o \mid B_1 \mid \dots \mid B_{j-1})$ is a matrix of $d \geqslant d_o$ rows and j marked boxes which is compatible with $\underline{e}$, then there exists a matrix of d rows and j+1 marked boxes $(B_o \mid B_1 \mid \dots \mid B_{j-1} \mid B_j)$ which is compatible with $\underline{e}$.

b) Let $(B_o \mid B_1 \mid \dots \mid B_{j-1})$ be a matrix of d rows and marked boxes which is compatible with $\underline{e}$, the index j verifying $n_k \neq h_{k+1}n_{k+1}+\dots+n_j$ for all $k<j$. Then there exists a matrix of d+1 rows and j+1 marked boxes compatible with $\underline{e}$ of type

$$\left(\begin{matrix}B_o\\0\end{matrix} \,\Big|\, \begin{matrix}B_1\\0\end{matrix} \,\Big|\, \dots \,\Big|\, \begin{matrix}B_{j-1}\\0\end{matrix} \,\Big|\, \begin{matrix}0\\1\end{matrix}\right).$$

Proof: a) We only must take into account that at most d_o-1 integers k verify $n_k > h_{k+1}n_{k+1}+\dots+h_{j-1}n_{j-1}+n_j$ and so the first column of B_j can be appropriately chosen and, moreover, it can be completed to the whole B_j whithout any difficulty.

b) It is evident.

3.3. Theorem.- For any curve with multiplicity sequence $\underline{e}$, $d_o \leqslant d(\vartheta) \leqslant d_1$, $d(\vartheta)$ being the number of dimension of ϑ. Conversely, if d is an integer such that $d_o \leqslant d \leqslant d_1$, then there exists an algebroid curve ϑ with multiplicity sequence $\underline{e}$ (and given algebracally field) such that $d=d(\vartheta)$.

Proof: Since d_o is the maximum of the satellitism orders of infinitely near points of ϑ (remark 1.3.) it is evident that $d_o \leqslant d(\vartheta)$. On the other hand, the inequality $d(\vartheta) \leqslant d_1$ follows from the fact that if $k < j$ are indices such that $n_k = h_{k+1} n_{k+1} + \ldots + n_j$ then according to (∗) the j-th marked row is one of the marked rows for some $j' < j$. Now take d with $d_o \leqslant d \leqslant d_1$ and consider the d first indices $j_1, \ldots, j_d$ verifying the condition in the definition of d_1. Using the part b) in the lemma one can easely construct a matrix with d rows, all of them marked, composed by j_d boxes which is compatible with $\underline{e}$. Now, by part a) in the lemma, applied $r-j_d$ times, we can construct a Hamburger-Noether matrix corresponding to a curve ϑ with $d(\vartheta)=d$.

3.4. Theorem.- Let d be an integer, $d_o \leqslant d \leqslant d_1$. Then for any curve ϑ with multiplicity sequence $\underline{e}$ and number of dimension d one has $d \leqslant \mathrm{Emb}(\vartheta) \leqslant n$, where $\mathrm{Emb}(\vartheta)$ denotes "embedding dimension". Conversely, if n' is an integer, $d \leqslant n' \leqslant n$, then there exists a curve ϑ with multiplicity sequence $\underline{e}$ and number of dimension d such that $\mathrm{Emb}(\vartheta)=n'$.

Proof: $d \leqslant \mathrm{Emb}(\vartheta)$ follows from the definition of d, and $\mathrm{Emb}(\vartheta) \leqslant n$ is a well-known fact. If $d \leqslant n' \leqslant n$, we can construct by theorem 3.3. a curve ϑ_o with multiplicity sequence $\underline{e}$ and $d=d(\vartheta_o)$ Now if in the procedure for constructing the Arf closure of ϑ_o (remark 2.6.) we use only n'-d steps, we obtain a curve ϑ with multiplicity sequence $\underline{e}$, with $d(\vartheta)=d$ and with $\mathrm{Emb}(\vartheta)=n'$ as desired.

REFERENCES.

[1] C.Arf. "Une interpretation algebrique de la suite des orders de multiplicité d'une branche algebrique". Proceedings of the London Mathematical Society (2), vol. 50(1949), pp. 256-287.

[2] A. Campillo. "Algebroid Curves in Positive Characteristic" Lecture Notes in Mathematics, vol. 813. Springer-Verlag.

[3] E.Casas."La proyección plana genérica de una rama de curva alabeada". Preprint.Dptº de Geometría. Universidad Central de Barcelona.

[4] P. Du Val. Note on Cahit Arf's "Une interpretation algébrique de la suite des orders de multiplicité d'une branche algébrique". Proceedings of the London Mathematica Society (2), vol. 50(1949),pp.288-294.

[5] J. Lipman. "Stable Ideals and Arf Rings".American Journal of Mathematics, vol. 93 (1971), pp. 649-685.

[6] F.Enriques-O.Chisini."Teoria geometrica delle equazioni". N.Jansichelli, Bologna, 1915-1924.Libre IV.

[7] O.Zariski."Algebraic Surfaces".Springer-Verlag-Berlin. New York (1972).

MODULI OF ALGEBROID PLANE CURVES

by

E. Casas

0.- Introduction. Let O be a smooth point on an algebraic surface S. the study of the formal (analytical) equivalence of irreducible algebroid (analytic) curves on S centered at O, or, equivalently, the study of the corresponding local moduli spaces (see S.1, pages 318 and 319, and, above all, Z.4) is still rich in open problems. Here we give a geometrical approach to the formal equivalence using the theory of infinitely near points. The infinitely near points in the succesive neighbourhoods of O lie on infinitely many projective lines and the group of formal transformations of S at O induces a group of homographies on every such projective line. The determination of these groups allows us to define a complete set of invariants for the irreducible algebroid curves on S centered at O.

We will take as base field the field C of complex numbers. However convergence questions are not considered here and thus we deal only with algebroid curves and formal transformations.

1.- Successive neighbourhoods of a point on a smooth surface. Let O be a smooth point on a smooth surface S. Although the points infintely near to O on S can be defined by abstraction as certain equivalence classes of branches centered at O (cf. W.1), here we will use blowing-ups to define them. The points on the first neighbourhood of O on S are the points of the exceptional curve $V_1(O)$ of the blow-up $S \longleftarrow S_1$ of O on S. The points of the i-th neighbourhood of O on S are defined recursively as the points that are in the first neighbourhood of a point in the (i-1)-th neighbourhood. Thus each infinitely near point of a neighbourhood of O is a simple point on a surface obtained from S by performing a succession of finitely many blowing-ups at infinitely near points in neighbourhoods of lo er order.

Let O_i be a point in the i-th neighbourhood of O and let S_i be the blow-up surface which it belongs to. We will say that an algebroid branch γ with center O goes through O_i with multiplicity n_i when the proper transform of γ on S_i is centered at O_i and its multiplicity there is n_i. We will also say that O_i is a point of the branch γ. In this way each branch is determined by the infinitely near points it contains. Conversely, each infinitely near point can be identified with the class of branches that go through it, which defines infinitely near points by abstraction (See W.1).

No matter wich way we choose to define the infintely near points they play a fundamental role in the intersection multiplicity of branches, as stressed in Noether's formula

$$\gamma.\tilde{\gamma} = \sum n_i m_i$$

where $\gamma.\tilde{\gamma}$ is the intersection muliplicity of the branches γ, $\tilde{\gamma}$, both centered at O, n_i, m_i are their respective multiplicities at an infinitely near point O_i (including O), and where the summation is extended over all points belogning to both branches (cf. W.1)

As it is customary we will call <u>proximate points</u> of an infinitely near point O_i the points in the first neighbourhood of O_i and also the points infinitely near to points in the first neighbourhood that belong to the corresponding proper transforms of the exceptional curve of the blowing-up of O_i. Since the multiplicity of any branch γ

at O_i equals the intersection multiplicity of the proper transform of γ with the exceptional curve of the blowing-up of O_i, we get the so called proximity relations: if O_i lies on a branch γ then $n_i = \sum n_s$, where n_s denotes the multiplicity of a point O_s on γ and where the sum is extended over all points on γ that are proximate to O_i.

The classical distinction between free and satellite infinitely near points, based on the kind of conditions imposed to the branches in order to go through them (cf. E.1, Libro 4º, Cap. 1) may be rephrased as follows: any infinitely near point O_i ($\neq O$) is proximate to the preceding point; O_i will be called satellite if in addition it is proximate to some O_j, $j < i-1$. Otherwise we will say that O_i is free. We refer to Z.3 and C.1. There are light differences in the definitions: in the classical texts the satellite points are said to be satellite of the last free point which precedes its, whereas in Z.3 free and satellite with respect to O stand respectively for free and satellite in the above sense.

Notice that all points in the first neighbourhood of O are free and that in the first neighbourhood of a point O_i infinitely near to O there are one or two satellite points according to whether O_i is free or satellite.

2.- Composition of a branch singularity. We call composition of a branch γ the sequence of multiplicities on γ of O and the successive infinitely near points on γ until the first smooth point (i. e., of multiplicity one) is reached. As we will see in greater detail below, the composition of a branch can be calculated from the characteristic exponents of its Puiseux series, and conversely, these can be obteined from the branch composition. Thus the equality of composition-s is one of the equivalent formulations (very likely the oldest) for the equisingularity of plane curves. (E.1, Libro 4º, Cap. I; Z.1 and Z.2 S.1).

To be more precise, let us consider an algebroid plane curve branch γ with center at 0 and let

$$y = \sum_{i=1}^{h} a_i x^i + \sum_{i=0}^{h_1} a_{1,i} x^{(\alpha'_1+i)/\alpha_1} + \ldots +$$

$$\sum_{i=0}^{h_{q-1}} a_{q-1,i} x^{(\alpha'_{q-1}+i)/\alpha_1\cdots\alpha_{q-1}} + \sum_{i=0}^{\infty} a_{q,i} x^{(\alpha'_q+i)/\alpha_1\cdots\alpha_q} \qquad [1]$$

be its Puiseux series, or, if we wish power series of a parameter t,

$$x = t^n$$

$$y = \sum_{i=1}^{h} a_i t^{ni} + \sum_{i=0}^{h_1} a_{1,i} t^{(\alpha'_1+i)\alpha_2\cdots\alpha_q} + \ldots + \sum_{i=0}^{h_{q-1}} a_{q-1,i} t^{(\alpha'_{q-1}+i)\alpha_q}$$

$$+ \sum_{i=0}^{\infty} a_{q,i} t^{\alpha'_q+i}$$

where $n = \alpha_1\cdots\alpha_q$ and where we assume that $(\alpha'_i,\alpha_i) = 1$, $a_{j,o} \neq 0$, $j = 1, \ldots, q$ and that

$$h < \alpha'_1/\alpha_1 < h+1 \text{ and } \alpha'_{j-1} + h_{j-1} < \alpha'_j/\alpha_j < \alpha_{j-1} + h_{j-1} + 1$$

for $j = 2,\ldots,q$, perhaps allowing some terms to vanish.

We recall that a branch determines its Puiseux series, once the local coordinates x, y are chosen, except for a change in the determination of $x^{1/n}$, i.e., but for a substitution of εt for t, where $\varepsilon^n = 1$.

We will say that α_j is the polydromy order of the series introduced by the characteristic term $a_{j,0} x^{\alpha'_j/\alpha_1\cdots\alpha_j}$.

In order to describe how the data above determine the

composition of the branch, write $\alpha'_1/\alpha_1 = n_1/n$, so that $\gcd(n_1,n) = \alpha_2 \ldots \alpha_q$, and perform the Euclidian algorithm to compute this greatest common divisor:

$$n_1 = hn + \nu_{1,1}$$
$$n = m_{1,1}\nu_{1,1} + \nu_{1,2}$$
$$\ldots$$
$$\nu_{1,s_1-2} = m_{1,s_1-1}\nu_{1,s_1-1} + \nu_{1,s_1}$$
$$\nu_{1,s_1-1} = m_{1,s_1}\nu_{1,s_1} \qquad (\nu_{1,s_1} = \alpha_2 \ldots \alpha_q)$$

Then on the branch γ in question we have:

h n-uple points $O, O_1, \ldots, O_{h-1}$

$m_{1,1}$ $\nu_{1,1}$-uple points $O_h, \ldots, O_{h+m_{1,1}-1}$

$m_{1,2}$ $\nu_{1,2}$-uple points $O_{h+m_{1,1}}, \ldots, O_{h+m_{1,1}+m_{1,2}-1}$

$\ldots$

m_{1,s_1} ν_{1,s_1}-uple points $O_{h+m_{1,1}+\ldots+m_{1,s_1-1}}, \ldots, O_{h+m_{1,1}+\ldots+m_{1,s_1}-1}$

Of all these points, $O_1, \ldots, O_h$ are free points and the remaining are satellite points.

Now proceed inductively as follows. Take the j-th characteristic exponent and set $\alpha'_j/\alpha_1 \ldots \alpha_j = n_j/n$, $\nu_{j,s_j} = \alpha_{j+1} \ldots \alpha_q$, so that

$$\nu_{j,s_j} = \gcd(\nu_{j-1,s_{j-1}}, n_j) - \gcd(n, n_1, \ldots, n_j)$$

Perform the Euclidian algorithm that computes this greatest common divisor,

$$n_j = (\alpha'_{j-1} + h_{j-1})\nu_{j-1,s_{j-1}} + \nu_{j,1}$$
$$\nu_{j-1,s_{j-1}} = m_{j,1}\nu_{j,1} + \nu_{j,2}$$
$$\ldots$$

$$\nu_{j,s_j-1} = m_{j,s_j}\nu_{j,s_j}$$

Then we obtain, after the last point in the (j-1)-step, the following points:

h_{j-1} $\nu_{j-1,s_{j-1}}$-uple points $O_0^{j-1},\ldots,O_{h_{j-1}-1}^{j-1}$

$m_{j,1}$ $\nu_{j,1}$-uple points $O_{h_{j-1}}^{j-1},\ldots,$ $O_{h_{j-1}+m_{j,1}-1}^{j-1}$

. . .

m_{j,s_j} ν_{j,s_j}-uple points $O_{h_{j-1}+m_{j,1}+\ldots+m_{j,s_j-1}}^{j-1},\ldots,O_{h_{j-1}+m_{j,1}+\ldots+m_{j,s_j}-1}^{j-1}$

Of all these points $O_0^{j-1},\ldots,O_{h_{j-1}}^{j-1}$ are free points and the remaining are satellite points.

After the last satellite point $O_{h_{q-1}+\ldots+m_{q,s_q}-1}^{q-1}$ we get simple free points O_i^q, $i \geq 0$.

The position of each free point O_i^j, $j=0,\ldots,q$, $i = 0,\ldots,h_j$ (take $O_i^0 = O_i$ and $h_q = \infty$) is determined, in relationship to the preceeding points, by the coefficient $a_{j,i}$ in the Puiseux series. Here by position we mean the following:

a) $i = 0$. The branches γ' through the satellite point before O_0^j and which have a free point after it, have a Puiseux series that coincides with the series [1] until the partial sum of degree (in x) $(\alpha'_{j-1}+h_{j-1})/\alpha_1\ldots\alpha_{j-1}$, both branches having the same characteristic exponent $\alpha'_j/\alpha_1\ldots\alpha_j$. Then in order that such a branch γ' goes through O_0^j it is necessary and sufficient that $a_{j,0}$ coincides with the corresponding coefficient in the branch γ', except for a factor which is an α_j-th root of unity. In fact one can choose the Puiseux series of γ' in such a way that the coefficient of the above characteristic term is $a_{j,o}$, using recursively the described indetermination.

b) $i \neq 0$. Since, as it is known, the branches that go through the last free point preceding O_i^j have a Puiseux series that coincides with our series up to the partial sum of degree $(\alpha'_j+i-1)/\alpha_1\cdots\alpha_j$ in x, the condition that such branches go through O_i^j is equivalent to the coincidence of the partial sums of degree $(\alpha'_j+i)/\alpha_1\cdots\alpha_j$, that is, the coincidence of the coefficient of $x^{(\alpha'_j+i)/\alpha_1\cdots\alpha_j}$ with $a_{j,i}$ and the non-appearance of a characteristic term in between (take $a_{0,i} = a_i$, $\alpha'_0 = 0$, $\prod_1^0 \alpha_j = 1$).

The conditions for a branch to go through a satellite point may be formulated in terms of the characteristic exponents. The precise form of these conditions need not be explicitized here. For proofs and additional details the reader is referred to E.1 (Libro 4, Cap.I) or S.1, and also to C.2, where a similar argument for surfaces is introduced.

3. Enriques' diagrams. It is extremely sugestive the way in which Enriques represented infinitely near points (again E.1, Libro 4, Chap I). The points lying on a curve branch with origin at O are represented as a sequence of points on a broken line in a plane, at equal distances between consecutive points, and according the following conventions: once O_{i-1} is represented, then the free points O_i in its first neighbourhood will be represented on arcs issuing from O_{i-1} with the same tangent at O_{i-1} as $O_{i-2}O_{i-1}$, and whose curvature corresponds to the position of O_i, that is, to the value of the corresponding coefficient in the Puiseux series. The proximate points to O_{i-2} that lie in successive neighbourhoods of O_{i-1} are placed on a line segment starting at O_{i-1} orthogonally to $O_{i-2}O_{i-1}$. These segments are taken alternatively to the rigth or to the left. The resulting

diagram contains curvilinear parts, on which the free points are represented, alternating with as many stair-like pieces, on which the satellite points lie, as characteristic exponents there are.

As an example let us consider the representation of a branch with characteristic exponents 25/7 and 173/42. The composition of such a branch is 42, 42, 42, 24, 18, 6, 6, 6, 6, 6, 6, 5, 1 and therefore its diagram looks like this

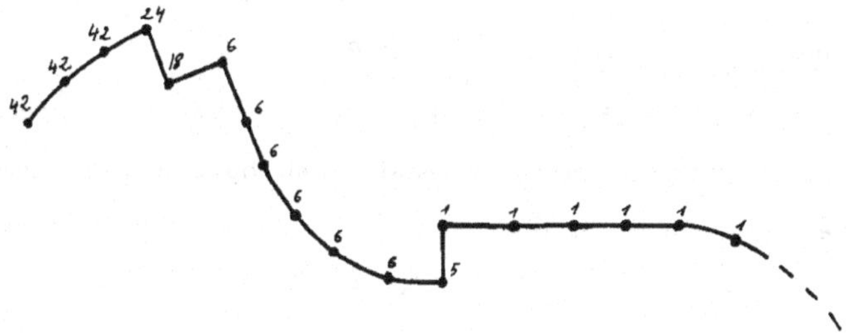

figure 1

As one can easily see, it is not necessary to indicate the multiplicities of the points on the diagram since it is enough to know the position of the first free simple point in order to know the composition and consequently the characteristic exponents. In fact the composition can be written using the proximity relations, and from the composition the characteristic exponents can be expressed as a continous fraction according to the division algorithms of the previous section.

Thus we see that the diagrams corresponding to branches in a given equisingularity class have the same structure, i. e., they can only differ in the curvature of corresponding curvilinear pieces. Since the formal equivalence of two equisingular branches is equivalent to the possibility of superposing the free points in their diagrams under a formal transformation, it is natural to ask how the formal transformations act on the infinitely near points.

4. Projective structure of the successive neighbourhoods. Let P be a point in the i-th neighbourhood of O. We will look at $V_1(P)$, the first neighbourhood of P, as a projective line. $V_1(P)$ is the exceptional line in the blowing-up of S_i at P and hence its points correspond projectively to the tangent directions to S_i at P.

From what we said in section 2, it turns out that the Puiseux series of the branches going through P and which have a free point on $V_1(P)$, can be chosen to be of the form

$$y = a_1x + \dots + a_r x^{\alpha/\beta} + a x^{\alpha'/\beta'} + \dots$$

where the exponents are as in [1] of section 2, the partial sum $a_1x + \dots + a_r x^{\alpha/\beta}$ is fixed, and a determines the position of the corresponding point on $V_1(P)$. Here we have $\alpha'/\beta' = (\alpha + 1)/\beta$ if P is free and that α'/β' is a well determined characteristic exponent if P is satellite, in which case we must assume $a \neq 0$. With these notations we have:

Proposition 1. If P is free, then the coefficients a can be taken as absolute coordinate of the point on $V_1(P)$ corresponding to the branch. If P is satellite, then we must take a^μ as absolute projective coordinate on $V_1(P)$, where μ is the polydromy order introduced by the characteristic term $ax^{\alpha'/\beta'}$.

Proof. As it is known, the point on $V_1(P)$ and the coefficient of the term of degree α'/β' depend only, for the branches under consideration, on the partial sum of degree α'/β' on the Puiseux series. Therefore it is enough to consider branches γ_a with Puiseux series

$$y = a_1x + \dots + a_r x^{\alpha/\beta} + ax^{\alpha'/\beta'}$$

where a is arbitrary (non zero if P is satellite) and the remaining coefficients are fixed.

For each value of a consider the point on $V_1(P)$ corresponding to γ_a. In this way we get an algebraic correspondence between the projective line with absolute coordinate a and $V_1(P)$. If P is free this correspondence is genericaly (1:1), by the section 2, hence birrational, hence a projectivity. If P is a satellite point the correspondence is (μ:1) so that now we can apply the same argument with the correspondence obtained by composing with the correspondence $a^{\mu} \leftrightarrow a$.

Remarks. 1) If P is free then the coordinate of the satellite point on $V_1(P)$ is ∞. If P is satellite then the coordinates of the two satellite points on $V_1(P)$ are 0 and ∞; the satellite point proximate to the point which precedes P has coordinate 0 or ∞ depending on the structure of the group of satellite points to which P belongs. If the characteristic exponent α'/β' is, with the notations of the section 2,

$$\alpha'/\beta' = \frac{1}{\alpha_1\cdots\alpha_{j-1}}\left[\alpha'_{j-1} + h_{j-1} + \cfrac{1}{m_{j,1} + \cfrac{1}{m_{j,2} + \cdots + \cfrac{1}{m_{j,s_j}}}}\right]$$

take

$$\alpha''/\beta'' = \frac{1}{\alpha_1\cdots\alpha_{j-1}}\left[\alpha'_{j-1} + h_{j-1} + \cfrac{1}{m_{j,1} + \cfrac{1}{m_{j,2} + \cdots + \cfrac{1}{m_{j,s_i} \pm \frac{1}{2}}}}\right]$$

where the ± sign is chosen in such a way that α''/β'' is greater than α'/β'. Now we can argue as in the previous proof taking instead of γ_a the branch given by

$$y = a_1x + \dots + a_r x^{\alpha/\beta} + ax^{\alpha'/\beta'} + x^{\alpha''/\beta''}$$

We get the same correspondence, which this time is easily extendable to $a = 0$. The selection of the sign, i. e., the parity of s_j, determines which of the two satellite points corresponds to the value $a = 0$.

2). The coordinate on the first neighbourhood of P described above depends not only on the local coordinates x, y but also on the partial sum chosen for the branches going through P. For this last choice there are a finite number of possibilities. For if in the Puiseux series [1] written in section 2 one takes $a_{j,i}$, $i \neq 0$, as a coordinate, then any change in the partial sum consists in substituting $\varepsilon x^{1/n}$ for $x^{1/n}$, where $\varepsilon^n = 1$, so that $a_{j,i}$ is replaced by $(\varepsilon^{(\alpha'_j+i)\alpha_{j+1}\cdots\alpha_q})a_{j,i}$. Similarly, when the coordinate under consideration is of the form $a_{j,0}^{\alpha_j}$, then it must be replaced by $\varepsilon^{\alpha'_j\alpha_j\cdots\alpha_q}a_{j,0}^{\alpha_j}$.

5.- <u>Transformations</u>. Given a smooth point O on a algebraic surface S, set $\mathcal{O}_O$ to denote the completion of the local ring of S at O. The automorphisms of $\mathcal{O}_O$ will be referred to as formal transformations of S at O. If we choose local coordinates x, y as before, then the formal transformations are automorphisms of $\mathbb{C}[[x,y]]$. Formal transformations act on the branches by transforming their equations. Our purpouse is to study the action these transformations induce on the succesive neighbourhoods of O.

More generally, given two simple points O, O' on two surfaces S, S' respectively, we call the isomorphisms between the completions of the corresponding local rings, formal transformations of O into O'. If $\Phi : \mathcal{O}_O \to \mathcal{O}_{O'}$ is such a transformation we will denote also Φ the induced bijection betwen the set of branches centered at O and the set of branches centered at O'. Since under this transformation the intersection multiplicity is invariant, we see, using the definition by abstraction of infinitely near points (as equivalence classes of branches defined inductively in terms of intersection multiplicities, cf. W.1) that Φ induces bijections between the succesive neighbourhoods of O and the corresponding successive neighbourhoods of O'. The resulting transformations will also be denoted Φ.

Theorem 1. Let Φ a formal transformation of O into O', O_i a point in the i-th neighbourhood of O and $O'_i = \Phi(O_i)$. Then

1) There exists a unique formal transformation Φ_i of O_i into O'_i, as points lying on the surfaces S_i and S'_i respectively, such that the following diagram

$$\begin{array}{ccc} \mathcal{O}_O & \xrightarrow{\ \Phi\ } & \mathcal{O}_{O'} \\ \downarrow & & \downarrow \\ \mathcal{O}_{O_i} & \xrightarrow{\ \Phi_i\ } & \mathcal{O}_{O'_i} \end{array}$$

is commutative, where the vertical morphisms are the monomorphisms which corresponds to the natural morphisms $S_i \to S$ and $S'_i \to S'$.

2) The transformation Φ defines a linear projectivity between $V_1(O_i)$ and $V_1(O'_i)$.

Proof: It is an easy consequence of the following lemma:

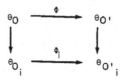

<u>Lemma 1</u>. With the notations of the previous theorem, Φ induces a projectivity $\Phi : V_1(O) \to V_1(O')$ and for all $O_1 \varepsilon V_1(O)$ there exists a unique Φ_1 that makes the following diagram commutative

$$\begin{array}{ccc} \theta_O & \xrightarrow{\Phi} & \theta_{O'} \\ \downarrow & & \downarrow \\ \theta_{O_1} & \xrightarrow{\Phi_1} & \theta_{O'_1} \end{array}$$

Assuming this lemma, the proof of 1) follows by induction. In fact the conmutativity of the diagrams implies that Φ and Φ_i induce the same transformations on the infinitely near points on which they can act. Then 2) follows from 1) and the first statement in the lemma.

<u>Proof of the lemma</u>. Let x, y local parameters at O and x',y' local parameters at O' and assume that

$$\Phi(x) = \sum_{i,j\geq 0} a_{i,j} x'^i y'^j \ , \quad \Phi(y) = \sum_{i,j\geq 0} b_{i,j} x'^i y'^j \ , \quad a_{0,0} = b_{0,0} = 0$$

Without loss of generality we may assume that $a_{0,1} = b_{1,0} = 0$, since we may perform a linear transformation on the parameters at O' in such a way that with respect the new parameters the condition is satisfied. Then we must have $a_{1,0} b_{0,1} \neq 0$. Now the branch whose equation is $\alpha x + \beta y = 0$ is transformed into the branch

$$\sum_{i,j\geq 0} (\alpha a_{i,j} + \beta b_{i,j}) x'^i y'^j = 0$$

whose tangent at the origin is $\alpha a_{1,0} x' + \beta b_{0,1} y' = 0$. Thus the induced transformation $V_1(O) \to V_1(O')$ is the projectivity given by

$$(\alpha, \beta) \longmapsto (\alpha a_{1,0}, \beta b_{0,1})$$

To see the second part of the lemma we may assume that O_1 corresponds to $y = 0$, performing a linear transformation of the parameters at O if necessary. Then O'_1 corresponds to $y' = 0$. Consecuently x, y/x are local parameters at O_1 and x', y'/x' are local parameters at O'_1. Now it suffices to see that Φ can be uniquely extended to an isomorphism between the rings $C[[x,y/x]]$ and $C[[x',y'/x']]$ and this follows from the following computation

$$\frac{\Phi(y)}{\Phi(x)} = \frac{\sum b_{i,j}x'^i y'^j}{\sum a_{i,j}x'^i y'^j} = \frac{\sum b_{i,j}x'^{i+j}(y'/x')^j}{\sum a_{i,j}x'^{i+j}(y'/x')^j} = \frac{\sum b_{i,j}x'^{i+j-1}(y'/x')^j}{\sum a_{i,j}x'^{i+j-1}(y'/x')^j} =$$

$$= (b_{0,1}/a_{1,0})(y'/x') + \dots$$

since the last denominator is a unit in $C[[x',y'/x']]$. Q.E.D.

Remark. The intrinsic character of the notions of proximity and satellitism guarantees that the projectivities in the theorem transform satellite points into satellite points and that the proximity relations are preserved.

If we considere in particular the group of formal transformations of O into itself that leave invariant successive points $O_1,\dots,O_i$ we get a group of homographic transformations of $V_1(O_i)$; this group will be denoted $G(O_i)$.

According the previous remark $G(O_i)$ is a group of affine transformations if we take the satellite point whose coordinate is ∞ as the improper point in $V_1(O_i)$, assuming of course that $O_i \neq O$. Moreover, the transformations in $G(O_i)$ are homotheties if O_i is a satellite point.

We will say that $G(O_i)$ acts transitively whenever its action on the open set of free points of $V_1(O_i)$ is transitive.

Now we can rephrase the problem of recognizing the formal equivalence between two branches in the following terms: given branches γ and γ' with the same composition, let us assume that they coincide up to the point O_i on the i-th neighbourhood of O and that they define free points on $V_1(O_i)$, say O_{i+1} and O'_{i+1}, respectively. If γ and γ' are formally equivalent then the transformation that takes γ into γ' induces an element of $G(O_i)$ which transforms O_{i+1} into O'_{i+1}. Conversely, if we can find such an element in the group $G(O_i)$, then the transformation which induces it allows us to replace γ by a formaly equivalent branch that has a higer contact with γ'. Proceeding in this way, we can decide in finitely many steps the formal equivalence of γ and γ', since, as it is known, and as we will see later, there are no obstructions to the formal equivalence from some neighbourhood of O onwards and moreover the order of this neighbourhood depends only on the equisingularity class under consideration.

Notice in particular that $G(O_i)$ acts transitively if O_i lies on a smooth branch, since any two smooth branches are formaly equivalent.

<u>6.- The "brute force" method</u>. Now we will analize the conditions imposed on the coefficients of the Puiseux series of two branches for them to be formaly equivalent. The method to reach such conditions is basicaly due to Zariski (vid. Z.4, especially pag. 39).

Since $G(O_i)$ acts transitively when O_i is in a smooth branch, there will be no loss of generality in assuming, given two equisingular branches γ_1, γ_2, that they admit Puiseux series of the form

$$\begin{cases} x = t^n \\ y = \sum_{i \geq 0} a_i t^{i+m} \end{cases} \qquad \begin{cases} x = t^n \\ y = \sum_{i \geq 0} b_i t^{i+m} \end{cases}$$

and such that they coincide until the last satellite point of the first group.

Let $\theta = \theta_O = C[[x,y]]$ be the completion of the local ring of S at O and let R_1 and R_2 be the local rings of γ_1 and γ_2 respectively, so that, since they are quotients of θ, we can write $R_1 = C[[\tilde{x},\tilde{y}]]$, $R_2 = C[[\hat{x},\hat{y}]]$, where $\tilde{x}$, $\tilde{y}$ (respectively $\hat{x}$, $\hat{y}$) are the remainder classes of x, y modulo the equation of γ_1 (resp. γ_2).

We can identify R_1 and R_2 with subrings of C[[t]], using the same parameter for both branches and the Puiseux series above. In this way C[[t]] appears to be the integral closure of R_1 and R_2 in its quotient field. Notice that another selection of the Puiseux series identify R_1 and R_2 to subrings of C[[t]] obteined from the former subrings by transforming them under an automorphism of the form $t \leftrightarrow \varepsilon t$, $\varepsilon^n = 1$.

Any isomorphism $\psi' : R_1 \to R_2$ can be extended in a unique way to an automorphism ψ of C[[t]]. This automorphism cleary satisfies $\psi(R_1) \subset R_2$. Conversely, if ψ is an automorphism of C[[t]] such that $\psi(R_1) \subset R_2$, then restricted to R_1 is an isomorphism of R_1 onto R_2. To see this it is enough to observe that the images of $\tilde{x}$, $\tilde{y}$ are series in t whose orders are n and m respectively, so that we will have

$$\psi(\tilde{x}) = \sum_{i,j \geq 0} c_{i,j} \hat{x}^i \hat{y}^j, \quad \psi(\tilde{y}) = \sum_{i,j \geq 0} d_{i,j} \hat{x}^i \hat{y}^j, \quad c_{0,0} = d_{0,0} = d_{1,0} = 0,$$

$$c_{1,0} d_{0,1} \neq 0$$

and hence these relations can be inverted, which shows that $\hat{x}$, $\hat{y}$ are in the image of ψ and consequently $\psi(R_1) = R_2$. In other words, the formal equivalence of the branches γ_1 and γ_2 is equivalent to the existence of an automorphism ψ of $C[[t]]$ such that $\psi(\tilde{x})$, $\psi(\tilde{y}) \in R_2$.

Set Γ to denote the common value semigroup of the branches γ_1 and γ_2,

$$\Gamma = \{ord_t f,\ f \in R_1\} = \{ord_t f,\ f \in R_2\} \subset N$$

That this semigroup is an equisingularity invariant is obvious, by Noether's formula, if we regard Γ as the set of intersection multiplicities of the branch under consideration with all algebroid curves centered at 0, and the equisingularity in terms of the composition.

The existence of a non-zero conductor for the extensions $R_i \to C[[t]]$, $i = 1,2$, is equivalent to the finiteness of $N - \Gamma$. The least integer c such that $c + i \in \Gamma$ for all $i \geq 0$ is usualy called conductor of Γ. The condition defining c is equivalent to say that the conductor for both extensions is $t^c C[[t]]$ (vid. Z.4, II.1 and II.2).

The only restrictions that the coefficients in one of our Puiseux series must satisfy is the non-vanishing of the characteristic coefficients. A sequence of coefficients satisfiying this condition will be said to be a admissible sequence of coefficients.Still another piece of notation: if $f = \sum_{i \geq 0} c_i t^i$ then we set $[f]_i = c_i$.

<u>Lemma 2</u>. For each $i \in \Gamma$ there exists a rational function $S_i(Z_0,\ldots,Z_{i-m-1},X,Y)$ such that for any admissible sequence of coefficients b_j we have that $S_i(b_0,\ldots,b_{i-m-1},t^n,\hat{y})$ is an element of order i of R_2 which satisfies the condition $[S_i(b_0,\ldots,b_{i-m-1},t^n,\hat{y})]_i = 1$.

Proof. It is enough to consider only those i in a generating set of Γ. Now a generating set of Γ is obtained, according to B.1, taking n and, for each group of satellite points on γ_2, the intersection multiplicity of γ_2 with a branch that contains the last free point preceding the group of satellites as a simple point. For $i = n$, $S_n = X$ will do. With respect to the remaining generators, observe that the required branches are obtained by taking as their Puiseux series the partial sums of the Puiseux series $y = y(x)$ of γ_2 once a characteristic term and all higher terms are dropped (cf. section 2). We will have series of the form

$$y = \sum_{j=0}^{v} b_j x^{\frac{j+m}{n}}$$

so that if $d = \mathrm{mcd}(n, j+m \mid b_j \neq 0)$, then we can take

$$\prod_{\varepsilon} \left(y - \sum_{j=0}^{v} b_j \varepsilon^{\frac{m+j}{d}} x^{\frac{m+j}{n}}\right)$$

where ε runs through all ε such that $\varepsilon^{n/d} = 1$, as the equation of one of these branches. Its intersection multiplicity i with γ_2 is given by

$$i = \mathrm{ord}_t \prod_{\varepsilon} \left[\sum_{j \geq 0} b_j t^{j+m} - \sum_{j=0}^{v} b_j \varepsilon^{\frac{m+j}{d}} t^{m+j}\right] > m+v$$

Since the coefficient of the initial term of this series is the product of the coefficients of the initial terms of its factors, which in turn are scalar multiples of characteristic coefficients, we see that the initial coefficient of our series has the form $\eta b_{j_0} \dots b_{j_u}$, where $\eta \neq 0$, $\eta \in C$ and the b_{j_k}'s are characteristic coefficients. Thus it is enough to take

$$S_i(Z_0,\dots,Z_{i-m-1},X,Y) = \frac{1}{nZ_{j_0}\dots Z_{j_u}} \prod_{\varepsilon}\left(Y - \sum_{j=0}^{v} Z_j \varepsilon^{\frac{m+j}{d}} X^{\frac{m+j}{n}}\right)$$

to fulfill the requirements of the lemma.

Let $f \in C[[t]]$. We are going to use the existence of the rational functions S_i to establish a finite recursive procedure to decide whether $f \in R_2$ or not. If $ord_t f = c$, there is nothing to worry about. Otherwise let us assume that the procedure has already been established for elements whose order is greater than i, where $i < c$, and that $ord_t f \geq i$. If $i \in \Gamma$ then we can find $g \in R_2$ such that $ord_t g = i$. Set $f' = f - \rho g$, where $\rho \in C$ is selected so that $ord_t f' > ord_t f \geq i$. Thus we can decide whether or not f' is an element of R_2, by induction, and hence whether or not $f \in R_2$. And if $i \notin \Gamma$ then we have $f \notin R_2$, or, if $f \in R_2$, we must have $[f]_i = o$, $ord_t f > i$ and the recursive procedure applies again.

Let us assume that ψ is a automorphism of $C[[t]]$, say $\psi(t) = t\left(\sum_{i \geq 0} \lambda_i t^i\right)$, $\lambda_0 \neq 0$. We call admissible the sequences of values of the λ's with $\lambda_0 \neq 0$. Then we have

<u>Proposition 2</u>. For each positive integer i such that $i + n \notin \Gamma$ there exists a rational function F_i in $2i + n - m + 1$ variables such that

$$F_i(\lambda_0,\dots,\lambda_{i-1},b_0,\dots,b_{i+n-m})$$

is defined for each admissible sequences of b's and λ's and

$$\alpha_i)\ \lambda_i = F_i(\lambda_0,\dots,\lambda_{i-1},b_0,\dots,b_{i+n-m}) \text{ , } i + n \notin \Gamma$$

are necessary and sufficient conditions in order that $\psi(\ddot{x}) \in R_2$.

<u>Proof.</u> We will assume that by an inductive process we have

constructed rational functions F_j, for $j < i$, $j+n \notin \Gamma$, and a rational expression $H_i(\lambda_0,\ldots,\lambda_{i-1},b_0,\ldots,b_{i-m+n-1},\hat{y},t^n)$ whose denominator contains only a product of characteristic coefficients, satisfying $ord_t(\psi(\tilde{x}) - H_i(\lambda,b,\hat{y},t^n)) \geq i+n$ if the preceding relations α_j, $j < i$, are verified, and such that the relation $\psi(\tilde{x}) \varepsilon R_2$ is equivalent to the verification of the relations α_j, for $j < i$, $j + n \notin \Gamma$, and the relation $\psi(\tilde{x}) - H_i(\lambda,b,\hat{y},t^n) \varepsilon R_2$. We show how to define F_i, if $i + n \notin \Gamma$, and H_{i+1}. To this end notice that in $\psi(\tilde{x}) = t^n\left[\sum_{i \geq 0} \lambda_i t^i\right]^n$ the coefficient of t^{i+n} contains a single term in which appears λ_i, namely $n\lambda_0^{n-1}\lambda_i$, and that the other terms involve only λ_j's with $j < i$. Thus if $i + n \notin \Gamma$, then we set $H_{i+1} = H_i$ and define F_i as the result of solving for λ_i the relation obtained by equating the coefficients of t^{i+n} in $\psi(\tilde{x})$ and $H_i(\lambda,b,\hat{y},t^n)$. The resulting F_i has the required properties since it will not involve any b_j with $j > i+n-m$ (such a b_j introduce t^{j+m}, with higer degree than t^{i+n}). And if $i+n \in \Gamma$ then we set

$$H_{i+1} = H_i + [\psi(\tilde{x}) - H_i]_{i+n} S_{i+n}$$

where S_{i+n} has been defined in lemma 2. It is easy to see that in this way the induction hypotesis is preserved. Since the relation $\psi(\tilde{x}) - H_i(\lambda,b,\hat{y},t^n) \varepsilon R_2$ is satisfied as soon $i+n = c$, the proof is complete.

With respect to $\psi(\tilde{y})$ we have

<u>Proposition 3</u>. For each positive integer i such that $i + m \notin \Gamma$ there exists a rational expression $G_i(\lambda_0,\ldots,\lambda_i,a_0,\ldots,a_{i-1},b_0,\ldots,b_{i-1})$, **defined for all admissible sequences** $\lambda_0,\ldots,\lambda_i$, $a_0,\ldots,a_{i-1},b_0,\ldots,b_{i-1}$, such that the relation $\psi(\tilde{y}) \varepsilon R_2$ is equivalent to the relations

$$\beta_i) \quad b_i = \lambda_0^i \frac{b_0}{a_0} a_i + G_i(\lambda_0,\ldots,\lambda_i,a_0,\ldots,a_{i-1},b_0,\ldots,b_{i-1}), \quad i+m \notin \Gamma.$$

Moreover λ_i appears in G_i with exponent 1 and non-zero coefficient for all admissible λ's, a's and b's.

Proof. We shall proceed as in the proof of proposition 2. From the expression

$$\psi(\tilde{y}) = \sum_{j \geq 0} a_j t^{m+j} \Big(\sum_{r \geq 0} \lambda_r t^r \Big)^{m+j}$$

we see that the relation $\psi(\tilde{y}) \varepsilon R_2$ is equivalent to the relation

$$\psi(\tilde{y}) - \lambda_0^m \frac{a_0}{b_0} \hat{y} \;\varepsilon\; R_2$$

Now let us assume by induction that this relation is equivalent to the relations β_j) for $j+m \notin \Gamma$, $j < i$, together with a relation of the form

$$\psi(\tilde{y}) - \lambda_0^m \frac{a_0}{b_0} \hat{y} - L_i(\lambda_0,\ldots,\lambda_{i-1},a_0,\ldots,a_{i-1},b_0,\ldots,b_{i-1},\hat{x},\hat{y}) \;\varepsilon\; R_2$$

where L_i is a rational expression whose denominator is a product of characteristic b's, without first degree terms in $\hat{x}$, $\hat{y}$, and such that $\operatorname{ord}_t E_i \geq m + i$, where E_i is the left term of the relation above, whenever the preceding relations $\beta_j (j < i,\ j+m \notin \Gamma)$ are verified.

We show how define G_i (if $i+m \notin \Gamma$) and L_{i+1} in such a way that the induction hypothesis is preserved. If $i+m \varepsilon \Gamma$ then, as it is easy to see, it is enough to set

$$L_{i+1} = L_i + \Big[\psi(\tilde{y}) - \lambda_0^m \frac{a_0}{b_0} \hat{y} - L_i \Big]_{m+i} S_{m+i}$$

Thus we may assume $i+m \notin \Gamma$. In this case the coefficient of t^{m+i} in E_i must vanish. Now in this coefficient there appears a unique term that contains a_i, namely $\lambda_0^{m+i} a_i$, and a unique term that contains λ_i, $m a_0 \lambda_0^{m-1} \lambda_i$; both terms come from $\psi(\tilde{y})$ and all other terms introduced by $\psi(\tilde{y})$ can involve λ_j and a_j only if $j < i$. From $\lambda_0^m \frac{a_0}{b_0} \hat{y}$ we get a unique term: $\lambda_0^m \frac{a_0 b_i}{b_0}$. The expression L_i contributes with terms that only involve λ_j, a_j, b_j with $j < i$. Notice that a term that involves a b_j which comes from the $\hat{y}$ in L_i, will affect a power of t not less than j+n+m, since $\hat{y}$ does not appear alone in L_i, so that it will affect our computation only if $i+m \geq j+n+m$, which implies that $j < i$. Altogether we get a relation of the form

$$\lambda_0^{m+i} a_i - \lambda_0^m \frac{a_0 b_i}{b_0} = G'_i(\lambda_0,\ldots,\lambda_i,a_0,\ldots,a_{i-1},b_0,\ldots,b_{i-1})$$

from which a relation of the form β_i follows inmediately. Since the induction hypothesis is preserved setting $L_{i+1} = L_i$ we can reach the end of the proof with the same argument used in proposition 2.

Propositions 1 and 2 imply immediately the following

<u>Theorem 2</u>. With the same notations as before, the formal equivalence of the branches γ_1 and γ_2 is equivalent to the existence of a solution with $\lambda_0 \neq 0$ of the simultaneous algebraic equations in $\lambda = (\lambda_0, \lambda_1, \ldots)$

$$\alpha_i)\ \lambda_i = F_i(\lambda_0,\ldots,\lambda_{i-1},b_0,\ldots,b_{i+n-m}), \qquad i \notin \Gamma - n$$

$$\beta_j)\ b_j = \lambda_0^j \frac{b_0}{a_0} a_j + G_j(\lambda_0,\ldots,\lambda_j,a_0,\ldots,a_{j-1},b_0,\ldots,b_{j-1}), \qquad j \notin \Gamma - m$$

where the F_i and G_j are rational functions with coefficients in C which are defined for all permissible sequences of a's, b's and λ's. Furthermore, λ_j appears effectively in G_j and only with exponent 1, for all permissible secuences of a's, b's and λ's.

7. First applications. First we see that due to the particular form of the equations in the first group above it is possible to eliminate the variables λ_i for all i such that $i+n \notin \Gamma$ and get an equivalent system in which only the remaining λ_i's and the relations comming from the β_j are present.

The theorem also implies the well known fact that the formal equivalence of branches is independent of the values of the coefficients a_i and b_i for $i+m \geq c$ (conductor of Γ). See for instance Z.4. Ignoring these coefficients we can represent each branch by a point in C^{c-m} (not lying in the hyperplanes corresponding to the vanishing of the characteristic terms). If we now eliminate the λ's in the algebraic equations provided by the theorem, we obtain the following

Corollary 1. The graph of the formal equivalence of branches in $C^{c-m} \times C^{c-m}$ is a constructible set.

Now we need to introduce some terminology. If P is an infinitely near point we will refer to branches that have P as a simple point and which have a free point in its first neighbourhood as the simplest branches through P. These branches are all in the same equisingularity class. The semigroup of the simplest branches through P will be called the semigroup of P.

Assume that P is a free point preceded by some satellite point (we exclude the case of points on a smooth branch). With a suitable selection of the local coordinates the Puiseux series of any branch γ through P can written in the form

$$x = t^n \qquad y = \sum_{j \geq 0} a_j t^{m+j} \quad , \quad m \notin (n)$$

which shows that the first characteristic exponent is m/n.

If a_j is the coefficient in the series of which the position of P depends (according to section 2) then we will say that P has relative level j with respect to γ, or with respect to the equisingularity type of γ.

If γ is one of the simplest branches through P we say that P has (absolute) level j.

We can easily see that the relative level of P with respect to γ is the sum of the multiplicities on γ of the free points on γ which precede P and which are preceded by some satellite point. In particular the relative level depends only on the equisingularity type of γ and on the order of the neighbourhood which contains P.

The action of G(P) can be shown to be transitive in many cases:

Corollary 2. Let P be a point infinitely near to 0 which is either a satellite point or a point preceded by satellite points. Let n the order of the simplest branches through P and m/n their first characteristic exponent; thus n and m are the smallest generators of the semigroup Γ of P. Then G(P) acts transitively in $V_1(P)$ except in the case when the level j of the points on $V_1(P)$ satisfies $j + n \notin \Gamma$ and $j + m \notin \Gamma$ [1].

Proof. Since there exist formal transformations that leave P fixed (at least the identity) the system of the equations α_i and β_i for $i < j$ is compatible if we take as $a_0, \dots, a_{j-1}$ the coefficients in the Puiseux

---

[1] Compare with the elimination rules on Z.4, Ch. III and Z.5, p.748.

series of the simplest branches through P and set $b_i = a_i$, $i = 1,\dots,j-1$. Let a_j, b_j be the coordinates of two arbitrary free points on $V_1(P)$. If $j+m \in \Gamma$ then β_j does not exist. If $j+m \notin \Gamma$ and $j+n \in \Gamma$ then α_j does not exist, λ_j is free and hence if we adjoint β_j to the system referred to above (where β_j is formed with the values a_j and b_j) we still get a compatible system. This is enough to guarantee the existence of formally equivalent branches going through the two selected points on $V_1(P)$, for if we choose any a_i, $i > j$, then we have freedom to select the corresponding b_i so that when we adjoint the relations α_i and β_i for $i > j$, the resulting system of equations is still compatible.

Thus we are especially interested in the case when the level j of the points in the first neighbourhood satisfies $j+n \notin \Gamma$, $j+m \notin \Gamma$. We call these first neighbourhoods <u>not necessarily transitive neighbourhoods</u>. In particular:

<u>Proposition 4</u>. If P is a satellite point and the group of satellites to which it belongs is preceded by some other such group, then the first neighbourhood of P is not necessarily transitive.

<u>Proof</u>. Let $x = t^n$, $y = \sum_{i \geq 0} a_i t^{m+i}$ be the Puiseux series of one of the simplest branches going through P. By hypothesis the coefficient a_j which determines the position of the point in the first neighbourhood of P satisfies $j > 0$ and is the coefficient of the last characteristic term. Therefore $j + m$ is not divisible by the greatest common divisor d of n and the $m + i$ with $i < j$ and $a_i \neq 0$. It is equivalent to say that d does not divide j, since d divides m due to the fact that $a_0 \neq 0$.

If now we compute the generators of Γ according B.1 (cf. section 6 above), the generators constructed with branches that do not pass through the last free point that precedes P are multiples of d; to see this it is enough to use Noether's formula for the intersection multiplicity and notice that all points in the branch entering in the formula have a multiplicity divisible by d. It is obvious that neither j + n nor j + m can be a linear combination of these generators of Γ.

It remains one more generator of Γ, namely the intersection multiplicity with the simplest branches going through the last free point preceding P. It will be enough to see that this multiplicity is greater than j + m, and hence also greater than j + n, to reach the conclusion. In fact the simplest branches through the last free point preceding P have order n/d and necesarily n/d > 1, because there are groups of satellite points preceding P. The Puiseux series of such a branch has the form

$$y = \sum_{i=0}^{j-1} a_i x^{\frac{m+i}{n}} + \ldots$$

where all the exponents of the terms which are not written have a common denominator n/d.

In order to see the claim let us writte the Puiseux series of our branch in the form

$$y = \sum_{i \geq 0} a_i x^{\frac{m+i}{n}}$$

Then the intersection multiplicity is obtained adding up the orders of the differences of determinations of the two series; there are n^2/d such diferences; of them n contribute with order $\frac{m+j}{n}$ and the remaining $n(\frac{n}{d} - 1)$ with orders which are non zero, so that the total sum is indeed greater than m + j.

Notice that as far as the action on a level j is concerned, the relations α_i, β_i, $i > j$ have no influence whatsoever. In fact, as in the proof of corollary 2, whith a suitable selection of the b_i, $i > j$, the relations α_i, β_i, $i > j$, can not alter the compatibility.

If $V_1(P)$ is not necessarily transitive and of level j, then the relations α_i, β_i, $i < j$, formed taking a's = b's = coordinates of P and the preceding points, define a quasi-affine variety in C^j of admissible solutions $\lambda = (\lambda_0,\ldots,\lambda_{j-1})$ (i.e. with $\lambda_0 \neq 0$), variety that we will denote Λ. In fact Λ has a natural structure of group variety which we can describe as follows: for every automorphism ψ of $C[[t]]$ let us consider the corresponding series $\sum \lambda_i t^i = \psi(t)/t$, so that the elements of Λ are the sequences of first j coefficients of the series which correspond to authomorphisms of $C[[t]]$ leaving P invariant, and conversely. Therefore the group structure of the set of automorphisms leaving P invariant induces a group structure on Λ which turns it into a group variety.

Let A(P) be the affine group of $V_1(P)$ (with the customary choice of the improper point). The relation α_j determines λ_j as a rational function on Λ. Thus if we substitute this rational function for λ_j in β_j we get an effective expression for the morphism of group varieties $\tau: \Lambda \longrightarrow A(P)$ which transforms $\lambda = (\lambda_0,\ldots,\lambda_{j-1}) \in \Lambda$ into the affine transformation induced by any automorphism whose corresponding series has $\lambda_0 + \ldots + \lambda_{j-1}t^{j-1}$ as a partial sum. It is obvious that $\mathrm{Im}\,\tau = G(P)$, and that the homothety ratio and the translation component of the element of G(P) appear as rational functions on Λ.

<u>8. Elementary branches.</u> We know that the action of G(P) is transitive when P is not preceded by satellite points and consequently we will

consider the case of points preceded by a single group of satellite points, and, among them, points lying on what we will call elementary branches:

Definition. An elementary branch is a branch which is formally equivalent to a branch of the form $x = t^n$, $y = t^m$, where $\gcd(n,m) = 1$ and $1 < n < m$.

Elementary branches are characterized by the fact that their module of differentials has maximum torsion, vid Z.5.

Lemma 3. Let Γ be the semigroup generated by n and m, where $\gcd(n,m) = 1$ and $1 < n < m$. Then $(\Gamma - n) \cap (\Gamma - m) = \Gamma \cup \{c-1\}$, where c-1 denotes the greatest element in the complementary of Γ. ($c = nm - n - m + 1$ as it is well known).

Proof: Any integer has a unique expression of the form $an + bm$ where $0 \leq b < n$. The elements in Γ are exactly those for which $a \geq 0$. Let now q be an integer not in Γ such that $q + n$ and $q + m$ belong to Γ. Then, if $q = an + bm$, $0 \leq b < n$, we will have $a < 0$. Since $q + n \in \Gamma$, we may conclude that $a + 1 \geq 0$ and hence that $a = -1$. Now from $q + m \in \Gamma$ we see that $-n + (b+1)m \in \Gamma$, so that we must have $b + 1 \geq n$, which implies $b = n - 1$ so that $q = -n + (n-1)m = nm - n - m = c - 1$.

Theorem 3. Let $O_0, \dots, O_i$ be successive free points on a elementary branch such that O_0 follows a satellite point. Let m/n be the characteristic exponent of the branch (which will be assumed written in irreducible form) and let Γ be the corresponding semigroup, which is generated by n and m. If $i + n \notin \Gamma$ and $i + m \notin \Gamma$, then $G(O_{i-1})$

leaves O_i invariant and acts transitively on the remaining free points. Otherwise $G(O_{i-1})$ acts transitively.

Proof. First notice that notations have been selected so that O_j has level j. Therefore the second statement is obvious from corollary 2.

Without loss of generality we may assume that the branch is given by $x = t^n$, $y = t^m$, and we are going to make explicit the process of obtaining the relations α, β for the transformation of the branch $x = t^n$, $y = t^m$ into the branch $x = t^n$, $y = t^m + b_i t^{m+i} + \ldots$. In fact these relations force λ_j to vanish for $j + n \notin \Gamma$, $j \leq i$ or $j + m \notin \Gamma$, $j < i$. Assume r is the least integer such that $\lambda_r \neq 0$ and $r + n \notin \Gamma$ or $r + m \notin \Gamma$; since in any case $r < i + n$ and $i + n \notin \Gamma$, the lemma above implies that if $j < r$ and $\lambda_j \neq 0$, then j is in Γ. Hence we can write

$$\psi(t) = t\Big(\sum_{j \in J_r} \lambda_j t^j + \lambda_r t^r + \ldots\Big)$$

where $J_r = \{ j \in \Gamma \mid j < r \}$.

If $r + n \notin \Gamma$ consider

$$\psi(\tilde{x}) = t^n\Big(\sum_{j \in J_r} \lambda_j t^j + \lambda_r t^r + \ldots\Big)^n$$

All terms arising from $\sum_{j \in J_r} \lambda_j t^j$ in this expression have exponents that belong to Γ. We may eliminate them subtracting suitable monomials in t^n and $t^m + b_i t^{m+i} + \ldots$ and in so doing we do not introduce new terms whose degree is lower than $m + i$. since $r \leq i$, $n + r < m + i$, the term $n\lambda_0^{n-1}\lambda_r t^{n+r}$ is unaffected. The hypothesis $n + r \notin \Gamma$ implies then that $\lambda_r = 0$.

If $r + m \notin \Gamma$ we consider

$$\psi(\tilde{y}) = t^m(\sum_{j\in J_r} \lambda_j t^j + \lambda_r t^r + \dots)^m$$

Since $r < i$, $m+r < m+i$, again the elimination of terms whose degree is in Γ does not affect the term $m\lambda_0^{m-1}\lambda_r t^{m+r}$, which also implies $\lambda_r = 0$.

Hence we have

$$\psi(t) = t(\sum_{j\in J_i} \lambda_j t^j + S(t))$$

with $\mathrm{ord}_t S(t) > i$. Now

$$\psi(\tilde{y}) = t^m(\sum_{j\in J_i} \lambda_j t^j + S(t))^m$$

In order to eliminate the term whose order is m we consider

$$\psi(\tilde{y}) - \lambda_0^m \hat{y} = t^m(\sum_{j\in J_i} \lambda_j t^j + S(t))^m - \lambda_0^m t^m - \lambda_0^m b_i t^{m+i} - \dots$$

This introduces a term of degree $m + i$. The successive eliminations of other terms whose degrees are in Γ do not affect the explicit terms. In particular $-\lambda_0^m b_i t^{m+i}$ remains. Eventually it will be the initial term and this forces b_i to vanish. So O_i is fixed.

Finally, in order to show that the action is transitive on $V_1(O_{i-1}) - O_i$ it is enough to observe that the formal transformation $\phi(x) = \lambda^n x$, $\phi(y) = \lambda^m y$ transforms the branch $x = t^n$, $y = t^m + a_i t^{m+i}$ into the branch $x = t^n$, $y = t^m + \lambda^i a_i t^{m+i}$.

9. The bifurcation invariant and the homothety ratio.

Definition: If γ is a singular branch, we will call bifurcation invariant of γ the relative level of the first free point on γ that does not lie on a elementary branch. We will denote it $s = s(\gamma)$. If γ is elementary we will set $s(\gamma) = \infty$.[1]

From the very definition it follows that $s(\gamma)$ is invariant under formal transformations. In fact (section 7) $s(\gamma)$ coincides with the sum of the multiplicities on γ of the free points of γ lying on elementary branches and which are preceded by some satellite point.

Given an equisingularity type it is easy to see that the invariant s can take only a finite number of values. For, unless γ is elementary, the point on γ of level $s(\gamma)$ lies in the first neighbourhood (necessarily non-transitive) of a point on an elementary branch, or it is the first free point after the second group of satellite points. If the branch has only one characteristic exponent, say m/n, m and n relatively prime, then either $s = \infty$ or $s + n \notin \Gamma$ and $s + m \notin \Gamma$, due to theorem 3. If there are more than one characteristic exponent let nd be the order of the branch and, as before, let m/n be the first characteristic exponent in irreducible form and let Γ be the semigroup generated by n, m. Then we have two posibilities, namely:

a) The point of level s precedes the second group of satellite points. In this case we will say that the <u>bifurcation</u> is <u>by a free point</u> and we must have $s/d + n \notin \Gamma$, $s/d + m \notin \Gamma$ in addition to the restriction imposed on s by the second characteristic exponent when the point of level s precedes the second group of satellite points.

b) s is the level of the first free point after the second group of satellite points. In this case the order of contact with the

---

(1) an equivalent invariant was considered by Zariski (Z.4, p. 32 and Z.5, pp. 785 and 786).

elementary branch is maximum and we will say that the bifurcation is by a satellite point. See figure 2 below.

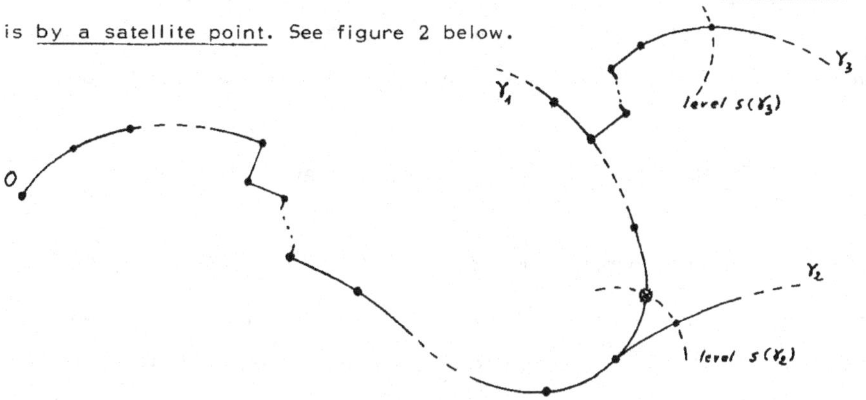

Figure 2. The branch γ_1 is assumed to be elementary and ⊗ represents a fixed point on γ_1. γ_2 bifurcates by a free point and γ_3 bifurcates by a satellite point.

Let γ be a singular branch, centered at O. Let O' denote the last point in the first group of satellite points on γ and let P be a point on γ, after O' and comming just before a free point. Then, if n is the multiplicity of O' on γ, we have:

Proposition 5. Let i be the level of the points of $V_1(P)$ relative to the equisingularity type of γ, and let μ be the polidromy order introduced by the corresponding characteristic term if P is a satellite point. Assume that ϕ is a formal transformation leaving P fixed. Then there exist parameters ρ_0, λ_0 which depend only on ϕ, and such that the homothety ratio of the transformation induced on the first neighbourhood of P equals $\rho_0^{-1}\lambda_0^{n+i}$, or $(\rho_0^{-1}\lambda_0^{n+i})^{\mu}$ whenever P is a satellite point.

Proof. Let S' be the surface on which O' lies and take local coordinates z_1, z_2 at O' so that the local equation of the exceptional locus of the morphism $S \to S'$ at O' be $z_1 z_2 = 0$. Since Φ leaves the satellite points fixed, the axes $z_i = 0$ will be fixed as well and thus

$$\Phi(z_1) = z_1 \delta(z_1, z_2) \qquad \Phi(z_2) = z_2 \rho(z_1, z_2)$$

where δ and ρ are invertible series.

Let $z_1 = t^n$, $z_2 = \sum_{j \geq 0} a_j t^{n+j}$ be the Puiseux series of the proper transform γ' of γ centered at O'. Then the infinitely near point whose position is determined by the coefficient a_i has precisely relative level i (with respect to γ and as point infinitely near to O). This is easly seen using what we observed after defining level, on section 7.

Let $\hat{z}_1$, $\hat{z}_2$ be the images of z_1, z_2 in the local ring R_2 of the transformed branch $\Phi(\gamma')$. We take C[[t]] as the integral closure of R_2 via the isomorphism induced by Φ^{-1} and the inclusion of the local ring R_1 of γ' in C[[t]] given by the Puiseux series above:

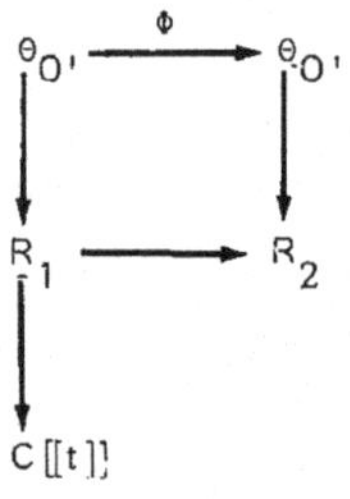

We have

$$\hat{z}_1 \delta(\hat{z}_1, \hat{z}_2) = t^n$$

$$\hat{z}_2 \rho(\hat{z}_1, \hat{z}_2) = \sum_{j \geq 0} a_j t^{n+j}$$

Now it is convenient to take $\tilde{t} = t\varepsilon^{-1}\delta(\hat{z}_1, \hat{z}_2)^{-1/n}$, where ε is an n-th root of unity and where we fix any determination of $\delta(\hat{z}_1, \hat{z}_2)^{-1/n}$. We have

$$\hat{z}_1 = t^n \quad , \quad \hat{z}_2 = \sum_{j\geq 0} a_j \rho^{-1}(\hat{z}_1, \hat{z}_2)\varepsilon^j \delta^{1+j/n}(\hat{z}_1, \hat{z}_2)\tilde{t}^{n+j}$$

If P is fixed then choosing ε in a suitable way we get that

$$\hat{z}_2 \equiv \sum_{h\geq 0} a_h \tilde{t}^{n+h} \mathrm{mod}\ \tilde{t}^{n+i} \qquad \dagger$$

so that we have

$$\hat{z}_2 \equiv \sum_{j\geq 0} a_j \rho^{-1}(\tilde{t}^n, \sum_{h=0}^{i-1} a_h \tilde{t}^{n+h})\varepsilon^j \delta^{1+j/n}(\tilde{t}^n, \sum_{h=0}^{i-1} a_h \tilde{t}^{n+h})\tilde{t}^{n+j}$$

mod $\tilde{t}^{n+i+1}$

From this we obtain an expression for the coefficient a_i of the Puiseux series of $\Phi(\gamma')$ like this:

$$\hat{a}_i = \rho_0^{-1}\varepsilon^i \delta_0^{\frac{n+i}{n}} a_i + H(a_0,\ldots,a_{i-1})$$

where $\rho_0 = \rho(0,0)$ and $\delta_0 = \delta(0,0)$. It is enough to take $\lambda_0 = \varepsilon\,\delta_0^{1/n}$.

Notice that if d is the greatest common divisor of n and all j such that $j < i$ and $a_j \neq 0$, then the condition $\dagger$ determines ε but for a d-th rooth of unity. This indetermination does not have any effect upon the conclusion of the proposition; for if P is not a satellite point then $a_i t^{n+i}$ is not a characteristic term and thus d divides i; and if P is a satellite point, then i is not divisible by d, but $i\mu$ is.

The preceding proposition can be applied to all points P coming before O'. If $P \neq O'$ we have a stronger result:

Corollary 3. With the same notations and hypothesis as above, and if $P \neq O'$ then the homothety ratio is λ_0^i or $\lambda_0^{i\mu}$ according to whether P is free or satellite, where again λ_0 is a parameter which depends only on Φ (and not on the level of P).

Proof. If $P \neq O'$ then the point of γ on $V_1(O')$ must be a fixed point of Φ. Since Φ has also the satellite points as fixed points we can conclude that the transformation induced by Φ on $V_1(O')$ is the identity. If we now apply the proposition for $P = O'$ we will have that $1 = \lambda_0^n \rho_0^{-1}$ and from this the claim follows inmediately.

Corollary 4. If the relative level of the points on $V_1(P)$ is greater than $s = s(\gamma)$, then the parameter λ_0 in corollary 3 satisfies $\lambda_0^s = 1$.

Proof. As before, no matter whether the bifurcation occurs by a free point or by a satellite point, in the first neigbourhood of level s there are three fixed points, so that the transformation induced at this implies that the homothety ratio at such a level must be one, i.e., $\lambda_0^s = 1$. If the bifurcation occurs by a satellite point then the same argument shows that $\lambda_0^{s\nu} = 1$, where ν is the polydromy order introduced by the second characteristic term; but here the selection of $\tilde{\gamma}$ guarantees that $a_s = \hat{a}_s$, so that we also get $\lambda_0^s = 1$.

Corollary 5. If the relative level of the points on $V_1(P)$ is greater than $s(\gamma)$, then there are only finitely many homothety ratios among the elements of G(P). Then G(P) turns out to be finite if its not transitive, or a group generated by finitely many homotheties and the translations if it is transitive.

Proof. That there are only finitely many homothety ratios among the elements of G(P) is a consequence of the previous corollary. It suffices then to notice that if G(P) contains a non trivial translation it must contain infinitely many. For then, by our observation at the end of section 7, G(P) will contain all translations.

The reader will observe that proposition 5, when the level i satisfies $i + m \notin \Gamma$ (which is always so if it is a non-transitive level), has an easier proof, as it follows from the existence of the relations β (theorem 2).

In particular the existence of non-trivial translations is ruled out in the following case:

Corollary 6. If P is a satellite point then G(P) is a finite group of homotheties except in the following two situations:

a) P belongs to the first group of satellite points.

b) P belongs to the second group of satellite points and all free points preceding it lie on a elementary branch.

Proof. By the fact that P is a satellite point, G(P) is a group of homotheties whose center is a satellite point. The cases we exclude are precisely those for which the level of the first neighbourhood of P is not higher than the bifurcation level. Thus we may apply corollary 5.

Remark: As far as the two cases excluded in the previous corollary, a) is dealt with in theorem 3, while in case b) G(P) is also the group of all homotheties with center at the point whose coordinate is zero. In fact this point is a satellite point and hence it is invariant. Without loss of generality we may assume that the simplest branches through P have Puiseux series $x = t^n$, $y = t^m + at^{m+s} + \ldots$, where a determines the point of the branch γ on the first neigbourhoo-

of P. Then it suffices to observe that any homothety may be induced by a transformation of the form $\Phi(x) = \lambda^n x$, $\Phi(y) = \lambda^m y$.

10. Semicanonical branches. Invariants. We fix a formal equivalence class of branches. In this class we will select finitely many branches which will be called semicanonical for the corresponding formal type. First we exclude the smooth case, in which we can take $y = 0$. Now let us begin choosing a branch in the class so that the elementary branch that has higest contact with it is $x = t^{n'}$, $y = t^{m'}$, with n' and m' relatively prime. If our class were a class of elementary branches, then this branch would be taken as semicanonical. Otherwise it determines the points on the branch until the point inmediately before the bifurcation level. On the first neighbourhood of this point acts the group of homotheties with center at the point determined by the elementary branch, or a satellite point (see theorem 3 and the remark above), according to whether the bifurcation occurs by a free point or by a satellite point. We thus see that without altering previous points, the coordinate of the point on the bifurcation level can be assumed to be 1. Beyond the bifurcation level we only encounter finite groups or transitive groups (corollary 5), and they are finite when P is a satellite point (corollary 6). Using suitable transformations it is possible to produce branches in the class such that the coordinate of the corresponding point on every transitive neighbourhood of level greater than s is zero. There are only finitely many branches of this type, since there are only finitely many non-transitive neighbourhoods and on each of them the orbits are also finite. These branches will be refered to as semicanonical branches of the class.

With a suitable choice of parameter each semicanonical branch will be represented by a Puiseux series of the form

$$x = t^n \quad , \quad y = t^m + t^{m+s} + \sum_j a_j t^{m+j}$$

where the summation is extended over the non-transitive levels greater than s. Such a series is determined by the branch but for a substitution of t by εt, where $\varepsilon^d = 1$, $d = \gcd(n,m,s)$. Notice that the first two coefficients of y are taken to be 1. These series will be called semicanonical series of the formal equivalence class.

Now we can define invariants of the class, which, together with s, will form a complete system of invariants.

Definition. Let γ be a branch, which we will assume singular and non-elementary. Let $j_1,\ldots,j_h$ be the non-transitive levels greater than $s = s(\gamma)$. Then we will say that the coefficients $a_{j_1},\ldots,a_{j_h}$ of the semicanonical series of the formal type of γ are the moduli of γ.

Remarks:

1.- We see from the definition that the system of moduli associated to each branch is multiform. Later we will deal with its polydromy.

2.- It is essential to notice that each modulus is defined for some branches and not defined for some others, according to whether the corresponding neighbourhood is not or is transitive. When we fix the invariant s, then all branches reach (if it exists) a first non transitive neghbourhood, which yields a first modulus for such branches. Beyond this point the transitivity or non-transitivity depends on the moduli already defined. In fact if we assume that $a_{j_1},\ldots,a_{j_r}$ are already defined for a branch and if i is the relative level of the next not necessarily transitive neghbourhood, then we look at the relations α, β which insure that the points of the semicanonical branch whose level is lower than i are fixed (in order to do that only the already determinated coefficients of the semicanonical series are needed). Once this is done, the local constancy of G_i in the relation β_i, taking into account the determination of λ_i by means of the α_i, decides the non-transitivity and consequently the existence of a modulus. In case such a modulus exists it is determined using the relations α, β that give a transformation of the branch γ into a semicanonical branch.

3.- Once we admit that some moduli may be definited for some branches, then the invariant s and the moduli form a complete system of invariants in the following sense: two equisingular branches γ and γ' are formally equivalent if and only if $s(\gamma) = s(\gamma')$ and the moduli of γ and γ' are defined for the same levels and they coincide.

4.- From the way the moduli are computed through the relations α, β it turns out that the moduli are algebraic functions of the coefficients of the Puiseux series of the branches for which they are defined, in the sense that they are obtained from them by means of a correspondence whose graph is a quasiprojective variety.

<u>Proposition 6</u>. Let P be a point lying on a semicanonical branch given by a series

$$x = t^n, \quad y = t^m + t^{m+s} + \sum_{r=1}^{h} a_{j_r} t^{m+j_r}$$

where all written terms are assumed to be non-zero. If $V_1(P)$ is non transitive and has relative level $i > s$, then the transformations of the form

$$\Phi(x) = \lambda^n x \quad , \quad \Phi(y) = \lambda^m y$$

with $\lambda^d = 1$, where d is the greatest common divisor of s and the j_r such that $j_r < i$, induce all elements in G(P). Consequenly G(P) is a group of homotheties with center at the point whose coordinate is zero. The homothety ratios are λ^i if P is free, or $\lambda^{i\mu}$ if P is satellite, where λ runs through the d-th roots of unity and where μ (in the case of a satellite point) is the order of polydromy introduced by the corresponding characteristic term.

Proof. We know that G(P) is a finite group of homotheties from corollaries 4 and 5. We will proceed using induction. If $i \leq j_1$, then a branch through a variable point of $V_1(P)$ has a series of the form

$$x = t^n \quad , \quad y = t^m + t^{m+s} + at^{m+i} + \ldots$$

and it is transformed into a branch with series

$$x = t^n \quad , \quad y = t^m + t^{m+s} + \lambda^i at^{m+i} + \ldots$$

by a transformation of the kind described in the statement (with d = s).

We see that these transformations induce all homotheties allowed by corollary 4 and hence induce all G(P).

Let us assume that the statement has already been proved for all levels lower than i and higher than s. In particular we will have a finite group of homotheties with center at the point whose coordinate is zero on each non-transitive neighbourhood of level lower than i and greater than s.

Consequently a transformation Φ which leaves P fixed must also leave invariant the points on γ whose levels j_r are lower than i. Since these points are disctint of the homothety centers we conclude that the homotheties induced by Φ are the identity. This and a suitable choice of the form of the transformed series adds to the condition $\lambda_0^s = 1$ the conditions $\lambda_0^{j_r} = 1$ (for $j_r < i$) which are to be fulfilled by the parameter λ_0 that determines the homothety ratio (λ_0^i or $\lambda_0^{i\mu}$ according to whether P is free or satellite) on the first neighbourhood of P. It is obvius that these homothety ratios are all present in those induced by the transformations described in the statement, so that these transformations induce the whole G(P).

The same kind of argument allows us to determine the homothety ratios arising on transitive neighbourhoods of higher level than the bifurcation level:

Corollary 7. Let P be a point on a branch γ and assume that P is either free or the last point in a group of satellite points. Let i the relative level of the points on $V_1(P)$ and suppose that $i > s(\gamma)$. Then the homothety ratios of the elements in G(P) are λ^i or $\lambda^{i\mu}$ according to whether P is free or satellite, where $\lambda^d = 1$, d is the greatest common divisor of s and the non transitive levels j, $s(\gamma) < j < i$, for which the center of the homotheties is not the point on γ, and μ is the polydromy order introduced by the corresponding characteristic term when P is a satellite point.

Corollary 8. If

$$x = t^n \quad , \quad y = t^m + t^{m+s} + \sum_{r=1}^{h} a_{j_r} t^{m+j_r}$$

is a semicanonical series then the remaining semicanonical series of the corresponding formal type are obtained from it by means of substitutions of the form $t^* = \lambda t$, $x^* = \lambda^n x$, $y^* = \lambda^m y$, where $\lambda^s = 1$ if $s = s(\gamma)$ is the corresponding bifurcation level. Thus the different determinations of the system of moduli of the corresponding branch are $(\lambda^{j_1} a_{j_1}, \ldots, \lambda^{j_h} a_{j_h})$ where $\lambda^s = 1$.

Proof. We first observe that among the substitutions described in the statement we find the substitutions $x^* = x$, $y^* = y$, $t^* = \lambda t$, with $1 = \lambda^{\gcd(n,m,s)}$ which act transitively on the semicanonical series of a given semicanonical branch. We need only check that the substitutions in the statement act transitively on the semicanonical branches. But this is obvious from the previous proposition since the transformations described there induce the whole group of transformations on each first neighbourhood for which a modulus is defined.

11. Some topological properties of the moduli space. Following Zariski (Z.4, Ch. III) we will define the topology in the moduli space using the coefficients in the short series.

Let us take a fixed equisingularity type. Then the corresponding moduli space M has as underlying set the set of formal equivalence clases of branches whose equisingularity type is the given one.

It is known (using for instance the transitivity on the neighbourhoods of relative level i with $i+m \in \Gamma$) that in each formal equivalence class there is a branch with Puiseux series (short series)

$$x = t^n \quad , \quad y = t^m + \sum a_i \, t^{m+i_j}$$

where m/n is the first characteristic exponent and where the summation is extended over all indices i_j such that $i_j+m\notin\Gamma$.

If ρ denotes the number of levels, relative to our equisingularity type, which do not belong to $\Gamma - m$, then the arrays of coefficients of the short series form an open set in C^ρ, namely the set $U = C^\rho - \bigcup_{k=2}^{g} H_k$, where each H_k denotes a hyperplane given by the vanishing of a characteristic coefficient and where g stands for the genus of the branches (the number of characteristic exponents in their series).

Now we get a surjection $U \to M$ when we associate to each short series its formal type. We topologize M with the quotient topology under this map, where U is endowed with the relative topology.

Next result (already used in Z.4) shows that we can do without some of the coefficients in the short series:

<u>Proposition 7</u>. If γ is a branch represented by a short series

$$x = t^n \quad , \quad y = t^m + \sum a_{i_j} t^{m+i_j}$$

then it is possible to determine a second short series

$$x = t^n \quad , \quad y = t^m + \sum b_{i_j} t^{m+i_j}$$

that represents a branch which is formally equivalent to γ, satisfies $b_{i_j} = 0$ if $i_j+n\in\Gamma$, and the remaining coefficients b_{i_j} are rational functions of the coefficients a_{i_j}; moroever, these rational functions are regular for every admissible value of the a_{i_j}.

<u>Proof</u>. Let i_r denote the least integer such that $i_r + n \in \Gamma$ and $a_{i_r} \neq 0$. Set $\tilde{a}_{i_j} = a_{i_j}$ if $i_j < i_r$ and $\tilde{a}_{i_r} = 0$. Now let us consider the

relations α, β arising in the transformation of γ into a branch $\tilde{\gamma}$ whose series has the form

$$x = t^n \quad , \quad y = t^m + \sum_{j=1}^{r} \tilde{a}_{i_j} t^{m+i_j} + \ldots$$

(the dots denote terms to be determined). It is obvious then that the system of the equations α_j, β_j, for $j < i_r$, admits the solution $\lambda_0 = 1$, $\lambda_1 = \ldots = \lambda_{i_r-1} = 0$. The equation α_{i_r} does not exist and we can determine λ_{i_r} as a regular function of the coefficients a_{i_j} in such a way that it satisfies β_{i_r}. Now in order to define $\tilde{\gamma}$, take, for $h > i_r$, $\lambda_h = 0$ if $h+n \in \Gamma$ and if $h+n \notin \Gamma$ define λ_h by solving for it in the relation α_h; and take, for $j > r$, $\tilde{a}_{i_j}$ to be the value determined by β_{i_j} if $i_j+m \notin \Gamma$, and zero otherwise. With this process the coefficient at the level i_r vanishes and the conclusion follows immediately by induction.

The bifurcation invariant determines on M a finite descending sequence of subspaces. Take

$$M_h = \{p \in M \mid s(p) \geq h\}$$

where h runs through all values of s for the given equisingularity type (the value $h = \infty$ is assumed to be included in the case of a single characteristic exponent).

We immediately get

Corollary 9. The subspaces M_h are closed in M.

Proof. Let γ be a branch that represents a limit point of M_h. We may assume that γ is given by a short series and that it exists a sequence of branches γ_j such that the γ_j represents points on M_h, are represented by short series, and converge to γ according to the topology of the coefficients. Furthermore, the previous proposition allows us to replace γ and the γ_j by formally equivalent branches $\tilde{\gamma}$ and $\tilde{\gamma}_j$, respectively, which still satisfy the same conditions as above and so that those of their coefficients whose level belong to $\Gamma - n$ are zero.

Under these circumstances we know that $s(\tilde{\gamma})$ and the $s(\tilde{\gamma}_j)$ are the levels of the first non-vanishing coefficients (after the term t^m) in the corresponding series. It is then clear that $s(\tilde{\gamma}) \geq s(\tilde{\gamma}_j) = h$, q.e.d.

In particular, in the case of a single characteristic exponent we see that M_∞ is reduced to a single point, corresponding to the elementary branches, and therefore this point is closed. It is known (Z.4, pages 31 and 33) that this is the only closed point in M. Later we will see that the situation is substantially different when there are two or more characteristic exponents.

Now we come to the basic result of this section. This result guarantees that the limit branch of a variable branch with constant formal type has the same formal type if and only if it has the same bifurcation invariant.

Let h_i, $i = 1,\dots,\nu$ be the values of the bifurcation invariant on the given equisingularity class. Assume they are ordered increasingly, so that

$$M = M_{h_1} \supsetneq M_{h_2} \supsetneq \cdots \supsetneq M_{h_\nu}$$

Then we have

<u>Theorem 4</u>. The subspaces $M_{h_i} - M_{h_{i+1}}$, for $i < \nu$, and the subspace

M_{h_ν} have no non-closed points in the relative topology.

Proof. Let $\{\gamma_r\}$ be a sequence of branches that have the same formal type, are given by short series and converge to γ. We claim that if $s(\gamma_r) = s(\gamma)$ then γ is formally equivalent to the γ_r. This will complete the proof.

We may assume, as in the proof of corollary 9, that in the series that give γ and the γ_r all coefficients whose level belong to $\Gamma - n$ are zero. Set $s = s(\gamma) = s(\gamma_r)$, and let

$$x = t^n \quad , \quad y = t^m + c(r)t^{m+s} + \sum a_{i_j}(r)t^{m+i_j}$$

$$x = t^n \quad , \quad y = t^m + ct^{m+s} + \sum a_{i_j} t^{m+i_j}$$

the series of γ_r and γ respectively, where the summation is extended over all indices i_j such that $i_j > s$, $i_j \notin (\Gamma - n)\cup(\Gamma - m)$, and where c and the $c(r)$ are non-zero, $c = \lim_r c(r)$ and $a_{i_j} = \lim_r a_{i_j}(r)$.

Let us take care of the coefficients c and $c(r)$ first. Suppose chosen a determination of the s-th root on an open set of C which contains c and the $c(r)$ and consider the transformations

$$\Phi : \Phi(x) = c^{-n/s}x \qquad \Phi(y) = c^{-m/s}y$$

$$\Phi_r : \Phi_r(x) = c(r)^{-n/s}x \qquad \Phi_r(y) = c(r)^{-m/s}y$$

The branches $\Phi(\gamma)$ and $\Phi_r(\gamma_r)$ satisfy the same conditions, respectively, as γ and γ_r do, and their coefficients of level s are equal to the unity. Thus we may assume, without loss of generality, that $c = c(r) = 1$, and hence that the series of γ_r and γ have the form

$$x = t^n \qquad y = t^m + t^{m+s} + \sum a_{i_j}(r)t^{m+i_j}$$

$$x = t^n \qquad y = t^m + t^{m+s} + \sum a_{i_j} t^{m+i_j}$$

where again the summation is extended over all indices which do not belong to $(\Gamma-n)\cup(\Gamma-m)$ and $a_{i_j} = \lim_r a_{i_j}(r)$.

Next we will proceed to get rid of the coefficients in the series giving γ_r which do not correspond to moduli, that is to say, those that determine the position of a point on a transitive neighbourhood. To avoid trivialities we will exclude, for each i_j such that the $a_{i_j}(r)$ are almost all zero, the branches γ_r such that $a_{i_j}(r) \neq 0$ (there are only finite many of them). In this way we can assume that the indices i_j over which the summation is extended still satisfy the further requirement that infinitely many $a_{i_j}(r)$ are non-zero.

Let i_q be the least of the indices i_j under consideration which in addition are levels of transitive neighbourhoods for the formal type of γ_r. Thus for $j < q$ the levels i_j correspond to non-transitive neighbourhoods and consequently the $a_{i_j}(r)$ are moduli for γ_r. Since the formal type of the γ_r is constant, $a_{i_j}(r)$ will take on only finitely many values and since $\lim_r a_{i_j}(r) = a_{i_j}$, we will have that for all $j < q$ there exists r_0 such that $a_{i_j}(r) = a_{i_j}$ for all $r \geq r_0$. Dropping finitely many branches in the sequence $\{\gamma_r\}$ we may further assume that $a_{i_j}(r) = a_{i_j}$ for all $j < q$.

The previous argument shows that if i_q does not exist then $\gamma_r = \gamma$ for all r in which case obviously γ and γ_r have the same formal type.

Otherwise the branches γ_r and γ have a common point P whose first neighbourhood has relative level i_q. By the definition of i_q the group G(P) acts transitively on $V_1(P)$.

Let $\Lambda \subset C^{i_q}$ be the group variety defined by the relations α_i, β_i, $i < i_q$, as we observed at the end of section 8, and whose points determine the transformations on $V_1(P)$ through the epimorphism $\tau : \Lambda \to G(P)$. Let Λ_0 be the subgroup of Λ defined by the condition $\lambda_0^{i_q} = 1$. Since $G(P)$ contains all translations, by corollary 5 and the hypothesis of transitivity, every translation is induced by an element of Λ_0. Therefore τ induces an epimorphism $\xi : \Lambda_0 \to C$ which assigns to each $\lambda \in \Lambda_0$ the translation component of $\tau(\lambda)$.

In particular we consider the fiber $\xi^{-1}(-a_{i_q})$, which is non-empty and does not contain any component of Λ_0 (otherwise the same would occur with all fibres, due to the action of Λ_0 itself on the set of fibers). Choose any $\lambda \in \xi^{-1}(-a_{i_q})$; then λ is a limit point of $\Lambda_0 - \xi^{-1}(-a_{i_q})$ and there is no difficulty in determining a sequence $\{\lambda(r)\}$ of points on Λ_0 which converges to λ and such that $\xi(\lambda(r)) = -a_{i_q}(r)$.

Set $\lambda = (\lambda_0, \ldots, \lambda_{i_q - 1})$, $\lambda(r) = (\lambda_0(r), \ldots, \lambda_{i_q-1}(r))$. Let us define $\tilde{a}_{i_j} = \tilde{a}_{i_j}(r) = a_{i_j}$ for $j < q$, set $\tilde{a}_{i_q} = \tilde{a}_{i_q}(r) = 0$ and consider the branches $\tilde{\gamma}, \tilde{\gamma}_r$ defined, respectively, by short series of the form

$$x = t^n \quad , \quad y = t^m + t^{m+s} + \sum \tilde{a}_{i_j} t^{m+i_j}$$

$$x = t^n \quad , \quad y = t^m + t^{m+s} + \sum \tilde{a}_{i_j}(r) t^{m+i_j}$$

where the coefficients after the i_q-th are to be determined. Now let us consider the relations α, β which translate the formal equivalence of γ and $\tilde{\gamma}$, and, for each r, of γ_r and $\tilde{\gamma}_r$. By construction we may take λ and $\lambda(r)$ in order to satisfy, in each case, the relations of level less than i_q. Then the relations α_{i_q} determine λ_{i_q} and $\lambda_{i_q}(r)$, which by construction give (with λ and $\lambda(r)$ respectively) a solution of the corresponding equation β_{i_q}. If $h > i_q$, set $\lambda_h = \lambda_h(r) = 0$ if

$h+n \in \Gamma$, and if $h+n \notin \Gamma$ take for λ_h and $\lambda_h(r)$ the values determined by the corresponding α_h, and, simultaneously take the indetermined coefficients of the above series according to the requirements of the succesive relations β. In this way the formal type of $\tilde{\gamma}$ and $\tilde{\gamma}_r$ is the same as for γ and γ_r respectively. From the fact that $\lambda(r) \to \lambda$ and because of the way $\tilde{\gamma}$ and $\tilde{\gamma}_r$ have been determined, it is clear that $\tilde{\gamma}_r \to \tilde{\gamma}$.

Finally let us apply the procedure introduced in the proof of proposition 7 to get rid of the coefficients whose level belongs to $\Gamma - n$ (notice that only levels higher than i_q are involved) in the series giving $\tilde{\gamma}$ and $\tilde{\gamma}_r$, without altering their partial sums of degree $m + i_q$. After this we obtain branches $\hat{\gamma}$ and $\hat{\gamma}_r$ formally equivalent to γ and γ_r respectively, satisfying the same conditions as γ and γ_r, but defined by series which lack the term of degree $m + i_q$ and whose previous terms coincide with those of the former series.

Iterating this for all the levels i_j under consideration and which correspond to a transitive neighbourhood we prove the claim and therefore the theorem.

Remark. Even through the moduli of a branch can be obteined from the coefficients of its Puiseux series through an algebraic correspondence, as we already pointed out, neverthless it is no possible in general to guarantee the continuity of the moduli with respect of the coefficients. If a such a continuous dependence were true the previous proof could be considerably shortened. In order to see what goes wrong with the continuity it is enough to observe, using section 10, that the vanishing of a modulus may, and sometimes it indeed does, modify the polydromy order of ulterior moduli.

Corollary 10. The moduli space of a given equisingularity type that has two or more characteristic exponents has no non-closed points.

<u>Proof.</u> The statement has already been proved for all points whose bifurcation invariant is maximum (by the theorem above and the corollary 9).

Let us now assume that $\{\gamma_r\}$ is a sequence of branches of constant formal type, that they are given by short series and that the sequence converges to a branch γ coefficientwise. Suppose that the formal type of γ is different from the formal type of the γ_r. We are going to get a contradiction and this will prove the corollary.

We have that $s(\gamma) > s(\gamma_r)$. As before there is no loss of generality in assuming that the first non-vanishing term beyond the degree m in the series giving the γ_r has degree $m + s$, where $s = s(\gamma_r)$. Write the series giving the γ_r in the form

$$x = t^n \quad , \quad y = t^m + c(r)t^{m+s} + \sum a_{i_j}(r)t^{m+i_j}$$

so that we will have $\lim_r c(r) = 0$. Let $(m + i_q)/n$ be the second characteristic exponent of the branches γ_r. Then $s < i_q$, since the case of a bifurcation by a satellite point, in which the maximum value of the bifurcation invariant occurs, has already been dealt with before.

Let ϕ_r be a formal transformation of the branch γ_r into a semicanonical branch. Then ϕ_r leaves fixed all points on γ_r until the point prior to the relative level s and it will induce on the first neighbourhood of relative level s a homothety of ratio $c(r)^{-1}$ since it transforms $c(r)$ into 1. Thus if we consider the relations α, β for the transformation of γ_r into a semicanonical branch, then the relation β_s will force λ_0 to satisfy $\lambda_0^s = c(r)^{-1}$.

If P is the last point in the second group of satellite points on

γ_r and if P' is the homologous point on the semicanonical branch $\Phi_r(\gamma_r)$, then the transformation $V_1(P) \to V_1(P')$ induced by Φ_r sends the origin and improper point respectively to the origin and improper point, since these points are satellite points. Consequently the term G_{i_q} in the relation β_{i_q} must vanish (once the previous relations are fulfilled) and hence β_{i_q} will be reduced to a relation of the form

$$\tilde{a}_{i_q} = \lambda_0^{i_q} a_{i_q}(r)$$

where $\tilde{a}_{i_q}$ is the modulus of level i_q (corollary 6).

Now when r varies the formal type of γ_r stays constant and therefore a_{i_q} can take only finitely many values. On the other hand when r goes to infinity, c(r) goes to zero, hence λ_0 goes to infinity, by the relation established before. By the relation β_{i_q} above we see that $\lim_r a_{i_q}(r) = 0$. But this is a contradiction since by hypothesis $\lim_r a_{i_q}(r)$ is equal to the second characteristic coefficient of γ which is necesarily non-zero.

Notice that in the proof above one does not reach a contradiction if it is allowed that γ has an equisingularity type different from the equisingularity type of γ_r. This suggests that if we were to take the distinct moduli spaces in a hypothetical union moduli space, topologized coefficientwise, then the closedness of points would be lost.

REFERENCES

B.1 H. Bresinsky. Semigroups corresponding to algebroid branches in the plane. Proc. Am. Math. Soc. 32-2, 1972.

C.1 E. Casas. La proyeccion plana genérica de una rama de curva alabeada. Coll. Math. XXIX, 2, 1978.

C.2 E. Casas. Singularidades de una hoja de superficie algebraica a partir de su serie de Puiseux. Coll. Math. XXIX, 2, 1978.

E.1 F. Enriques - O. Chisini. Teoria geometrica delle equazioni ... N. Zanichelli, Bologna, 1915-1924.

S.1 J.G. Semple - G.T. Kneebone. Algebraic curves. Oxford University Press, 1959.

W.1 B.L. Van der Waerden. Infinitely near points. Ind. Mat., 12, 1950.

Z.1 O. Zariski. Studies in equisingularity I. Equivalent singularities of plane algebroid curves. Am. J. Math. XC, 3, 1968.

Z.2 O. Zariski. Studies in equisingularity III. Saturation of local rings and equisingularity. Am. J. Math. XC, 3, 1968.

Z.3 O. Zariski. General theory of saturation and of saturated local rings II. Am. J. Math. XCIII, 4, 1971.

Z.4 O. Zariski. Le problème des modules pour les branches planes. Cours donné au Centre de Matématiques de l'Ecole Polytechnique, Paris, Octobre - Novembre 1973.

Z.5 O. Zariski. Characterization of plane algebroid curves whose module of diferentials has maximum torsion. Proc. Nat. Acad. of Science U.S.A., 56, 3, 1966.

Dep. Geometría y Topología
Facultad de Matemáticas
Universidad de Barcelona

INVARIANTS TOPOLOGIQUES DE GERMES D'APPLICATIONS STABLES ET FINIES

André GALLIGO

I - Position du problème

Rappelons que deux germes d'applications analytiques

$$f \text{ et } g : (\mathbf{C}^n,o) \longrightarrow (\mathbf{C}^p,o)$$

sont dits équivalents s'il existe deux germes d'isomorphismes h et k de $(\mathbf{C}^n,o)$ et de $(\mathbf{C}^p,o)$ tels que $g = k \circ f \circ h$; que f est dit stable si tout déploiement

$$F : (\mathbf{C}^n,o)\times(\mathbf{C}^k,o) \longrightarrow (\mathbf{C}^p,o)\times(\mathbf{C}^k,o)$$

de f est équivalent au déploiement trivial de f, $f(\underline{x},\underline{t}) = (f(\underline{x}),\underline{t})$; que les germes d'applications analytiques stables sont classifiés à équivalence près par leurs algèbres associées

$$Q(f) = \mathbf{C}\{x_1,\dots,x_n\}/(f_1,\dots,f_p) .$$

Cette classification étant insuffisante R.Thom et J.Mather ont étudié l'équivalence topologique : dans la définition précédente remplacer isomorphisme par homéomorphisme. Un problème se pose alors naturellement : comment "lire" sur leurs algèbres associées que deux germes d'applications stables sont topologiquement équivalents? Plus précisément :

« Décrire des classes C de germes d'applications stables et des invariants numériques I des algèbres associées Q tels que si f f et $g : (\mathbf{C}^n,o) \longrightarrow (\mathbf{C}^p,o)$ appartiennent à C et si $I(Q(f) \neq I(Q(g))$, f ne soit pas topologiquement équivalent à g.»

I est alors un invariant topologique.

Nous nous restreindrons ici au cas des germes d'applications finies $f : (\mathbf{C}^n,o) \longrightarrow (\mathbf{C}^p,o)$ avec $n \leqslant p$.

Par ouverture de la stabilité on sait que si $\underline{f} : U \subset \mathbf{C}^n \longrightarrow V \subset \mathbf{C}^p$ désigne un représentant de f pour tout x voisin de 0 dans U, le germe de $\underline{f}$ en x que nous notons f_x est un germe d'application stable. Son algèbre associée $Q(f_x)$ est dite voisine de $Q(f)$.

II - Résultats

1. On dit que f est Σ_i, $i \in \mathbf{N}$, si f est de corang i, autrement dit i est la dimension de plongement de l'algèbre associée $Q(f)$ qui admet donc la présentation :

$$Q(f) = \mathbf{C}\{x_1,\dots,x_i\}/I \quad \text{avec} \quad I \subset (\mathcal{M}_i)^2$$

I désignant un idéal et $\mathcal{M}_i$ l'idéal maximal.

R. May puis J.Damon ont démontré que Σ_i est un invariant topologique pour tous les germes d'applications stables finis.

2. On dit que f est $\Sigma_i(j)$ si f est Σ_i et si $j \in \mathbf{N}$ est le corang de la dérivée seconde intrinsèque de f, autrement dit

$$1+i+j = \dim_{\mathbf{C}} Q(f)/\mathcal{M}_Q^3$$

$\mathcal{M}_Q$ désignant l'idéal maximal de Q.

J.Damon a démontré que $\Sigma_i(j)$ est un invariant topologique pour les germes d'applications stables, finis, Σ_i tels que $p-n \geq \frac{i(i-1)}{2}$.

3. $\delta = \dim_{\mathbf{C}} Q(f)$ est un invariant topologique pour les germes d'applications stables finis des deux classes Σ_2 et D.A.T.
On dit que f est D.A.T. (discrete algebra type) s'il n'existe qu'un nombre fini de type à isomorphisme près d'algèbres voisines de Q(f) ayant même Σ_i.

4. On appelle fonction d'Hilbert-Samuel de Q la fonction

$$\ell \in \mathbf{N} \longmapsto h(\ell) = \dim_{\mathbf{C}}(Q(f)/\mathcal{M}_Q^{\ell+1}).$$

J.Damon a démontré que si f est D.A.T. la fonction d'Hilbert-Samuel de Q(f) est un invariant topologique. Plus particulièrement si (n,p) sont dans les "bonnes dimensions" définies par J.Mather [M], Q(f) lui-même est un invariant topologique.

5. Si f est Σ_2 notons $Q(f) = \mathbf{C}\{x,y\}/I$ et $\nu(Q)$ l'ordre minimal des éléments de l'idéal I. Alors f étant Σ_2 si $p-n \geq \nu(Q)-1$ la fonction d'Hilbert-Samuel est un invariant topologique.

III - Aperçu des méthodes

1) Revient à démontrer par récurrence sur le nombre entier i que si f est Σ_i et g est Σ_k avec $k \leq i$ alors f et g ne sont pas topologiquement équivalents si $k \neq i$. On utilise les ensembles $\Sigma_j(g)$, $j \leq i$, des points x voisins de 0 dans $\mathbf{C}^n$ tels que g_x soit Σ_j. Alors le type à homéomorphisme près de :

$$\Sigma_{i-1}(g) \cup \Sigma_i(g) = \mathbf{C}^n \setminus (\Sigma_0(g) \cup \dots \cup \Sigma_{i-2}(g))$$

est un invariant topologique de g par hypothèse de récurrence et $\Sigma_{i-1}(g) \cup \Sigma_i(g)$ est vide si $k < i-1$, est une variété topologique si $k=i-1$, n'est pas une variété topologique si k=i. Ce dernier résultat étant obtenu en calculant des groupes d'homologie.

2) Est un peu plus délicat mais procède de la même méthode.

3) Consiste à démontrer que δ est égal au nombre maximum de points d'une fibre de f en un point voisin de 0, qui est un invariant topologique évident. Les algèbres voisines de Q(f) s'obtenant par déformation, on est ramené à déformer une algèbre artinienne de colongueur δ, qui représente un point épais de $\mathbf{C}^i$, en δ points simples. Ceci est toujours possible si la dimension de plongement de Q(f) est 2 mais est, en général, impossible si elle excède 2. La démonstration consiste à construire de telles déformations pour une certaine classe d'algèbres artiniennes puis à montrer que les algèbres D.A.T. soit y appartiennent soit se déforment platement en algèbres de cette classe.

L'outil essentiel est l'escalier d'un idéal I de $\mathbf{C}\{x_1,\dots,x_i\}$:
En translatant parallèlement à lui-même un hyperplan de $\mathbf{R}^i$ on définit un bon ordre de $\mathbf{N}^i$, ceci permet d'associer à toute série s son plus petit exposant exp $s \in \mathbf{N}^i$ et de former l'ensemble E(I) des exp s des séries s d'un idéal I de $\mathbf{C}\{x_1,\dots,x_i\}$. Il existe alors un plus petit ensemble fini F appelé l'escalier de I tel que

$$E(I) = \bigcup_{\alpha \in F} (\alpha + \mathbf{N}^i).$$

Si $Q = \mathbf{C}\{\underline{x}\}/I$ et $\delta = \dim_{\mathbf{C}} Q < \infty$, le complémentaire de E(I) dans $\mathbf{N}^i$ contient δ points dits sous l'escalier. Lorsqu'on considère la direction d'hyperplans de pente $(1,\dots,1)$ et que l'on effectue des changements de coordonnées linéaires génériques, on associe à I l'escalier générique qui a la particularité que ses "marches" sont toutes de hauteur 1 ; c'est un invariant analytique de l'algèbre Q qui en dimension 2 est équivalent à la fonction d'Hilbert-Samuel de Q.

On définit des générateurs de I adaptés à l'escalier qui permettent de décrire précisément les déformations de Q.

4) Utilise la classification des algèbres D.A.T. d'abord en trois classes $\Sigma_2, \Sigma_{r(i)}$, $r > i$, i=o ou 1, $\Sigma_{3(2)}$, puis en de nombreuses sous-classes jusqu'à arriver à la classification par la fonction d'Hilbert-Samuel. Il s'agit de les distinguer topologiquement grâce aux invariants topologiques précédents mais aussi par des raisonnements en cascade utilisant la structure topologique de sous-ensembles de $\mathbf{C}^n$ analogues au $\Sigma_j(g)$ décrit en 1) mais où le type Σ_j est remplacé par l'appartenance à une sous-classe étudiée au cran précédent.

5) On commence par démontrer la lissité des strates d'Hilbert-Samuel :

$$\Sigma_h(f) = \{x \in \mathbf{C}^n / Q(f_x) \text{ admet } h \text{ comme fonction H.S.}\}.$$

Ceci s'obtient à l'aide d'un résultat analogue de Briançon-Iarrobino sur Hilb{x,y} que l'on remonte "par éclatement" dans l'espace des jets $\Sigma_2(n,p)$.

La démonstration de 5) procède ensuite par récurrence sur δ. On définit la fonction d'Hilbert-Samuel déformée

$$\tilde{h}_Q(\ell) = \max\{h_{Q'}(\ell) \,/\, Q' \text{ voisine de } Q \text{ mais de colongueur } \delta-1\}$$

qui se représente aussi par un escalier générique.

Passer de h_Q à $\tilde{h}_Q$ revient, sur l'escalier générique correspondant, à enlever un carreau à la dernière marche qui n'est pas de longueur 1 :

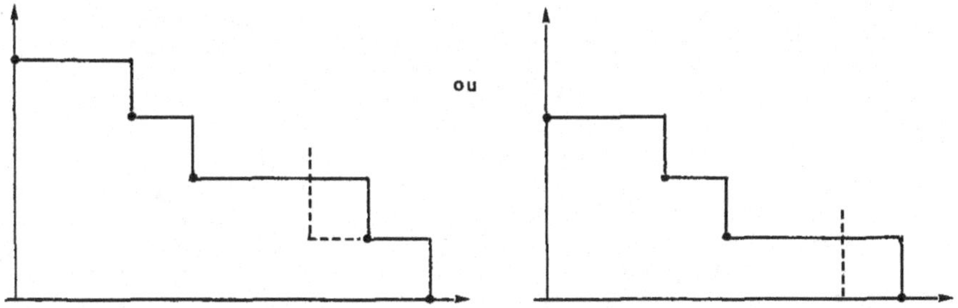

Pour pouvoir repasser de $\tilde{h}_Q$ à h_Q il faut connaître en outre le nombre $\alpha(Q)$ des dernières marches de h_Q. Par hypothèse de récurrence $\tilde{h}_Q$ est un invariant topologique.

Pour relier $\alpha(Q)$ à $\tilde{h}_Q$ on décrit explicitement, à l'aide des générateurs de I adaptés à l'escalier générique, toutes les déformations de Q(f) en une algèbre Q' ayant $\tilde{h}_Q$ comme fonction d'Hilbert-Samuel. Ceci permet de décrire l'incidence des strates Σ_{h_Q} et $\Sigma_{\tilde{h}_Q}$ et de calculer le type topologique de $\overline{\Sigma}_{\tilde{h}_Q}$ dont la cohomologie locale redonne $\alpha(Q)$:

$$\dim_{\mathbf{Q}} H^{\bullet}\left(\overline{\Sigma}_{h_Q}, \mathbf{Q}\right)_{\text{loc}} = 2\alpha(Q)-1.$$

Bibliographie

Briançon J. *Description de* $\mathrm{Hilb}^n\,\mathbf{C}\{x,y\}$. Invent.Math. 41, (1977)

Briançon J. et Galligo A. *Déformation d'un point de* $\mathbf{R}^2$ *ou* $\mathbf{C}^2$. Astérisque 7 et 8, (1973)

Damon J. *Investigating the topological stratification...* Proc.11th Col. Brazil, Math. Soc., (1977)

Damon J. *Topological Properties of D.A.T. I*: Adv. in Math Supp. Ser., vol.5, (1979)
II: Amer.J. Math.101, n°6, (1979)

Damon J. et Galligo A. *A topological invariant...* Invent. Math. 32, (1976)

Galligo A. *A propos du théorème de préparation*. Lecture Notes, n°409, (1974)

Galligo A. *Stabilité et théorème de division*. Ann. Inst. Fourier, T.24-2, (1979)

Mather J. *Stability of* C^∞*-mappings I à VI* (surtout) *VI*: The nice dimensions, Lecture Notes, n°192, (1970).

Mather J. *How to stratify a mapping and jet spaces*. Lecture Notes, n°535, (1975)

Damon J. et Galligo A. *The Hilbert-Samuel partition of* Σ_2. (à paraître).

SINGULARITES ISOLEES ET SECTIONS PLANES DE VARIETES DETERMINANTIELLES

M. GIUSTI

M. MERLE

Première partie

SINGULARITES ISOLEES ET NUAGES DE NEWTON

M. GIUSTI

Soit $(X,0)$ un germe d'espace analytique complexe. Si on choisit un longement $(X,0) \hookrightarrow (\mathbb{C}^n,0)$ et des générateurs $f_1,\ldots,f_p$ de l'idéal définissant n représentant du germe dans un voisinage ouvert de l'origine, on peut lui ssocier p sous-ensembles de $\mathbb{N}^n$ (ou nuages de Newton) comme suit :

$$f_i = \sum_{n\in\mathbb{N}^n} f_{i\alpha}\, x^\alpha \qquad (x^\alpha = x_1^{\alpha_1}\ldots x_n^{\alpha_n})$$

$$N_i = \{\alpha \in \mathbb{N}^n \mid f_{i\alpha} \neq 0\} \qquad i = 1,\ldots,p \quad .$$

n se pose alors le problème d'existence suivant :

Etant donnés p sous-ensembles, non vides, $N_1,\ldots,N_p$ de $\mathbb{N}^n$ ne contenant pas l'origine (ou nuages de points), existe-t-il une singularité isolée codimension maximale p (donc intersection complète) admettant $N_1,\ldots,N_p$ mme nuages de Newton ?

Un cas particulier important de ce problème consiste à se donner des uages contenus dans des hyperplans de $\mathbb{N}^n$; les variétés considérées admettent ne action de $\mathbb{C}^*$, et c'est la variante suivante :

Etant donnés p + n entiers $(d_1,\ldots,d_p\, ;\, a_1,\ldots,a_n)$, existe-t-il une ingularité isolée de codimension p définie par p polynômes quasi-homogènes de degrés $d_1,\ldots,d_p$ pour les poids $a_1,\ldots,a_n$ donnés aux variables ?

Pour ce dernier problème, et dans le cas des hypersurfaces, différents auteurs ont envisagé diverses conditions nécessaires portant sur certaies séries de Poincaré associées (cf. quelques références dans Arnold [1]) u sur certains graphes associés (cf. Orlik-Randell [5]).

Toujours dans le cas particulier des hypersurfaces, Kouchnirenko [3] donne des conditions nécessaires et suffisantes pour le problème général. Nous proposons ci-dessous une méthode complètement différente pour traiter le cas des intersections complètes ; nous donnerons des conditions nécessaires et suffisantes d'existence permettant de caractériser les nuages des singularités isolées d'intersections complètes, ainsi qu'un algorithme effectif pour les vérifier.

Je tiens à remercier M. Merle des nombreuses et positives discussions que j'ai eues avec lui sur ce travail.

1. NOTATIONS ET RESULTATS.

Remarquons tout d'abord qu'on peut se restreindre à l'étude des nuages finis, grâce à l'existence d'un jet suffisant pour une application définissant une singularité isolée d'intersection complète (cf. Mather [4]) et au fait qu'être isolée et intersection complète est une condition ouverte.

Sans perte de généralité, on supposera également dans toute la suite qu'aucun nuage n'est entièrement constitué de points à distance 1 de l'origine ; sinon on se ramène à une situation de p - 1 nuages dans $\mathbb{N}^{n-1}$.

1.1 Pour toute partie I de $\{1,...,n\}$, on note $|I|$ le cardinal de I et $\mathbb{N}^I$ le I-plan de coordonnées de $\mathbb{N}^n$. Etant donné un couple ordonné $I \subset K$ de $\{1,...,n\}$, on définit alors l'épaississement de $\mathbb{N}^I$ dans $\mathbb{N}^K$ comme l'ensemble des points de $\mathbb{N}^K$ à distance au plus 1 de $\mathbb{N}^I$. On le note $\mathbb{N}^{I,K}$.

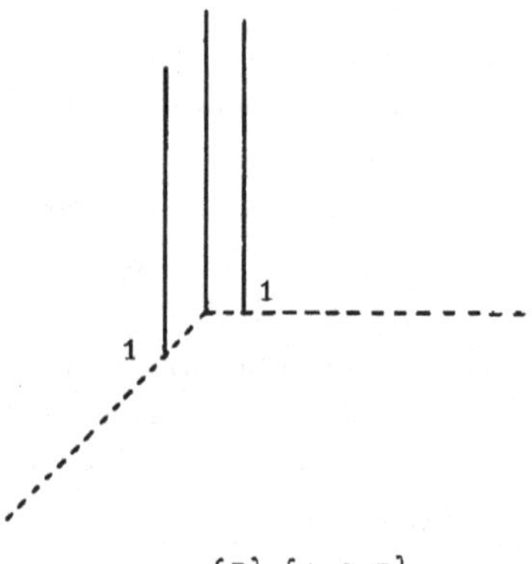

$\mathbb{N}^{\{3\},\{1,2,3\}}$

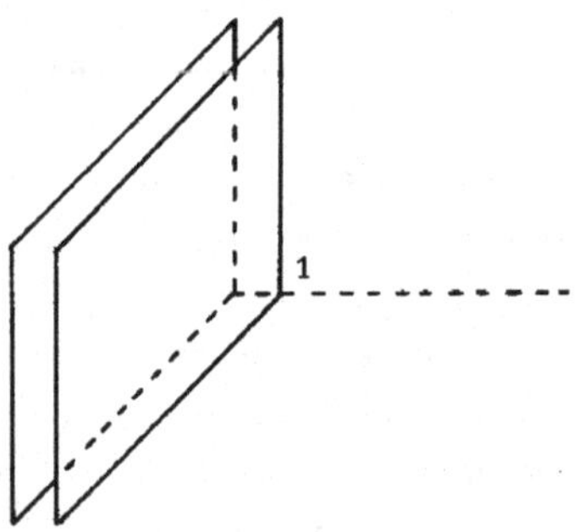

$\mathbb{N}^{\{1,3\},\{1,2,3\}}$

$\mathbb{N}^{I,K}$ est une réunion disjointe de translatés de $\mathbb{N}^I$ ($\mathbb{N}^{I,K} = \mathbb{N}^I \coprod_{k \in K-I} \mathbb{N}^{I,I\cup\{k\}}$) et c'est une fonction croissante de I et de K.

1.2 Maintenant, étant donnés p sous-ensembles non vides $N_1,\ldots,N_p$ de $\mathbb{N}^n - \{0\}$, considérons pour tout couple ordonné $I \subset K$ de parties de $\{1,\ldots,n\}$ les nuages qui coupent l'épaississement $\mathbb{N}^{I,K}$:

$$E(I,K) = \{i \in \{1,\ldots,p\} \mid N_i \cap \mathbb{N}^{I,K} \neq \emptyset\}$$

et l'entier :

$$D(I,K) = |K| - |E(I,K)| \quad .$$

$E(I,K)$ hérite des propriétés de croissance de $\mathbb{N}^{I,K}$ (et $E(I,K) = \coprod_{k \in K-I} E(I,I \cup \{k\})$), alors que $D(I,K)$ n'est plus que fonction décroissante de I.

Dans le cas où les nuages considérés sont les nuages de Newton d'un germe X, on peut interpréter $E(I,I)$ et $D(I,I)$ (notés dorénavant $E(I)$ et $D(I)$ par simplification) comme suit : le premier correspond aux équations ne s'annulant pas identiquement sur $\mathbb{C}^I$. Le second est une dimension virtuelle de $X \cap \mathbb{C}^I$: c'est une minoration de la dimension de $X \cap \mathbb{C}^I$ (celle qu'il aurait s'il était intersection complète). Nous verrons plus loin que sous certaines hypothèses de généricité, $X \cap \mathbb{C}^I$ est lisse et localement intersection complète de dimension $D(I)$ en tout point de $X \cap (\mathbb{C}^*)^I$.

Si K est différent de I, $E(I,K)$ et $D(I,K)$ n'ont pas une interprétation aussi claire. Néanmoins soit $m(I,K)$ et $\mathfrak{U}(K)$ les idéaux définissant respectivement $\mathbb{C}^I$ et $X \cap \mathbb{C}^K$ dans $\mathbb{C}^K$. L'idéal tronqué à l'ordre un $\mathfrak{U}(K) + m^2(I,K)/m^2(I,K)$ n'a pas habituellement une interprétation très précise, mais ici la dimension virtuelle du germe qu'il définit dans $\mathbb{C}^K$ est $D(I,K)$.

Quoiqu'il en soit, nous sommes maintenant en mesure de formuler le théorème d'existence suivant :

1.3 <u>Théorème</u> : <u>Etant donnés</u> p <u>nuages</u> $N_1,\ldots,N_p$ <u>de</u> $\mathbb{N}^n$, <u>il existe une singularité isòlée d'intersection complète de codimension</u> p <u>dans</u> $(\mathbb{C}^n,0)$, <u>les admettant pour nuages de Newton si et seulement si</u> :

<u>Toute partie non vide</u> I <u>de</u> $\{1,\ldots,n\}$ <u>vérifiant</u> :

(i) $D(I) > 0$

(ii) $E(I) < p$

possède la propriété $*(I)$:

$$*(I) \qquad \sup_{\substack{K\supset I \\ E(I,K)<p}} D(I,K) + D(I) \leq n-p+1 \quad .$$

1.4 Remarque : La résolution du problème d'existence ne dépend que des points des nuages donnés dont la distance aux hyperplans de coordonnées de $\mathbb{N}^n$ est au plus 1 ; plus précisément il suffit de connaître l'application $|E|$.

1.5 Variante quasi-homogène du théorème.

Un germe de $(\mathbb{C}^n,0)$ défini par p polynômes est dit quasi-homogène de type $(d_1,\dots,d_p ; a_1,\dots,a_n)$ si ses nuages de Newton sont contenus dans des hyperplans parallèles d'équations :

$$\sum_{j=1}^{n} a_j \alpha_j = d_i \qquad (i = 1,\dots,p) \quad .$$

Soient $\Gamma(I)$ le semi-groupe de $\mathbb{N}$ engendré par les a_i $(i \in I)$, et $\Gamma(I,K)$ la partie $\Gamma(I) \cup \bigcup_{k\in K-I} (a_k + \Gamma(I))$ de $\Gamma(K)$. Alors :

$$E(I,K) = \{i \in 1,\dots,p\} \mid d_i \in \Gamma(I,K)\} \quad .$$

1.6 Remarque : Nous verrons que la propriété $*(I)$ implique en fait qu'un germe générique pour ses nuages de Newton est lisse et localement intersection complète en tout point de $(\mathbb{C}^*)^I$, ce qui est automatiquement satisfait si l'une des deux conditions (i) ou (ii) n'est pas vérifiée (cf. 3.3). Aussi appellerons-nous désormais parties non triviales de $\{1,\dots,n\}$ tout I vérifiant (i) et (ii).

Enfin s'il n'y a aucune partie non triviale, le problème a évidemment une solution ; par exemple c'est le cas bien connu où chaque nuage coupe tous les axes de coordonnées de $\mathbb{N}^n$.

1.7 Corollaires du théorème.

1.7.1 Il existe une singularité d'intersection complète de codimension p si et seulement si toute partie non triviale I vérifie l'affaiblissement de $*(I)$:

$$D(I) \leq n-p \qquad \text{(cf. 3.3)} \quad .$$

Voyons maintenant quelques conditions nécessaires d'existence :

.7.2 Tout I non trivial doit vérifier :

$$2\,D(I) \leq n - p + 1 \quad .$$

.7.3 Soit N la réunion des nuages $N_1,\ldots,N_p$. Une deuxième condition nécessaire d'existence est alors la suivante :

Si $\mathbb{N}^I$ ne coupe pas N, au moins $|I| + p - 1$ de ses épaississements dans es $|I| + 1$-plans de $\mathbb{N}^n$ coupent N.

Cette condition traduit simplement le fait suivant : si p polynômes léfinissent une singularité isolée d'intersection complète, p combinaisons inéaires génériques en font autant.

.7.4 3ème condition nécessaire d'existence (conséquence de 1.7.2) :

Le cardinal de toute partie non triviale est majoré par $\frac{n+p-1}{2}$.

Cette remarque borne le cardinal des parties où il suffit de vérifier a propriété * pour répondre au problème. (Nous donnerons à la fin de ce traail un algorithme pour vérifier *).

. EXEMPLES.

.1 Cas des courbes.

La condition nécessaire et suffisante s'écrit : pour tout I non triial, on a :

$$*(I) \qquad \sup_{\substack{K \supset I \\ E(I,K)<p}} D(I,K) = 1 \quad .$$

n utilisant les propriétés de monotonie de D, il suffit même de vérifier * ur les parties non triviales minimales pour l'inclusion.

Exemple 1 :

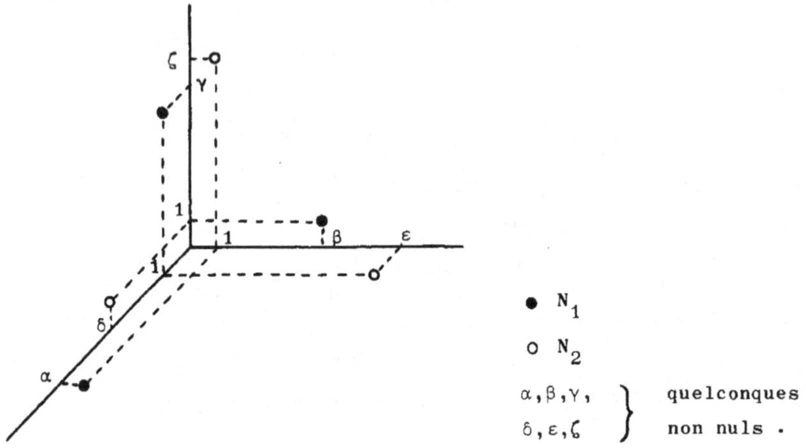

Les 3 parties non triviales sont {1}, {2} et {3}. Si K est tout {1,2,3} les épaississements des axes coupent les deux nuages et sont donc éliminés. Si |K| vaut 2, chaque épaississement d'axe ne coupe qu'un nuage, et D(I,K) vaut 1.

Pour a générique, les équations

$$\begin{cases} a_1\, x^{\alpha}y + a_2\, y^{\beta}z + a_3\, z^{\gamma}x = 0 \\ a_4\, x^{\delta}z + a_5\, z^{\varepsilon}y + a_6\, y^{\zeta}x = 0 \end{cases}$$

définissent une singularité isolée d'intersection complète.

2.2 Cas des surfaces.

Condition nécessaire et suffisante d'existence : Toute partie non triviale vérifie :

$$*(I) \qquad \begin{cases} D(I) = 1 \\ \sup\limits_{\substack{K\supset I \\ E(I,K)<p}} D(I,K) \leq 2 \end{cases}$$

Pour les mêmes raisons de monotonie que précédemment, il suffit encore de vérifier * sur les parties non triviales minimales pour l'inclusion.

Exemple 2 : Pour tout $\underline{a}$, les deux équations : (α, α', β, β', γ et $\delta \geq 2$)

$$a_1\, xy + a_2\, z^{\gamma} + a_3\, t^{\delta} = 0$$

$$a_4\, xy + a_5\, x^{\alpha} z + a_6\, y^{\beta} z + a_7\, x^{\alpha'} t + a_8\, y^{\beta'} t = 0$$

ne définissent pas une singularité isolée d'intersection complète.

En effet, $\{3,4\}$ n'est pas trivial, mais $D(\{3,4\},\{1,2,3,4\}) = 3$.

2.3 Cas des hypersurfaces.

Dans ce cas, les I non triviaux se caractérisent par la condition $E(I) = \emptyset$, et la condition $*(I)$ devient :

$$\sup_{\substack{K \supset I \\ E(I,K)=\emptyset}} |K| + |I| \leq n \ .$$

D'où la condition nécessaire et suffisante d'existence :

Si $\mathbb{N}^I$ ne coupe pas N, au moins $|I|$ de ses épaississements dans les $|I| + 1$-plans de coordonnées le font.

Remarquons que c'est la condition nécessaire 1.7.3 qui devient dans ce cas suffisante.

Il est intéressant de comparer cette condition nécessaire et suffisante à celle donnée par Kouchnirenko qui est en apparence plus restrictive :

Pour tout I, trivial ou non, au moins $|I|$-épaississements de $\mathbb{N}^I$ dans les $|I| + 1$-plans de coordonnées coupant N. ($\{1,\dots,n\}$ est évidemment exclu.) (cf. [3], 1.13, Remarque (ii)).

Exemple 3 : Existe-t-il une singularité isolée d'hypersurface quasi-homogène de type (265 ; 1, 24, 33, 58) ?

Ce type satisfait à une condition nécessaire d'existence consistant à vérifier que

$$\prod_{j=1}^{n} \frac{z^{d-a_j} - 1}{z^{a_j} - 1}$$

est un polynôme (exemple de V.M. Izlev cité dans Arnold [1]).

Il est aisé de voir que 241, 232 et 207 n'appartenant pas au semi-groupe engendré par 24 et 23, un seul épaississement de $\mathbb{N}^{\{2,3\}}$ coupe N et la réponse est négative.

§ 3. GERMES GENERIQUES POUR LEURS NUAGES DE NEWTON.

Une idée naturelle est de considérer l'ensemble des germes admettant $N_1,\dots,N_p$ comme nuages de Newton, et de calculer pour ceux qui sont intersections complètes (à reconnaître) la dimension générique (à préciser) de leur lieu singulier.

On exclut immédiatement le cas trivial $n < p$.

3.1 Considérons l'espace vectoriel $Y = \prod_{i=1}^{p} \mathbb{C}^{|N_i|}$, et l'application :

$$F : (\mathbb{C}^n,0) \times Y \longrightarrow (\mathbb{C}^p,0) \times Y$$

$$(x,y) \longmapsto (F_1(x,y),\dots,F_p(x,y),y)$$

où $F_i(x,y) = \sum_{\alpha\in \mathbb{N}_i} y_{i\alpha}\, x^\alpha \quad (x^\alpha = x_1^{\alpha_1} \dots x_n^{\alpha_n})$.

Les fibres $X_y = F^{-1}(0,y)$ décrivent ainsi tous les germes de $\mathbb{C}^n$ admettant $N_1,\dots,N_p$ comme nuages de Newton.
Notons Σ_y le lieu critique de l'application définissant X_y , obtenu en annulant les p-mineurs de la matrice jacobienne

$$\mathfrak{J} = [\mathfrak{J}_{ij}]_{\substack{i=1,\dots,p\\ j=1,\dots,n}} \qquad (\mathfrak{J}_{ij} = \frac{\partial F_i}{\partial x_j}) \ .$$

X_y est un germe de singularité isolée d'intersection complète si et seulement si $X_y \cap \Sigma_y$ se réduit à l'origine. En utilisant la partition canonique $\coprod_{I\subset\{1,\dots,n\}} (\mathbb{C}^*)^I$ de $\mathbb{C}^n$, il suffit d'étudier l'intersection de $X_y \cap \Sigma_y$ avec $\mathbb{C}^I$, et ce pour un y générique.

En un point x de $\mathbb{C}^I$, $\mathfrak{J}_{ij}$ s'écrit $\sum_{\alpha \in \frac{\partial N_i}{\partial x_j} \cap \mathbb{N}^I} \alpha_j\, y_{i\alpha}\, x^\alpha$, où $\frac{\partial N_i}{\partial x_j}$ est le j-ième nuage dérivé de N_i :

$$\frac{\partial N_i}{\partial x_j} = \{\alpha \in \mathbb{N}^n \mid \alpha + (0,\dots,0,\underset{\substack{\uparrow\\ \text{j-ème place}}}{1},0,\dots,0) \in N_i\} \ .$$

D'où l'introduction naturelle de l'ensemble :

$$Z(I) = \{(i,j) \in \{1,\dots,p\} \times \{1,\dots,n\} \mid \frac{\partial N_i}{\partial X_j} \cap \mathbb{N}^I = \emptyset\} .$$

Ainsi $\mathfrak{J}_{ij}$ s'annule identiquement sur $\mathbb{C}^I$ si et seulement si (i,j) est dans $Z(I)$.

Remarquons que sous l'hypothèse faite au début du § 1 sur nos nuages :

$$E(I) = \{i \in \{1,\dots,p\} \mid \{i\} \times I \not\subset Z(I)\}$$

et que $\mathfrak{J}$ restreinte à $\mathbb{C}^I$ se met sous la forme :

$$\mathfrak{J}(I) = \begin{array}{r|c|c|} \multicolumn{1}{r}{} & \multicolumn{1}{c}{I} & \multicolumn{1}{c}{} \\ \cline{2-3} E(I) & \mathfrak{J}'(I) & \\ \cline{2-3} & 0 & \mathfrak{J}''(I) \\ \cline{2-3} \end{array}$$

$\mathfrak{J}'(I)$ n'est pas autre chose que la matrice jacobienne des équations définissant $X_y \cap \mathbb{C}^I$.

Démontrons maintenant un lemme de transversalité ad hoc qui va nous permettre de préciser la généricité dont nous avons besoin dans Y.

3.2 Lemme de transversalité : Soient V un germe de $(\mathbb{C}^n,0)$ lisse en dehors des hyperplans de coordonnées, et $(W,0)$ un germe de $(\mathbb{C}^p,0)$ dont l'ouvert des points lisses est partout dense. Alors il existe un ouvert de Zariski non vide dans Y tel que, pour tout y dans cet ouvert, soit la fibre $F^{-1}(W \times \{y\}) \cap V \cap (\mathbb{C}^*)^n$ est vide, soit sa codimension dans $V \cap (\mathbb{C}^*)^n$ est celle de W. De plus si W est lisse, cette fibre l'est aussi.

Ce lemme de tranversalité à la Bertini-Sard appartient évidemment à une famille nombreuse et ancienne (cf. par exemple Kleiman [2]). Excluons le cas trivial où $V \cap (\mathbb{C}^*)^n$ est vide, et considérons les deux restrictions π_1 et π_2 à $F^{-1}(W \times Y) \cap V \cap (\mathbb{C}^*)^n$ des projections canoniques. π_1 est une fibration de fibre $W \times \mathbb{C}^{\dim Y - p}$. D'autre part, s'il n'existe pas d'ouvert dense de Zariski de Y où la fibre $\pi_2^{-1}(y)$ est vide, l'image de π_2 contient un ouvert dense de Zariski au-dessus duquel $\pi_2^{-1}(y)$ est lisse en tout point de l'ouvert partout dense des points lisses de $F^{-1}(W \times Y) \cap V \cap (\mathbb{C}^*)^n$ et :

$$\dim \pi_2^{-1}(y) + \dim Y = \dim F^{-1}(W\times Y)\cap V\cap(\mathbb{C}^*)^n$$

$$= \dim W + \dim Y - p + \dim (V\cap(\mathbb{C}^*)^n) \quad .$$

D'où la conclusion voulue : pour y générique, on a l'égalité

$$\operatorname{Codim}_{V\cap(\mathbb{C}^*)^n} \pi_2^{-1}(y) = \operatorname{Codim}_{\mathbb{C}^p} W$$

qui reste vraie quand $\pi_2^{-1}(y)$ est vide avec les conventions habituelles.

3.3 <u>Corollaire</u> : <u>lemme du rejet dans les ténèbres extérieures</u> :
<u>Il existe un ouvert dense de Zariski dans Y au-dessus duquel X_y est lisse et localement intersection complète en dehors des hyperplans de coordonnées.</u>

Il suffit d'appliquer le lemme précédent à la situation $V=\mathbb{C}^n$ et $W=\{0\}$.

En particulier, il existe un ouvert de Zariski non vide $\Omega(I)$ de Y tel que, pour tout y dans cet ouvert, $X_y\cap(\mathbb{C}^*)^I$ est lisse, intersection complète de dimension D(I) ; si D(I) est négatif ou nul, X_y n'a pas de points dans $(\mathbb{C}^*)^I$. Dans le cas contraire, la matrice jacobienne $\mathfrak{J}'(I)$ est de rang maximum E(I) en tout point de $X_y\cap(\mathbb{C}^*)^I$. D'autre part si E(I) vaut p, $\mathfrak{J}(I)$ lui-même est a fortiori de rang maximum en tout point de $X_y\cap(\mathbb{C}^*)^I$: X_y est donc lisse et localement intersection complète en tout point de $(\mathbb{C}^*)^I$.

Ces dernières remarques justifient les interprétations de E(I) de D(I) données dans 1.2 ainsi que la définition des parties non triviales dans 1.6 : si I est trivial, X est lisse et localement intersection complète en tout point de $(\mathbb{C}^*)^I$.

3.4 <u>Démonstration du théorème</u> : Définissons pour tout I non vide la $p\times n$-matrice m(I) :

$$m_{ij}(I) = \begin{cases} 0 & \text{si } \dfrac{\partial N_i}{\partial X_j}\cap \mathbb{N}^I = \emptyset \quad (\text{i.e. } (i,j)\in Z(I)) \\ \\ z_{ij} & \text{si } \dfrac{\partial N_i}{\partial X_j}\cap \mathbb{N}^I \neq \emptyset \quad (\text{i.e. } (i,j)\notin Z(I)) \end{cases}$$

à coefficients dans $\mathbb{C}[z_{ij}]_{(i,j)\notin Z(I)}$.

L'annulation des p-mineurs de m(I) définit une section d'une variété déterminantielle générique par un certain $(np-|Z(I)|)$-plan de coordonnées ; appelons la $\mathcal{D}(Z(I))$.

Pour accéder à la codimension générique de $X_y \cap \Sigma_y \cap (\mathbb{C}^*)^I$ dans $X_y \cap (\mathbb{C}^*)^I$, on aimerait appliquer le lemme de transversalité 3.2 à la situation des $np - |Z(I)$ nuages $\frac{\partial N_i}{\partial X_j} \cap \mathbb{N}^I$ $((i,j) \notin Z(I))$ de $\mathbb{N}^I$, avec $W = \mathcal{D}(Z(I))$ et $V = X_{y_o} \cap \mathbb{C}^I$ $(y_o \in \Omega(I))$

Mais ceci nous est interdit à cause de la dépendance commune en y de $\Sigma_y \cap \mathbb{C}^I$ et $X_y \cap \mathbb{C}^I$: rien ne prouve en effet que l'ouvert où la dimension de $\Sigma_y \cap X_{y_o} \cap (\mathbb{C}^*)^I$ est générique contienne y_o.

Cependant dans le cas où I est non trivial, le rang de $\mathfrak{Z}(I)$ en un point de $X_y \cap (\mathbb{C}^*)^I$ $(y \in \Omega(I))$ n'est maximum d'après 3.3 que si le rang de $\mathfrak{Z}''(I)$ est $p - |E(I)|$.

Ce n'est certainement pas le cas si D(I) est strictement plus grand que $n - p$. Dans le cas contraire, soit Σ''_y le germe obtenu en annulant les $(p - |E(I)|)$-mineurs de $\mathfrak{Z}''(I)$:

$$(X_y \cap \Sigma_y) \cap \mathbb{C}^I = (X_y \cap \mathbb{C}^I) \cap \Sigma''_y \quad .$$

Introduisons $Z''(I) = Z(I) \cap ((\{1,\dots,p\} - E(I)) \times (\{1,\dots,n\} - I))$ et $\mathcal{D}(Z''(I))$ la variété déterminantielle définie par l'annulation des $(p - |E(I)|)$-mineurs de $m''(I)$.

Il est maintenant agréable de constater que $X_y \cap \mathbb{C}^I$ et Σ''_y sont indépendants dans le sens suivant : Y est somme directe des deux sous-espaces vectoriels

$$Y' = \sum_{i \in E(I)} \mathbb{C}^{|N_i|} \times \{0\}$$

$$Y'' = \{0\} \times \sum_{i \notin E(I)} \mathbb{C}^{|N_i|}$$

$$y = (y', y'') \quad , \quad y' \in Y' \quad , \quad y'' \in Y''$$

et

$$X_y \cap \mathbb{C}^I = X_{y'} \cap \mathbb{C}^I$$

$$\Sigma''_y = \Sigma''_{y''} \quad .$$

Soit $\Omega'(I)$ l'ouvert partout dense de Y' projection de $\Omega(I)$. On peut maintenant appliquer le lemme de transversalité à la situation des

$(n - |I|)(p - |E(I)|) - |Z''(I)|$ nuages $\frac{\partial N_i}{\partial X_j} \cap \mathbb{N}^I$ de $\mathbb{N}^I$ $((i,j) \notin Z''(I))$, avec

$$V = X_{y'} \cap \mathbb{C}^I \quad (y' \in \Omega'(I)) \quad \text{et} \quad W = \mathcal{D}(Z''(I)) \quad .$$

Il existe donc un ouvert dense de Zariski $\Omega''(I)$ dans Y'' tel que, pour tout y''

dans cet ouvert, la codimension de $X_{y'} \cap \Sigma_{y''} \cap (\mathbb{C}^*)^I$ dans $X_{y'} \cap (\mathbb{C}^*)^I$ est la codimension $C(I)$ de $\mathcal{A}(Z''(I))$ dans $\mathbb{C}^{(n-|I|)(p-|E(I)|)-|Z''(I)|}$.
Voilà qui motive l'étude, effectuée en collaboration avec M. Merle, des sections des variétés déterminantielles génériques par des plans de coordonnées, dont il suffit maintenant de connaître la codimension pour achever la démonstration du théorème :

3.5 Lemme : Pour tout I non trivial,

$$C(I) = \mathrm{Sup}(0\,,\, n-p+1 - \sup_{\substack{K\supset I \\ E(I,K)<p}} D(I,K)) \quad .$$

En effet, si pour tout I non trivial la propriété $*(I)$ est vérifiée, $X_y \cap \Sigma_y \cap (\mathbb{C}^*)^I$ est vide pour tout y dans $\Omega'(Y) \times \Omega''(Y)$; si ce n'était pas le cas, sa dimension serait d'après le lemme de transversalité 3.2 :

$$\dim X_y \cap (\mathbb{C}^*)^I - C(I) \leq$$

$$\leq \dim X_y \cap (\mathbb{C}^*)^I + \sup_{\substack{K\supset I \\ E(I,K)<p}} D(I,K) - (n-p+1) \qquad \text{(lemme 3.5)}$$

$$\leq D(I) + \sup_{\substack{K\supset I \\ E(I,K)<p}} D(I,K) - (n-p+1) \qquad \text{(lemme 3.3)}$$

$$\leq 0 \qquad \text{contradiction} \quad .$$

La même conclusion restant vraie pour un I trivial (cf. 3.3), la condition suffisante est démontrée, puisque les seuls I exclus $(D(I) > n-p)$ ne vérifient pas $*(I)$.
Réciproquement, si un I non trivial ne vérifie pas $*(I)$, on a pour tout y :

$$\dim X_y \cap \Sigma_y \cap \mathbb{C}^I \geq \dim X_y \cap \mathbb{C}^I + \dim \Sigma_y \cap \mathbb{C}^I - |I|$$

$$\geq D(I) - C(I)$$

$$= \mathrm{Sup}(D(I)\,,\, D(I) + \sup_{\substack{K\supset I \\ E(I,K)<p}} D(I,K) - (n-p+1))$$

$$> 0 \quad ,$$

et X_y n'est pas une singularité isolée d'intersection complète de codimension p dans $(\mathbb{C}^n,0)$.

<u>Démonstration de 3.5</u> : La codimension C(I) se calcule à partir de la taille t(Z"(I)) de la matrice m"(I) (cf. 2ème partie, 1.1) :

$$t(Z''(I)) = \sup_{\substack{\emptyset \neq J \subset \{1,\dots,n\}-I \\ \emptyset \neq J' \subset \{1,\dots,p\}-E(I) \\ J\times J' \subset Z''(I)}} (|J| + |J'|)$$

grâce à la formule (2ème partie, 1.3) :

$$C(I) = \mathrm{Sup}(0\,,\, n - |I| - \mathrm{Sup}(p - E(I)\,,\, t(Z''(I))) + 1) \quad .$$

Or on lit très bien l'application E sur la matrice m(I) :

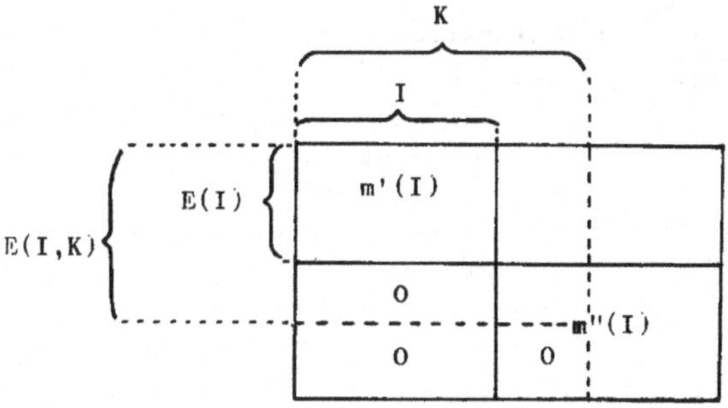

m(I)

et on déduit aisément 3.5.

Pour vérifier * il suffit donc de construire l'ensemble partiellement ordonné des Z(I) $(K \supset I \Rightarrow Z(I) \subset Z(K))$ et de vérifier que toute partie non triviale I de cardinal $|I| \leq \frac{n+p-1}{2}$ (cf. 1.7.4) satisfait :

1) $2\,D(I) \leq n - p + 1$
2) $t(Z''(I)) \leq (n - p + 1 - 2\,D(I)) + (p - |E(I)|)$.

*
* *
*

BIBLIOGRAPHIE

[1] V.I. Arnold : Normal forms of functions in neighbourhoods of degenerate critical points, Uspechi Mat. Nauk 29, 2 (176), (1974), 11-49 et Russian Math. Surveys 29, 2 (1974), 10-50 (cf. p. 21).

[2] S. Kleiman : The transversality of a general translate, Compositio Math. 28 (1974), 287-297.

[3] A.G. Kouchnirenko : Polyèdres de Newton et nombres de Milnor, Inv. Math. 32 (1976), 1-31.

[4] J. Mather : Stability of C^{∞} mappings -III- : finitely determined map-germs, Publications I.H.E.S. No 35 (1968), 127-156.

[5] P. Orlik, R. Randell : The structure of weighted homogeneous polynomials Proc. of Symposia in Pure Mathematics, "Several complex variables", vol. XXX, part I (1977), 57-64.

Centre de Mathématiques
Ecole Polytechnique
F-91128 PALAISEAU Cedex

"Laboratoire Associé au
C. N. R. S. n° 169"

Deuxième partie

SECTIONS DE VARIETES DETERMINANTIELLES PAR LES PLANS DE COORDONNEES

M. GIUSTI

M. MERLE

Etant donnés deux entiers n et p ($n \geqslant p$) et un corps algébriquement clos k, on considère dans l' espace vectoriel $M_k(n,p)$ des $n \times p$-matrices à coefficients dans k la sous-variété D des matrices singulières, sur laquelle s'annule l'idéal $\mathfrak{J}$ engendré par les p-mineurs.

Soit un morphisme $V \xrightarrow{\varphi} M_k(n,p)$ (V variété affine sur k) ; les premiers résultats généraux sur $\varphi^*(\mathfrak{J})$ ont été obtenus à notre connaissance par F.S. Macaulay ("The algebraic theory of modular systems" 1916 [8] theorem 53) dans le cas où V est un espace affine sur k : la codimension de $\varphi^{-1}(D)$ dans V est majorée par n-p+1. Si elle est égale à n-p+1, $\varphi^*(\mathfrak{J})$ définit une variété de Cohen-Macaulay, qui est par ailleurs génériquement lisse donc réduite.

$\mathfrak{J}$ est donc l'idéal des fonctions s'annulant sur D, qu'on appellera désormais <u>variété déterminantielle générique</u>.

Dans le même esprit, plusieurs auteurs dont Eagon, Eagon-Northcott, Buchsbaum-Rim,... [3,5,1] ont prolongé l'étude aux idéaux déterminantiels d'une matrice à coefficients dans un anneau, en introduisant un complexe de Koszul généralisé.

Dans ce travail nous étudions la section de D par un plan de coordonnées de $M_k(n,p)$. Pour ce faire nous n'utilisons pas les techniques précédentes mais la construction naturelle d'une résolution des singularités de D qui permet une étude plus géométrique.

On trouve vraisemblablement cette idée pour la première fois dans le livre de T.G. Room ("The geometry of determinantal loci" 1938 [11] . Elle est reprise entre autres par I.R. Porteous [9] , F. Ronga [10] , G. Kempf [7] ,.. .

Signalons enfin que Eagon-Hochster [4], et De Concini-Eisenbud-Procesi [2] démontrent directement le caractère réduit de l'anneau quotient par $\mathfrak{J}$.

Eagon-Hochster considèrent d'ailleurs certaines sections de D par des plans de coordonnées particuliers.

1 - Notations et énoncé des résultats.

1.1 Soit Z un sous-ensemble de $\{1,\dots,n\} \times \{1,\dots,p\}$.
Sa taille t(Z) est le demi-périmètre maximum d'un rectangle non vide $I \times J$ contenu dans $Z (I \subset \{1,\dots,n\}$, $J \subset \{1,\dots,p\}$). Si L(Z) est le plan de coordonnées de $M_k(n,p)$ défini par les équations : $X_{ij} = 0$ $((i,j) \in Z)$ nous nous intéressons à la section D(Z) de D par L(Z). Nous appellerons désormais Z une configuration de zéros. D admet une stratification canonique par le rang ;

$$D = \bigcup_{i=o}^{p-1} D_i$$

(où D_i est la strate des matrices de rang exactement i.)

De plus, les $\overline{D_i}$ (i=o,....,p-1) forment une filtration croissante. Maintenant soient r(Z) le rang maximum des éléments de D(Z), et $D_i(Z)$ (i=o,....,r(Z)) la partition par le rang de D(Z).

1.2 Proposition : Les $\overline{D_i(Z)}$ (i=o,...,r(Z)) forment encore une filtration strictement croissante de D(Z).

Ce résultat n'est pas du tout général ; pour une section par un plan quelconque il peut devenir inexact comme le montre le contre-exemple suivant :

Considérons la sous-variété de k^2 défini par l'annulation du déterminant de la matrice :

$$\begin{bmatrix} x & 0 & 0 \\ 0 & x & 0 \\ 0 & 0 & y \end{bmatrix}$$

$\overline{D_2(Z)}$ ne contient pas $\overline{D_1(Z)}$.

1.3 Théorème. <u>La codimension c(Z) de D(Z) dans L(Z) et le rang générique r(Z) des éléments de D(Z) ne dépendent que de la taille de la configuration des zéros.</u>

Ils sont donnés de manière explicite par les formules suivantes :

$$\begin{cases} c(Z) = n-p+1 - [\operatorname{Inf}(t(Z),n+1) - \operatorname{Inf}(t(Z),p)] \\ r(Z) = \operatorname{Inf}(p-1, n+p-t(Z)). \end{cases}$$

Quand tous les p-mineurs s'annulent identiquement sur L(Z), la codimension est évidemment nulle ; plus généralement on a le

1.4 Théorème. <u>La codimension ne dépend en fait que de l'ensemble des p-mineurs ne s'annulant pas sur L(Z).</u>

La formule explicite sera donnée au chapitre 3.

1.5 Corollaire. <u>Les trois assertions suivantes sont équivalentes</u> :

(i) <u>La codimension est maximale</u>

(ii) <u>$t(Z) \leq p$</u>

(iii) <u>Aucun des p-mineurs ne s'annule identiquement sur L(Z).</u>

Dans ce cas D(Z) est une variété de Cohen-Macaulay [8,..]

1.6 Dans tout ce paragraphe nous excluons les configurations de zéros qui contiennent des rectangles de largeur maximale p.
Nous élaborons alors un procédé permettant de déterminer les composantes irréductibles de D(Z) ce qui nous permet de donner un critère d'irréductibilité puis de décider quand D(Z) est de Cohen - Macaulay.

1.6.1 Théorème. D(Z) est irréductible si et seulement si la taille n'est pas l'intervalle [p,n].

1.6.2 Théorème. D(Z) est de Cohen - Macaulay si et seulement si la taille n'est pas dans l'intervalle [p+1,n].

Taille	Codimension	Rang générique	Nombre de p-mineurs non identiquement nuls	Observations
$0 \leq t(Z) < p$	Maximale $n-p+1$	$p-1$	Maximal $\binom{n}{p}$	Cohen-Macaulay irréductible
$t(Z) = p$				Cohen-Macaulay réductible
$p < t(Z) \leq n$	$n - t(Z) + 1$	$p-1$	non nul non maximal	Non Cohen-Macaulay réductible
$n < t(Z) \leq n+p$	0	$n+p-t(Z)$	0	Lisse

1.7. Question : l'idéal des fonctions s'annulant sur D(Z) est-il encore engendré par les p-mineurs ?

2 - Modification et stratification canoniques d'une variété déterminantielle.

A la variété déterminantielle D est associé un objet extrêmement simple : la famille des systèmes linéaires associés.

Soit donc

$$S = \{(x,\lambda) \in M_k(n,p) \times \mathbb{P}^{p-1} \mid \operatorname{Ker} x \ni \lambda\}$$

munie des deux restrictions des projections canoniques :

$$\begin{array}{ccc} S & \xrightarrow{\pi_2} & \mathbb{P}^{p-1} \\ {\scriptstyle \pi_1}\downarrow & & \\ M_k(n,p) & & \end{array}$$

$\pi_1^{-1}(D_p)$ est vide, et la restriction de π_1 à $\pi_1^{-1}(D_i)$ est une fibration sur D_i, de fibre $\mathbb{P}^{p-i-1}$ $(0 \leq i \leq p-1)$.

L'image de S par π_1 n'est autre que D, et ces deux variétés ont même dimension car :

$$\begin{aligned} \dim S &= \sup_{0 \leq i \leq p-1} \dim \pi_1^{-1}(D_i) \\ &= p-1 + \sup_{0 \leq i \leq p-1} (\dim D_i - i) \\ &= \dim D_{p-1} \\ &= \dim D \end{aligned}$$

($\dim D_i$ est une fonction strictement croissante de i)

Examinons maintenant comment cette situation passe à la section par $L(Z)$. Si la partition de $D(Z)$ par le rang vérifie 1.2, on a :

$$\dim D(Z) = \dim S(Z) - p + r(Z) + 1 .$$

Ceci cesse d'être vrai pour une section quelconque de la variété déterminantielle générique. (Reprendre l'exemple donné dans 1.2).

<u>Démonstration de 1. 2.</u> Soit x_o un point de $D_{i-1}(Z)$ $(1 \leq i \leq r(Z))$. Considérons la suite x_j $(0 \leq j \leq p)$ de points voisins de x_o, définie par récurrence comme suit : on obtient x_j en générisant dans $L(Z)$ la $j^{\text{ème}}$ colonne de x_{j-1}.
Le rang de x_j est donc au plus celui de x_{j-1} majoré de un, et comme celui de x_p est $r(Z)$, un des éléments de la suite est dans $D_i(Z)$.
Le point est évidemment que ces générisations successives sont indépendantes, ce qui serait d'ailleurs assuré par des hypothèses plus générales.

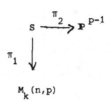

3 - Etude de la variété S(Z).

Ce paragraphe est consacré à l'étude de la variété S(Z), principalement à l'aide de la projection $\pi_2 : S(Z) \to \mathbb{P}^{p-1}$.

3.1 Calcul de dimensions.

Si l'on stratifie $\mathbb{P}^{p-1}$ par les plans de coordonnées, la restriction de π_2 au dessus de chaque strate est une fibration vectorielle.

Soit J un sous-ensemble non vide de $\{1,\dots,p\}$. Notons

$$F(J) = \{ i \in \{1,\dots,n\} \ ; \ \forall j \in J, (i,j) \in Z \}$$

$$\Sigma_J = \{ \lambda \in \mathbb{P}^{p-1} \ ; \ \forall j \in J \ \ \lambda_j \neq 0 \ , \ \forall j \notin J \ \ \lambda_j = 0 \}$$

$$S_J = \pi_2^{-1}(\Sigma_J).$$

Pour tout $\lambda \in \Sigma_j$, $\pi_2^{-1}(\lambda)$ est un produit de n sous-espaces vectoriels de k^p, définis dans L(Z) par les équations

$$\forall i \in \{1,\dots,n\} - F(J) \quad \sum_{j \in J} \lambda_j x_{ij} = 0$$

$\pi_2^{-1}(\lambda)$ est de codimension $n - F(J)$ dans L(Z), codimension constante sur Σ_J.

3.1.1 $S_J = \pi_2^{-1}(\Sigma_J)$ est donc de dimension $np - |Z| - n + |F(J)| + |J| - 1$.

$(S_J)_{\emptyset \neq J \subset \{1,\dots,p\}}$ étant une partition finie de S , nous venons de déterminer la dimension de S égale à

$$np - |Z| - [n - \sup_{\emptyset \neq J \subset \{1,\dots,p\}} (|J| + |F(J)|) + 1]$$

La projection π_1, au-dessus de l'ouvert $D_{r(Z)}$ (des matrices singulières de L(Z) de rang maximum r(Z)) a des fibres isomorphes à $\mathbb{P}^{p-r(Z)-1}$.
La dimension de $D_{r(Z)}$, donc de D (voir 1.2) est égale à

$$np - |Z| - [n - \sup_{J \neq \emptyset} (|J| + |F(J)|) + p - r(Z)] .$$

Compte tenu de la définition de la taille (1.1) nous avons montré le

3.1.2 <u>Théorème</u>. Soit Z une configuration de zéros et $t(Z)$ sa taille. La codimension $c(Z)$ de la variété $D(Z)$ des matrices singulières ayant cette configuration de zéros est donnée par

$$c(Z) = n - \sup(p, t(Z)) + p - r(Z).$$

3.1.3 <u>Remarque</u> : $c(Z)$ ne dépend en fait que de la taille $t(Z)$ car $r(Z)$ s'exprime lui aussi en fonction de $t(Z)$. En effet :

- Lorsque $c(Z) \neq 0$ (i.e. $D(Z) \neq L(Z)$) $r(Z) = p-1$ (voir 1.2),

$$t(Z) \leq n \quad \text{et} \quad c(Z) = n - \sup(p, t(Z)) + 1$$

- Par contre, si $c(Z) = 0$, la formule 3.1.2 nous montre que

$$r(Z) = n + p - t(Z). \quad (t(Z) > n).$$

3.1.4 <u>Corollaire</u> : Si aucun des p-mineurs ne s'annule identiquement sur $L(Z)$, la codimension $c(Z)$ est maximum, égale à $n-p+1$.

<u>Démonstration</u> : Il suffit de montrer que $t(Z) \leq p$. Si tel n'était pas le cas, il existerait un ensemble $J \subset \{1,\ldots,p\}$ tel que $|J| + |F(J)| > p$.

Les lignes de $F(J)$ seraient alors de rang $p - |J| < |F(J)|$ et il existerait donc un mineur identiquement nul, ce qui est contraire à l'hypothèse. ∎

3.1.5 <u>Exemple</u> : Lorsque L n'est plus un plan de coordonnées, le corollaire 3.1.4 (et donc le théorème 3.2.1 qui suit) peuvent être faux. Considérons l'exemple (que nous a indiqué D. Eisenbud) des matrices singulières de la forme :

$$\begin{pmatrix} x & 0 \\ 0 & x \\ y & z \end{pmatrix}$$

D est de codimension 1 dans k^3. Cependant aucun des 2-mineurs ne s'annule identiquement (ils sont même linéairement indépendants).

3.2 Soit $I \subset \{1,\dots,n\}$. Nous noterons désormais x_I la matrice $|I| \times p$ extraite de x, construite sur les lignes de I et Δ_I son déterminant lorsque $|I| = p$.

3.2.1 <u>Théorème</u> : La codimension $c(Z)$ de la variété déterminantielle $D(Z)$ ne dépend que de l'ensemble $\{(i_1\dots i_p) \in \{1,\dots,n\}^p \; ; \; \Delta_{i_1\dots i_p} \not\equiv 0\}$ des p-mineurs non identiquement nuls sur $L(Z)$.

<u>Démonstration</u> : Si $c(Z) = 0$ le résultat est trivial. Si $c(Z) \neq 0$ reconsidérons un instant la formule qui donne la codimension

$$c(Z) = n - \sup(p, t(Z)) + 1$$

où, rappelons-le, $t(Z)$ est le demi-périmètre maximum d'un rectangle non vide de Z.

Pour $i \in \{1,\dots,n\}$, notons Z_i l'ensemble $\{j \in \{1,\dots,p\} ; (i,j) \in Z\}$ et U_i le complémentaire de Z_i dans $\{1,\dots,p\}$.

$t(Z)$ peut alors se calculer par :

$$t(Z) = \sup_{\substack{\emptyset \neq I \subset \{1,\dots,n\} \\ \left|\bigcap_I Z_i\right| \neq \emptyset}} |I| + \left|\bigcap_I Z_i\right| = \sup_{\substack{\emptyset \neq I \subset \{1,\dots,n\} \\ \left|\bigcup_I U_i\right| \neq p}} p + |I| - \left|\bigcup_I U_i\right|$$

Pour tout $I \subset \{1,\dots,n\}$ notons

f la fonction $I \longmapsto n - p + 1 - |I| + \left|\bigcup_I U_i\right|$

g la fonction $I \longmapsto n - p + 1 - |I| + r(I)$, où $r(I)$ est le rang d'un x_I générique.

Remarquons que $c(Z)$ est égal à $\displaystyle\operatorname{Inf}_{\left|\bigcup_I U_i\right| < p} f(I)$ et que la fonction g s'exprime uniquement à l'aide de l'ensemble non vide

$$\{(i_1,\dots,i_p) \in \{1,\dots,n\}^p \quad ; \quad \Delta_{i_1,\dots,i_p} \not\equiv 0\}$$

des mineurs non identiquement nuls sur $L(Z)$. En effet

$$r(I) = \operatorname*{Sup}_{\Delta_{i_1,\dots,i_p} \not\equiv 0} \left|(i_1,\dots,i_p) \cap I\right| .$$

Le théorème résulte alors du lemme suivant :

3.2.2 Lemme

$$\operatorname*{Inf}_{\left|\bigcup_I U_i\right| < p} f(I) = \operatorname*{Inf}_{r(I) < p} g(I) .$$

Démonstration : Comme $r(I)$ est majoré par $\left|\bigcup_I U_i\right|$, le deuxième terme minore le premier.

Réciproquement, si le deuxième est égal à $g(I_o)$, deux cas sont possibles :

- Le rang d'un x_{I_o} générique est maximum (et nécessairement égal à $|I_o|$). Alors

$$g(I_o) = n - p + 1 = f(\emptyset)$$

- le rang d'un x_{I_o} générique n'est pas maximum : nous pouvons lui appliquer le théorème 3.1.2 et en déduire l'existence d'un rectangle non vide $J_o \times H_o$ de Z, inclus dans $I_o \times \{1,\ldots,p\}$, maximal et de demi-périmètre $p + |I_o| - r(I_o)$: Donc

$$g(I_o) = f(J_o) .$$

Ceci termine la démonstration.

3.3 La transformée stricte de $D(Z)$ par π_1 ($c(Z) \neq 0$).

Le morphisme π_1 est un isomorphisme au-dessus de l'ouvert partout dense D_{p-1} des matrices de rang $p-1$. Nous appelons transformée stricte T de D par π_1 l'adhérence $\overline{\pi_1^{-1}(D_{p-1})}$ et nous voulons caractériser les composantes de S qui sont dans T.

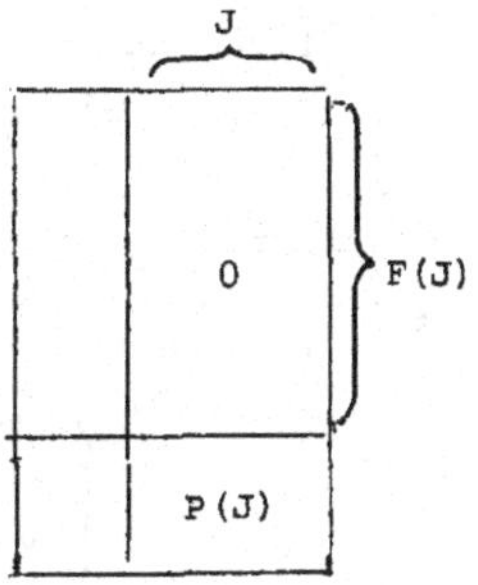

Soit $\emptyset \neq J \subset \{1,\ldots,p\}$. Notons $P(J)$ la matrice construite sur les colonnes J et sur les lignes $\{1,\ldots,n\} - F(J)$, et $P_R(J)$ la sous-matrice construite sur les lignes R ($R \subset \{1,\ldots,n\} - F(J)$).

Nous définissons comme suit un nouvel invariant ne dépendant que de J et de la configuration de zéros Z :

$$\tau(J) = \operatorname*{Inf}_{\substack{R \subset \{1,\dots,n\} - F(J) \\ |R| = |J| - 1}} t(P_R(J))$$

où $t(P_R(J))$ est la taille de la configuration de zéros induite par Z sur la matrice $P_R(J)$;

3.3.1 Proposition : Une condition nécessaire et suffisante pour que S_J soit inclue dans la transformée stricte T est que $\tau(J)$ ne dépasse pas $|J|-1$

Démonstration : S_J est inclue dans T si et seulement si $\pi_1(S_J)$ contient au moins une matrice de rang $p-1$.
La proposition va résulter des deux lemmes suivants :

3.3.2 Lemme : Soient q sous-espaces vectoriels $F_1,\dots,F_q$ d'un espace vectoriel F. Les deux conditions suivantes sont équivalentes :

(i) Il existe un système libre de q vecteurs $v_1,\dots,v_q$ vérifiant

$$\forall i \in \{1,\dots,q\} \quad v_i \in F_i$$

(ii) $\forall\, I \subset \{1,\dots,q\} \qquad \dim \sum_{i \in I} F_i \geqslant |I|$

Démonstration :

i) $\Rightarrow$ ii) est clair

ii) $\Rightarrow$ i) se montre par récurrence sur q. ■

Définissons $\mathcal{H}(Z)$ comme l'ensemble des parties J non vides de $\{1,\dots,p\}$ qui vérifient : $\tau(J) \leqslant |J| - 1$.

3.3.3 Lemme : Les deux conditions suivantes sont équivalentes

i) $\pi_1(S_{\{1,\dots,p\}})$ contient des matrices de rang $p-1$

ii) $\{1,\dots,p\} \in \mathcal{H}(Z)$.

Démonstration : Etant donnés une configuration de zéros Z et un point $\lambda \in \Sigma_{\{1,\dots,p\}}$, la fibre $\pi_2^{-1}(\lambda)$ est définie par les relations suivantes :

$$\begin{cases} \forall (i,j) \in Z \quad (\text{i.e. } j \in Z_i) \quad x_{ij} = 0 \\ \forall i \in \{1,\dots,n\} \ \sum_{j=1}^{p} \lambda_j x_j = 0 \,. \end{cases}$$

Notons E_i le sous-espace vectoriel de k^p défini par

$$\forall j \in Z_i \quad x_j = 0$$

et Λ l'hyperplan de k^p dual de la droite λ.

Avec ces notations on a : $\pi_2^{-1}(\lambda) = \prod_{i=1}^{n} (\Lambda \cap E_i)$

Pour qu'une matrice x appartenant à $\pi_1(\pi_2^{-1}(\lambda))$ (c'est-à-dire une matrice dont la collection $(x_i)_{1 \leq i \leq n}$ des vecteurs lignes vérifie $\forall i \in \{1,\ldots,n\}$ $x_i \in \Lambda \cap E_i$) soit de rang $p-1$ il faut et il suffit d'après 3.3.2 que :

$$\exists R \subset \{1,\ldots,n\}, |R| = p-1 \; ; \; \forall I \subset R, \dim \sum_{i \in I} \Lambda \cap E_i \geq |I|$$

Prenons alors λ générique dans $\Sigma_{\{1,\ldots,p\}}$ (donc dans $\mathbb{P}^{p-1}$). Il est facile de voir, par récurrence sur $|I|$, que

$$\sum_{i \in I} \Lambda \cap E_i = \Lambda \cap (\sum_{i \in I} E_i)$$

En conséquence $\dim \sum_{i \in I} \Lambda \cap E_i = (\dim \sum_{i \in I} E_i) - 1 = p - |\bigcap_{i \in I} Z_i| - 1.$

En résumé $\pi_1(S_{\{1,\ldots,p\}})$ rencontre D_{p-1} si et seulement si

$$\exists R \subset \{1,\ldots,n\}, |R| = p-1 \; ; \; \forall I \subset R, |I| + |\bigcap_{i \in I} Z_i| \leq p-1 ,$$

ce qui est l'assertion du lemme.

<u>Fin de la démonstration de 3.3.1</u> :

Appliquons la modification de Room à la variété D' des matrices singulières $(n-|F(J)|) \times |J|$ ayant la configuration de zéros Z' induite par Z dans $(\{1,\ldots,n\} - F(J)) \times J$.

On obtient alors une variété $S'_J = (\pi'_2)^{-1}(\Sigma'_J)$.

Il est clair que S_J est le produit de S'_J par un espace affine.

En appliquant le lemme 3.3.3 à S'_J on termine la démonstration de 3.3.1. ∎

Nous possédons maintenant un critère permettant de caractériser les éléments de la partition $(S_J)_{\emptyset \neq J \subset \{1,\dots,p\}}$ qui sont dans T.

Nous voulons maintenant déterminer les composantes irréductibles de T (et donc de D).

3.3.4 <u>Remarque</u> : Si $I \in \mathcal{H}(Z)$ et $J \subset I$, alors $J \in \mathcal{H}(Z)$.

3.3.5 <u>Lemme</u> : $S_I \subset \overline{S_J}$ lorsque les deux conditions suivantes sont réalisées

i) $I \subset J$ (i.e. $\Sigma_I \subset \overline{\Sigma_J}$)

ii) $\forall K, I \subset K \subsetneq J$, $|K| + |F(K)| < |J| + |F(J)|$.

<u>Démonstration</u> : Il suffit de montrer l'inclusion au voisinage d'un point ξ de l'image inverse d'un point de Σ_I. Les seules parties de $\pi_2^{-1}(\overline{\Sigma_J})$ qui coupent un voisinage V suffisamment petit de ξ sont les $S_K (I \subset K \subset J)$.

$V \cap \pi_2^{-1}(\overline{\Sigma_J})$ est donc de dimension égale à celle de S_J

soit $np - |Z| + |J| - 1 - (n - |F(J)|)$.

D'autre part $V \cap \pi_2^{-1}(\overline{\Sigma_J})$ est définie dans $L(Z) \times \mathbf{P}^{|J|-1}$

par les $n - |F(J)|$ équations $\sum_{j \in J} \lambda_j x_{ij} = 0$ $(\forall i \in \{1,\dots,n\} - F(J))$.

$\pi_2^{-1}(\overline{\Sigma_J})$ est donc, au voisinage de ξ , une intersection complète, ce qui prouve que $S_I \subset \overline{S_J}$ au voisinage de ξ , et donc partout puisque S_I est irréductible.

3.3.6 Décrivons maintenant un procédé qui donne les composantes irréductibles de T.

Considérons l'ensemble $\mathcal{H}(Z) = \{J \subset \{1,\dots,p\}; \tau(J) \leq |J| - 1\}$.

Une composante irréductible de T est nécessairement l'adhérence d'une partie S_J avec $J \in \mathcal{H}(Z)$.

- Les composantes de dimension maximum sont associées aux éléments J de $\mathcal{H}(Z)$ tels que $|J| + |F(J)| = \sup(p, t(Z))$.

Remarquons alors que pour de tels J , $\pi_2^{-1}(\overline{\Sigma_J})$ est intersection complète et $\pi_1(\pi_2^{-1}(\overline{\Sigma_J}))$ est une réunion de composantes de dimension maximum. (C'est même le produit d'un espace affine par une variété algébrique $D'_J(Z)$. $D'_J(Z)$ est la variété des matrices $(n - |F(J)|) \times J$ singulières ayant la configuration de zéros induite par Z sur $p(J)$. $D'_J(Z)$ est de codimension maximum $n - |J| - |F(J)| + 1$ et c'est une variété de Cohen - Macaulay)

- Pour obtenir les composantes de T de dimension immédiatement inférieure à la dimension maximum, on considère l'ensemble $\mathcal{K}_1(Z)$ obtenu en retirant à $\mathcal{K}(Z)$ les parties J déjà nommées et toutes leurs sous-parties.

Les éléments $J_1 \in \mathcal{K}(Z)$ tels que $|J_1| + |F(J_1)|$ est maximum dans $\mathcal{K}_1(Z)$ donnent les composantes recherchées , etc ...

3.3.7 Démontrons maintenant le critère d'irréductibilité et déterminons quand D(Z) est de Cohen - Macaulay. Excluons le cas trivial où $t(Z) > n$ $(D(Z) = L(Z))$. Deux cas restent possibles :

- Si $t(Z) < p$, le procédé ci-dessus ne fournit qu'une composante irréductible de D(Z) ; comme D(Z) est de Cohen - Macaulay, D(Z) est irréductible.

- Si t(Z) est dans l'intervalle $[p,n]$, il existe alors des ensembles J , non vides, de $\mathcal{K}(Z)$, différents de $\{1,\dots,p\}$ tels que $|J| + |F(J)| = t(Z)$.
Pour un tel J , $\pi_1(\overline{S_J})$ est une composante de D de dimension maximum.

Soit $j \notin J$. Parmi les sous-ensembles contenant j , il y en a aussi dans $\mathcal{K}(Z)$ (ne serait-ce que $\{j\}$ lui-même). Prenons en un, K, maximal pour l'inclusion. $\pi_1(\overline{S_J})$ est une composante de D nécessairement distincte de $\pi_1(\overline{S_J})$.

Examinons maintenant si D est de Cohen - Macaulay (C.M.).
Si la taille est égale à p , D(Z) est de C.M.

Si D(Z) n'est pas équidimensionnel, il ne peut être de C.M. ;

Supposons donc $p < t(Z) \leq n$ et D équidimensionnel.

Nous allons montrer que D (équidimensionnel de dimension d) n'est pas de Cohen-Macaulay en mettant en évidence deux sous-variétés D_1 et D_2 de D de dimension d telles que :

$$D_1 \cup D_2 = D \quad , \quad D_1 \cap D_2 \text{ de codimension au moins deux dans D.}$$

Soit $J \subset \{1,2,\dots,p\}$ une partie maximale pour l'inclusion parmi celles qui vérifient $|J| + |F(J)| = t(Z)$. J est différent de $\{1,2,\dots,p\}$ puisque $t(Z) > p$.

Pour toute partie K qui rencontre J et qui vérifie $|K|+|F(K)| = t(Z)$ il est facile de voir que $|J\cup K| + |F(J\cup K)| = t(Z)$ ce qui montre que $K \subset J$.

Les sous-ensembles K de $\{1,2,\dots,p\}$ vérifiant $|K|+|F(K)| = t(Z)$ sont associés aux composantes de S de dimension maximum d. Nous venons de voir que pour un tel K, K est inclus soit dans J, soit dans son complémentaire.

Posons alors $S_1 = \bigcup_{K\subset J} \overline{S}_K$ et $S_2 = \bigcup_{K\cap J = \emptyset} \overline{S}_K$

$S_1 \cup S_2$ contient toute composante de S de dimension d, d'autre part $S_1 \cap S_2 = \emptyset$ ($\pi_2(S_1) \subset \overline{\Sigma}_J$ et $\pi_2(S_2) \subset \overline{\Sigma}_{\complement J}$).

Soient $D_1 = \pi_1(S_1)$ et $D_2 = \pi_1(S_2)$. Appelons D_1' et D_2' les transformées strictes (voir 3.3) de D_1 et D_2. Nous avons alors

$$D_1 \cap D_2 = \pi_1(\pi_1^{-1}(D_1) \cap D'_2)$$

$$\text{avec} \quad \pi_1^{-1}(D_1) = D_1' \cup E_1$$

Comme $S_1 \cap S_2 = \emptyset$, $D_1' \cap D_2' = \emptyset$ et l'on a $D_1 \cap D_2 = \pi_1(E_1 \cap D_2')$

Supposons alors que $D_1 \cap D_2$ soit de dimension d-1, et plaçons nous dans un voisinage V d'un point générique d'une composante de dimension d-1 de $D_1 \cap D_2$. Il faut alors que

dim $E_1 \cap D_2' =$ d-1, et nécessairement E_1 contient une composante de dimension d qui rencontre D_2'. Cette composante est nécessairement dans S_2.

Symétriquement nous avons une composante exceptionnelle E_2 de dimension d contenue dans S_1 telle que π_1 $(E_2 \cap D_1') = D_1 \cap D_2$.

On voit donc que l'image inverse de $D_1 \cap D_2 \cap V$ par π_1 a au moins deux composantes de dimension d qui se projettent toutes deux surjectivement sur $D_1 \cap D_2 \cap V$.

Or l'image inverse d'un point quelconque de D est un espace projectif On a donc une contradiction.

$D_1 \cap D_2$ est de codimension au moins deux dans D. Comme $D_1 \cup D_2 = D$, d'après un théorème de Hartshorne ([6] corollaire 3.9. p.46) D ne peut être de Cohen - Macaulay ∎

BIBLIOGRAPHIE

[1] D.A. BUCHSBAUM - D.S. RIM. A generalized Koszul complex.II Depth and multiplicity - Transactions American Mathematical Society, Vol. 111 (1964), p. 197-224.

[2] C. DE CONCINI - D. EISENBUD - C. PROCESI : Young diagrams and determinantal varieties. Inv. Math. 56 (1980) p. 129-173

[3] J.A. EAGON. Ideals generated by the subdeterminants of a matrix. Thesis. University of Chicago (1961).

[4] J.A. EAGON - M. HOCHSTER : Cohen Macaulay rings,Invariant theory and the generic perfection of determinantal loci. Amer. J. Math. 93 (1971) p.1020-1058

[5] J.A. EAGON - D.G. NORTHCOTT. Ideals defined by matrices and a certain complex associated with them. Proceedings of the Royal Society, A, Vol. 269 (1967), p. 147 - 172.

[6] A. GROTHENDIECK. Local cohomology - Lecture Notes in Mathematics n° 41, Springer - Verlag.

[7] G. KEMPF. On the geometry of a theorem of Riemann. Ann. Math. 98 (1973) p. 178 - 185.

[8] F.S. MACAULAY. The algebraic theory of modular systems. Cambridge Tracts. Vol. 19 (1916).

[9] I.R. PORTEOUS. Simple singularities of maps. Proceedings of Liverpool Singularities Symposium. Lecture Notes in Mathematics n°192. Springer-Verlag.

[10] F. RONGA. Le calcul des classes duales aux singularités de Boardman d'ordre deux. Comm. Math. Helv. Vol. 47 -1- (1972) p. 15-35.

[11] T.G. ROOM. The geometry of determinantal loci. Cambridge University Press, Cambridge (1938).

ON THE TOPOLOGY OF COMPLEX ALGEBRAIC MAPS

by

M. Goresky and R. MacPherson

In this largely expository note we give some homological properties of algebraic maps of complex algebraic varieties which are rather surprising from the topological point of view. These include a generalisation to higher dimension of the invariant cycle theorem for maps to curves.

These properties are all corollaries of a recent deep theorem of Deligne, Gabber, Beilinson, and Bernstein which is stated in §2 . This theorem involves intersection homology and the derived category. One of our objects here is to popularize it by giving corollaries involving only ordinary homology. For this reason some readers may wish to begin with §3.

§1. Intersection homology.

For any complex algebraic variety V , let $D_c^b(V)$ be the algebraically constructible bounded derived category of the category sheaves of $\mathbb{Q}$-module on V . (Objects of $D_c^b(V)$ are bounded complexes of sheaves of $\mathbb{Q}$-modules on V that are cohomologically locally constant on the strata for some stratification of V by complex algebraic submanifolds; see ([GM2], §1.11).

If $\underline{\underline{S}}^\cdot \in D_c^b(V)$ and $U \subset V$, $\mathcal{H}^k(U,\underline{\underline{S}}^\cdot)$ (resp. $\mathcal{H}_c^k(U,\underline{\underline{S}}^\cdot)$) denotes the hypercohomology (resp. hypercohomology with compact supports) of the restriction of $\underline{\underline{S}}^\cdot$ to U .

If $p \in V$, let $\mathcal{D}_p^o$ be the "open disk" of points at distance less than ε from p , where distance is the usual Euclidean distance using some local analytic embedding of a neighborhood of P in $\mathbb{C}^N$. For $\underline{\underline{S}}^\cdot \in D_c^b(V)$, $\mathcal{H}^k(\mathcal{D}_p^o,\underline{\underline{S}}^\cdot)$ and $\mathcal{H}_c^k(\mathcal{D}_p^o,\underline{\underline{S}}^\cdot)$ are independent of the choices for small enough ε .

A <u>local system</u> on a space U is a locally constant sheaf of $\mathbb{Q}$-module on U .

Definition-Proposition ([G M2], §4.1).

Let V be a complex algebraic variety of pure dimension n , U be a nonsingular Zariski open and dense subvariety, and L be a local system on U . Then there is an object $\underline{\underline{IC}}^{\cdot}(V,L)$ in $D_c^b(V)$ called the sheaf of intersection homology chains on V with coeficients in L , which is defined up to canonical isomorphism in $D_c^b(V)$ by the following properties :

1) $\underline{\underline{IC}}^{\cdot}(V,L)$ restricted to U is L[n] .

2) V can be stratified by strata $\{S_\ell\}$ where S_ℓ has dimension ℓ , so that if $p \in S_\ell$,

a) $\mathcal{H}^k(\mathcal{D}_p^o, \underline{\underline{IC}}^{\cdot}(V,L)) = 0$ unless k is a dimension marked \$ in the figure below.

b) $\mathcal{H}_c^k(\mathcal{D}_p^o, \underline{\underline{IC}}^{\cdot}(V,L)) = 0$ unless k is a dimension marked £ in the figure below.

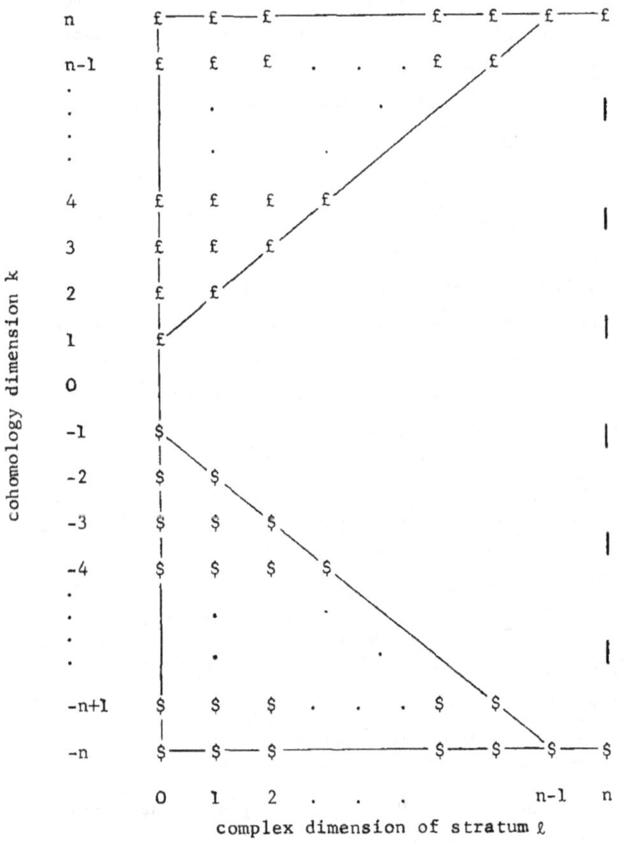

Remarks.

1. The regions of marks and in the figure are sharp in the sense that if they are made smaller, existence fails and if they are made bigger, uniqueness fails.

2. $\underline{\underline{IC}}^{\cdot}(V,L)$ is independent of U in the sense that if L and L' agree where they are both defined, then $IC^{\cdot}(V,L)$ is equivalent in $D_c^b(V)$ to $\underline{\underline{IC}}^{\cdot}(V,L')$.

3. If L is the constant sheaf $\mathbb{Q}_U$, then $\underline{\underline{IC}}^{\cdot}(V,L)$ is denoted $\underline{\underline{IC}}^{\cdot}(V)$. It is a purely topological invariant of V .

Example. If V is nonsingular, $\underline{\underline{IC}}^{\cdot}(V)$ is $\mathbb{Q}_V[n]$.

§2. The decomposition theorem.

The following theorem was conjectured in [ΓM] §2.10 . It has been proved by P. Deligne, O. Gabber, A. Beilinson, and I. Bernstein. ([D4]) .

Theorem. Let $f : X \to Y$ be a proper projective map of complex algebraic varieties. Then there exist closed subvarieties $V_\alpha \subset Y$ and local systems L_α , and integers ℓ_α such that there is an equivalence in $D_c^b(Y)$

$$**) \qquad Rf_*\underline{\underline{IC}}^{\cdot}(X) \approx \bigoplus_\alpha i_*^\alpha\, \underline{\underline{IC}}^{\cdot}(V_\alpha,L_\alpha)\,[\ell_\alpha]$$

where $i^\alpha : V_\alpha \hookrightarrow Y$ is the inclusion.

Remarks.

1. A decomposition **) of $Rf_*\underline{\underline{IC}}^{\cdot}(X)$ can be found with the restriction that the varieties V_α are irreducible and the local systems L_α are indecomposable. Under this restriction the objects $i_*^\alpha\underline{\underline{IC}}^{\cdot}(V_\alpha,L_\alpha)[\ell_\alpha]$ are indecomposable in $D_c^b(Y)$ in the sense that whenever $i_*\underline{\underline{IC}}^{\cdot}(V_\alpha,L_\alpha)[\ell_\alpha] = \underline{\underline{S}}^{\cdot} \oplus \underline{\underline{T}}^{\cdot}$ then either $\underline{\underline{S}}^{\cdot}$ or $\underline{\underline{T}}^{\cdot}$ is equivalent to zero in $D_c^b(Y)$ ([GM2] , §4.1 , corollary 2) . Also with this restriction, the list of summands $i_*^\alpha\underline{\underline{IC}}\ (V_\alpha,L_\alpha)$ is uniquely determined. (We do not know if the category $D_c^b(Y)$ has such unique decompositions into indecomposables in general).

2. There are generalisations of the Poincaré duality theorem and the hard Lefschetz theorem relative to f . (They specialize to the classical theorems when X is nonsingular and Y is a point).

Let $Loc(V,\ell)$ be the direct sum of the L_α for those α such that $V = V_\alpha$ and $\ell = \ell_\alpha$. Then Poincaré duality ([GM2] , §5.3 and §1.6) says that there is an isomorphism.

$$\mathrm{Loc}(V,\ell) = \mathrm{Hom}(\mathrm{Loc}(V,-\ell)),\mathbb{Q}) \ .$$

Hard Lefschetz ([BB]) says that there exists a map $\Lambda : \mathrm{Loc}(V,\ell) \to \mathrm{Loc}(V,\ell+2)$ for all ℓ such that for $\ell > 0$

$$\Lambda^{\ell} : \mathrm{Loc}(V,-\ell) \xrightarrow{\simeq} \mathrm{Loc}(V,\ell)$$

is an isomorphism.

3. Although the theorem is a purely topological result about complex varieties, the proof uses characteristic p techniques. Such a decomposition exists for any complex which is pure in the sense of [D2] , §6.2. The complex $\underline{\underline{IC}}^{\cdot}(X)$ is pure by a result of Gabber [D3] and Rf_* preserves purity by [D2] , §6.2.

4. If X is nonsingular and of complex dimension n , then

$$\mathcal{H}^k_c Rf_* \underline{\underline{IC}}^{\cdot}(X) \simeq \mathcal{H}^k_c Rf_* \mathbb{Q}_X[n] \simeq H^{n+k}_c(X) \simeq H_{n-k}(X)$$

where $H_{n-k}(X)$ is the ordinary homology of X with rational coefficients. So the splitting $**)$ gives rise to a decomposition of the homology of X . Using the numbering of dimensions of [GM1] and [CGM] ,

$$H_k(X) \approx \bigoplus_{\alpha} IH_{k_\alpha}(V_\alpha, L_\alpha)$$

where $k_\alpha = k - n + \dim V_\alpha + \ell_\alpha$. For an example where this decomposition is worked out in detail, see [BM] .

5. If $f : X \to Y$ is a resolution of singularities of Y , then one of the terms in the decomposition $**)$ of $Rf_*\mathbb{Q}_X$ will be $\underline{\underline{IC}}^{\cdot}(X)$. Thus the intersection homology of X is contained in the ordinary homology of any resolution of X .

Example. Suppose $Y \subset \mathbb{CP}^{n+1}$ is the cone with vertex p over a nonsingular variety in $\mathbb{CP}^n$. Suppose $f : X \to Y$ is the blow-up of Y at p , D is the exceptional divisor, and $c_1(N) \in H^2(D)$ is the first Chern class of its normal bundle. Then

$$\mathrm{Loc}(Y,0) = \mathbb{Q}_{Y-P}$$

$$\mathrm{Loc}(p,\ell) = \mathbb{Q}_p \otimes (\mathrm{Image} \cap c_1(N))$$

and all of the other $\mathrm{Loc}(V,\ell)$ are zero.

The stalk at p of the cohomology sheaf $\underline{\underline{H}}^i(Rf_*\mathbb{Q}_X)$ is $H^i(D)$. It splits in pieces $H^i(\underline{\underline{IC}}^\cdot(Y,\mathbb{Q}))_p$ and $H^i(i_*^p\underline{\underline{IC}}^\cdot(p,\mathbb{Q}))_p$ as $H^i(D)$ splits into primitive and non-primitive cohomology.

§3. Resolutions.

Let X be a nonsingular complex algebraic variety and let $f : X \to Y$ be a proper projective algebraic map. For any point $p \in Y$, let $\mathcal{D}_p \subset Y$ be the set of points of distance at most ε from p and let $S \subset Y$ be the set of points of distance exactly ε from p. (Here "distance" means the usual Euclidean distance with respect to some local analytic embedding of a neighborhood of p in $\mathbb{C}^N$). Let M be $f^{-1}(\mathcal{D}_p)$, B be $f^{-1}(S)$, and $\overline{f} : B \to S$ be the restriction of f.

$$\begin{array}{ccccc} B & \underset{i}{\subset} & M & \subset & X \\ {\scriptstyle \overline{f}}\downarrow & & \downarrow & & \downarrow{\scriptstyle f} \\ S & \subset & \mathcal{D}_p & \subset & Y \end{array}$$

It is well known from stratification theory that for small enough ε, M is a compact manifold with boundary B and the topological type of the pair (M,B) is independent of the choices.

Let $K \subset H_*(B)$ be the kernel of $i_* : H_*(B) \to H_*(M)$. In this section and the next, we address the following

Question. To what extent is K determined by the data $\overline{f} : B \to S$ (and the fact that f is algebraic) ?

Remarks. 1. Of the information in the long exact sequence in homology for the pair (M,B), $K \subset H_*(B)$ is the only part that could be determined by these data since blowing up a point in $f^{-1}(p)$ will change $H_*(M)$.

2. Just from the topological fact that B is the boundary of the manifold M, we see that K is a maximal isotropic subspace for the intersection pairing on $H_*(B)$; i.e. $K = K^\perp$. (See [Do], prop. 9.6, p. 305). In particular $\dim K = \frac{1}{2} \dim H_*(B)$.

<u>Corollary 1</u>. If Y is an n dimensional variety with an isolated singular point at p, and $f : X \to Y$ is a resolution of singularities (so $\overline{f} : B \to S$ is a homeomorphism), then

$$K = H_n(B) \oplus H_{n+1}(B) \oplus \dots \oplus H_{2n-1}(B)$$

<u>Proof</u>. In $\mathcal{D}_p$ the decomposition **) of the theorem has the form $\underline{\underline{IC}}^{\cdot}(Y)$ plus terms concentrated at p. Only the term $\underline{\underline{IC}}^{\cdot}(Y)$ effects K. So the problem becomes the following : Which cycles in $H_*(S)$ are boundaries in $IH(\mathcal{D}_p)$? Here we use the interpretation of intersection homology of [GM1] and [CGM] (proved to be equivalent in [GM2]).

$\mathcal{D}_p$ is topologically a cone with base S and vertex p. Any cycle Z in S is the boundary of its cone to p. The cone is allowable as a chain in $\underline{\underline{IC}}^{\cdot}(B)$ if (and only if) the dimension of Z is at least n ([CGM] , §2.1). So $H_n(S) \oplus H_{n+1}(S) \oplus \dots$ is in K. Since it is a maximal isotropic subspace of $H_*(S)$, it is all of K.

<u>Examples.</u> The simplest example is a node (or normal crossing) of a curve. Topologically the picture is like this :

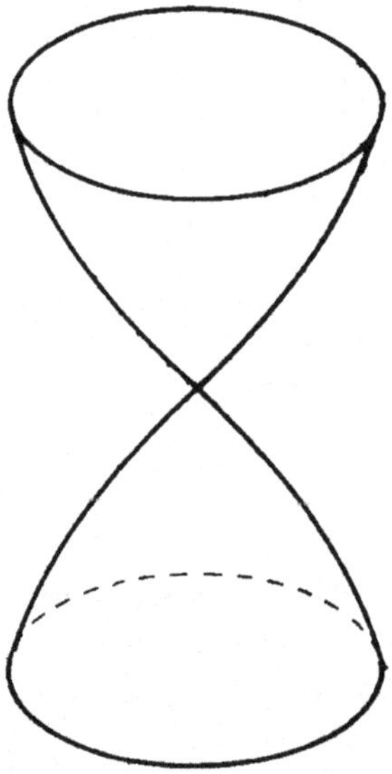

The corollary asserts that the resolution must be the figure on the left rather than the one on the right (as may be seen by several classical arguments).

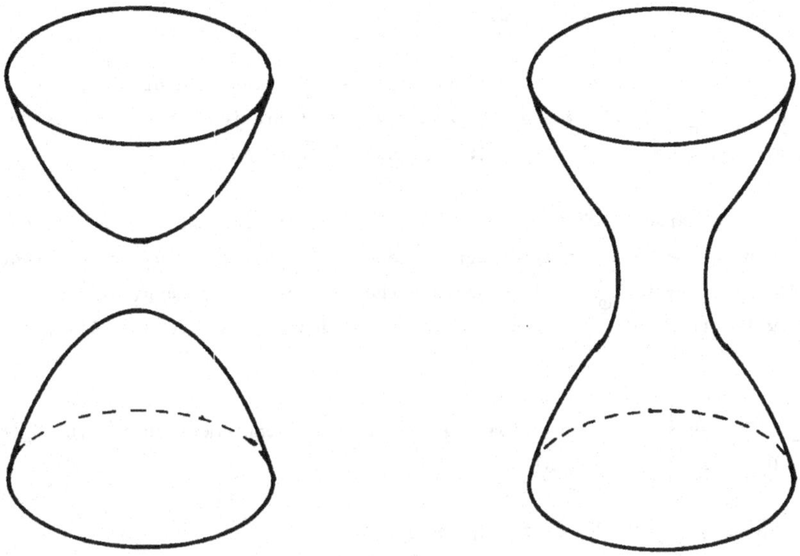

For surfaces, the corollary follows from Grauert's blowing down criterion. For dimension three and more, it seems new.

§4. <u>Generalized invariant cycle theorem.</u>

Given $\overline{f} : B \to S$ as in the last section, we will construct a subspace $J \subset H_*(B)$. Choose a Whitney stratification of S by strata $\{U_\varphi\}$ of odd dimension such that $\overline{f}$ restricted to the inverse image of U_φ is a topological fibration onto U_φ for each φ . (This can be done : the U may be taken to be restrictions to S of a stratification of Y by complex manifolds with the similar fibration property. See [H] and [T] , p. 276). Choose a triangulation T of S so that each stratum U_α is a union of interiors of simplices. (See [G]). Let R be the union of all simplices Δ of the barycentric subdivision of T such that for all U_φ , $\dim (\Delta \cap U_\varphi) < \frac{1}{2} \dim U_\varphi$.

<u>Definition</u> . $J \subset H_*(B)$ is the image of the map

$$H_*(\overline{f}^{-1}(R)) \to H_*(B)$$

<u>Lemma</u>. J is independent of the choices ($\{U_\varphi\}$ and T) in its construction. It is

a maximal isotropic subspace of $H_*(B)$.

Example. If $\bar{f}$ is a topological fibration and S is a manifold of dimension $2m-1$, then $J = F_{m-1} H_*(B)$ where F_s denotes the fibration of $H_*(B)$ of the Leray spectral sequence for $\bar{f}$. (See [S], p. 473-4, theorem 1).

Remarks. For any "perversity" function p and any stratified map $B \to S$, we may similarly define a "perverse Leray filtered piece" $J_p \subset H_*(B)$ by using the inequality $\dim(\Delta \cap U_\varphi) \leq p(\dim U_\varphi)$. We conjecture that if $p(c)$ and $c-p(c)$ are both nondecreasing functions of c , then J_p is independent of the choices. If $p(c) = s$, then $J_p = F_s$.

Corollary 2. The subspace K is always a vector space complement to J in $H_*(B)$. (That is $J \cap K = \{0\}, J + K = H_*(B)$).

Proof. We decompose $H_*(B)$ as in §2 remark 4

$$H_*(B) = \bigoplus_\alpha IH_*(S \cap V_\alpha, L_\alpha)$$

Arguing as in the proof of corollary 1, we see that

$$K = \bigoplus_\alpha IH_{a_\alpha}(S \cap V_\alpha, L_\alpha) \oplus \ldots \oplus IH_{2a_\alpha - 1}(S \cap V_\alpha, L_\alpha)$$

where $a_\alpha = \dim_{\mathbb{C}} V_\alpha$. We claim that

$$J = \bigoplus_\alpha IH_o(S \cap V_\alpha, L_\alpha) \oplus \ldots \oplus IH_{a_\alpha - 1}(S \cap V_\alpha, L_\alpha)$$

From this, corollary 2 and the lemma clearly follow.

To establish the claim, let R° be an open regular neighborhood of R . Then

$$J = \text{image}(H_*(\bar{f}^{-1}(R')) \to H_*(B)) = \bigoplus_\alpha \text{Image}(IH_*(R^\circ \cap V_\alpha, L_\alpha) \to IH_*(S \cap V_\alpha, L_\alpha))$$

We will show that for all α .

$$IH_o(S \cap V_\alpha, L_\alpha) \oplus \ldots \oplus IH_{a_\alpha - 1}(S \cap V_\alpha, L_\alpha) = \text{Image}(IH_*(R^\circ \cap V_\alpha, L_\alpha) \to IH_*(S \cap V_\alpha, L_\alpha))$$

The inclusion $\subset$ follows from [GM1] §3.4 plus the fact that $R^{\bar{m}} \subset R \cap V_\alpha$ for $i \leq a_\alpha - 1$ since V_α is a union of strata U_φ. The inclusion $\supset$ then follows from the fact that J is self-annihilating under the intersection pairing. This fact may be seen by using stratified general position [M] to find a homeomorphism $h : V_\alpha \to V_\alpha$ isotopic to the identity such that $h(R \cap V_\alpha) \cap (R \cap V_\alpha)$ is empty.

Remark. For general f, corollary 2 is the best possible result on K from the data $\bar{f} : B \to S$, except for integrality considerations. For example, for $f : T \times T \to T$ where T is a curve of genus one, there is an automorphism of the topological fibration $\bar{f}$ taking any complement to $J \cap H_1(B, \mathbb{Z})$ in $H_1(B, \mathbb{Z})$ as a $\mathbb{Z}$-module to any other complement.

Example. If Y is a curve, corollary 2 is equivalent to the invariant cycle theorem:

> Let $\varphi : F \subset B$ be the inclusion of a fiber and let $\mu : H_*(F) \to H_*(F)$ be the monodromy map. Then the composed map
>
> $$H_{i+2}(M,B) \to H_{i+1}(B) \xrightarrow{\varphi^*} H_i(F)$$
>
> is a surjection to the kernel of $(1-\mu)$, i.e. to the invariant cycles (see [C], introduction).

This follows from the Wang exact sequence for the fibration $\bar{f} : B \to S$ over a circle ([S], p. 456, Cor. 6)

$$H_{i+1}(F) \xrightarrow{\varphi_*} H_{i+1}(B) \xrightarrow{\varphi^*} H_i(F) \xrightarrow{1-\mu} H_i(F)$$

and the fact that $J \cap H_{i+1}(B)$ is the image of φ_*

§5. Leray Spectral Sequence.

Corollary 3. Let $f : X \to Y$ and $f' : X' \to Y$ be two proper projective maps of non-singular complex algebraic varieties to Y. Then if $R^i f_* \mathbb{Q}_X$ and $R^i f_* \mathbb{Q}_{X'}$ are isomorphic for all i, then $H_*(X)$ and $H_*(X')$ are isomorphic. In fact the whole Leray spectral sequences of f and f' coincide.

Proof. By dévissage, the triples V_α, L_α, ℓ_α occuring in the decomposition **) may be determined from the $R^i f_* \mathbb{Q}_X$. Then $H_*(X)$ will be a direct sum of hypercohomology of the factors $\underline{\underline{IC}}^\cdot(V_\alpha, L_\alpha)$ with a dimension shift depending on ℓ_α. Similarly, the Leray spectral sequence for f will decompose into a sum of spectral sequences for the hypercohomology of each $\underline{\underline{IC}}^{\cdot}(V_\alpha, L_\alpha)$ (see [GM2], §1.2).

Examples 1. If the $R^i f_* \mathbb{Q}_X$ are all locally constant sheaves, then all the V_α wil be Y, and the $\underline{\underline{IC}}^{\cdot}(V_\alpha, L_\alpha)$ will be locally constant sheaves so their spectral sequences will degenerate at E^2. (This case was a result in [D1]).

2. For an example of a V such that the spectral sequence for $\underline{\underline{IC}}^{\cdot}(V)$ does not degenerate at E^2, take any surface S with a curve C such that $H_1(C) \to H_1(S)$ is not injective. Blow up enough points on C then blow down its reduced transform.

Institut des Hautes Etudes Scientifiques
35 route de Chartres
91440 Bures-sur-Yvette (France)

June 1981

IHES/M/81/32

REFERENCES

[BB] A.A. Beilinson - I.N. Bernstein : Appendix to "La conjecture de Weyl II" of P. Deligne : ℓ-adic modules and weight filtrations, preprint, Moscow 1981 (in Russian).

[BM] W. Borho - R. MacPherson : Représentations des groupes de Weyl et homologie d'intersection pour les variétés nilpotentes, Comptes Rendus de l'Acad. Sci. Paris, 1981.

[C] C.H. Clemens : Degeneration of Kähler manifolds, Duke Math. Journal 44 (1977), 215 - 290.

[CGM] J. Cheeger - M. Goresky - R. MacPherson : L^2-cohomology and intersection homology for singular algebraic varieties, proceedings of year in differential geometry, I.A.S., S. Yau, ed, (1981) Annals of Math. Studies, Princeton.

[ΓM] S. Gelfand - R. MacPherson : Verma modules and Schubert-cells : a dictionary, IHES preprint, 1980.

[D1] P. Deligne : Théorème de Lefschetz et critères de dégénérescence de Suites Spectrales, Publ. Math. IHES 35 (1968) 107 - 126.

[D2] P. Deligne : La conjecture de Weil, II, Publ. Math. IHES 52(1980) 137 - 252.

[D3] P. Deligne : Pureté de la cohomologie de MacPherson-Goresky, d'après un exposé de O. Gabber, rédigé par P. Deligne, IHES preprint, Fév. 1981.

[D4] P. Deligne et al. : To appear in Proceedingsof C.I.R.M. conference "Analyse et Topology sur les Espaces Singuliers", Marseille-Luminy 1981.

[Do] A. Dold : Lectures on Algebraic Topology, Springer Verlag (1972)

[G] M. Goresky : Triangulation of stratified objects, Proc. Am. Math. Soc. 72 (1978) 193 - 200.

[GM1] M. Goresky - R. MacPherson : Intersection Homology Theory, Topology 19(1980) 135 - 162.

[GM2] M. Goresky - R. MacPherson : Intersection Homology II, preprint 1981.

[H] R. Hardt : Semi-algebraic local-triviality in semi-algebraic mappings, Am. Jour. of Math. 102 (1980) 191 - 302.

[M] C. McCrory : Stratified general position, Algebraic and Geometric Topology, p. 142 - 146 Springer Lect. Notes in Math. 664 (1978).

[S] Spanier, Algebraic Topology, McGraw-Hill, 1966.

[T] R. Thom, Ensembles et morphismes stratifiés, Bull. Amer. Math. Soc. 75 (1969) 240 - 284.

SINGULARITES DES SCHEMAS DE HILBERT PONCTUELS

Jean-Michel GRANGER

Dans cette note, on se propose d'étudier les déformations plates des sous - espaces de dimension zéro de $\mathbb{C}^r$, et des germes de tels sous espaces ou " points épais " en $0 \in \mathbb{C}^r$. Ceci nous conduit à considérer les schémas de Hilbert ponctuels $\mathrm{Hilb}^n \mathbb{C}^r$ et $\mathrm{Hilb}^n \mathcal{O}_r$ $(\mathcal{O}_r = \mathbb{C}\{x_1, \ldots, x_r\})$ paramétrant ces sous espaces et ces " points " de longueur fixée n . Les problèmes d'existence de certains types de déformations s'expriment en termes de relations d'incidence entre sous - ensembles de $\mathrm{Hilb}^n \mathbb{C}^r$ et $\mathrm{Hilb}^n \mathcal{O}_r$. De la solution de ces problèmes, on peut déduire des résultats sur la géométrie des schémas de Hilbert ponctuels (irréductibilité, composantes irréductibles, détermination de lieux singuliers). Les deux principaux outils utilisés ici sont les suivants :

- Une présentation du germe de $\mathrm{Hilb}^n \mathbb{C}^r$ en un point en terme de platificateur local ([7]) d'un morphisme analytique convenable noté φ . Dans cette présentation $\mathrm{Hilb}^n \mathcal{O}_r$ apparait comme un lieu de ramification maximum de φ .

- Un théorème de connexité pour les germes d'espaces analytiques dû à A. GROTHENDIECK ([5]) . Dans [3] et [11] , T. Gaffney et R. Lazarsfeld utilisent ce théorème pour obtenir une minoration de la dimension des lieux de ramifications de morphismes analytiques ou algébriques.

En appliquant une méthode analogue à des restrictions appropriées de φ , on obtient une minoration de la dimension du germe (E_ν, z) où z est une intersection complète et $E_\nu \subset \mathrm{Hilb}^n \mathcal{O}_r$ paramètre les points d'ordre au moins ν (degré minimum d'une hypersurface de plongement).

- Pour $r = 2$, on en déduit une bonne propriété de décroissance pour la filtration de $\mathrm{Hilb}^n \mathcal{O}_2$ par les E_ν $(E_\nu = \overline{E_\nu - E_{\nu+1}})$ et le fait que E_2 est le lieu singulier de $E_1 = \mathrm{Hilb}^n \mathcal{O}_2$, réduit.

Dans [9] , A. IARROBINO montre un résultat analogue pour un type particulier d'idéaux : les intersections complètes dont le gradué associé est aussi une intersection complète. Parmi les exceptions qu'il met en évidence, signalons le cas d'une famille contenue dans l'intérieur de $\overline{L}$:

<u>EXEMPLE</u> . - Soit $H(2, 2, 2)$ l'ensemble des intersections complètes $I = (f, g, h)$ telles que :

1/ $\nu(f) = \nu(g) = \nu(h) = 2$

2/ $\dim \mathcal{O}_3/_I = 8$.

On a alors $\dim H(2, 2, 2) = 12$, $\dim \overline{L} = 14$, et on peut montrer que $\overline{L}$ est un voisinage de $H(2, 2, 2)$ (voir [4] , partie 2, § V) .

b/ <u>Questions sur les strates d'Hilbert-Samuel</u>

- Lorsque $r \geq 3$, la description des Z_T est beaucoup plus compliquée que dans le cas $r = 2$ (cf. Proposition 3) et les questions suivantes demeurent ouvertes.

- Calcul ou encadrement précis de la dimension de Z_T .

- Détermination des composantes irréductibles, des composantes connexes de Z_T (quand les Z_T sont-elles irréductibles, ou connexes ?) .

- Conditions pour que Z_T soit lisse, ou détermination du lieu singulier de Z_T .

Le problème des déformations d'un idéal I sur des idéaux de fonction d'Hilbert Samuel donnée $T' \neq T(I)$ est encore plus ardu même lorsque $r = 2$. En terme de schéma de Hilbert, on peut s'intéresser en premier lieu à des conditions nécessaires ou suffisantes pour que

$$Z_T \subset \overline{Z}_{T'} \text{ , où } Z_T \text{ et } Z_{T'}$$

sont des strates d'Hilbert Samuel. Les trois conditions nécessaires suivantes fournissent une première indication assez utile :

1/ $\dim Z_T < \dim Z_{T'}$

2/ Pour tout j , $t_o + \ldots + t_j \geq t'_o + \ldots + t'_j$

3/ $r(T) \geq r(T')$ (nombre minimum de générateurs d'un idéal dans Z_T)

On trouvera dans [4] divers calculs de relations d'incidence $Z_T \subset \overline{Z}_{T'}$ et des contre exemples montrant que les conditions 1/ 2/ 3/ ci dessus ne sont pas suffisantes.

– Pour $r \geq 3$, on obtient un résultat de nature opposée (la minoration obtenue étant en général croissante par rapport à ν) : les intersections complètes d'ordre $\nu \geq 2$ sont " presque toutes " non alignables (§ 1) .

Ces résultats énoncés au § 1 constituent une partie de ma thèse ([4]) , où on trouvera des démonstrations plus détaillées.

§ 1 ENONCE DES RESULTATS

Un point z_I de longueur n de support $\{ 0 \}$ dans $\mathbb{C}^r$ est déterminé par un idéal de définition I de $\mathcal{O}_r = \mathbb{C}\left\{ x_1, \ldots, x_r \right\}$ et la platitude d'une déformation z_{I_t} de z_I équivaut à la condition suivante :

$$n_t = \sum_{i=1}^{k} \dim_{\mathbb{C}} \mathcal{O}_{z_{I_t}, p_i} \quad \text{est constant où} \quad \left\{ p_1, \ldots, p_k \right\} \text{ est le support}$$

de la fibre z_{I_t} .

On dit que z_I est <u>lissifiable</u> si il admet une déformation en n points simples et qu'il est <u>alignable</u> si il admet une déformation à support constant dont la fibre générale z_{I_t} $(t \neq 0)$ est de dimension de plongement un (i.e. d'algèbre locale isomorphe à $\mathbb{C}\{x\}/_{(x^n)}$) . Parmi les déformations à support constant, on peut aussi chercher les déformations d'un type particulier faisant intervenir les invariants suivants ($\mathfrak{m}$ désigne l'idéal maximal de $\mathbb{C}\left\{ x_1, \ldots, x_r \right\}$) .

- L'<u>ordre</u> $\nu(I)$ de l'idéal I défini par $I \subset \mathfrak{m}^{\nu(I)}$, $I \not\subset \mathfrak{m}^{\nu(I)+1}$.
- La <u>fonction d'Hilbert Samuel</u> de I :

$$T(I) = (t_o, \ldots, t_j, \ldots) \quad , \quad t_j = \dim_{\mathbb{C}} \frac{I + \mathfrak{m}^j}{I + \mathfrak{m}^{j+1}} \quad .$$

La considération des déformations à $T(I)$ constant conduit à la notion de <u>strate d'Hilbert Samuel</u> relative à T , $Z_T \subset \mathrm{Hilb}^n \mathcal{O}_r$ paramétrant les idéaux I tels que $T(I) = T$.

Le cadre naturel pour étudier ces problèmes est celui des schémas de Hilbert ponctuels (cf. § 2) :

- $\mathrm{Hilb}^n \mathbb{C}^r$, espace analytique paramétrant les sous espaces de dimension zéro et de longueur n de $\mathbb{C}^r$.
- $\mathrm{Hilb}^n \mathcal{O}^r$, sous espace de $\mathrm{Hilb}^n \mathbb{C}^r$ paramétrant les " points " de support $\{ 0 \}$ de $\mathbb{C}^r$ (ou les idéaux de colongueur n de $\mathcal{O}_r$) .

Rappelons que d'après [2], ou [6], ces espaces sont connexes.

Lorsque $r = 2$, on a les résultats suivants :

THEOREME 1 . - (FOGARTY [2], HARTSHORNE) . L'espace $\mathrm{Hilb}^n \mathbb{C}^2$ est lisse, réduit, connexe de dimension $2n$ et tout point de $\mathbb{C}^2$ est lissifiable.

THEOREME 2 . - (BRIANÇON [1]) . $\mathrm{Hilb}^n \mathcal{O}_2$ est irréductible de dimension pure $n-1$, et tout point de $\mathbb{C}^2$ est alignable.

Notons $W \subset \mathrm{Hilb}^n \mathcal{O}_r$ l'ouvert formé par les sous espaces constitués de n points simples de $\mathbb{C}^r$ et $L \subset U \subset \mathrm{Hilb}^n \mathcal{O}_2$ les ouverts formés respectivement par les points alignés (de dimension de plongement un) et intersections complètes. On a évidemment :

$$L \subset U \subset \overline{W} \cap \mathrm{Hilb}^n \mathcal{O}_r \subset \mathrm{Hilb}^n \mathcal{O}_r .$$

D'après les résultats précédents, $\overline{L} = \mathrm{Hilb}^n \mathcal{O}_2$.

Par contre, pour $r \geq 3$, L et U ont des adhérences distinctes sauf pour des valeurs exceptionnelles de n, ainsi que $\mathrm{Hilb}^n \mathcal{O}_r$ et $\overline{W} \cap \mathrm{Hilb}^n \mathcal{O}_r$ (cf § 4) .

La démonstration du Théorème 2 utilise une description précise des strates d'Hilbert - Samuel :

PROPOSITION 3 . - (BRIANÇON [1], IARROBINO [8])

1/ Lorsque $r = 2$, la strate d'Hilbert - Samuel Z_T est lisse connexe de dimension :

$$\dim Z_T = n - \nu - \sum_{j \geq \nu} \delta(j)(\delta(j)-1)/2$$

où $\delta(j) = t_{j-1} - t_j$.

2/ La strate Z_T est de dimension $n - \nu$ si et seulement si Z_T contient des intersections complètes qui forment alors un ouvert dense $U_T = Z_T \cap U$ de Z_T .

Considérons les sous ensembles Z_ν et E_ν de $\mathrm{Hilb}^n \mathcal{O}_r$ constitués respectivement par les idéaux d'ordre ν et $\geq \nu$ ($E_\nu = \bigcup_{\nu' \geq \nu} Z_{\nu'}$) . La démonstration du Théorème 2 consiste à établir que $\mathrm{Hilb}^n \mathcal{O}_2 = \overline{Z}_1$ (adhérence de Z_1) et à remarquer que d'après la Proposition 3 , Z_1 est lisse connexe de dimension $n-1$ (noter que lorsque $r = 2$, on a : $L = Z_1$) .

Nous démontrons ici un résultat plus précis (conjecturé dans [1]) :

THEOREME 4 . - Si $r = 2$, l'ensemble E_ν des idéaux d'ordre $\geq \nu$ est égal à l'adhérence dans $\text{Hilb}^n \mathbb{C}\{x, y\}$ de l'ensemble Z_ν des idéaux d'ordre ν.

Ce théorème résulte de l'inégalité : $\dim(E_\nu, z_I) \geq n - \nu$, et nous donnons en même temps (§ 3 Proposition 8) une minoration de $\dim(E_\nu, z_I)$ pour $r \geq 3$ dont nous tirons quelques conséquences, concernant l'inclusion stricte $\overline{L} \subsetneq \overline{U}$ au § 4.

Le Théorème 4 permet de déterminer le lieu singulier de l'espace réduit sous-jacent à $\text{Hilb}^n \mathbb{C}\{x, y\}$:

THEOREME 5. – Le lieu singulier de $\text{Hilb}^n \mathbb{C}\{x, y\}_{red}$ est l'ensemble $E_2 = \overline{Z}_2$ des idéaux d'ordre au moins deux.

On obtient un résultat analogue mais a priori plus faible (car concernant une structure non réduite sur $\text{Hilb}^n \mathcal{O}_2$) en évaluant la dimension de l'espace tangent de Zariski en un point de $\text{Hilb}^n \mathcal{O}_2$. Pour $r \geq 3$, on trouve par cette méthode des composantes irréductibles entièrement singulières dans $\text{Hilb}^n \mathcal{O}_r$, $r \geq 3$, n assez grand.

§ 2 SCHEMAS DE HILBERT PONCTUELS ET PLATIFICATEURS

Le schéma de Hilbert $\text{Hilb}^n X$ d'un espace analytique X est caractérisé par les propriétés suivantes :

1/ L'espace réduit sous jacent est l'ensemble des sous espaces de dimension zéro et de longueur n de X.

2/ Il existe un sous espace $\widetilde{Z}$ de $Z \times X$ tel que la projection $p : \widetilde{Z} \longrightarrow Z$ soit plate finie de fibre $p^{-1}(z) = \{z\} \times z$.

3/ Pour tout espace analytique T et tout $\widetilde{T} \subset T \times X$ tel que la projection $\widetilde{T} \longrightarrow T$ soit plate finie à fibres de longueur n, il existe un unique morphisme analytique $\lambda : T \longrightarrow Z$, tel que : $\widetilde{T} = (\lambda \times \text{Id})^{-1}(\widetilde{Z})$.

Le germe de $\text{Hilb}^n X$ en un point z est caractérisé par une propriété analogue à 3/ où on remplace les espaces et les applications par des germes. Si z est réductible, ce germe admet une décomposition naturelle en produit, ce qui permet de se restreindre au cas où z_{red} est égal à $\{0\} \subset \mathbb{C}^r$.

Dans la proposition suivante, on donne une présentation du germe de $\text{Hilb}^n \mathbb{C}^r$ en un point z_I en terme de platificateur local ([7]) d'un germe de morphisme convenable. Soit $(f_1, \ldots, f_p)$ un système de générateurs de I et

$\{ \bar{e}_0, \ldots, \bar{e}_{n-1} \}$ une base sur $\mathbb{C}$ de $\mathcal{O}_r/_I$, telle que $e_0 = 1$, $e_i \in \mathfrak{m}$ si $i \geq 1$.

On note : $F_i(\underline{x} ; \underline{a}) = f_i(\underline{x}) + \sum_{j=1}^{n-1} a_{ij} e_j(\underline{x})$

où $\underline{x} = (x_1, \ldots, x_r)$

$$\underline{a} = (a_{i,j} ; 1 \leq i \leq p ; 1 \leq j \leq n-1) \in \mathbb{C}^{p(n-1)}$$

et φ l'application : $\mathbb{C}^{r+p(n-1)} \longrightarrow \mathbb{C}^{p+p(n-1)} = \mathbb{C}^{pn}$ définie par $\varphi(\underline{x}, \underline{a}) = (F_1, \ldots, F_p ; \underline{a})$.

PROPOSITION 6 . - ([4]) Le germe de $\mathrm{Hilb}^n \mathbb{C}^r$ en z_I est isomorphe au platificateur local du germe de l'application φ en 0 .

Dans cet isomorphisme le point $(u_1, \ldots, u_p, \underline{a})$ paramètre le sous espace (de longueur n) d'équations :

$$\begin{cases} F_1(\underline{x} ; \underline{a}) = u_1 \\ \quad \cdots\cdots \\ F_p(\underline{x} ; \underline{a}) = u_p \end{cases} .$$

Le germe de $\mathrm{Hilb}^n \mathcal{O}_r$ est alors défini par les conditions suivantes :

1/ $u_1 = \ldots = u_p = 0$

2/ La fibre de φ au-dessus de $(0, \underline{a})$ est de support $\{ 0 \}$ et de longueur n .

La condition 2/ montre en particulier que φ est un isomorphisme au dessus de $\mathrm{Hilb}^n \mathcal{O}_r$.

§ 3 DEMONSTRATION DES THEOREMES 4 ET 5 .

a/ Le théorème de connexité

THEOREME 7 . - Soit (X, x) un germe d'espace analytique complexe irréductible de dimension n et (Y, x) un germe de sous espace défini par p équations $f_1, \ldots, f_p \in \mathcal{O}_{X,x}$. Alors Y est connexe en dimension $\geq n-p-1$.

Dire que Y est connexe en dimension $\geq n-p-1$ signifie que si $W \subset Y$ et $\dim W < n-p-1$, $Y-W$ est connexe. Nous utiliserons le Théorème 5 sous la forme suivante :

COROLLAIRE . - Sous les hypothèses du Théorème 5, supposons que $Y = Y_1 \cup Y_2$, où Y_1 et Y_2 sont des réunions de composantes irréductibles distinctes. Alors $\dim Y_1 \cap Y_2 \geq n - p - 1$.

Le théorème démontré par GROTHENDIECK ([5] Th. XIII.2.1.) concerne en fait des spectres d'anneaux locaux complets. Le Théorème 7 s'en déduit facilement. L'énoncé analogue pour les espaces analytiques réels serait faux, ainsi que les théorèmes 4 et 5 (comparer à [8], 5.B.) .

b/ Démonstration du Théorème 4

Les résultats de [1] permettent d'abord de se ramener facilement au cas des intersections complètes :

LEMME . - Soit $U_\nu \subset \mathrm{Hilb}^n \mathcal{O}_2$ l'ensemble des intersections complètes d'ordre ν. On a pour tout $\nu \geq 1$: $Z_{\nu+1} \subset \bar{Z}_\nu \cup \bar{U}_{\nu+1}$ donc $E_\nu = \bar{Z}_\nu \cup \bar{U}_{\nu+1}$.

Soit $I = (f_1, \dots, f_r)$ une intersection complète dans $\mathcal{O}_r$.

PROPOSITION 8 . - La dimension de E_ν en z_I est au moins égale à :

$$(n-1)(r-1) + f(\nu, r)$$

où $f(\nu, r) = \nu^r - r\binom{\nu+r-1}{r} + r - 1$.

Lorsque $r = 2$, la Proposition 8 donne :

$$\dim(E_\nu, z_I) \geq n - \nu .$$

Donc : $\dim(E_\nu, z_I) = n - \nu$, d'après la Proposition 3 . Si $z_I \in E_{\nu+1} = E_\nu - Z_\nu$, on a donc par un argument évident de dimension $z_I \in \bar{Z}_\nu$, ce qui démontre le Théorème 4 .

Démonstration de la Proposition 8 .

Dans le cas particulier où $p = r$, le germe d'application φ de la Proposition 6 est plat et on a donc : $(\mathrm{Hilb}^n \mathbb{C}^r, z_I) \approx (\mathbb{C}^{nr}, 0)$.

Supposons de plus que $e_0, \dots, e_{n-1}$ soient des monômes et que $\{e_0, \dots, e_{N-1}\}$ soit l'ensemble des monômes de degré $\leq \nu$ de $\mathcal{O}_r$ $\left(N = \binom{\nu+r-1}{r}\right)$.

Soit $\varphi' : S = (\mathbb{C}^{r + r(n-N)}, 0) \longrightarrow S$, le germe d'application obtenu par restriction de φ de la façon suivante :

$$\varphi'(\underline{x}, \underline{a}') = (G_1, \ldots, G_r, \underline{a}')$$

$$G_i(\underline{x}, \underline{a}') = f_i(\underline{x}) + \sum_{j=N}^{n-1} a_{i,j}\, e_j(\underline{x})$$

$$\underline{a}' = (a_{i,j} \; ; \; 1 \leq i \leq r \; ; \; N \leq j \leq n-1) \in \mathbb{C}^{r(n-N)} \; .$$

Soit $W_o = \{0\} \times \mathbb{C}^{r(n-N)} \subset S$ paramétrant, dans un voisinage de z_I les sous schémas de $\mathbb{C}^r$ contenant à l'origine un point d'ordre ν (donc intersections complètes d'ordre $\geq \nu^r$) .

Notons $R^k(\varphi') = \left\{ m \in S \; ; \; \deg_m \varphi' \geq k+1 \right\}$. On a $W_o \subset R^{\nu^r - 1}(\varphi')$, et d'autre part, les points de $E_\nu = (\mathrm{Hilb}^n \mathcal{O}_r, z_I) \cap W_o$ sont déterminés par la condition : $\deg_{(0, \underline{a}')} (\varphi') = n$ (cf les remarques suivant la Proposition 6). Ainsi :

$$(E_\nu, z_I) = W_o \cap R^{n-1}(\varphi') \qquad (1) \; .$$

On construit alors par récurrence sur k une suite de sous espaces W_k de W_o satisfaisant aux conditions suivantes :

1/ $W_o \supset \ldots \supset W_k \supset \ldots \supset W_{n-\nu^r}$, et W_k est irréductible.

2/ $W_k \subset R^{\nu^r + k - 1}(\varphi')$

3/ $\mathrm{codim}_S W_{k+1} \leq \mathrm{codim}_S W_k + 1$.

La condition 2/ donne, d'après (1) :

$$W_{n-\nu^r} \subset (E_\nu, z_I) \; .$$

L'inégalité de la Proposition 8 résulte alors immédiatement de la condition 3 .

Pour construire W_{k+1} lorsque $0 \leq k < n-\nu^r$, on considère l'application :

$$\varphi' \times \varphi' \; : \; W_k \times S \longrightarrow S \times S$$

et $\Delta'_k = (\varphi' \times \varphi')^{-1} (\Delta_S)$ où Δ_S est la diagonale de S . D'après le Théorème 7 on voit facilement que Δ'_k est connexe en dimension $\geq \dim W_k - 1$. On distingue alors deux cas :

1/ Si la diagonale Δ_k de W_k est contenue dans l'adhérence de $\Delta'_k - \Delta_k$, on peut prendre $W_{k+1} = W_k$.

2/ Sinon $\Delta_k \cap \overline{\Delta'_k - \Delta_k}$ est de dimension au moins égale à $\dim W_k - 1$ d'après le corollaire du Théorème 7 . Une composante irréductible, de dimension maximum de cette intersection fournit après projection isomorphe sur W_k , le germe W_{k+1} cherché □

c/ Pour terminer ce paragraphe, nous esquissons la démonstration du Théorème 5 . Le cas $n = 3$ où $Z_2 = \{ \mathfrak{m}^2 \}$ peut être traité séparément. Pour $n \geq 4$, Z_2 contient un sous ensemble dense Z_2' formé par les idéaux isomorphes à l'un des suivants :

$$I_{p,q} = (x^p + y^q, xy) , \quad p + q = n \quad .$$

D'après ce Théorème 4 , il reste donc à montrer que $\mathrm{Hilb}^n \mathbb{C}\{x, y\}_{red}$ est singulier au point z paramétrant $I_{p,q}$. On considère pour cela la présentation du germe $(\mathrm{Hilb}^n \mathbb{C}^2, z)$ fournie par la Proposition 6 et on en extrait une déformation transverse à la strate Z_T de z . Cette strate est lisse de dimension $n - 2$ (Proposition 3) et on constate que la trace de $\mathrm{Hilb}^n \mathcal{O}_2$ dans cette section transverse est un germe de courbe singulière isomorphe à la courbe plane $C_{p,q}$ d'équation $x^p + y^q = 0$. On peut alors montrer que le germe de $\mathrm{Hilb}^n \mathcal{O}_2$ en z est isomorphe à :

$$(Z_T \times C_{p,q}, (z, 0)) \quad .$$

§ 4 QUELQUES REMARQUES SUR $\mathrm{Hilb}^n \mathcal{O}_r$, $r \geq 3$ ET SUR LES STRATES D'HILBERT SAMUEL

a/ Dans [10] IARROBINO a montré que pour $r \geq 3$, et n assez grand, $\mathrm{Hilb}^n \mathcal{O}_r$ contient des points non alignables et même non lissifiables. Cela résulte d'une minoration de la dimension :

$$\dim \mathrm{Hilb}^n \mathcal{O}_r \geq a(r)\, n^{2-2/r}$$

où $a(r)$ ne dépend pas de n , et du fait que pour r fixé, cette minoration est d'ordre supérieur à $\dim \overline{L} = (r-1)(n-1)$ et à $\dim \overline{W} = rn$.

En examinant à quelle condition on a $f(\nu, r) \geq 0$ (Proposition 8) , on obtient par le même argument l'existence au voisinage de " presque toute " intersection complète de familles de points où les points alignables ($\in \overline{L}$) ne sont pas denses :

<u>PROPOSITION 9</u> . - Soit $r \geq 3$, $\nu \geq 2$ tels que (ν, r) soit distinct de (2,3), (3,3) ou (2,4) . Pour tout idéal I d'ordre ν , on a : $\dim(E_\nu, z_I) \geq \dim L$ et z_I admet une déformation sur les idéaux non alignables.

Donnons pour conclure le premier de ces contre exemples concernant des intersections complètes : soient f et g des polynômes homogènes de degré 3 et premiers entre eux, et I l'idéal engendré par f et g. On a :

$$T_o = T(I) = (1, 2, 3, 2, 1) .$$

Les fonctions d'Hilbert - Samuel d'ordre deux possibles d'après les conditions 1/ 2/ 3/ sont :

$$T_1 = (1, 2, 2, 2, 1, 1)$$
$$T_2 = (1, 2, 2, 1, 1, 1, 1)$$
$$T_3 = (1, 2, 1, 1, 1, 1, 1, 1)$$

On trouve très facilement une déformation dans Z_{T_1} et par un calcul plus compliqué dans Z_{T_3}. Pour fixer les idées si $f = y^3 + c\,x^3$, $g = x^2 y$, $c \neq 0$ ces déformations sont respectivement :

$$F = y^3 + c\,x^3 + t\,x\,y \;, \quad G = g = x^2 y$$

et

$$\begin{cases} F = y^3 + c\,x^3 + t(y - mx)(y - \frac{5}{8} m\,x) \\ G = x^2 y - t/_{6m^2}(y - mx)(y - 4\,mx), \quad c = -\frac{m^3}{4} \end{cases} .$$

Par contre, l'idéal I n'admet pas de déformation dans Z_{T_2} dès que l'espace vectoriel $< f, g >$ ne contient pas de cube $(ax + by)^3$, ce qui est une condition générique sur $< f, g >$. On peut noter que les idéaux isomorphes à $(y^3 + cx^3, x^2y)$, $c \neq 0$, remplissent cette condition et forment un ouvert dense de Z_{T_o}.

REFERENCES

[1] J. BRIANÇON Description de $\mathrm{Hilb}^n\, \mathbb{C}\{x, y\}$. Inventiones Math. n° 41 pp 45 - 89

[2] J. FOGARTY Algebraic families on an algebraic surface. Amer. J. of Math., n° 10 , pp 511 - 521, 1968

[3] T. GAFFNEY Multiple points and ramification loci for finite maps (Notes - Northeastern U.)

[4] M. GRANGER Géométrie des schémas de Hilbert ponctuels. Thèse-Université de Nice

[5] A. GROTHENDIECK Cohomologie locale des faisceaux cohérents et théorèmes de Lefschetz locaux et globaux. SGA 2, North Holland, 1968

[6] R. HARTSHORNE Connectedness of the Hilbert scheme. Publ. Math. de l'I.H.E.S. n° 29 pp 261 - 304 (1966)

[7] H. HIRONAKA, M. LEJEUNE, B. TEISSIER Platificateur local en géométrie analytique et applatissement local. Singularités à Cargèse, Astérisque, n° 7 et 8 , p 441 - 463, 1973

[8] A. IARROBINO Punctual Hilbert schemes. Memoirs of the Am. Math. Society, Vol 10, n° 188, 1977

[9] A. IARROBINO Complete intersection algebras having no deformations to $k[x]/(x^n)$ (preprint)

[10] A. IARROBINO Reducibility of the families of 0 - dimensional schemes on a variety. Inventiones Math. n° 15 pp 72-77, 1972

[11] R. LAZARSFELD Branched coverings of projective spaces. Thesis. Brown University

J.M. GRANGER
I.M.S.P.
Université de Nice
MATHEMATIQUES
Parc Valrose
06034 NICE CEDEX

ON DEFORMATION OF CURVES AND A FORMULA OF DELIGNE*

Gert-Martin Greuel

ontents

bstract: We study deformations of germs of reduced complex curve ingularities and of singular projective curves in some $P^n(\mathbb{C})$. In both ases a deformation is topologically trivial iff the Milnor numbers of he singularities are constant during the deformation. The Milnor umber also occurs naturally in the degree of the singular Todd class f Baum-Fulton-MacPherson and in a formula of Deligne concerning the imension of the base space of the semiuniversal deformation. Some pplications of this fact are given in particular to the non-smoothbility of certain curves.

. Milnor number and Hirzebruch-Riemann-Roch formula

.1. Let $(C, 0) \subset (\mathbb{C}^n, 0)$ be the germ of a reduced complex curve ingularity. We recall some basic facts about the Milnor number $= \mu(C, 0)$ (cf. [B-G]) and deduce some results about families of ompact curves. Moreover we fix the notations.

et $\mathcal{O} = \mathcal{O}_{C,0}$ be the local ring of $(C, 0)$ with maximal ideal $\mathfrak{m}$ and

$$n : (\bar{C}, \bar{0}) \to (C, 0)$$

) This is a modified version of [G_2]. The author gratefully acknowledges the financial support of the Deutsche Forschungsgemeinschaft and of the Stiftung Volkswagenwerk for a visit to the IHES, during which this paper was written.

the normalization. If $(C, 0) = \bigcup_{i=1}^{r} (C_i, 0)$ is the decomposition into irreducible components (branches) we have $(\bar{C}, \bar{0}) = \coprod_{i=1}^{r} (\bar{C}_i, \bar{0}_i)$, the disjoint union of smooth germs, and

$$\bar{\mathcal{O}} := n_* \mathcal{O}_{(\bar{C}, \bar{0})} = \prod_{i=1}^{r} \mathbb{C}\{t_i\}.$$

Moreover we consider:

$\bar{\mathfrak{m}} = \prod_{i=1}^{r} \bar{\mathfrak{m}}_i$, the Jacobson radical of the semilocal ring $\bar{\mathcal{O}}$,

$\bar{\mathfrak{m}}_i = t_i \, \mathbb{C}\{t_i\}$, the maximal ideal of $\mathbb{C}\{t_i\}$,

$K = \prod_{i=1}^{r} \mathbb{C}\{\{t_i\}\}$, the total ring of fractions of $\bar{\mathcal{O}}$,

$\mathbb{C}\{\{t_i\}\}$ = quotient field of the convergent power series ring $\mathbb{C}\{t_i\}$,

$\mathcal{C} = \mathrm{Ann}_{\bar{\mathcal{O}}} (\bar{\mathcal{O}} / \mathcal{O})$, the conductor ideal of $\mathcal{O} \subset \bar{\mathcal{O}}$,

$\Omega = \Omega^1_{(C, 0)}$, the holomorphic (Kählerian) 1-forms on $(C, 0)$

$\bar{\Omega} = n_* \, \Omega^1_{(\bar{C}, \bar{0})}$, $\Omega^1_{(\bar{C}, \bar{0})}$ = holomorphic 1-forms on $(\bar{C}, \bar{0})$,

$T^1_{(C, 0)} = \mathrm{Ext}^1_{\mathcal{O}} (\Omega, \mathcal{O})$, space of first order infinitesimal deformations of $(C, 0)$,

$\omega = \mathrm{Ext}^{n-1}_{\mathcal{O}_{\mathbb{C}^n,0}} (\mathcal{O}, \Omega^n_{\mathbb{C}^n,0})$, the Grothendieck dualizing module of $(C, 0)$.

We can describe ω also as

$$\omega = \{\alpha \in \bar{\Omega} \otimes_{\mathcal{O}} K \mid \mathrm{res}(f\alpha) = 0 \text{ for all } f \in \mathcal{O}\},$$

where

$$\mathrm{res}(\beta) := \sum_{i=1}^{r} \mathrm{res}\,(\beta_i)$$

for each $\beta = (\beta_1, \ldots, \beta_r) \in \bar{\Omega} \otimes K = \prod_{i=1}^{r} \mathbb{C}\{\{t_i\}\} dt_i$, (cf. [Se]).

C, 0) is called smoothable if the base space (S, 0) of the semi-
niversal deformation (cf. [Gr], [Tj]) contains a component over which
ne generic fiber is smooth. Such a component of (S, 0) is called a
noothing component for (C, 0). The Zariski tangent space of (S, 0)
s isomorphic to $T^1_{(C, 0)}$ and (C, 0) is said to be unobstructed if
im (S, 0) = $\dim_{\mathbb{C}} T^1_{(C, 0)}$, i.e. if (S, 0) is smooth.

.2. We shall consider the following numerical invariants of (C, 0):

= $m(C, 0) = e_{\mathfrak{m}}(\mathcal{O})$	: multiplicity of the local ring $\mathcal{O}$,
= $\delta(C, 0) = \dim_{\mathbb{C}}(\overline{\mathcal{O}}/\mathcal{O})$	: δ-invariant,
= $r(C, 0) = \dim_{\mathbb{C}}(\overline{\mathcal{O}}/\overline{\mathfrak{m}})$	: number of branches,
= $t(C, 0) = \dim_{\mathbb{C}}(\omega/\mathfrak{m}\omega)$	: Cohen-Macaulay type,
= $c(C, 0) = \dim_{\mathbb{C}}(\overline{\mathcal{O}}/\mathcal{C})$	: multiplicity of the conductor,
= $\tau(C, 0) = \dim_{\mathbb{C}} T^1_{(C,0)}$	: Tjurina number,
= $\mu(C, 0) = \dim_{\mathbb{C}}(\omega/d\mathcal{O})$	: Milnor number.

ote that in the definition of μ the map $d : \mathcal{O} \to \omega$ is defined to be
ne composition of the exterior derivation $d : \mathcal{O} \to \Omega$ and the canonical
orphism $\Omega \to \bar{\Omega} \hookrightarrow \omega$ ("canonical class").

roposition (cf. [B-G], 1.2.1):

$$\mu = 2\delta - r + 1$$

n particular, $\mu = 0$ iff (C, 0) is smooth.

<u>1.3.</u> The importance of μ comes from the fact that it controls the topology in a flat family of reduced curves. μ is the number of virtua vanishing cycles in a small deformation $f : (X, 0) \to (\mathbb{C}, 0)$ of $(C, 0) = (f^{-1}(0), 0)$. These cycles actually appear for smoothable curves in a nearby smooth fiber. In order to make this statement precise, we have to choose a good representative for f.

Let $B \subset \mathbb{C}^n$ be a small open ball of radius ε with center 0 and $D \subset \mathbb{C}$ be a small disc with radius δ and center 0. For sufficiently small B and D we may assume that a representative of $f : (X, 0) \to (\mathbb{C}, 0)$ is given by the following commutative diagram

$$\begin{array}{ccc} X & \overset{i}{\hookrightarrow} & B \times D \\ {\scriptstyle f}\downarrow & \swarrow {\scriptstyle \Pi} & \\ D & & \end{array}$$

where i is a closed immersion and Π is the projection on the second factor. Such a representative with $0 < \delta << \varepsilon$ sufficiently small will be called a <u>good representative</u> of f. We use the following notations $(t \in D)$:

$$C_t \times \{t\} = f^{-1}(t),$$

$$\mu_t = \mu(C_t) = \sum_{x \in C_t} \mu(C_t, x),$$

$$\delta_t = \delta(C_t) = \sum_{x \in C_t} \delta(C_t, x),$$

$$r'_t = r'(C_t) = \sum_{x \in C_t} (r(C_t, x) - 1),$$

$$b^i_t = b^i(C_t) = \dim_{\mathbb{C}} H^i(C_t, \mathbb{C}).$$

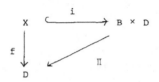

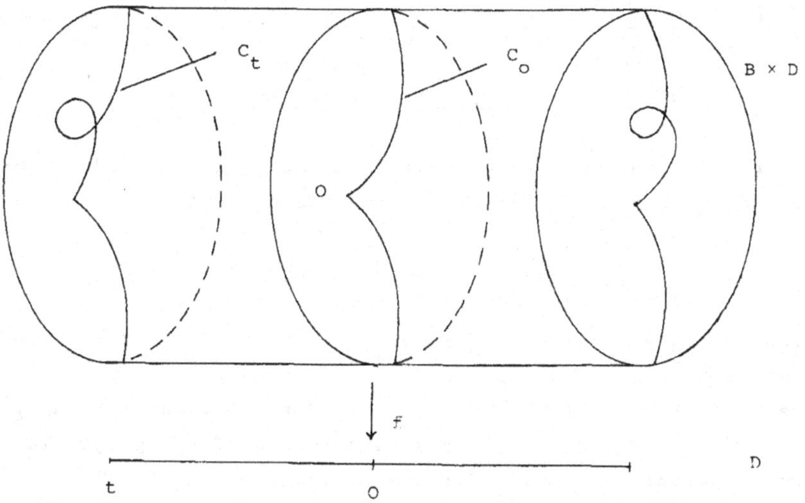

Figure 1

heorem (cf. [B-G], 4.2.2, 4.2.4, 5.2.2) *Let* $f : X \to D$ *be a good epresentative of a deformation of* $(C, 0)$. *Then for all* $t \in D$:

1) (a) $b_t^o = 1$ (i.e. C_t *is connected*),

(b) $\mu_o - \mu_t = b_t^1$,

(c) $\mu_o - \mu_t \geqslant \delta_o - \delta_t \geqslant 0$.

2) *The following conditions are equivalent*:

(a) μ_t *is constant for all* $t \in D$,

(b) δ_t *and* r_t' *are constant for all* $t \in D$,

(c) $b_t^1 = 0$ *for all* $t \in D$,

(d) C_t *is contractible for all* $t \in D$.

3) *Let* $\sigma : D \to X$ *be a section of* f. *If* $\mu(C_t, \sigma(t))$ *is constant for all* $t \in D$, *then* $C_t - \sigma(t)$ *is smooth*. *If* $C_t - \sigma(t)$ *is smooth, the following conditions are equivalent*.

(a) $\mu(C_t, \sigma(t))$ *is constant for all* $t \in D$,

(b) $\delta(C_t, \sigma(t))$ *and* $r(C_t, \sigma(t))$ *are constant for all* $t \in D$,

(c) $b_t^1 = 0$ *for all* $t \in D$,

(d) $f : X \to D$ *admits a weak simultaneous resolution* (cf. $[T_1]$),

(e) <u>The topological pairs</u> (B, C_o) <u>and</u> (B, C_t) <u>are homeomorphic</u> for all $t \in D$,

(f) $f : X \to D$ <u>is a topologically trivial fibration</u>.

Note that for plane curves even stronger statements can be made. For comments on this theorem in connection with the concept of equisingularity see [B-G] and [B-G-G].

<u>1.4.</u> We consider now families of global curves. Let $f : X \to T$ be a proper, flat family of reduced projective curves over some open and connected subset $T \subset \mathbb{C}$. I.e. X is a closed two dimensional analytic subspace of $P^n(\mathbb{C}) \times T$ and f is the restriction of the projection on the second factor, such that $f : X \to T$ is flat and $f^{-1}(t) = C_t \times \{t\}$ is a reduced curve for each $t \in T$. Let μ_t, δ_t, r_t' and b_t^i have the same meaning as in 1.3. We do not assume that b_t^o, the number of connected components of C_t, is equal to one. The following example shows that for a global analogue of theorem 1.3 we have to replace b_t^1 by the topological Euler characteristic

$$\chi_{top,\, t} = \chi_{top}(C_t) = b_t^o - b_t^1 + b_t^2 .$$

<u>Example</u>: Let C_t be the family of quadrics in $P^2(\mathbb{C})$ defined by $x^2 - y^2 - tz^2 = 0$:

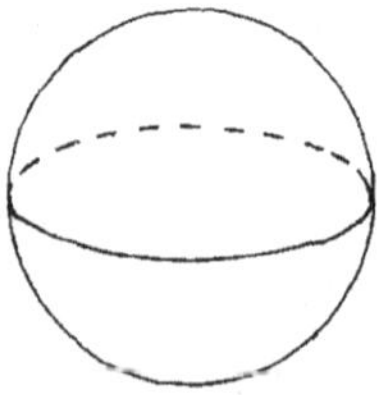

$C_{t \neq 0} \approx S^2$

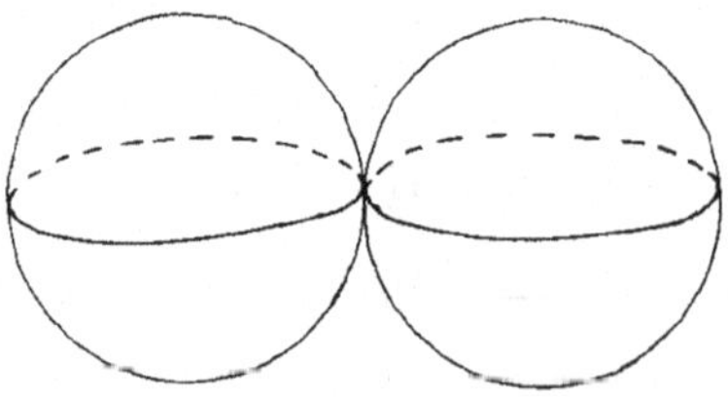

$C_o \approx S^2 \vee S^2$

<u>Figure</u> 2

$\mu_t = 0$, $b_t^1 = 0$, $\chi_{top,\, t} = 2$ if $t \neq 0$ and $\mu_o = 1$, $b_o^1 = 0$, $\chi_{top,\, 0} = 3$. Hence $\mu_o - \mu_t = \chi_{top,\, 0} - \chi_{top,\, t} \neq b_o^1 - b_t^1$.

heorem: Let $f : X \to T$ be a flat family of reduced projective curves s above.

1) (a) b_t^o is constant for all $t \in T$ (and equal to the number of connected components of X),

(b) $\mu_t - \mu_{t'} = \chi_{top,\, t} - \chi_{top,\, t'}$ for all $t, t' \in T$.

2) The following conditions are equivalent:

(a) μ_t is constant for all $t \in T$,

(b) δ_t and r_t' are constant for all $t \in T$,

(c) $\chi_{top,\, t}$ is constant for all $t \in T$,

(d) b_t^o, b_t^1, b_t^2 are constant for all $t \in T$.

3) If the number of singular points of C_t is independent of $t \in T$, then the conditions (2) (a),...,(d) are moreover equivalent to

(e) $f : X \to T$ admits a weak simultaneous resolution,

(f) the topological pairs $(P^n(\mathbb{C}), C_t)$ and $(P^n(\mathbb{C}), C_{t'})$ are homeomorphic for all $t, t' \in T$,

(g) $f : X \to T$ is a topological locally trivial fibration.

roof: (1) (a) follows easily by topological arguments from 1.3 (1) (a). t follows also from the fact, that $f_* f^{-1}\mathcal{O}_T$ is a locally free sheaf of rank equal to the number of connected components of X) and $(f_* f^{-1}\mathcal{O}_T)_t \cong H^o(C_t, \mathbb{C}) \otimes \mathcal{O}_{T,t}$.

1) (b) follows from the local statement 1.3 (1) (b) by a Mayer-ietoris argument. More elegantly it is a consequence of the formula $\chi_{alg,\, t} = \chi_{top,\, t} - \mu_t$ (cf. prop. 1.5) and the fact, that $\chi_{alg,\, t}$ s constant.

2) The equivalence of (a) and (b) follows from prop. 1.2 and theorem .3 (1) (c), the equivalence of (a) and (c) from (1) (b). The mplication (d) $\Longrightarrow$ (c) is trivial. "δ_t = constant" implies normalization in family" (cf. [T_2]), hence b_t^2 = number of irreducible omponents of C_t = number of connected components of $\bar{C}_t$ (normalization of C_t) is constant. Therefore (b) and (c) together with (1) (a) mply (d).

(3) is an easy consequence of the local theorem; the assumption implies that the restriction $f : \Sigma \to T$ (Σ = critical set of f) is an unramified covering, therefore the equivalence of (a) and (e) follows from 1.3 (3). In order to see that (a) implies (f) and (g) let U be some (arbitrary small) open tubular neighbourhood of Σ in $P^n \times T$. By the Ehresmann fibration theorem, $f : X - X \cap U \to T$ is a C^∞ locally trivial subfibration of the projection $P^n \times T - U \to T$. Pasting these diffeomorphisms together with the local homeomorphisms inside U given by 1.3 (3) we get the desired global homeomorphisms.

1.5. We show how the Milnor number occurs naturally in the general Riemann-Roch formula for singular curves.

Let $C \subset P^n(\mathbb{C})$ be a reduced compact curve, not necessarily irreducible or connected. For any sheaf $\mathcal{F}$ on C s.t. $H^i(C,\mathcal{F}) \neq 0$ only for finitely many i and $\dim_{\mathbb{C}} H^i(\mathbb{C},\mathcal{F}) < \infty$ for all i let

$$\chi(\mathcal{F}) = \sum_i (-1)^i \dim_{\mathbb{C}} H^i(C,\mathcal{F})$$

denote the Euler characteristic of $\mathcal{F}$. We write $h^i(C) = \dim_{\mathbb{C}} H^i(C,\mathcal{O}_C)$ and $\chi_{alg}(C) = h^o(C) - h^1(C)$ for the Euler characteristic of the structure sheaf $\mathcal{O}_C$. Note that $h^o(C) = b^o(C)$ and

$$\mu(C) = \sum_{x \in C} \mu(C, x) = \dim_{\mathbb{C}} H^o(C, \omega_C/d\mathcal{O}_C) = \chi(\omega_C/d\,\mathcal{O}_C).$$

It is convenient to introduce the <u>virtual topological Euler characteristic</u>

$$\chi_\tau(C) := \chi_{top}(C) - \mu(C).$$

<u>Proposition:</u> <u>For any reduced projective curve</u> C, $2\chi_{alg}(C) = \chi_\tau(C)$.

<u>Proof:</u> Take Euler characteristics of the two exact sequences

$$0 \to \mathbb{C}_C \to \mathcal{O}_C \xrightarrow{d} d\,\mathcal{O}_C \to 0,$$

$$0 \to d\mathcal{O}_C \to \omega_C \longrightarrow \omega_C/d\mathcal{O}_C \to 0$$

and notice $\chi_{alg}(C) = -\chi(\omega_C)$ by duality.

Let $\mathcal{L}$ be any invertible sheaf on C, $n : \bar{C} \to C$ the normalization and $\bar{\mathcal{L}} = n_* n^*\mathcal{L}$, where $n^*\mathcal{L}$ denotes the analytic primage sheaf. The degree of $\mathcal{L}$ is defined to be

$$\deg \mathcal{L} = c_1(\mathcal{L})\,[C].$$

Here $c_1(\mathcal{L}) \in H^2(C, \mathbb{Z})$ denotes the first Chern class of $\mathcal{L}$ and $[C] = \sum_i [C_i] \in H_2(C, \mathbb{Z})$ the homology class defined by C ($[C_i]$ = image of the fundamental class of the irreducible component C_i of C in $H_2(C, \mathbb{Z})$, $i = 1,\ldots,b^2(C)$). The Riemann-Roch formula for the non-singular curve $\bar{C}$ says

$$\chi(n^*\mathcal{L}) = \chi(\mathcal{O}_{\bar{C}}) + \deg(n^*\mathcal{L}).$$

From the exact sequence

$$0 \to \mathcal{L} \to \bar{\mathcal{L}} \to \bar{\mathcal{L}}/\mathcal{L} \to 0$$

we obtain $\chi(n^*\mathcal{L}) = \chi(\bar{\mathcal{L}}) = \chi(\mathcal{L}) + \delta$. Since $\deg(n^*\mathcal{L}) = c_1(n^*\mathcal{L})([\bar{C}]) = n^*c_1(\mathcal{L})([\bar{C}]) = c_1(\mathcal{L})(n_*[\bar{C}]) = c_1(\mathcal{L})[C] = \deg\mathcal{L}$, we deduce the (well-known) Riemann-Roch formula for the singular curve C:

$$\chi(\mathcal{L}) = \chi_{alg}(C) + \deg(\mathcal{L}).$$

Replacing χ_{alg} by $\frac{1}{2}\chi_\tau$ we obtain

<u>Theorem</u> (Hirzebruch-Riemann-Roch): <u>For any reduced projective curve</u> C <u>and any invertible sheaf</u> $\mathcal{L}$ <u>on</u> C,

$$\chi(\mathcal{L}) = c_1(\mathcal{L})[C] + \tfrac{1}{2}\chi_\tau(C).$$

I consider this to be the appropriate formulation of a Hirzebruch-Riemann-Roch formula in the following sense: the left hand side depends on analytical data while the right hand side depends only on the topological bundle ($c_1(\mathcal{L})$), on the topology of C ($\chi_{top}(C)$) and on

$\mu(C)$ which reflects the topology of the singularities (in the sense of 1.4). But note that μ is not a topological invariant of C as can be seen by taking a curve C with cuspidal singularities: C is homeomorphic to its normalization $\bar{C}$ while $\mu(C) > \mu(\bar{C}) = 0$.

It is also interesting to interpret the singular Riemann-Roch theorem of Baum-Fulton-MacPherson (cf. [B-F-M]) with the help of μ. Recall that

$$\chi(\mathcal{L}) = \varepsilon(ch(\mathcal{L}) \cap \tau(C))$$

where $ch(\mathcal{L}) = 1 + c_1(\mathcal{L})$ is the Chern character of $\mathcal{L}$, $\tau(C) = \tau_o(C) + \tau_1(C)$, $\tau_i(C) \in H_{2i}(C, \mathbb{Q})$ is the Todd homology class, $\cap : H^p(C, \mathbb{Z}) \times H_q(C, \mathbb{Z}) \to H_{q-p}(C, \mathbb{Z})$ is the cap product and ε is induced by the mapping of C to a point. Therefore

$$\begin{aligned} \chi(\mathcal{L}) &= \varepsilon(\tau_o(C) + c_1(\mathcal{L}) \cap \tau_1(C)) \\ &= \varepsilon(\tau_o(C)) + c_1(\mathcal{L})[C], \end{aligned}$$

since $\tau_1(C) = [C]$. Hence $\varepsilon(\tau_o(C)) = \frac{1}{2}(\chi_{top}(C) - \mu(C))$. Define

$$\underline{\mu}(C) := \sum_{x \in C} \mu(C, x)x \in H_o(X, \mathbb{Z})$$

to be the <u>Milnor divisor</u> of C, which is concentrated in the singular points of C. Let $c_o(C) \in H_o(C, \mathbb{Z})$ be the 0-th singular homology Chern class of MacPherson (cf. [MP]). Since $\varepsilon(c_o(C)) = \chi_{top}(C)$, we obtain

$$\tau_o(C) = \frac{1}{2}(c_o(C) - \underline{\mu}(C)).$$

Hence the homology Todd classes are completely understood for reduced complex curves.

. A formula of Deligne

.1. We keep the notation of 1.1. Let $\Theta = \mathrm{Hom}_{\mathcal{O}}(\Omega, \mathcal{O})$ and $= \mathrm{Hom}_{\bar{\mathcal{O}}}(\bar{\Omega}, \bar{\mathcal{O}})$. Since any derivation of $\mathcal{O}$ has a unique extension o $\bar{\mathcal{O}}$ we have the inclusion $\Theta \subset \bar{\Theta}$. Let

$$m_1 = m_1(C, 0) = \dim_{\mathbb{C}} \bar{\Theta}/\Theta.$$

heorem (Deligne, [D], 2.27, 2.34)

1) Let $(C, 0)$ be a reduced smoothable curve singularity and $(E, 0)$ smoothing component of the semiuniversal base. Then

$$\dim (E, 0) = 3\delta - m_1 .$$

2) Let $(C, 0)$ be an arbitrary reduced curve singularity and $\mathfrak{a} \subset \bar{\mathcal{O}}$ any ideal contained in the conductor $\mathcal{C}$. Let $\bar{G} = \mathrm{Aut}(\bar{\mathcal{O}}/\mathfrak{a})$ nd $G \subset \bar{G}$ the stabilizor of $\mathcal{O}/\mathfrak{a}$. Then

$$m_1 = r + \dim_{\mathbb{C}} \bar{G}/G.$$

lthough this statement is purely local, Deligne proves it by globaliz-ng the smoothing family. It would be interesting to know a local proof. e reformulate Deligne's result in order to get formulas which are better uited to concrete calculations and deduce estimates which contain only basic" invariants. For quasihomogeneous curves this formula reduces to particularly elegant one, containing only the Milnor number and the ohen-Macaulay type. For any reduced curve singularity $(C, 0)$ we et

$$e := e(C, 0) := 3\delta - m_1.$$

.2. For the proofs as well as for concrete calculations it is conven-ent to use the language of fractional ideals. For $\mathcal{O}$-ideals $\mathfrak{X}, \mathfrak{L} \subset K$ which contain a non-zero divisor there is an isomorphism

$$\mathrm{Hom}_{\mathcal{O}}(\mathfrak{b}, \mathfrak{a}) \cong \mathfrak{a} : \mathfrak{b} = \{x \in K \mid x\mathfrak{b} \subset \mathfrak{a}\}$$

where $x \in \mathfrak{a} : \mathfrak{b}$ corresponds to the multiplication with x. We set $\mathfrak{a}^{-1} = \mathcal{O} : \mathfrak{a}$. Let $\mathfrak{g}$ be any further $\mathcal{O}$-ideal, containing a non-zero divisor, then:

(a) $(\mathfrak{a} : \mathfrak{b}) : \mathfrak{g} = \mathfrak{a} : (\mathfrak{b} \cdot \mathfrak{g}) = (\mathfrak{a} : \mathfrak{g}) : \mathfrak{b}$,

(b) $\dim_{\mathbb{C}}(\mathfrak{b}/\mathfrak{a}) = \dim_{\mathbb{C}}(\omega : \mathfrak{a}/\omega : \mathfrak{b})$ if $\mathfrak{a} \subset \mathfrak{b}$,

(c) $\omega : \omega = \mathcal{O}$.

While (a) is trivial, (b) and (c) concern local duality (cf. [H-K]).

We now identify some $\mathcal{O}$-modules with fractional ideals in K. After a choice of local uniforming parameters $t_1, \dots, t_r$ we define an $\bar{\mathcal{O}}$-isomorphism

$$\phi : \bar{\Omega} \otimes K \xrightarrow{\cong} K ,$$

$$dt := (dt_1, \dots, dt_r) \mapsto t := (t_1, \dots, t_r),$$

which maps $\bar{\Omega}$ isomorphically onto $\bar{\mathfrak{m}}$. Let $j : \Omega \to \bar{\Omega}$ denote the canonical morphism and

$$M := \phi \circ j(\Omega) \subset K.$$

Of course $M \cong \Omega/T\Omega$, where $T\Omega$ denotes the torsion submodule of Ω. M depends on the choice of the local parameters and is determined only up to multiplication by a unit of $\bar{\mathcal{O}}$.

<u>Lemma</u>: <u>For a quasihomogeneous curve singularity there exist local parameters</u> $t_1, \dots, t_r$ <u>such that</u> ϕ <u>induces an isomorphism</u> $\Omega/T\Omega \xrightarrow{\cong} \mathfrak{m}$ (<u>i.e.</u> $M = \mathfrak{m}$).

<u>Proof</u>: By definition of "quasihomogeneous" there exist local coordinates of $(\mathbb{C}^n, 0)$, $x_1, \dots, x_n$, and positive integers (weights) $w_1, \dots, w_n$ such that the equations of $(C, 0) \subset (\mathbb{C}^n, 0)$ are homogeneous with respect to these weights. Each branch C_i of C can be parametrized by

$$t_i \mapsto (t_i^{w_1} p_i^1, \ldots, t_i^{w_n} p_i^n)$$

where $p_i = (p_i^1, \ldots, p_i^n) \in C_i - \{0\}$. The inclusion $\mathcal{O} \subset \bar{\mathcal{O}}$ is deter mined by

$$x_\nu \mapsto (x_\nu(t_1), \ldots, x_\nu(t_r)) = (t_1^{w_\nu} p_1^\nu, \ldots, t_r^{w_\nu} p_r^\nu)$$

$\nu = 1, \ldots, n$. If $g \in \mathcal{O}$ is (quasi-) homogeneous of degree q then, by the chain rule and the Euler relation,

$$t_i \frac{dg}{dt_i} (x(t_i)) = \sum_{\nu=1}^{n} \frac{\partial g}{\partial x_\nu} (x(t_i)) w_\nu x_\nu (t_i) = qg(x(t_i)).$$

Hence $\phi(dg) = qg$ which proves $\phi(d\mathcal{O}) = \mathfrak{m}$. Since M is the $\mathcal{O}$-ideal generated by $\phi(d\mathcal{O})$, $M = \mathfrak{m}$.

Remark: Kunz-Ruppert prove in [K-R], Satz 2.1, that $j\Omega \cong \mathfrak{m} \Longleftrightarrow \mathcal{O}$ quasihomogeneous.

2.3. $\phi : \bar{\Omega} \otimes K \to K$ induces an embedding of ω (resp. $d\mathcal{O}$) into K. We call the image again ω (resp. $d\mathcal{O}$). Then

$$t^{-1}\omega = \{g \in K \mid \text{res}\,(fg) = 0 \quad \forall\, f \in \mathcal{O}\}.$$

For what follows it is convenient to choose a different embedding of ω into K.

Lemma-notation: There is an $\bar{\mathcal{O}}$ generator f of $\mathcal{C}$ such that, if we define

$$\tilde{\omega} := ft^{-1}\omega,$$

the following holds:

$$\mathcal{C} = \tilde{\omega} : \bar{\mathcal{O}}, \qquad \mathcal{O} \subset \tilde{\omega} \subset \bar{\mathcal{O}}.$$

In particular, $\tilde{\omega} \cong \omega$ is a dualizing module and $\tilde{\omega} = \mathcal{O}$ iff $\mathcal{O}$ is Gorenstein.

Proof: It is obvious that

$$t^{-1}\omega : \bar{\mathcal{O}} = \{g \in K \mid \text{res}\,(hg) = 0 \;\forall\, h \in \bar{\mathcal{O}}\} = \bar{\mathcal{O}}.$$

(Note that this is just the finite duality theorem $\bar{\mathfrak{m}} = \bar{\Omega} = \text{Hom}_{\mathcal{O}}(\bar{\mathcal{O}}, \omega)$ for $(\bar{C}, 0) \to (C, 0)$ of Hartshorne ([H])). Hence $gt^{-1}\omega : \bar{\mathcal{O}} = \mathcal{C}$ for any $\bar{\mathcal{O}}$-generator g of $\mathcal{C}$. Furthermore

$$\omega : (\omega\bar{\mathcal{O}}) = (\omega : \omega) : \bar{\mathcal{O}} = \mathcal{O} : \bar{\mathcal{O}} = \mathcal{C} = gt^{-1}\omega : \bar{\mathcal{O}} = \omega : g^{-1}t\bar{\mathcal{O}},$$

whence $gt^{-1}\omega\bar{\mathcal{O}} = \bar{\mathcal{O}}$ and $gt^{-1}\omega$ must contain a unit u of $\bar{\mathcal{O}}$. Set $f = u^{-1}g$, then $1 \in t^{-1}f\omega = \tilde{\omega}$ and $\mathcal{O} \subset \tilde{\omega}$.

2.4. Consider the inclusions

(α) $\mathcal{C} \subset \mathfrak{m} \subset \mathcal{O} \subset \tilde{\omega} \subset \bar{\mathcal{O}} \subset t^{-1}\omega \subset f^{-1}\bar{\mathcal{O}}$,

(β) $\mathcal{C} \subset \mathfrak{m} \subset \bar{\mathfrak{m}} \subset \omega \subset f^{-1}\bar{\mathfrak{m}}$,

(γ) $\mathcal{C} \subset d\mathcal{O} \subset M \subset \tilde{\omega}M \subset \bar{\mathcal{O}}M \subset \bar{\mathfrak{m}} \subset \bar{\mathcal{O}}$,

(δ) $\mathcal{C} \subset \tilde{\omega} : \bar{\mathfrak{m}} \subset M^{-1} \subset \tilde{\omega} : M \subset \bar{\mathcal{O}} \subset \bar{\mathcal{O}} : \bar{\mathfrak{m}}$.

These inclusions are easy to verify. E.g. $\mathcal{C} = t\tilde{\omega} : \bar{\mathfrak{m}}$ and the fourth inclusion in (δ) follows from $\tilde{\omega} : (\tilde{\omega} : M) = M \supset \mathcal{C} = \tilde{\omega} : \bar{\mathcal{O}}$. Using the duality between $\omega/\bar{\Omega}$ and $\bar{\mathcal{O}}/\mathcal{O}$ we obtain:

(a) $\dim_{\mathbb{C}} \omega/\bar{\Omega} = \dim_{\mathbb{C}} \omega/\bar{\mathfrak{m}} = \dim_{\mathbb{C}} \bar{\mathcal{O}}/\mathcal{O} = \delta$,

(b) $\dim_{\mathbb{C}} \omega/\mathfrak{m} = \dim_{\mathbb{C}} \omega/\bar{\mathfrak{m}} + \dim_{\mathbb{C}} \bar{\mathfrak{m}}/\mathfrak{m} = 2\delta - r + 1 = \mu$,

(c) $\dim_{\mathbb{C}} \bar{\mathcal{O}}/\tilde{\omega} = \dim_{\mathbb{C}} f^{-1}\bar{\mathfrak{m}}/\omega = \dim_{\mathbb{C}} f^{-1}\bar{\mathfrak{m}}/\bar{\mathfrak{m}} - \dim_{\mathbb{C}} \omega/\bar{\mathfrak{m}} = c - \delta$,

(d) $\dim_{\mathbb{C}} \tilde{\omega}/\mathcal{O} = \dim_{\mathbb{C}} \bar{\mathcal{O}}/\mathcal{O} - \dim_{\mathbb{C}} \bar{\mathcal{O}}/\tilde{\omega} = 2\delta - c$.

We define

$$d_o := \dim_{\mathbb{C}} \bar{\Omega}/j\Omega = \dim_{\mathbb{C}} \bar{\mathfrak{m}}/M,$$

$$d_1 := \dim_{\mathbb{C}} \tilde{\omega}:M/M^{-1} = \dim_{\mathbb{C}} \tilde{\omega}M/M .$$

<u>emma</u>: <u>For any reduced curve singularity</u>,

i) $m_1 = c - d_o + d_1$,

ii) $d_o - d_1 = \dim_{\mathbb{C}} \bar{\mathfrak{m}}/\tilde{\omega}M = m - r + \dim_{\mathbb{C}} \bar{\mathcal{O}}M/\tilde{\omega}M$.

<u>roof</u>: (ii) follows from (γ), noticing that $m = \dim_{\mathbb{C}} \bar{\mathcal{O}}/\bar{\mathcal{O}}M$. Since $_o - d_1 = \dim_{\mathbb{C}} M^{-1}/\omega : \bar{\mathfrak{m}}$, we see from ($\delta$) that $m_1 = \dim_{\mathbb{C}} \bar{\mathcal{O}} : \bar{\mathfrak{m}}/M^{-1} =$ $\text{im}_{\mathbb{C}} \bar{\mathcal{O}} : \bar{\mathfrak{m}}/\tilde{\omega} : \bar{\mathfrak{m}} - d_o + d_1$. The result follows from the equalities $\bar{\mathcal{O}} : \mathfrak{m}/\tilde{\omega} : \bar{\mathfrak{m}} = \bar{\mathcal{O}}/\tilde{\omega} : \bar{\mathcal{O}} = \bar{\mathcal{O}}/\mathcal{C}$.

<u>.5.</u> <u>Theorem</u>: <u>Let</u> (C, O) <u>be a reduced curve singularity, then</u>:

1)
$$
\begin{aligned}
e &= 3\delta - c + d_o - d_1 \\
&= 3\delta - c + m - r + \dim_{\mathbb{C}} \bar{\mathcal{O}}M/\tilde{\omega}M \\
&= \mu + 2\delta - c - d_1 - \dim_{\mathbb{C}}(\Omega/d\mathcal{O} + T\Omega) \\
&= \mu + t - 1 + \dim_{\mathbb{C}} \bar{\mathcal{O}}/\tilde{\omega}M - \dim_{\mathbb{C}} \bar{\mathcal{O}}/\tilde{\omega}\mathfrak{m} \\
&= \mu + t - 1 + \dim_{\mathbb{C}} \mathrm{Hom}_{\mathcal{O}}(\Omega, \mathcal{O})/\mathrm{Hom}_{\mathcal{O}}(\bar{\mathcal{O}}, \mathcal{O}) \\
&\qquad - \dim_{\mathbb{C}} \mathrm{Hom}_{\mathcal{O}}(\mathfrak{m}, \mathcal{O})/\mathrm{Hom}_{\mathcal{O}}(\bar{\mathcal{O}}, \mathcal{O}).
\end{aligned}
$$

2) <u>The following inequalities hold</u> (<u>for singular</u> (C, O)):

$$\delta \leq \delta + t - 1 + m - r \leq 3\delta - c + m - r \leq e \leq \mu + 2\delta - c \leq 3\delta - r < 3\delta.$$

3) <u>If</u> (C, O) <u>is quasihomogeneous, then</u> $e = \mu + t - 1$.

<u>emarks:</u> (1) By the theorem of Deligne, e is the dimension of a smoothing component if (C, O) is smoothable. Notice also that the heorem gives formulas for $\dim \bar{G}/G = 3\delta - e - r$.

(2) 2.5 (3) was first proved by Buchweitz [B] in the special case of onomial curves.

Proof: (1) The first two formulas follow from $e = 3\delta - m_1$ and lemma 2.4. From the exact sequence

$$0 \to \bar{\Omega}/d\mathcal{O} \to \omega/d\mathcal{O} \to \omega/\bar{\Omega} \to 0$$

we obtain $d_o = \dim_{\mathbb{C}} \bar{\Omega}/d\mathcal{O} - \dim_{\mathbb{C}} M/d\mathcal{O} = \mu - \delta - \dim_{\mathbb{C}} M/d\mathcal{O}$. Therefore, using the lemma below,

$$\begin{aligned} e &= \mu + t - 1 + \dim_{\mathbb{C}} \tilde{\omega}\mathfrak{m}/\mathfrak{m} - \dim_{\mathbb{C}} \tilde{\omega}M/d\mathcal{O} \\ &= \mu + t - 1 + \dim_{\mathbb{C}} \mathcal{O}/\tilde{\omega}M - \dim_{\mathbb{C}} \bar{\mathcal{O}}/\tilde{\omega}\mathfrak{m} && (*) \\ &= \mu + t - 1 + \dim_{\mathbb{C}} M^{-1}/\tilde{\omega} : \bar{\mathcal{O}} - \dim_{\mathbb{C}} \mathfrak{m}^{-1}/\tilde{\omega} : \bar{\mathcal{O}} && (**). \end{aligned}$$

(*) follows since $\dim_{\mathbb{C}} \bar{\mathcal{O}}/d\mathcal{O} = \dim_{\mathbb{C}} \bar{\mathcal{O}}/\mathfrak{m}$ (cf. 2.4 (b)) and (**) implies the last equality of the theorem since $\tilde{\omega} : \bar{\mathcal{O}} = \mathcal{C} = \bar{\mathcal{O}} : \mathcal{O}$ by lemma 2.3.

(2) follows immediately from (1), and (3) from $M = \mathfrak{m}$ (lemma 2.2.).

Lemma: $2\delta - c = t - 1 + \dim_{\mathbb{C}} \tilde{\omega}\mathfrak{m}/\mathfrak{m}$.

Proof: $t = \dim_{\mathbb{C}} \tilde{\omega}/\mathfrak{m}\tilde{\omega} = \dim_{\mathbb{C}} \tilde{\omega}/\mathfrak{m} - \dim_{\mathbb{C}} \tilde{\omega}\mathfrak{m}/\mathfrak{m} = 2\delta - c + 1 -$

$\dim_{\mathbb{C}} \tilde{\omega}\mathfrak{m}/\mathfrak{m}$ (2.4 (d)) (cf. [H-K], Satz 3.6).

Corollary: Let (C, O) be a reduced, smoothable curve singularity.

(1) If (C, O) is quasihomogeneous, then

$$\tau \geq \mu + t - 1$$

and equality holds iff (C, O) is unobstructed.

(2) If (C, O) is Gorenstein, then $e \leq \mu$. If (C, O) is irreducible and Gorenstein, then $e = \mu$ iff (C, O) is quasihomogeneous.

roof: (1) and the first statement of (2) are obvious. Consider (M_C, O), the monomial curve of (C, O), which is also Gorenstein. ence

$$e(M_C, O) = \mu(M_C, O) = \mu(C, O) \geqslant e(C, O).$$

(M_C, O) can be smoothed over that irreducible component (E_{M_C}, O) ver which (C, O) sits (openness of versality). (C, O) is quasi-omogeneous iff it is isomorphic to (M_C, O). If (C, O) is not uasihomogeneous it can be specialized to (M_C, O) along a $\mathbb{C}$-orbit in M_C. But this implies $e(C, O) < \dim (E_{M_C}, O) = \mu(C, O)$ by the principle of economy" (cf. [T_2] for complete intersections and Po] for the general case).

.6. Remarks: (1) Cor. 2.5 (1) gives a useful criterion to decide hether smoothable curves are obstructed.

2) Cor. 2.5 (2) can be generalized: Assume (C, O) irreducible nd $t(M_C, O) \leqslant 2$. Then $e \leqslant \mu + t - 1$ and $e = \mu + t - 1$ iff $C, O)$ is quasihomogeneous (since in this case $t(C, O) = t(M_C, O)$).

3) Again in Cor 2.5 (2) suppose that (C, O) is irreducible and oreover unobstructed. Then $\tau \leqslant \mu$ and $\tau = \mu$ iff (C, O) is uasihomogeneous. This statement can be considered as a partial eneralization of a theorem of Saito [Sa]. Since $\tau = \dim_{\mathbb{C}} T\Omega$ by ocal duality, this can also be considered as a generalization of a heorem of Zariski [Z] about plane curves. For a different proof see aldi [W].

4) The equality $\tau = \mu$ holds for quasihomogeneous complete inter-ections of any dimension > 0 (cf. [G_1]). Related results for surfaces ere proved by Wahl in [Wa].

roblems: (1) Does the inequality $e \leqslant \mu + t - 1$ hold for all or at east for all smoothable curves?

2) Is the assumption "irreducible" in Cor 2.5 (2) necessary?

Example: The following example shows that the assumption "Gorenstein" in Cor 2.5 (2) is necessary.

$\mathcal{O} = \mathbb{C}\{t^5, t^6, t^7 + t^8\}, \mu = 2\delta = 12,\ c = 10,\ m = 5.$ Since $\mathfrak{m}^{-1} = \mathbb{C} + t^5\mathbb{C}\{t\} \neq t^5\mathbb{C}\{t\} = M^{-1}$, $\mathcal{O}$ is not quasihomogeneous and $t = \dim_{\mathbb{C}} \mathfrak{m}^{-1}/\mathcal{O} = 2$. $\tilde{\omega} = \langle 1, t-t^2, t^5, t^6, t^7, t^8, t^{10}\mathbb{C}\{t\}\rangle_{\mathbb{C}}$,

$\bar{\mathcal{O}}M = \tilde{\omega}M$, $j\Omega = d\mathcal{O}$, $d_o = 6$, $d_1 = 2$ whence $e = 12$. Since $\mathcal{O}$ defines a curve in $\mathbb{C}^3$ it is not obstructed and smoothable. We obtain $e = \tau = \mu = 12$.

Note that for the corresponding monomial curve $\mathcal{O} = \mathbb{C}\{t^5, t^6, t^7\}$, $\mu = 12$, $t = 2$, $e = \mu + t - 1 = 13$.

We computed $\tilde{\omega}$ with the help of the following useful lemma.

Lemma: (Delorme [De]). Let $\mathcal{O} \subset \mathbb{C}\{t\}$ be the local ring of an irreducible curve with semigroup $\Gamma = v(\mathcal{O})$ where v denotes the t-adic valuation. Let $N \subset \mathbb{C}\{t\}$ be any $\mathcal{O}$-module with $\mathcal{O} \subset N$ and $v(N) = \{\nu \in \mathbb{Z} \mid c - 1 - \nu \notin \Gamma\}$. Then $N = \tilde{\omega}$. In particular, $\tilde{\omega} = \langle t^\nu \mathbb{C}\{t\} \mid c - 1 - \nu \notin \Gamma \rangle_{\mathbb{C}}$ for quasihomogeneous $\mathcal{O}$.

Proof: Let f be any generator of $\mathcal{C}$. Then $-1 \notin v(f^{-1}N)$ and $f^{-1}N \subset f^{-1}\tilde{\omega}$ (by 2.3); hence $\mathcal{O} \subset N \subset \tilde{\omega}$. Since $\dim_{\mathbb{C}} N/\mathcal{O} = 2\delta - c = \dim_{\mathbb{C}} \tilde{\omega}/\mathcal{O}$ (2.4 d), the lemma follows.

3. Applications and examples

3.1. The following criterion for non smoothability is a local version of a similar criterion of Mumford ([M]).

Proposition: *Let (C, O) be a reduced curve singularity and let $(X, O) \overset{\sigma}{\leftrightarrows} (T, O)$ be a deformation of (C, O) together with a section σ. Let $f : X \to T$ be a sufficiently small representative of $(X, O) \to (T, O)$ and $C_t = f^{-1}(t)$ for $t \in T$. Assume:*

(i) *$(C_t, \sigma(t))$ is not isomorphic to (C_o, O), C_t is singular in $\sigma(t)$ and $C_t - \sigma(t)$ is non singular.*

(ii) *T is irreducible and $\dim T \geqslant e(C, O)$.*

Then there is an analytic open dense subset $T_o \subset T$, such that $(C_t, \sigma(t))$ is not smoothable if $t \in T_o$.

Proof: We may assume that T is a closed subspace of the semi-universal base S of (C, O). Let $T_o = T - A \cap T$, where A is the union of all smoothing components of (C, O) in S. Either A is empty or $\dim A = e \leqslant \dim T$. T cannot be a component of A since there are no smooth fibers over T. Hence T_o is dense in T and T must lie in some component S' of S over which are no smooth fibers. By openness of versality no fibers over T_o are smoothable.

3.2. We want to apply this proposition to the examples of Mumford and Pinkham in order to extend their results. We consider first the examples of Mumford.

Let $\bar{\mathcal{O}} = \mathbb{C}\{t\}$, $0 < k \leqslant d$ two integers and V a $\mathbb{C}$-vectorspace in $\bar{\mathcal{O}}$ such that $\bar{m}^d \supset V \supset \bar{m}^{2d}$ and $\dim_{\mathbb{C}} V/\bar{m}^{2d} = k$. Let (C_V, O) be the irreducible curve singularity with local ring $\mathcal{O}_V = \mathbb{C} + V$. Of course $\delta_V := \delta(C_V, O) = 2d - k - 1$. This family of curves is parametrized by the Grassmannian $G(k, d)$ of the k-dimensional subspaces $\bar{V} := V/\bar{m}^{2d}$ in $\bar{m}^d/\bar{m}^{2d}$.

Mumford [M] considered the case $d = 2k$ and he showed by globalizing the curves in this family that, for generic V and sufficiently big

k, $(C_V, 0)$ is not smoothable. Using Proposition 3.1 and the formula $e = \mu + t - 1$ we can easily treat the case for general k and d, giving moreover bounds for these numbers.

$G(k, d)$ is stratified into locally closed submanifolds G_ν, $\nu = (\nu_1, \dots, \nu_k)$, $d \leq \nu_1 < \nu_2 < \dots < \nu_k \leq 2d - 1$, corresponding to the possible semigroups

$$\Gamma_\nu = \{0, \nu_1, \nu_2, \dots, \nu_k\} \cup \{\lambda \in \mathbb{N} \mid \lambda \geq 2d\}.$$

It is not difficult to see that $\dim G_\nu = 2kd - \sum_{i=1}^{k} (\nu_i + i)$.
The open stratum of maximal dimension $k(d - k)$ is $G_{max} = G_{(d, d+1, \dots, d+k-1)}$. Unfortunately we cannot apply prop. 3.1 directly to the family $(C_V, 0)$, $\bar{V} \in G(k, d)$, since there might be isomorphic curves in this family. Therefore consider in G_ν the vector space $\bar{V}_\nu$ which corresponds to the monomial curve $(C_\nu, 0)$ with semigroup Γ_ν. Since any isomorphism between $(C_\nu, 0)$ and $(C_V, 0)$ is induced by multiplication of the parameter t with a unit $\in \bar{\mathcal{O}}^*$, a transversal slice T_ν through the orbit of $\bar{\mathcal{O}}^*$ (or of $\bar{\mathcal{O}}/\bar{m}^d \setminus \bar{m}/\bar{m}^d$) in $(C_\nu, 0)$ contains no curves isomorphic to $(C_\nu, 0)$. Since $\mathbb{C}^* + \bar{m}^{2d-\nu_1}$ stabilizes V_ν, we obtain

$$\dim T_\nu \geq \dim G_\nu - (2d - \nu_1) + 1.$$

We see from 3.1. that the generic irreducible curve with semigroup Γ_ν is not smoothable if $e(C_\nu, 0) \leq \dim T_\nu$, i.e. if $t(C_\nu, 0) \leq (2k - 6)(d + 1) - \sum_{i=2}^{k} (\nu_i + i) + 9$. Of course $t(C_\nu, 0)$ can be computed in all cases. We do this (using lemma 2.6) only for

$$\Gamma_\nu = \Gamma_{max} = \{0, d, d+1, \dots, d+k-1, \lambda \geq 2d\}.$$

In this case $t(C_\nu, 0) = d - k$.

We deduce that the generic irreducible curve singularity with semigroup Γ_{max} is not smoothable if $(d - k - 3)(k - 6) \geq 14$. E.g. taking $d = 17$ and $k = 9$, the curve in $(\mathbb{C}^9, 0)$ with parametrization

$$x_0 = t^{17}, \; x_i = t^{17+i} + \sum_{j=1}^{8} a_{ij} t^{25+j} \qquad (i = 1, \dots, 8)$$

is not smoothable for generic $a_{ij} \in \mathbb{C}^{64}$.

3.3. Let (C, O) be the curve singularity consisting of r lines in $(\mathbb{C}^n, O)$. Consider the d-tuple Veronese embedding

$$v_d : P^{n-1} \to P^{N-1}, \qquad N = \binom{n+d-1}{d},$$

which sends $x = (x_1 : \ldots : x_n)$ to all monic monomials in $x_1, \ldots, x_d$ of degree d; $v_d(x) = (v_d(x)_1 : \ldots : v_d(x)_N)$.

Definition: r points $p_1, \ldots, p_r$ in P^{n-1} are said to be in general position, if for all $d \geqslant 1$, $v_d(p_1), \ldots, v_d(p_r)$ span a linear subspace of P^{N-1} of dimension $\min\{r, N\} - 1$. The points are said to be in generic position, if each subset of $\{p_1, \ldots, p_r\}$ is in general position. We say that r lines in $(\mathbb{C}^n, O)$ are in general (resp. generic) position, if this is the case for the corresponding points in P^{n-1}.

Note that $v_d(p_1), \ldots, v_d(p_r)$ span a subspace of P^{N-1} of dimension q iff $p_1, \ldots, p_r$ impose q independent conditions on the space of homogeneous polynomials of degree d. If d_o is defined by $\binom{n+d_o-2}{d_o-1} < r \leqslant \binom{n+d_o-1}{d_o}$ then $p_1, \ldots, p_r$ are in general position iff $v_d(p_1), \ldots, v_d(p_r)$ span a subspace of maximal dimension in P^{N-1} for $d = 1, \ldots, d_o$; therefore we have to check only finitely many conditions. For $r = n, n+1$ "general position" means just that the lines are minimally embedded, while "generic position" means that each n of them are linear independent.

Lemma: Let $(C, O) \subset (\mathbb{C}^n, O)$ be the curve singularity consisting of r lines in general position. Let d be defined by $\binom{n+d-1}{d} < r \leqslant \binom{n+d}{d+1}$. Then

$$\delta(C, O) = (d+1)r - \binom{n+d}{d}.$$

If $r \leqslant n$ then $t(C, O) = r-1$. If $n < r \leqslant \frac{n(n+1)}{2}$ and if the lines

<u>are in generic position, then</u> $t = r - n$.

<u>Proof</u>: Let P_ν denote the space of homogeneous polynomials of degree ν, $\bar{\mathcal{O}}_\nu = \prod_{i=1}^{r} \mathbb{C} \cdot t_i^\nu$ and $\mathcal{O}_\nu = \mathcal{O} \cap \bar{\mathcal{O}}_\nu$. Consider the canonical mappings

$$P_\nu \twoheadrightarrow \mathcal{O}_\nu \hookrightarrow \bar{\mathcal{O}}_\nu .$$

If $L_i = \{t_i p_i \mid t_i \in \mathbb{C}\}$, $p_i \in \mathbb{C}^n - \{0\}$, $i = 1,\ldots,r$, are the r lines, then $\phi_\nu : P_\nu \to \bar{\mathcal{O}}_\nu$ is defined by

$$P_\nu \ni g \longmapsto (g(p_1)t_1^\nu,\ldots,g(p_r)t_r^\nu) \in \bar{\mathcal{O}}_\nu .$$

The monic monomials of degree ν constitute a basis of P_ν and $\bar{\mathcal{O}}_\nu$ has $t_1^\nu,\ldots,t_r^\nu$ as basis. With respect to these bases ϕ_ν is given by the matrix

$$A_\nu = \begin{pmatrix} v_d(p_1) \\ \vdots \\ v_d(p_r) \end{pmatrix} = \begin{pmatrix} v_d(p_1)_1,\ldots,v_d(p_1)_N \\ \vdots \\ v_d(p_r)_1,\ldots,v_d(p_r)_N \end{pmatrix}$$

with r lines and $N = \binom{n+\nu-1}{\nu}$ columns. By the definition of general position, $\operatorname{rk} A_\nu = \min(r, N)$. Hence

$$P_\nu \xrightarrow{\cong} \mathcal{O}_\nu \qquad \text{if } 0 \leqslant \nu \leqslant d,$$

$$\mathcal{O}_\nu \xrightarrow{\cong} \bar{\mathcal{O}}_\nu \qquad \text{if } \nu \geqslant d+1.$$

Therefore $\delta = \sum_{\nu=0}^{d} \dim_{\mathbb{C}} \bar{\mathcal{O}}_\nu/\mathcal{O}_\nu = (d+1)r - \sum_{\nu=0}^{d} \binom{n+\nu-1}{\nu} = (d+1)r - \binom{n+d}{d}$.

In order to compute $t = \dim_{\mathbb{C}} \mathfrak{m}^{-1}/\mathcal{O}$, consider $\mathfrak{m}_\nu^{-1} = \{g \in \bar{\mathcal{O}}_\nu \mid g\mathcal{O}_1 \subset \mathcal{O}_{\nu+1}\}$. Of course

$$t = \sum_{\nu \geqslant 0} \dim_{\mathbb{C}}(\mathfrak{m}_\nu^{-1}/\mathcal{O}_\nu).$$

From what we have shown above, it follows easily that $\mathfrak{m}_\nu^{-1} = \mathcal{O}_\nu$ if $\nu \geqslant d+1$ or $\nu < d-1$ and $\mathfrak{m}_d^{-1} = \bar{\mathcal{O}}_d$. In general it is rather difficult to see what $\mathfrak{m}_{d-1}^{-1}$ is. If $d = 0$, $t = r - 1$ follows immediately. If $d = 1$, we claim that $\mathfrak{m}_o^{-1} = \mathcal{O}_o$. We can describe $\mathfrak{m}_o^{-1}$ explicitly by

$$\mathfrak{m}_o^{-1} = \{(g_1,\dots,g_r) \in \mathbb{C}^r \mid \exists\ C \in M(n \times n, \mathbb{C}) \text{ s.t. } GA_1 = A_1 C\}.$$

Here $G \in M(r \times r, \mathbb{C})$ is the diagonal matrix with diagonal elements $g_1,\dots,g_r$, where $M(p \times p, \mathbb{C})$ is the set of $p \times p$-matrices with complex coefficients. Let $M = \mathbb{C}^n \subset \mathbb{C}^r$ be the subspace spanned by the coordinates $i_1,\dots,i_n$, let $\Pi_M : \mathbb{C}^r \to M$ denote the projection, $G_M : M \to M$ the restriction of G to M and $A = \Pi_M \circ A_1 : \mathbb{C}^n \to M$. By the hypothesis of generic position, A is an isomorphism. Hence $(g_1,\dots,g_r) \in \mathfrak{m}_o^{-1}$ implies that $G_M = A\,C\,A^{-1}$, i.e. $g_{i_1},\dots,g_{i_n}$ are the eigenvalues of C. Varying $M \subset \mathbb{C}^r$ we see that $g_1 = \dots = g_r$ whence $\mathcal{O}_o = \mathfrak{m}_o^{-1}$.

<u>3.4.</u> For fixed r and n we consider the family of reduced curves consisting of r lines in $\mathbb{C}^n$ through 0 which is parametrized by r distinct points in P^{n-1}. If the lines are in general position then δ is constant and we can consider these lines as fibers of a flat holomorphic mapping over a smooth $r(n-1)$-dimensional base. In order to apply the criterion 3.1 for non smoothability, we have to take a small transversal slice T through the orbit of the projective linear group $PGL(n-1, \mathbb{C})$ which operates on that base. Now $\dim T = r(n-1) - (n^2 - 1)$ and $e = \mu + t - 1 = 4r - 3n - 2$ if $n < r \leqslant \frac{n(n+1)}{2}$ and $e \leqslant 3\delta - r = (3d+2)r - 3\binom{n+d}{d}$ if $\binom{n+d-1}{d} < r \leqslant \binom{n+d}{d+1}$. From 3.1 we deduce the following extension of a result of Pinkham ([P], Theorem 11.10):

<u>Theorem:</u> <u>Almost all curve singularities</u> $(C, 0) \subset (\mathbb{C}^n, 0)$ <u>consisting of</u> r <u>lines in general position</u>, $n < r \leqslant \frac{n(n+1)}{2}$ (<u>resp</u>. $d \geqslant 2$ <u>and</u> $\binom{n+d-1}{d} < r \leqslant \binom{n+d}{d+1}$), <u>are not smoothable if</u> $(r-n-2)(n-5) \geqslant 7$ (<u>resp</u>. $r(n-3-3d) + 3\binom{n+d}{d} \geqslant n^2 - 1$). <u>E. g.</u> $(C, 0)$ <u>is not</u>

<u>smoothable if</u> r <u>is in the interval of the following table</u>:

n	6	7	8	9	10
r	[15, 21]	[13, 30]	[13, 72]	[13, 193]	[14, 419]

<u>Remarks:</u> 1) Pinkham proves this theorem for the range $n < r \leqslant 2n - 1$ using his method of negative grading. Strange to say that he gets the same bounds for non smoothability in this range.

2) It is known that n and $n + 1$ lines in $(\mathbb{C}^n, 0)$ are always smoothable (cf. [P], 11.13) as well as arbitrary many lines in $(\mathbb{C}^2, 0)$ or $(\mathbb{C}^3, 0)$. Eisenbud has shown (unpublished) that $n + 2$ lines in $(\mathbb{C}^n, 0)$ are smoothable; but $n + 3$ lines are in general not smoothable if $n \geqslant 12$ after the theorem. Note that for fixed n, the theorem shows the existence of non smoothable curves of lines only for r within some finite interval (which is growing with n). Also we obtain nothing for $n = 4, 5$.

<u>Problem:</u> Do there exist for $n \geqslant 4$ non smoothable curves of r lines if r tends to infinity? Do there exist smoothable ones?

Any smoothing of the singularity at the origin can be extended to a smoothing of the global curve consisting of r lines in $P^n(\mathbb{C})$ through 0. Such a smoothed curve would have genus δ which we computed above. But notice that such a smoothing may not be possible in the same P^n, in general the embedding of the smoothed curve can be much bigger.

<u>3.5.</u> The formula of Deligne cannot only be used to show that certain curve singularities are not smoothable but also that the semi-universal base for a smoothable curve is not smooth, namely if $e < \tau$. We want to apply this to n and $n + 1$ lines in $(\mathbb{C}^n, 0)$.

Using standard techniques for the computation of T^1 (cf. [R - V]) and the results of 3.1. we deduce easily:

Lemma: Let $(C, O) \subset (\mathbb{C}^n, O)$ be the curve singularity consisting of r lines in general position. If $d \geq 0$ and $r \leq \binom{n+d}{d+1}$ then $T^1_\nu = 0$ for all $\nu \geq d$.

Remark: Since (C, O) is homogeneous, $T^1 = T^1_{(C,O)}$ carries a natural grading (cf. [P]) with ν-th graded piece equal to T^1_ν. Notice that (C, O) is negatively graded if $r \leq \frac{n(n+1)}{2}$, which was proved by Pinkham if $r \leq 2n - 1$ ([P], Theorem 11.1, cf. also [R - V], Theorem 4.9.). After [P], Prop. 11.2 this implies that each curve singularity with r smooth branches, $r \leq \frac{n(n+1)}{2}$, and r different tangents in general position is isomorphic to its tangent cone.

Proposition: Let (C, O) be n linear independent lines in $(\mathbb{C}^n, O)$ and let S be the semi-universal base of (C, O). Then:

(1) S is pure $(2n - 3)$-dimensional.

(2) $n = 2 \Longrightarrow T^1 = T^1_{-2}$, $\dim T^1 = 1$

$n \geq 3 \Longrightarrow T^1 = T^1_{-1}$, $\dim T^1 = n(n-2)$.

In particular is (C, O) obstructed iff $n \geq 4$.

Proof: All deformations of (C, O) were described in [B - G], Prop. 7.2.6. These are all ordinary multiple points, hence smoothable So (1) follows from $\dim S = \mu + t - 1$ (Th. 2.5) and lemma 3.3. Using that the ideal of (C, O) in $(\mathbb{C}^n, O)$ is generated by $x_i x_j$, $1 \leq i < j \leq n$, if x_i are the coordinates of $\mathbb{C}^n$, it is not difficult to obtain (2).

3.6. In order to do the same for $n + 1$ lines in $(\mathbb{C}^n, O)$ we first have to classify all possible deformations and then to compute the graded pieces of T^1. This was done in some detail in [G_2]. We state here only the result.

Let A_k denote the plane curve singularity with equation $x^2 + y^{k+1} = 0$ and L^n_r the curve singularity consisting of r lines in $(\mathbb{C}^n, 0)$ in general position. $A_k \times L^n_r$ denotes a curve singularity which is isomorphic to the union of $A_k \subset (\mathbb{C}^2 \times \{0\}, 0)$ and $L^n_r \subset (\{0\} \times \mathbb{C}^n, 0)$ in $(\mathbb{C}^2 \times \mathbb{C}^n, 0)$.

Proposition: Let $(C, 0) \subset (\mathbb{C}^n, 0)$ be $n + 1$ linear independent lines (i.e. of type L^n_{n+1}) and $F : X \to S$ a small representative of the semiuniversal deformation of $(C, 0)$. Then:

(1) All, except at most one, singular points of $X_t = F^{-1}(t)$, $t \in S$, are of type L^p_p. If $x \in X_t$ is singular and not of type L^p_p then it is either of type L^p_{p+1} or of type $A_2 \times L^p_p$ or of type $A_3 \times L^p_p$ (for some p).

(2) All deformations of $(C, 0)$ are smoothable and S is pure dimensional. If the $n + 1$ lines have the property that there are s subsets of n linear independent lines then $t = n + 2 - s$ and therefore $\dim S = 2n + 3 - s$.

(3) If all subsets of n lines are linear independent then the following holds:

$$n = 2 \Longrightarrow \quad T^1 = T^1_{-1} \oplus T^1_{-2} \oplus T^1_{-3}, \ \dim T^1_{-1} = \dim T^1_{-3} = 1, \ \dim T^1_{-2} = 2$$

$$n = 3 \Longrightarrow \quad T^1 = T^1_{-1} \oplus T^1_{-2}, \ \dim T^1_{-1} = 3, \quad \dim T^1_{-2} = 2,$$

$$n \geqslant 4 \Longrightarrow \quad T^1 = T^1_{-1}, \quad \dim T^1 = \frac{n(n-1)}{2} .$$

In particular is $(C, 0)$ obstructed iff $n \geqslant 5$.

References:

[B-F-M] : Baum, P., Fulton, W., MacPherson, R.: Riemann-Roch for Singular Varieties, IHES, Publ. Math., 45, 101 - 145 (1975)

[B-G-G] : Briancon, J., Galligo, A., Granger, M.: Déformations équisingulières des germes de courbes gauches réduites, preprint of the Dep. of Math., Univ. of Nice, France.

[B] : Buchweitz, R.: On Deformation of Monomial curves, in: Seminaire Demazure-Pinkham-Teissier 1976/77, SLN 777, Springer Verlag (1980).

[B-G] : Buchweitz, R.-O., Greuel, G.-M.: The Milnor Number and Deformations of Complex Curve Singularities, Inv. Math. 58, 241 - 281 (1980).

[D] : Deligne, P.: Intersections sur les surfaces reguliêrs, SGA 7 II, 1 - 38, SLN 340, Springer Verlag (1973).

[De] : Delorme, C.: Quelques remarques sur les courbes monomiales, preprint.

[Gr] : Grauert, H.: Über die Deformationen isolierter Singularitäten analytischer Mengen, Inv. Math. 15, 171 - 198 (1972).

[G_1] : Greuel, G.-M.: Dualität in der lokalen Kohomologie isolierter Singularitäten, Math. Ann. 250, 157 - 173 (1980).

[G_2] : Greuel, G.-M.: Deformationen spezieller Kurvensingularitäten und eine Formel von Deligne, Teil II der Habilitationsschrift, Bonn 1979.

[H] : Hartshorne, R.: Residues and Duality, SLN 20, Springer Verlag (1966)

[H-K] : Herzog, J., Kunz, E.: Der kanonische Modul eines Cohen-Macaulay Rings, SLN 238, Springer Verlag (1971)

[K-R] : Kunz, E., Ruppert, W.: Quasihomogene Singularitäten algebraischer Kurven, manuscripta math. 22, 47 - 61 (1977).

[MP] : MacPherson, R.: Chern Classes of Singular Varieties, Ann. of Math. 100, 423 - 432 (1974)

[M] : Mumford, D.: Pathologies IV, Am. J. of Math. 97, 847 - 849 (1975).

[P] : Pinkham, H.: Deformation of Algebraic Varieties with G_m-Action, Asterisque 20 (1974)

[Po] : Pourcin, G.: Déformation de singularité isolées, Asterisque 16, 161 - 173 (1974)

[R-V] : Rim, D.S., Vitulli, M.: Weierstrass Points and Monomial Curves, Journ. of Alg. 48, 454 - 476 (1977).

[Sa] : Saito, K.: Quasihomogene isolierte Singularitäten von Hyperflächen, Inv. Math. 14, 123 - 142 (1971).

[Se] : Serre, J.-P.: Groupes algebriques et corps des classes, Hermann (1959).

[T_1] : Teissier, B.: Résolution simultanée I, II, in: Séminaire Demazure-Pinkham-Teissier 1976/77, SLN 777, Springer Verlag (1980)

[T_2] : Teissier, B.: The Hunting of Invariants in the Geometry of Discriminants, in P. Holme (edit.), Real and Complex Singularities, Oslo 1976, Northholland (1978).

[Tj] : Tjurina, G.N.: Locally Flat Deformations of Isolated Singularities of Complex Spaces, Math. of the USSR-Izvestia 3, 967 - 999 (1969).

[Wa] : Wahl, J.: Smoothings of Normal Surface Singularities, preprint.

[W] : Waldi, R.: Deformation von Gorenstein-Singularitäten der Kodimension 3, Math. Ann. 242, 201 - 208 (1979)

[Z] : Zariski, O.: Characterization of Plane Algebroid Curves whose Module of Differentials has Maximum Torsion, Proc. Nat. Acad. Sci. 56, 781 - 786 (1966).

G.-M. Greuel
Math. Institut der Universität
Wegeler Str. 10
D - 53 Bonn.

DROITES EN POSITION GENERALE DANS L'ESPACE PROJECTIF

Robin HARTSHORNE

André HIRSCHOWITZ

§ 0. INTRODUCTION.

On va démontrer le théorème suivant (on travaille sur un corps de base k, algébriquement clos, de caractéristique quelconque).

THEOREME 0.1. : Soit Y une réunion de r droites disjointes en position générale dans l'espace projectif $\mathbb{P}^N$, $N \geq 3$.
Alors pour tout $n \geq 0$, l'application naturelle

$$\rho(n) : H^o(\mathbb{P}^N, \mathcal{O}_{\mathbb{P}^N}(n)) \longrightarrow H^o(Y, \mathcal{O}_Y(n))$$

est de rang maximum.

Ici on dit qu'un morphisme $\rho : V \longrightarrow W$ d'espaces vectoriels est de rang maximum s'il est injectif ou surjectif ou bijectif. La signification de l'expression en position générale est qu'il existe un ouvert de Zariski non-vide U dans l'espace qui paramètre les réunions de r droites dans $\mathbb{P}^N$, disons $U \subseteq G(1,N)^r$, G étant la variété Grassmannienne, tel que pour tout Y correspondant à un point y de U, l'assertion du théorème soit vraie.

En langage géométrique, le théorème dit que des droites en positio générale imposent des conditions indépendantes sur les hypersurfaces d'un degré donné n qui les contiennent.

Un cas spécial du théorème est le suivant :

si

$$\binom{N+n}{N} \leq r(n+1),$$

alors il existe une réunion de r droites dans $\mathbb{P}^N$ qui n'est pas contenue dans une hypersurface de degré n.
En effet, ces nombres sont les dimensions des espaces vectoriels figurant dans le théorème. Cette inégalité entraine donc que $\rho(n)$ est injectif, et par conséquent que son noyau $H^o(\mathbb{P}^N, \mathcal{I}_Y(n))$ est nul. Par suite il n'y a pas d'hypersurface de degré n contenant Y.

Pour des petites valeurs de N, r, n, on retrouve des résultats de géométrie projective classique. Par exemple, dans $\mathbb{P}^3$, trois droites disjointes sont contenues dans une unique surface quadrique. Quatre droites en position générale ne sont pas contenues dans une surface quadrique. Cinq droites en position générales ne sont pas contenues dans une surface cubique. Le premier cas qui semble dépasser les moyens de la géométrie classique dans $\mathbb{P}^3$ est le cas $n=4$, $r=7$. Dans ce cas ci, l'énoncé, dont nous ne connaissons pas de démonstration par des méthodes classiques, dit que sept droites en position générale ne sont pas contenues dans une surface quartique.

Dans notre démonstration, on remarque d'abord que la condition sur un schéma Y de vérifier l'énoncé du théorème est une condition ouverte sur le schéma de Hilbert des sous-schémas fermés de $\mathbb{P}^N$. Donc pour démontrer le théorème, il suffit d'exhiber un schéma Y, correspondant à un point y du schéma de Hilbert qui est dans la fermeture de l'ensemble de points correspondants aux réunions disjointes de droites, et qui vérifie l'énoncé du théorème. Notre preuve, qui se fait par récurrence sur N et n, consiste en un choix convenable de tels schémas Y, qui en général auront des points singuliers et des éléments nilpotents.

Notre motivation originelle pour ce travail était le problème sur les fibrés vectoriels de rang 2 sur $\mathbb{P}^3$ [2,§5] , de l'existence de tels fibrés ayant la "cohomologie naturelle". Il se trouve que le résultat démontré ici n'entraine pas le résultat voulu pour les fibrés vectoriels. Néanmoins, des méthodes analogues ont aussi conduit à la solution de ce problème-là (voir l'article en préparation [3]). Signalons aussi l'article [4] où l'un de nous a résolu un problème analogue pour les courbes rationnelles dans $\mathbb{P}^3$.

§ 1. PREMIERES REDUCTIONS.

Notons d'abord que l'application $\rho(n)$ du théorème est de rang maximum si et seulement si $H^0(\mathcal{I}_Y(n)) = 0$ ou $H^1(\mathcal{I}_Y(n)) = 0$. Donc, quand on fait varier Y dans une famille plate, d'après les théorèmes de semi-continuité de cohomologie, cette condition est une condition ouverte sur la famille des Y.

D'autre part, pour un Y donné, d'après le théorème de Serre, il y a un entier $n_o = n_o(Y)$ tel que pour tout $n \geq n_o$, $H^1(\mathcal{I}_Y(n)) = 0$, donc $\rho(n)$ est surjectif. Pour n et r fixé, la famille des Y possible est une famille bornée, donc en appliquant encore une fois les théorèmes de semi-continuité, on peut choisir l'entier n_o indépendamment de Y. Autrement dit, pour N et r fixé, il existe un entier n_o tel que pour tout $n \geq n_o$ et tout Y réunion de r droites disjointes en $\mathbb{P}^N$, l'application $\rho(n)$ soit surjective. De ce fait, il n'y a qu'un nombre fini de valeurs de n pour lesquelles il faut choisir Y avec soin, et la propriété que, pour tout n, $\rho(n)$ est de rang maximum, est une propriété ouverte sur la famille des Y. Donc pour démontrer le théorème, il suffit de trouver, pour tout N, r, n, un schéma Y avec $\rho(n)$ de rang maximum.

Supposons maintenant qu'on ait choisis N, n, et r tels que

$$\binom{N+n}{N} = r(n+1),$$

et qu'on ait trouvé Y tel que $\rho(n)$ soit bijectif. Alors pour tout $r' > r$, en ajoutant d'autres droites à Y, on obtient un schéma Y' avec $\rho(n)$ injectif. D'autre part, pour $r'' < r$, en retirant des droites de Y, on obtient un schéma Y" avec $\rho(n)$ surjectif. Donc pour démontrer le théorème pour N et n donné et pour tout r, il suffit de le démontrer pour l'unique entier r qui donne l'égalité ci-dessus.

Malheureusement, cet entier n'existe pas toujours. C'est pourquoi nous allons considérer des schémas Y qui sont réunions de droites et d'un ensemble de points alignés, pour obtenir une égalité analogue. En particulier, on va démontrer le théorème suivant.

<u>THEOREME 1.1.</u> : Pour chaque $N \geq 3$ et $n \geq 0$, soit

$$r = \left[\frac{1}{n+1} \binom{N+n}{N}\right]$$

et

$$q = (n+1)\left(\frac{1}{n+1}\binom{N+n}{N} - r\right)$$

où [ ] dénote la partie entière d'un nombre rationnel.
Alors il existe un schéma $Y \subseteq \mathbb{P}^N$ qui consiste en une réunion disjointe de r droites et de q points alignés sur une droite qui n'intersecte pas les autres droites, tel que l'application naturelle

$$\rho(n) : H^0(\mathbb{P}^N, \mathcal{O}_{\mathbb{P}^N}(n)) \longrightarrow H^0(Y, \mathcal{O}_Y(n))$$

soit bijective.

D'après la discussion ci-dessus, ce théorème entraine le théorème (0.1). En effet, pour $r' > r$, on ajoute des droites à Y, la première étant la droite qui contient les q points, et on obtient une réunion de droites Y' avec $\rho(n)$ injectif. Pour $r'' \leq r$, on retire les q points et, au besoin, des droites, et on obtient une réunion de droites Y" avec $\rho(n)$ surjectif.

La démonstration se fait par récurrence sur N et n, dans les paragraphes suivants.

<u>§ 2. LE CAS N=3.</u>

Notons H_n l'assertion du théorème (1.1) pour N=3 et un $n \geq 0$ donné. En explicitant la définition de r et q on trouve

a) $r = \frac{1}{6}(n+2)(n+3)$, $q=0$, si $n \equiv 0,1 \pmod 3$

b) $r = \frac{1}{6}(n+1)(n+4)$, $q=\frac{1}{3}(n+1)$, si $n \equiv 2 \pmod 3$.

On va démontrer les assertions H_n par récurrence sur n. D'abord nous traitons les cas n=0,1,2,3 à la main.

H_o. r=1, q=0. Y est une droite, et il faut démontrer que $H^o(\mathcal{O}_{\mathbb{P}^3}) \longrightarrow H^o(\mathcal{O}_Y)$ est bijectif, ce qui est évident.

H_1. r=2, q=0. Il faut montrer qu'il existe deux droites dont la réunion n'est pas contenue dans un plan, ce qui est évident.

H_2. r=3, q=1. De façon analogue, il faut montrer qu'il existe une réunion Y de trois droites disjointes et un point qui n'est contenue dans aucune surface quadrique. On sait que trois droites disjointes sont contenues dans une unique surface quadrique. Il suffit donc de prendre le point hors de la quadrique.

H_3. r=5, q=0. Il faut montrer qu'il existe une réunion Y de cinq droites qui n'est pas contenue dans une surface cubique. On considère la surface quadrique non singulière Q qui contient les droites L_1, L_2, L_3. On prend les droites L_4, L_5 en position générale dans $\mathbb{P}^3$, de façon que leur intersection avec Q consiste en quatre points P_1, P_2, P_3, P_4 en position générale sur Q. Alors si une surface cubique F contient Y, ou bien F est la réunion de Q avec un plan H qui contient L_4 et L_5 (ce qui est impossible - voir H_1), ou bien l'intersection de F et Q est une courbe C sur Q, de type (3,3), qui contient L_1, L_2, L_3 et les quatre points P_1, P_2, P_3, P_4. Une telle courbe C doit être la réunion de L_1, L_2, L_3 avec trois droites L'_1, L'_2, L'_3 de l'autre famille de droites sur Q, donc C ne peut pas contenir les quatre points P_i en position générale. Donc Y n'est pas contenue dans une surface cubique.

La démonstration dans le cas général est calquée sur celle du cas n=3. On fixe une surface quadrique non singulière Q, et on prend pour Y une réunion de droites dont une partie est située sur Q. Alors on montre d'abord qu'une surface F de degré n contenant Y contient aussi Q. Donc F est réunion de Q et d'une surface F' de degré n-2 qui contient la partie de Y non située sur Q. Ici on peut appliquer l'hypothèse de récurrence H_{n-2}.

Cette stratégie marche facilement pour $n \equiv 0,2 \pmod 3$. Mais pour $n \equiv 1 \pmod 3$ on ne peut pas ajuster les conditions convenablement sans considérer certaines spécialisations de Y qui sont des schémas avec éléments nilpotents.

On va commencer avec la partie simple de la démonstration.

PROPOSITION 2.1. : Si $n \equiv 0$ ou $2 \pmod 3$ et $n \geq 3$, alors H_{n-2} entraine H_n.

DEMONSTRATION. Prenons d'abord $n=3k$ avec $k \geq 1$. Pour démontrer H_n, il faut trouver une réunion de $r = \frac{1}{2}(k+1)(3k+2)$ droites Y qui n'est pas contenue dans une surface F de degré n. Fixons une surface quadrique non singulière Q. On prend Y = Y' U Y" où Y' est la réunion de 2k+1 droites dans une des deux familles de droites sur Q, et Y" est la réunion de $\frac{1}{2}k(3k+1)$ droites qui intersectent Q transversalement.

Supposons alors qu'une surface F de degré n=3k contient Y. Si F ne contient pas Q, alors l'intersection $F \cap Q$ est une courbe C de type (3k,3k) sur Q, qui contient Y' et les k(3k+1) points de $Y'' \cap Q$. Donc C est la réunion de Y' et d'une courbe C' de type (k-1,3k) qui contient les k(3k+1) points. Notons maintenant que

$$\dim H^{0}(\mathcal{O}_Q(k-1,3k)) = k(3k+1).$$

Ceci se démontre en utilisant l'isomorphisme $Q \cong \mathbb{P}^1 \times \mathbb{P}^1$ et le fait que $\mathcal{O}_Q(a,b) \cong p_1^* \mathcal{O}_{\mathbb{P}^1}(a) \otimes p_2^* \mathcal{O}_{\mathbb{P}^1}(b)$. Donc si les k(3k+1) points $Y'' \cap Q$ sont en position générale, il n'y a pas de telle courbe C'.

On conclut que F contient Q, donc F est réunion de Q et d'une surface F' de degré n-2 qui contient Y". Mais d'après H_{n-2}, il existe une telle réunion Y" de droites qui n'est pas contenue dans une surface de degré n-2. Donc la même chose est vraie pour Y" réunion de droites en position générale, ce qui a pour conséquence que l'ensemble de points $Y'' \cap Q$ est en position générale sur Q, ce qu'on a supposé plus haut.

On conclut finalement que Y n'est pas contenue dans une surface F de degré n, ce qui démontre H_n dans le cas $n \equiv 0 \pmod 3$.

Supposons maintenant que n=3k+2 avec $k \geq 1$. Il faut trouver Y réunion de $r = \frac{1}{2}(k+1)(3k+6)$ droites et $q = k+1$ points alignés tel que Y ne soit pas contenue dans une surface F de degré n=3k+2. De façon analogue, on prend $Y = Y' \cup Y''$ où Y' est une réunion de 2k+2 droites dans une famille sur Q et k+1 points sur Q, alignés sur une droite de la même famille. On prend Y" une réunion de $\frac{1}{2}(k+1)(3k+2)$ droites en position générale, non sur Q.

Si F est une surface de degré n=3k+2 qui contient Y mais ne contient pas Q, alors $F \cap Q$ est une courbe C de type (3k+2,3k+2) sur Q qui contient les 2k+2 droites et les k+1 points de Y' plus les (k+1)(3k+2) points en position générale de $Y'' \cap Q$. Donc C est réunion des 2k+2 droites et d'une courbe C' de type (k,3k+2) qui contient les k+1+(k+1)(3k+2) points. Or les conditions imposées aux k+1 points de Y' n'empèchent pas de répartir ces (k+1)(3k+3) points sur k+1 droites de la famille considérée plus haut, à raison de 3k+3 par droite. Dans ce cas, la courbe C' contient ces k+1 droites, ce qui est incompatible avec son bidegré.

On conclut que F contient Q, donc il existe une surface F' de degré n-2=3k qui contient Y", ce qui contredit H_{n-2}. Donc F n'existe pas, et la proposition est démontrée.

Il nous reste à démontrer le cas $n \equiv 1 \pmod 3$. Pour ceci nous allons utiliser certains schémas avec éléments nilpotents et il serait dangereux de continuer à employer le langage géométrique utilisé jusqu'ici. Nous allons donc passer au langage des schémas.

DEFINITION. Soient H et Y deux sous-schémas de $\mathbb{P}^N$. On définit l'intersection schématique $H \cap Y$ de H et Y comme le schéma qui correspond au faisceau d'idéaux $\mathcal{I}_H + \mathcal{I}_Y$. Si H est une hypersurface, définie localement par une équation f, on définit le schéma résiduel $Z = \mathrm{res}_H Y$ par le faisceau d'idéaux

$$\mathcal{I}_Z = f^{-1}\ker(\mathcal{I}_{Y,\mathbb{P}^N} \longrightarrow \mathcal{I}_{Y\cap H,H}) ,$$

de façon qu'il y a une suite exacte

$$0 \longrightarrow \mathcal{O}_Z(-d) \xrightarrow{f} \mathcal{O}_Y \longrightarrow \mathcal{O}_{Y\cap H} \longrightarrow 0$$

où $d = \deg H$.

EXEMPLE 2.1.1. Si on considère dans $\mathbb{P}^3$ une famille plate de sous-schémas dont le schéma général est une réunion de deux droites disjointes, et où les deux droites se rencontrent en un point dans une fibre spéciale, alors le schéma spécial est une conique dégénérée avec éléments nilpotents au point singulier. (Voir [1, III, 9.8.4] pour une situation analogue). On peut vérifier ceci par un calcul simple. Posons par exemple Y_t dans $\mathbb{A}^3$ réunion des deux droites $y = z = 0$ et $x = z-t = 0$. Alors l'idéal de Y_t et

$$(y,z) \cap (x,z-t) = (xy,xz,y(z-t),z(z-t)).$$

Si $t=0$ on obtient l'idéal (xy,xz,yz,z^2) qui représente la réunion des droites $x=0$ et $y=0$ dans le plan $z=0$, plus un point immergé à l'origine.

Notons Y le schéma Y_0 ci-dessus, et considérons l'intersection schématique de Y avec un plan $H : y=x$ qui est transverse aux deux droites. On obtient que $H \cap Y$ est le schéma d'idéal $(y-x,x^2,xz,z^2)$.

Ceci est un point triple dans le plan H, c'est à dire un schéma supporté par un point $P \in H$ et dont l'anneau structural est $\mathcal{O}_{P,H}/\mathcal{M}^2_{P,H}$. C'est un schéma de longueur 3.

On voit tout de suite que, dans ce cas, le schéma résiduel $res_H Y$ a pour idéal (z,xy), qui est une conique dégénérée réduite dans le plan $z=0$.

Les mêmes considérations s'appliquent naturellement à l'intersection d'un tel schéma Y avec une surface non singulière transverse aux droites.

PROPOSITION 2.2. Si $n=3k+1$ avec $k \geq 2$. Alors $H_{3k-3} \Rightarrow H'_{3k-1} \Rightarrow H_{3k+1}$, où H'_{3k-1} est l'assertion suivante :

(H'_{3k-1}) : Il existe un schéma $Y \subseteq \mathbb{P}^3$ qui est réunion de $\frac{1}{2}(k-1)(3k-2)$ droites et 2k coniques dégénérées ayant leurs points singuliers sur une surface quadrique non singulière Q et tel que l'application naturelle

$$\rho(3k-1) : H^0(\mathcal{O}_{\mathbb{P}^3}(3k-1)) \longrightarrow H^0(\mathcal{O}_Y(3k-1))$$

soit bijective. (Ici on appelle conique dégénérée une réunion de deux droites distinctes qui se rencontrent en un point).

DEMONSTRATION. $H'_{3k-1} \Rightarrow H_{3k+1}$. Il faut trouver Y une réunion de $\frac{1}{2}(k+1)(3k+4)$ droites tel que $\rho(3k+1)$ soit bijective. Comme la condition est ouverte sur l'ensemble des Y, il suffit de trouver une spécialisation d'une réunion de droites disjointes qui ait la propriété voulue. Fixons une surface quadrique non singulière Q. Nous allons prendre $Y = Y' \cup Y''$, où Y' est la réunion de 2k+1 droites dans une famille sur Q, et Y" est la réunion de $\frac{1}{2}(k-1)(3k-2)$ droites en position générale et de 2k coniques dégénérées avec éléments nilpotents aux points singuliers, qui soient limites de paires de droites disjointes, comme dans l'exemple (2.1.1), et telles que leurs points singuliers soient situés sur Q, en position générale sur Q. Alors Y est spécialisation d'une réunion de $\frac{1}{2}(k+1)(3k+4)$ droites disjointes.

Notons $Y''\cap Q$ l'intersection schématique de Y" avec Q. Comme dans l'exemple (2.1.1), le schéma résiduel $res_Q Y''$ est égal à Y''_{red}. Ecrivons la suite exacte de restriction de $\mathbb{P}^3$ à Q et la suite associée pour le schéma résiduel de l'intersection $Y''\cap Q$. On obtient un diagramme exacte

$$\begin{array}{ccccccccc} 0 & \to & H^o(\mathcal{O}_{\mathbb{P}^3}(3k-1)) & \to & H^o(\mathcal{O}_{\mathbb{P}^3}(3k+1)) & \to & H^o(\mathcal{O}_Q(3k+1)) & \to & 0 \\ & & \downarrow \rho(3k-1) & & \downarrow \rho(3k+1) & & \downarrow \alpha(3k+1) & & \\ 0 & \to & H^o(\mathcal{O}_{Y''_{red}}(3k-1)) & \to & H^o(\mathcal{O}_Y(3k+1)) & \to & H^o(\mathcal{O}_{Y\cap Q}(3k+1)) & \to & 0 \end{array}$$

La flèche verticale à gauche n'est autre que $\rho(3k-1)$. Le schéma Y''_{red} satisfait aux hypothèses de H'_{3k-1}, donc en appliquant H'_{3k-1}, on trouve que $\rho(3k-1)$ est bijectif. D'autre part, chacune des coniques dégénérées avec nilpotents de Y" intersecte Q en un point triple (2.1.1) et deux points simples, tandis que les droites de Y" intersectent Q en deux points chacune. Donc l'intersection schématique $Y\cap Q$ est la réunion de 2k+1 droites dans une famille plus 2k points triples et $3k^2-k+2$ points simples. D'après le lemme (2.3) ci-dessous, l'application naturelle $\alpha(3k+1)$ est bijective. On en conclut que $\rho(3k+1)$ est bijectif, ce qui démontre H_{3k+1}.

$H_{3k-3} \Rightarrow H'_{3k-1}$. Il faut trouver Y une réunion de $\frac{1}{2}(k-1)(3k-2)$ droites et 2k coniques dégénérées avec leurs points singuliers sur Q, tel que $\rho(3k-1)$ est bijectif. Les tels schémas Y forment une famille irréductible, et comme plus haut, il suffit de trouver un schéma spécial de ce type ayant les propriétés voulues. Donc nous allons spécialiser Y de façon qu'une des deux droites de chaque conique dégénérée soit contenue dans Q, ces droites appartenant à la même famille de droites sur Q, et de plus, on va faire rentrer une des droites simples de Y dans Q, toujours dans la même famille de droites sur Q. Ici il faut supposer $\frac{1}{2}(k-1)(3k-2) > 0$, ce qui explique l'hypothèse $k \geq 2$ dans la proposition.

Considérons alors l'intersection $Y \cap Q$ et le schéma résiduel $Y'' = \mathrm{res}_Q Y$. On obtient comme précédemment un diagramme :

$$\begin{array}{ccccccccc}
0 & \rightarrow & H^o(\mathcal{O}_{\mathbb{P}^3}(3k-3)) & \rightarrow & H^o(\mathcal{O}_{\mathbb{P}^3}(3k-1)) & \rightarrow & H^o(\mathcal{O}_Q(3k-1)) & \rightarrow & 0 \\
 & & \downarrow \rho(3k-3) & & \downarrow \rho(3k-1) & & \downarrow \alpha(3k-1) & & \\
0 & \rightarrow & H^o(\mathcal{O}_{Y''}(3k-3)) & \rightarrow & H^o(\mathcal{O}_Y(3k-1)) & \rightarrow & H^o(\mathcal{O}_{Y\cap Q}(3k-1)) & \rightarrow & 0
\end{array}$$

Le schéma résiduel Y" est la réunion des 2k droites issues des coniques dégénérées de Y qui ne sont pas dans Q, plus les $\frac{1}{2}(3k^2-5k)$ droites simples de Y situées hors de Q. Donc Y" est une réunion de $\frac{1}{2}k(3k-1)$ droites sans aucune restriction sur leur position. D'après l'assertion H_{3k-3} l'application $\rho(3k-3)$ est bijective.

L'intersection $Y \cap Q$ est la réunion de 2k+1 droites d'une famille sur Q plus $2k+(3k^2-5k) = 3k^2-3k$ points en position générale sur Q. D'après le lemme (2.3) dans le cas trivial $t=d=q''=0$, $\alpha(3k-1)$ est bijectif. On en conclut que $\rho(3k-1)$ est bijectif, ce qui démontre la proposition.

<u>LEMME 2.3.</u> : Soit Y une réunion de r droites d'un même système sur une surface quadrique non singulière Q avec q points simples, q" points simples sur une droite du même système, d points doubles et t points triples (au sens de 2.1.1), le tout en position générale sur Q.

Soit $n \geq 0$ vérifiant :

1°) $r(n+1) + q + q'' + 2d + 3t = (n+1)^2$

2°) $t+d \leq n+1$

3°) $q'' \leq n+1$

4°) si $r < n$ alors $t \leq \frac{n+1-q''}{2} + (n-r-1)\left[\frac{n+1}{2}\right]$, sinon $t=0$.

Alors l'application naturelle

$$\alpha(n) : H^0(\mathcal{O}_Q(n)) \longrightarrow H^0(\mathcal{O}_Y(n))$$

est bijective.

DEMONSTRATION. Si $r \geq n$, l'énoncé est évident. Supposons donc $r < n$. Il faut montrer que Y n'est pas contenu dans une courbe de type (n,n) sur Q, et il suffit de le montrer pour une spécialisation de Y. Choisissons n-r droites $L_1,\ldots,L_{n-r}$ non contenues dans Y mais dans le même système que les droites de Y. Les conditions 1) 2) 3) et 4) assurent qu'on peut trouver des entiers non négatifs t_i, d_i, q_i vérifiant

$$q_1 \geq q''$$

$$2t_i + d_i + q_i = n+1 \text{ pour } i=1,\ldots,n-r,$$

$$t_1+\ldots+t_{n-r}=t,\ d_1+\ldots+d_{n-r} \leq d,\ q_1+\ldots+q_{n-r} \leq q+q''.$$

On peut donc mettre t_i points triples, d_i points doubles, et q_i points simples de Y sur chaque droite L_i de façon que la direction tangente à chaque point double n'est pas dans L_i. Si C est une courbe de type (n,n) sur Q qui contient Y, alors C contient les r droites de Y, et le nombre d'intersection de C avec chaque droite L_i est au moins n+1, donc C contient aussi les droites $L_1,\ldots,L_{n-r}$. Par suite, C est la réunion de n droites dans chacun des deux systèmes sur Q. Les n droites du second système doivent contenir ensemblistement les t points triples, les $d_1+\ldots+d_{n-r}$ points doubles situés sur les L_i, et schématiquement les $d - (d_1+\ldots+d_{n-r})$ points doubles et les $q+q''-q_1-\ldots-q_{n-r}$ points simples non situés sur les droites L_i. Mais par construction

$$t + d_1+\ldots+d_{n-r}+2(d-d_1-\ldots-d_{n-r}) + q+q'' - q_1 -\ldots- q_{n-r} = n+1$$

et cette inclusion est impossible puisque les points triples et doubles sur les L_i peuvent être pris en position générale. Le lemme est démontré.

REMARQUE : Les hypothèses du lemme 2.3 ne sont certainement pas les lus faibles possibles. Signalons (exemple : n=2, r=q=q"=d=0, t=3) que es conditions 1) 2) 3) du lemme sont insuffisantes. Ici, comme au §3 our $H''_{n,N}$, il semble difficile de formuler un énoncé naturel.

ROPOSITION 2.4. Le théorème (1.1) est vrai dans le cas N=3.

EMONSTRATION. Nous avons déjà vérifié que H_o, H_1, H_2, H_3 sont vrais. 'autre part, dans les propositions (2.1) et (2.2) nous avons vérifié es implications $H_{n-2} \Rightarrow H_n$ pour $n \equiv 0,2 \pmod 3$, $n \geq 3$, et $_{n-4} \Rightarrow H'_{n-2} \Rightarrow H_n$ pour $n \equiv 1 \pmod 3$, $n \geq 7$. Pour compléter la récur-ence il suffit de vérifier H_4. On peut remarquer que la démonstration e $H'_{n-2} \Rightarrow H_n$ dans (2.2) marche aussi pour le cas n=4, c'est à dire $'_2 \Rightarrow H_4$. Malheureusement l'assertion H'_2 est fausse (une réunion de eux coniques dégénérées est toujours contenue dans une surface quadri-ue !), et il faut donc trouver une autre démonstration de H_4.

Pour ceci il faut trouver une réunion Y de sept droites telle que (4) soit bijectif. Fixons(*) une surface quadrique non singulière Q et pécialisons Y de la façon suivante. On prend $Y = \cup L_i$, $i=1,\dots,7$, ù L_1, L_2, L_3 sont trois droites d'une famille sur Q, et L_4, L_5, L_6, $_7$ ne sont pas sur Q. En plus, L_4 et L_5 se rencontrent en point P_1 vec éléments nilpotents sur Q, et L_5 et L_6 se rencontrent en un point $_2$ avec éléments nilpotents sur Q. Enfin L_7 ne rencontre pas les autres roites. En utilisant le même diagramme que dans (2.2), remarquons que 'intersection schématique $Y \cap Q$ est la réunion des droites L_1, L_2, L_3 lus 2 points triples et 4 points simples. Les hypothèses du lemme 2.3) sont alors vérifiées et $\alpha(4)$ est bijectif. D'autre part le sché a résiduel $res_Q Y$ est la réunion de L_4, L_5, L_6, L_7 avec la structure

(*) La construction suivante a été proposée par Bernard ANGENIOL pendant l'exposé oral, à LA RABIDA.

de schéma réduite. Pour ce schéma Z il faut vérifier que $\rho(2)$ est bijectif. Il suffit de montrer que Z n'est pas contenu dans une surface quadrique. De fait, les droites L_4, L_6, L_7 sont contenues dans une unique surface quadrique Q'. Si L_5 était aussi contenue dans Q', alors $L_5 \cap L_7$ serait non vide, ce qui est impossible. On en conclut que $\rho(2)$ est bijectif pour Z et donc du diagramme de (2.2), que $\rho(4)$ est aussi bijectif. Ceci termine la démonstration de (1.1) pour le cas N=3.

§ 3. LE CAS $N \geq 4$.

Il s'agit de démontrer le théorème (1.1) pour $N \geq 4$. Pour un n donné, notons $H_{n,N}$ l'énoncé de (1.1). La validité de $H_{0,N}$ et $H_{1,N}$ pour tout $N \geq 4$ est élémentaire, et laissée au lecteur. Nous allons envisager deux autres énoncés. Pour les formuler nous introduisons les entiers s et p, liés à r et q de façon évidente et définis par les relations

$$\binom{N+n}{N} = (n+1)s-p, \quad 0 \leq p \leq n.$$

$(H'_{n,N})$. On suppose $N \geq 4$ et $n \geq 1$. Il existe un schéma Y dans $\mathbb{P}^N$ qui est réunion de p coniques dégénérées, avec leurs points singuliers dans un hyperplan $H = \mathbb{P}^{N-1}$, et de s-2p droites simples, et tel que l'application $\rho(n)$ correspondante soit bijective.

$(H''_{n,N})$. On suppose $N \geq 3$ et $n \geq 2$. Soient r', q', q", d, t des entiers non négatifs vérifiant

1) $r'(n+1) + q' + q'' + 2d + 3t = \binom{N+n}{N}$

2) $t + d < n$

3) $q'' \leq n$

4) $q' \geq \text{Max}\,(q''-2,\ n-q''-1)$.

Alors il existe un schéma Y dans $\mathbb{P}^N$ qui est réunion de r' droites, q' points simples, q" points alignés, d points doubles et t points triples et tel que l'application $\rho(n)$ correspondante soit bijective. Ici un point double est un schéma de longueur 2 concentré en un point, et un point triple, comme dans (2.1.1), est un schéma dans un plan H de la forme $\mathcal{O}_{P,H}/\mathfrak{m}^2_{P,H}$.

Pour démontrer le théorème (1.1) nous allons établir les implications suivantes :

a) $H'_{n-1,N} + H''_{n,N-1} \Rightarrow H_{n,N}$ pour $N \geq 4$

b) $H'_{n-1,N} + H''_{n,N-1} \Rightarrow H'_{n,N}$ pour $N \geq 4$

c) $H_{n-1,N} + H''_{n,N-1} \Rightarrow H''_{n,N}$ pour $N \geq 4$

d) $H_{n-2,3} + (2.3) \Rightarrow H''_{n,3}$

DEMONSTRATION de a). $H'_{n-1,N} + H''_{n,N-1} \Rightarrow H_{n,N}$ pour $n \geq 2$, $n \geq 4$. Soient r et q comme dans (1.1) pour N et n donné. Il faut trouver Y réunion de r droites et q points alignés telle que $\rho(n)$ soit bijectif. Fixons un hyperplan $H = \mathbb{P}^{N-1}$ dans $\mathbb{P}^N$. Nous allons prendre pour Y une spécialisation d'un tel schéma, sous la forme $Y = Y' \cup Y''$ où Y' est une réunion de r' droites dans H et q points alignés dans H, et Y" est une réunion d'un certain nombre de coniques dégénérées avec éléments nilpotents aux points singuliers, ayant leurs points singuliers dans H, et d'autres droites non contenues dans H, de sorte que Y''_{red} satisfasse aux hypothèses de $H'_{n-1,N}$. On considère l'intersection schématique de Y avec H, et son schéma résiduel, qui est Y''_{red}. En utilisant une suite exacte comme dans la démonstration de (2.2), on peut appliquer $H'_{n-1,N}$ à Y''_{red}. Pour terminer la démonstration, il faut voir que $Y \cap H$ satisfait aux hypothèses de $H''_{n,N-1}$. De fait, $Y \cap H$ est la réunion de r' droites, q"=q points alignés, avec $q'' \leq n$ par définition de q, un nombre t de points triples égal au nombre de coniques dégénérées dans l'énoncé $H'_{n-1,N}$, qui satisfait donc $t < n$, et finalement un nombre q' de points en position générale, où q' est le nombre de droites simples dans $H'_{n-1,N}$.

Ce nombre q' est égal à s(n-1,N) - 2p(n-1,N) donc

$$q' \geq \frac{1}{n} \binom{N+n-1}{N} - 2(n-1).$$

Pour vérifier la condition 4) de $H''_{n,N-1}$, il suffit de montrer que

$$\frac{1}{n}\binom{N+n-1}{N} - 2(n-1) \geq n-1, \text{ autrement dit : } \binom{N+n-1}{N} \geq 3n(n-1).$$

Un calcul élémentaire montre que cette inégalité est vérifiée pour $n=2,3,4$, $N \geq 5$; et pour $n \geq 5$, $N \geq 4$.

Pour $n=2,3,4$, $N=4$ on vérifie directement que $q' \geq n-1$.

<u>DEMONSTRATION de</u> b). $H'_{n-1,N} + H''_{n,N-1} \Rightarrow H'_{n,N}$ pour $n \geq 2$, $n \geq 4$.

L'idée et la même que pour a). On fixe un hyperplan H. Si $H'_{n-1,N}$ recquiert plus de coniques dégénérées que $H'_{n,N}$, on crée de nouvelles coniques dégénérées avec éléments nilpotents comme plus haut. Si $H'_{n-1,N}$ recquiert moins de coniques dégénérées que $H'_{n,N}$, on laisse entrer une des droites de chaque conique supplémentaire dans H. Alors, comme plus haut, on a $Y = Y' \cup Y''$ où $Y' \subseteq H$ est une réunion de droites, et Y" est une réunion de droites et de coniques dégénérées, éventuellement avec éléments nilpotents. Le lemme ci-dessous assure que les nombres de droites intervenant dans la construction précédente ne sont pas négatifs. On applique $H''_{n,N-1}$ au schéma résiduel de l'intersection $Y \cap H$. On applique $H''_{n,N-1}$ à l'intersection $Y \cap H$, qui est une réunion de droites, de points triples correspondant aux nouvelles coniques dégénérées, de points doubles correspondant aux anciennes coniques dégénérées, donc avec $d+t < n$, et finalement q' points correspondant aux droites simples de $H'_{n-1,N}$ qui ne font pas partie de coniques dans Y. On a $q' = s_{n-1} - 2p_{n-1}$ dans la première construction et $q' = s_{n-1} - p_{n-1} - p_n$ dans la seconde. Pour vérifier la condition 4) de $H''_{n,N-1}$, il suffit de montrer que

$$\frac{1}{n}\binom{N+n-1}{N} - (2n-1) \geq n-1$$

autrement dit

$$\binom{N+n-1}{N} \geq n(3n-2)$$

Un calcul élémentaire montre que cette inégalité est vérifiée pour $n=2$, $N \geq 7$; pour $n=3,4$, $N \geq 5$; et pour $n \geq 5$, $N \geq 4$. Pour les autres cas, on vérifie directement que $q' \geq n-1$.

LEMME 3.1. Même lorsque $q_{n-1} < q_n$, on a, pour $n \geq 2$, $N \geq 4$,

$$s_n - p_n \geq s_{n-1} - p_{n-1}$$

DEMONSTRATION. : L'inégalité à montrer s'écrit encore

$$\frac{1}{n+1}\binom{N+n}{N} - \frac{np_n}{n+1} \geq \frac{1}{n}\binom{N+n-1}{N} - \frac{(n-1)p_{n-1}}{n}$$

Les estimations sur p_n et p_{n-1} nous réduisent à prouver

$$n\binom{N+n}{N} - n^3 \geq (n+1)\binom{N+n-1}{N},$$

ou encore

$$n\binom{N+n-1}{N-1} - n^3 \geq \binom{N+n-1}{N}$$

soit

$$\frac{N-1}{N}\binom{N+n-1}{N-1} \geq n^2.$$

Le premier membre de cette inégalité étant croissant en N, il suffit de la vérifier pour N=4, ce qui se fait élémentairement. Le lemme est démontré.

DEMONSTRATION de c). $H_{n-1,N} + H''_{n,N-1} \Rightarrow H''_{n,N}$ pour $n \geq 2$, $N \geq 4$. Soit Y comme dans l'énoncé $H''_{n,N}$. On fixe l'hyperplan H et on garde hors de H certaines des r' droites et certains des q' points de façon à satisfaire les hypothèses de $H_{n-1,N}$. Il résulte des conditions 1) 2) et 3) de $H''_{n,N}$ et de l'inégalité

$$4n-3 + \left[\frac{1}{n} \binom{N+n-1}{N}\right] \leq \binom{N+n-1}{N-1}$$

que r' et q' sont suffisamment grands pour permettre la construction précédente, compte tenu qu'une droite est équivalente à n+1 points. L'inégalité ci-dessus se traite comme dans la démonstration du lemme 3.1. (en fait on a $n^3 \geq 4n-3$ pour $n \neq 2$). De plus la condition 4) de $H''_{n,N}$ permet de créer les alignements exigés par $H_{n-1,N}$. On met dans H les autres composantes de Y. Alors on applique $H_{n-1,N}$ au schéma résiduel $res_H Y$, et on applique $H''_{n,N-1}$ à $Y \cap H$. Pour vérifier que $Y \cap H$ satisfait aux hypothèses de $H''_{n,N-1}$, la seule chose non-évidente est la condition 4). Pour celle-ci, il suffit de montrer que $H_{n-1,N}$ a au moins n-1 droites, c'est à dire que

$$\left[\frac{1}{n} \binom{N+n-1}{N}\right] \geq n-1.$$

Pour ceci il suffit de vérifier

$$\frac{1}{n} \binom{N+n-1}{N} - 1 \geq n-1,$$

autrement dit

$$\binom{N+n-1}{N} \geq n^2.$$

Cette inégalité est vérifiée pour tout $n \geq 2$, $N \geq 4$.

<u>DEMONSTRATION de d).</u> $H_{n-2,3}$ + (2.3) $\Rightarrow$ $H''_{n,3}$ pour $n \geq 2$. Soit Y comme dans l'énoncé $H''_{n,3}$. On fixe une quadrique non singulière Q et on garde hors de Q certaines des r' droites et certains des q' points de façon à satisfaire les hypothèses de $H_{n-2,3}$. Il résulte des conditions 1) 2) et 3) de $H''_{n,3}$ et de l'inégalité élémentaire

$$4n-3 + 2 \left[\frac{1}{n-1} \binom{n+1}{3}\right] \leq (n+1)^2$$

que r' et q' sont suffisamment grands pour permettre la construction précédente, compte tenu qu'une droite est équivalente à n+1 points. De plus, la condition 4) de $H''_{n,3}$ permet de créer les alignements exigés par $H_{n-2,3}$. On fait rentrer dans Q tout le reste. On applique $H_{n-2,3}$ au schéma résiduel $res_Q Y$ et on applique le lemme (2.3) à $Y \cap Q$. La seule difficulté consiste à vérifier l'hypothèse 4) de (2.3). Soient $\bar{r}$, $\bar{q}$, $\bar{q}''$, $\bar{d}$, et $\bar{t}$ les nombres pour lesquels il faut vérifier cette hypothèse. On a $\bar{t} = t$, $\bar{d} = d$, $\bar{q}'' = q''$ et $\bar{q} \geq 2 \left[\frac{1}{n-1} \binom{n+1}{3}\right]$. Si la condition 4) n'est pas vérifiée, la condition 1) donne

$$\bar{r}(n+1)+2\left[\frac{1}{n-1} \binom{n+1}{3}\right] + \bar{q}'' + 3\left(1 + \left[\frac{n+1-\bar{q}''}{2}\right] + (n-\bar{r}-1)\left[\frac{n+1}{2}\right]\right) \leq (n+1)^2.$$

Comme $\bar{q}'' + 3\left[\frac{n+1-\bar{q}''}{2}\right]$ est minoré par n, on obtient

$$\bar{r}(n+1) - \left[\frac{n+1}{2}\right]) \leq (n+1)^2 - (n-1) \left[\frac{n+1}{2}\right] - \left[\frac{n(n+1)}{3}\right] - n - 3.$$

Des considérations élémentaires montrent que pour $n \geq 6$, cette inégalité implique

$$\bar{r} \leq n-4$$

auquel cas 4) résulte de 2). Pour n=2,3,4,5, on trouve respectivement

$$\bar{r}=0,\ \bar{r} \leq 1,\ \bar{r} \leq 2,\ \bar{r} \leq 2,$$

auquel cas 4) résulte de $\bar{t} \leq n-1$ sauf si

$$n=4,\ \bar{r}=2,\ \bar{t}=3,\ \bar{q}''=4, \bar{d}=0,\ \bar{q}=2.$$

Nous allons montrer que la conclusion du lemme 2.3 reste vraie dans ce cas.

Il faut montrer que Y n'est pas contenu dans une courbe de type (4,4) sur Q, et il suffit de le montrer pour une spécialisation de Y. Choisissons trois droites L_1, L_2, L_3 dans un système de Q. On met L_1 et L_2 dans Y, les $\bar{q}''$ points alignés sur L_3 avec un des $\bar{q}$ points. Si C est une courbe de type (4,4) sur Q qui contient Y, alors C contient L_1, L_2, L_3 et les trois droites de l'autre système passant par les trois points triples. Le reste de C doit donc être constitué par une conique passant par les trois points triples et le dernier point, mais une telle conique n'existe pas en général.

FIN DE LA DEMONSTRATION DE (1.1).

Pour terminer, notons que $H'_{1,N}$ est vrai pour tout $N \geq 4$ (facile) et que les assertions $H_{n,3}$ pour $n \geq 0$ ne sont autres que les assertions H_n déjà prouvées au paragraphe 2. Donc la récurrence commence bien, et les assertions a), b), c), d) suffisent pour démontrer le théorème.

REFERENCES.

[1] R. HARTSHORNE, Algebraic Geometry, Graduate texts in mathematics, 52, Springer Verlag (1977).

[2] R. HARTSHORNE, On the classification of algebraic space curves, in Fibrés vectoriels et équations differentielles, (Nice 1979) Progress in Math. 7, Birkhäuser, Boston (1980) 83 - 112.

[3] R. HARTSHORNE, A. HIRSCHOWITZ, Cohomology of a general instanton bundle, manuscrit, Berkeley (1981).

[4] A. HIRSCHOWITZ, Sur la postulation générique des courbes rationnelles. Acta Math. 146 : 3-4 (1981). pp. 209-230.

Robin HARTSHORNE
Department of Mathematics
University of California
BERKELEY CA 94720.

André HIRSCHOWITZ
I.M.S.P.
Université de Nice
06034 - NICE CEDEX

LIMITES D'ESPACES TANGENTS ET TRANSVERSALITE DE VARIETES POLAIRES

J.P.G. HENRY

M. MERLE

§ 1 - INTRODUCTION.

Le premier résultat sur la transversalité de variétés polaires relatives aux plans qui servent à les définir est dissimulé dans [10], Th. 1, p. 269. Il peut s'écrire ainsi : soit f germe d'application analytique

$$f : \mathbb{C}^n,0 \longrightarrow \mathbb{C},0$$

avec $f^{-1}(0)$ à singularité isolée à l'origine, on définit Γ_{z_n} la courbe polaire relative pour l'hyperplan $z_n = 0$, par

$$\Gamma_{z_n} = \text{clôture de } \{x/T_x\ f^{-1}(f(x)) \text{ contenu dans } (z_n = 0)\}$$

$$\Gamma_{z_n} = \{x/\frac{\partial f}{\partial z_1} = \dots = \frac{\partial f}{\partial z_{n-1}} = 0\} \quad .$$

Le résultat de B. Teissier de 1973 est que, si $z_n = 0$ est un hyperplan assez général, alors Γ_{z_n} est transverse à $z_n = 0$. La démonstration utilise des dépendances intégrales d'idéaux et le théorème est employé pour relier la suite μ^* (des $\mu^{(n-i)}$) aux multiplicités de variétés polaires : $\mu^{(n-i)}$, défini comme le nombre de Milnor de l'intersection de $X = f^{-1}(0)$ par i hyperplans génériques, est aussi la multiplicité d'une variété polaire. Dans le cas ci-dessus, on a

$$e(\Gamma_{z_n}, z_n) = e(\frac{\partial f}{\partial z_1}, \dots, \frac{\partial f}{\partial z_{n-1}}, z_n) = \mu^{(n-1)}(f^{-1}(0)) \quad ,$$

la transversalité implique que c'est aussi la multiplicité de la courbe polaire Γ_{z_n}.

Le progrès suivant, en 1978, est encore limité au cas des singularités isolées, mais dans le cas intersection complète et non plus hypersurface. Le lemme clef de [3] (3.1) peut s'exprimer comme suit :

Soit Z donné par $f_1 = f_2 = \dots = f_{p-1} = 0$ germe à l'origine de singularité isolée d'espace analytique complexe, on suppose que $X = f_p^{-1}(0) \cap Z$ est une intersection complète à singularité isolée et on regarde la variété polaire Γ_{z_n} relative à

$$f_p : Z \longrightarrow \mathbb{C}$$

pour l'hyperplan $z_n = 0$. Γ_{z_n} est définie par $f_1 = \dots = f_{p-1} = 0$ et annulation des mineurs $\frac{\partial(f_1,\dots,f_p)}{\partial(z_{i_1},\dots,z_{i_p})}$, $i_1,\dots,i_p \in [1,n-1]$. Alors, si l'hyperplan défini par $z_n = 0$ est assez général, Γ_{z_n} est transverse à $z_n = 0$ et sa multiplicité est $\mu^{(n-1)}(f_1,\dots,f_p) + \mu^{(n-1)}(f_1,\dots,f_{p-1})$.

La proposition 1 de cet article admet comme corollaires des généralisations des résultats cités ci-dessus, la seule restriction portant sur la dimension de la base de l'application servant à définir les variétés polaires, qui doit être un. On retrouve également des résultats sur la transversalité aux variétés polaires absolues, démontrées dans [6].

§ 2 - LIMITES D'ESPACES TANGENTS.

K désignera le corps des réels ou celui des complexes.

Proposition 1 : Soient X un espace sous-analytique de dimension $d \geq 2$ de K^{n+2}, contenant l'origine, et Y une droite contenant ce point ; X^o désigne la partie lisse de X. Nous pouvons supposer qu'au voisinage de l'origine est donné un plongement du couple (X,Y) dans $(K^{n+1} \times K, 0)$ qui identifie Y avec $\{0\} \times K$. Avec ces hypothèses et pour tout hyperplan L transverse à Y et à X à l'origine, il existe un ouvert dense U de la grassmanienne $G(k,n+1)$ des $(k+1)$-plans de K^{n+2} $(k \geq n-d+3)$ contenant Y tel que, pour tout H dans U, on ait la propriété (P) suivante :

(P) Pour toute suite de couples $(x_i, H_i)_{i\in\mathbb{N}}$ avec $x_i \in H_i \cap X^o$, $(x_i)_{i\in\mathbb{N}}$ convergeant vers 0, $(H_i)_{i\in\mathbb{N}}$ dans $G(k,n+1)$, convergeant vers H, la limite de $((T_{x_i} X^o) \cap L)_{i\in\mathbb{N}}$ si elle existe, est coupée transversalement par H.

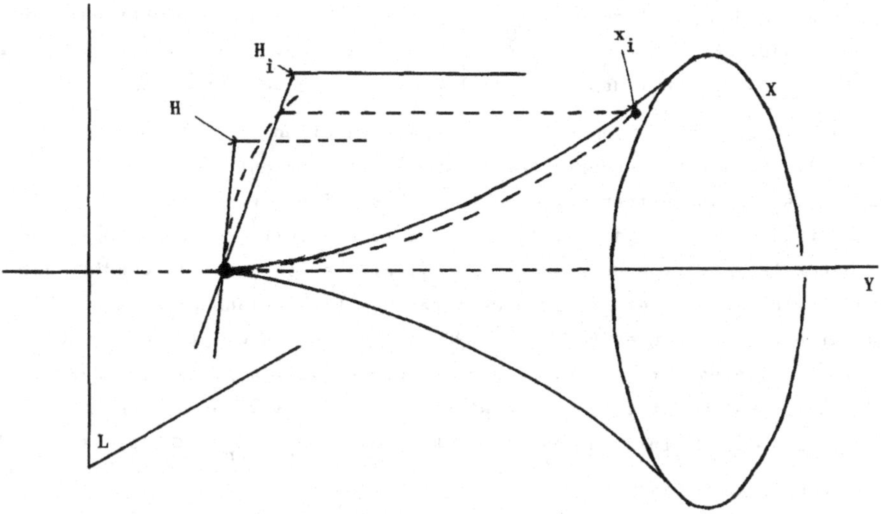

<u>Remarques</u> : (1) L'hypothèse de sous-analyticité est là pour assurer la validité du "lemme des petits chemins".

(2) La proposition porte en fait sur une propriété de la projection de direction Y : la propriété (P) se lit dans L.

(3) L'hypothèse $\dim Y \leq 1$ est essentielle, comme le montre l'exemple du § 4. Il est cependant possible de la remplacer par une hypothèse sur le morphisme de projection de X sur Y induit par la projection de $K^{n+1} \times K \to K$: on impose des conditions du type de [5] th. 1, p. 242. Cette version plus générale sera publiée ultérieurement. Mais aucune hypothèse d'incidence sur (X,Y) n'est nécessaire et on ne suppose pas $Y \subset X$.

(4) L'idée de cette proposition nous est venue en analysant un résultat de V. Navarro ([8], lemme (4.6) et [9] th. 3.12).

Nous remercions J.L. Brylinski pour ses suggestions.

<u>Démonstration</u> : 1/ Montrons que l'ensemble U des (k+1)-plans vérifiant la propriété (P) est un ouvert.

On considère pour cela la sous-variété V de $X^o \times \mathrm{Grass}(k,n+1) \times \mathrm{Grass}(d-1,n+1)$ définie par

$$V = \{(x,H,T)/H \ (k+1)\text{-plan contenant } Y,\ x \in X^o \cap H,\ T = T_x X \cap L\} \quad .$$

Prenons l'adhérence $\overline{V}$ de V dans $X \times \mathrm{Grass}(k,n+1) \times \mathrm{Grass}(d-1,n+1)$ et considérons la fibre $\overline{V}_o$ de $\overline{V}$ au-dessus de 0 pour la projection $\overline{V} \to X$. En intersectant $\overline{V}_o$

avec le fermé Z des couples (H,T) tels que dim $T \cap H \geq d-n+k$, on obtient l'ensemble des couples (x_i, H_i) avec x_i dans $H_i \cap X^o$, x_i convergeant vers 0, H_i vers H et tels que la limite T des $(T_{x_i} X^o) \cap L$ coupe H avec une dimension excédentaire. On voit que le complémentaire U de la projection de $\overline{V}_o \cap Z$ dans Grass(k,n+1) (projection propre) est l'ouvert qui convient.

2/ Nous allons montrer maintenant que U n'est pas vide. On reprend une technique inaugurée dans [11], 2.14, corollaire 2 et utilisée dans [2] lemme, pour montrer le même type de résultat. On stratifie -ici pour la condition (w) de Verdier- la variété des sections de X par des (k+1)-plans passant par Y. La condition (w) de Verdier ([14], p. 296 (1.4)) pour un couple de strates (Z,Λ) exprime que dans tout voisinage de t_o appartenant à Λ, la distance du plan $T_z Z$ à $T_t\Lambda$ est majorée, à une constante multiplicative près, par la distance de z à Λ. On définit un ouvert de (k+1) plans $U_{r_1 \dots r_k, s} \subset \mathrm{Grass}(k+1,n+1)$ associé au morphisme

$$K^{(n+1-k)k} \times K^k \times K \longrightarrow K^{n+1} \times K$$

défini par

$$((a_i^o), \dots, (a_i^{n-k}), z_1, \dots, z_k, u) \longrightarrow (x_o, \dots, x_n, y)$$

$$x_o = \sum_{i=1}^{i=k} a_i^{(0)} z_i^{r_i}$$

$$\dots$$

$$x_{n-k} = \sum_{i=1}^{i=k} a_i^{(n-k)} z_i^{r_i}$$

$$x_{n-k+1} = z_1^{r_1}$$

$$x_n = z_k^{r_k}$$

$$y = u^s \quad ,$$

et $X_{r_1,\dots,r_{k,s}}$ est l'image réciproque de $X \subset K^{n+1} \times K$ par ce morphisme. Il existe alors un ouvert dense de $\Lambda = K^{(n+1-k)\times k} \times \{0\} \times \{0\}$ tel que $X_{r_1,\dots,r_k,s}$ vérifie la condition de Verdier le long de cet ouvert $\Lambda^{\bullet}$. Ceci définit, en associant à chaque $(\underline{a}_i^o), \dots, (\underline{a}_i^{n-k})$ le (k+1)-plan d'équations :

• La condition $k \geq n-d+3$ assure que Λ est adhérente à $X_{r_1,\dots,r_k,s} \setminus \Lambda$.

$$x_o = \sum_{i=1}^{i=k} a_i^{(0)} x_{n-k+i}$$

...

$$x_{n-k} = \sum_{i=1}^{i=k} a_i^{(n-k)} x_{n-k+i}$$

un ouvert dense de (k+1)-plans de $K^{n+1} \times K$ passant par Y ; nous noterons $U_{r_1,\dots,r_k,s}$ cet ouvert. Montrons que tout k+1 plan H, dans l'intersection U de ces ouverts denses, pour $r_1,\dots,r_k,s$ parcourant $\mathbb{N}^{k+1}$, vérifie la propriété (P). On peut supposer, quitte à changer les coordonnées $x_o,\dots,x_{n-k}$, que le (k+1)-plan H est le plan défini par $x_o = \dots = x_{n-k} = 0$.
Prenons un disque d'épreuve (i.e. un morphisme analytique $\gamma : (\mathbb{D},0) \to (X,0)$) tel que la courbe image dans X, ait sa projection sur L tangente à $L \cap H$, ce qui s'écrit en paramétrant la courbe par $x_o = x_o(t),\dots,x_n = x_n(t)$, $y = y(t)$ et en notant v la valuation naturelle de $K\{t\}$

$$\inf(v(x_o),\dots,v(x_{n-k})) > \inf(v(x_{n-k+1}),\dots,v(x_n)) .$$

On notera $v(x_{n-k+i}) = r_i$, $v(y) = s$.
On peut alors remonter ce disque dans $X_{r_1,\dots,r_k,s}$ d'au moins une façon. Si $(\eta_o,\dots,\eta_n,\zeta)$ sont les coordonnées d'une normale à X au point $(x_o,\dots,x_n,y)$ alors

$$(z_1^{r_1}\,\eta_o,\dots,z_k^{r_k}\,\eta_o,\dots,z_1^{r_1}\,\eta_{n-k},\dots,z_k^{r_k}\,\eta_{n-k},\dots,\dots$$

$$\dots,r_i z_i^{r_i-1}(\eta_{n-k+i} + \sum_{j=0}^{j=n-k} a_i^j\,\eta_j),\ s u^{s-1}\zeta)$$

sont les coordonnées d'une normale à $X_{(r_1,\dots,r_k,s)}$ au point correspondant.
La condition de Verdier s'écrit : pour $1 \le i \le k$ et $0 \le j \le n-k$

$$v(z_i^{r_i}\,\eta_j) \ge \inf_{1\le i\le k} (v(z_i),v(u)) + \inf v((z_i^{r_i-1}(\eta_{n-k+i} + \sum_{j=0}^{j=n-k} a_i^j\,\eta_j),\ v(s u^{s-1}\zeta)) .$$

Comme $v(z_i) = v(u) = 1$, on a, pour $1 \le i \le k$, $0 \le j \le n-k$,

$$(*) \qquad v(x_{n-k+i}\,\eta_j) \ge \inf_{1\le\ell\le k} (v(x_{n-k+\ell}(\eta_{n-k+\ell} + \sum_{j=0}^{j=n-k} a_\ell^j\,\eta_j)),v(y\,\zeta)) .$$

Nous allons montrer que dans le terme de droite $v(y\zeta)$ ne peut être un minimum strict. Ceci résulte de ce que le chemin est pris dans X, la tangente au chemin est donc "orthogonale" à la normale $(\eta_o,\eta_1,\ldots,\eta_n,\zeta)$ ce qui s'écrit (en notant $x_i' = (x_i \circ \gamma)'$) :

$$\sum_{i=0}^{i=n} x_i' \eta_i + y'\zeta = 0 \quad .$$

Compte-tenu de

$$x_j = \sum_{i=1}^{i=k} a_i^j x_{n-k+i}$$

pour $0 \leq j \leq n-k$ et des inégalités (*) et en remarquant que $v(x_i') = v(x_i) - 1$, $v(y') = v(y) - 1$, on voit que $v(y\zeta)$ ne peut être minimum strict dans le terme de droite de (*). Il existe alors un indice i tel que

$$v(x_{n-k+i}\,\eta_j) \geq v(x_{n-k+i}(\eta_{n-k+i} + \sum_{j=0}^{j=n-k} a_i^j \eta_j))$$

pour $0 \leq j \leq n-k$, d'où pour $0 \leq j \leq n-k$:

$$v(\eta_j) \geq v(\eta_{n-k+i}) \quad .$$

On a donc vu que pour tout chemin dont la projection sur L est une courbe tangente à H et tout champ de vecteurs normaux à X le long de ce chemin, la limite de leurs projections sur L est un vecteur qui a une composante non nulle dans H. Ceci équivaut à la propriété (P) pour le k-plan H. U convient donc, et, étant intersection dénombrable d'ouverts denses, il est dense dans Grass (k+1,n+1) $(U = \cap\, U_{r_1,\ldots,r_k,s})$.

Remarque : Il est assez facile de voir qu'on peut avoir un énoncé un peu plus général de la proposition en considérant non pas des (k+1)-plans H passant par Y, mais des drapeaux de plans contenant $Y, D_1 \supset D_2 \supset \ldots D_i \supset \ldots D_n \supset Y$ et en demandant que le (n+2-i)-plan D_i vérifie la propriété (P).

§ 3 - TRANSVERSALITE AUX VARIETES POLAIRES.

La proposition précédente va permettre de démontrer des propriétés de transversalité aux variétés polaires des plans qui servent à les définir, aussi bien dans le cas absolu que dans le cas relatif pour reprendre la terminologie de [12]. Pour un cadre général concernant les variétés polaires on renvoie le lecteur à [6] et à [12].

Corollaire 1 (transversalité pour les "polaires absolues" cf. [6] (5.1.2)) : Soit $X \subset \mathbb{C}^{N+1}$ un germe de sous-analytique en 0, de dimension d. Alors pour presque tout drapeau $D_1 \supset D_2 \supset \ldots \supset D_N$, si on note $P_X(D_{k+1})$ la variété polaire associée au N-k - plan D_{k+1}, définie par

$P_X(D_{k+1}) =$ Clôture de $\{x \in X^o$ partie lisse de $X / T_x X^o \cap D_{k+1}$ est de dimension $\geq d-k\}$

alors D_k coupe $P_X(D_{k+1})$ transversalement.

Démonstration : En utilisant la proposition 1, si le drapeau $(D_1, \ldots, D_N)$ est assez général, on peut supposer que D_k et D_{k+1} vérifient la propriété (P), avec $Y = \{0\}$. Si $P_X(D_{k+1})$ ne coupait pas D_k transversalement, il existerait une courbe Γ dans $X^o \cup \{0\}$ tangente à D_k en 0. Désignons par λ la tangente à cette courbe. Montrons d'abord que λ ne peut être contenue dans D_{k+1}. En effet, le long de Γ, $T_x X^o$ coupe D_{k+1} suivant un plan de dimension $\geq d-k$, et la limite de $T_x X^o$ le long de Γ ne coupe pas D_{k+1} transversalement (pour $k \leq d-2$). Si λ était contenue dans D_{k+1}, ceci contredirait la propriété (P) pour D_{k+1}. Supposons maintenant que λ soit dans D_k, alors (λ, D_{k+1}) engendre D_k. Mais $T = \lim_{x \in \Gamma\ x \to 0} T_x(X^o)$ contient λ, alors comme on a $\dim(T \cap D_{k+1}) \geq d-k$, on a $\dim(T \cap D_k) \geq d-k+1$, et on aurait alors une contradiction à la propriété (P) pour D_k. ■

On s'intéresse maintenant aux variétés polaires relatives, quand la base est lisse de dimension 1.

Corollaire 2 (transversalité pour les "polaires relatives") : On suppose X germe en 0 d'espace analytique de dimension d dans K^N, et $f : X \to K$ analytique. Pour un drapeau $D_1 \supset D_2 \supset \ldots D_N$ de K^N, on définit

$$P_{X_f}(D_k) = \text{clôture } \{x \in X^o / \dim T_x X_{f(x)} \cap D_k \geq d-k\} .$$

Alors, pour presque tout drapeau, $P_{X_f}(D_k)$ est transverse à D_k.

Démonstration : On construit $Z \subset X \times K \subset K^N \times K$ le graphe de f

$$Z = \{(x, f(x) \mid x \in X\} .$$

On a alors une correspondance bijective entre les (N-k)-plans de K^N et les (N-k+1)-plans de $K^N \times K$ contenant $Y = \{0\} \times K$. Si $D_k \times Y = H$ vérifie la propriété (P) pour Z et $L = K^N \times \{0\}$, nous

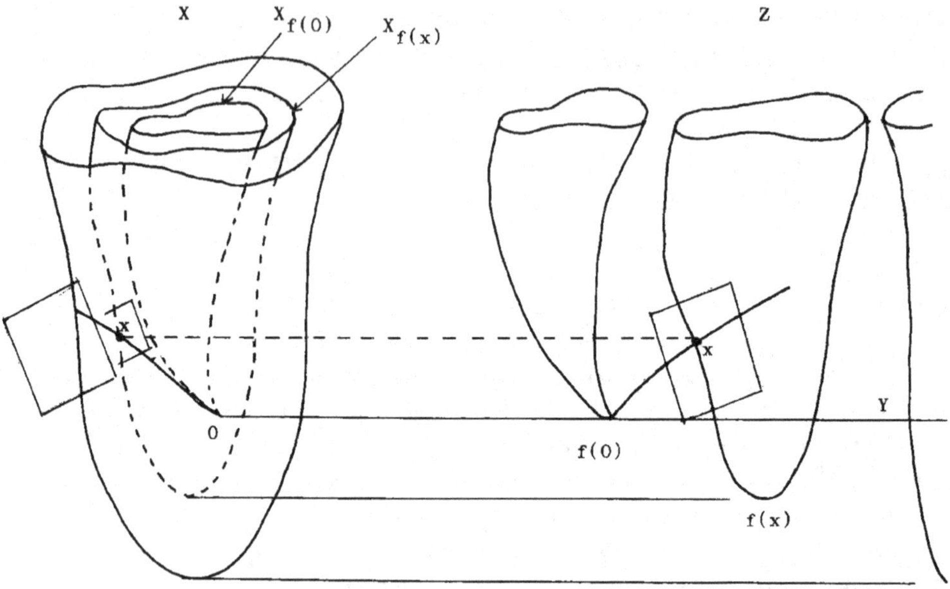

allons montrer que $P_{X_f}(D_k)$ est transverse à D_k. S'il n'en était pas ainsi, il existerait une courbe Γ dans $P_{X_f}(D_k)$, avec $\Gamma - \{0\} \subset X^o$, tangente à D_k. On remonte cette courbe dans le graphe et on y obtient une courbe dont la projection sur L est tangente à $H \cap L = D_k$! Or le long de cette courbe $T_x Z$ coupé avec la direction de L n'est autre que $T_x X$ qui coupe D_k avec la dimension au moins $d-k$, la limite $\lim_{x \to 0} T_x Z \cap L$ vérifie encore la même propriété, ce qui contredit (P) pour $D_k \times Y = H$.

Corollaire 3 (B. Teissier, cf. [12], lemme 3) : Si (X^o, Y) avec Y de dimension 1, $Y \subset X$, satisfait les conditions de Whitney, alors $P_X(D_2)$ (où $D_2 \subset K^{N+1}$ est un plan de codimension 2 générique) est vide.

Démonstration : On peut supposer D_2 transverse à Y et que l'hyperplan engendré par D_2 et Y vérifie (P). Si $P_X(D_2)$ n'était pas vide, on pourrait trouver une courbe γ, avec $\gamma - \{0\} \subset X^o$, et $\gamma \subset P_X(D_2)$. Soit T la limite des plans tangents à X^o le long de γ. T doit intersecter D_2 avec une dimension au moins $d-1$, puisque γ est dans $P_X(D_2)$. T doit contenir Y (condition a) de Whitney), l doit contenir la limite ℓ des normales $\overline{y\cdot\pi(y)}$ (y sur γ, $\pi(y)$ projection de sur Y) (condition b) de Whitney). Supposons que $\ell \subset D_2$. On aurait alors une ontradiction avec le fait que $(D_2 + Y)$ vérifie (P). D_2 ne contient donc pas ℓ. $_2$, Y et ℓ sont donc trois espaces supplémentaires, T devrait contenir Y et ℓ, t couper D_2 avec la dimension $d-1$, ce qui est contradictoire avec $\dim T = d$. $X^{(D_2)}$ est donc vide. ∎

Toujours avec Y de dimension 1, on obtient que la condition (a) de hitney passe aux sections hyperplanes génériques. On retrouve pour $\dim Y = 1$ ue les conditions de Whitney passent aux sections hyperplanes génériques, fait onjecturé par Teissier dans [10], résultat de Briançon-Speder pour les familles 'hypersurfaces à singularité isolée [11], généralisé par Navarro [8] aux sous-nalytiques et Teissier dans le cas où Y est de dimension quelconque, mais X ıalytique complexe [12].

4 - EXEMPLE.

Considérons l'hypersurface analytique Z de $\mathbb{C}^2 \times \mathbb{C}^2$ définie par :

$$x_1^2 + x_2^2 + y_1x_1 + y_2x_2 = 0 \quad .$$

otons $Y = \{(x_1,x_2,y_1,y_2) \; ; \; x_1 = 0, \; x_2 = 0\}$ et $X = Z - Y$. Le couple (X,Y) vérifie a condition de frontière. Par contre si H_λ est un hyperplan contenant Y d'équaion $x_2 - \lambda x_2 = 0$ et γ_λ l'arc analytique de X défini par :

$$x_1 = \lambda t^2 \quad , \quad x_2 = t^2 \quad , \quad y_1 = t \quad , \quad y_2 = -\lambda t - (1+\lambda^2)t^2 \quad ,$$

e long de γ_λ les coordonnées projectives de T_xX sont :

$$\eta_1 = 2x_1 + y_1 = t + 2\lambda t^2 \quad , \quad \eta_2 = 2x_2 + y_2 = -\lambda t + (1-\lambda^2)t^2 \quad ,$$

$$\zeta_1 = x_1 = \lambda t^2 \quad , \quad \zeta_2 = x_2 = t^2 \quad ,$$

t la limite de T_xX a donc pour coordonnées projectives $(1 \; ; -\lambda \; ; 0 \; ; 0)$ qui sont elles de H_λ.

Pour tout hyperplan H_λ contenant Y il existe donc une courbe conteue dans H_λ qui donne naissance à une limite d'espace tangents contenue dans $_\lambda$.

BIBLIOGRAPHIE

[1] J. Briançon, J.P. Speder : Les conditions de Whitney impliquent μ^*-constant, Ann. Inst. Fourier, 26 (1976), 153-163.

[2] J. Briançon, J.P. Henry, J.P. Speder : Les conditions de Whitney en un point sont analytiques, Note aux C.R. Acad. Sc. Paris, t. 282 (1976), p. 279.

[3] M. Giusti, J.P. Henry : Minorations de nombres de Milnor, Bull. S.M.F., 108 (1980), 17-45.

[4] J.P.G. Henry, M. Merle : Sections planes, limites d'espaces tangents et transversalité de variétés polaires, Note aux C.R. Acad. Sc. Paris, t. 291 (1980), p. 291.

[5] H. Hironaka : Stratification and flatness in real and complex singularities, Proc. of the Nordic Summer School on real and complex singularities, Oslo 1976, Noordhoff 1977, p. 199-265.

[6] Lê D.T. et B. Teissier : Variétés polaires locales et classes de Chern des variétés singulières, (à paraître dans Annals of Maths.).

[7] V. Navarro : Sobre la topologie local y global de las intersecciones completas, Thèse, Universidad Politecnica de Barcelona, E.T.S.I.I., Diagonal 657, 1978.

[8] V. Navarro : Conditions de Whitney et sections planes, Inventiones Math. 61, 3 (1980), 199-266.

[9] V. Navarro, D. Trotman : Whitney regularity and generic wings, Ann. Inst. Fourier 31, 2 (1981), 87-111.

[10] B. Teissier : Cycles évanescents, sections planes et conditions de Whitney, in Singularités à Cargèse 1972, Astérisque 7-8 (1973), 285-362.

[11] B. Teissier : The hunting of invariants in the geometry of discriminants, Proc. of the Nordic Summer School on real and complex singularities, Oslo 1976, Noordhoff 1977, p. 565-675.

[12] B. Teissier : Variétés polaires locales et conditions de Whitney, Note aux C.R. Acad. Sc. Paris, t. 291 (1980), p. 799.

[13] B. Teissier : Variétés polaires -I-, Inventiones Math. 40 (1977), 267-292.

[14] J.L. Verdier : Stratifications de Whitney et théorème de Bertini-Sard, Inventiones Math. 36 (1976), 295-312.

Centre de Mathématiques
Ecole Polytechnique
F-91128 PALAISEAU Cedex

BETWEEN EQUIMULTIPLICITY AND NORMAL FLATNESS

M. Herrmann und U. Orbanz

Introduction. Zariski's paper "Reduction of the singularities of algebraic three dimensional varieties" [Z] is to be seen as one starting point for the resolution of singularites in all dimensions. It suggested to get a desingularisation of a given variety X by blowing up smooth centers contained in the singular locus of X. To be able to prove that this process will finally improve the situation, the first thing needed is a measure for the singularity. Secondly we would like to have some condition for a smooth subvariety D of X, which allows to conclude that blowing up X along D will actually improve the singularity. It is clear that such a condition will be closely related to the way of measuring the singularity.

For surfaces (in characteristic 0, embedded in a three dimensional nonsingular variety), this program was carried out by Zariski in the above cited paper, and the result was used to resolve the singularities of an algebraic three dimensional variety by local uniformisation. The measure for the singularity in Zariski's proof (and in every surface proof afterwards except that by Lipman [Li_1]) was the multiplicity, and one condition for a regular curve to be a permissible center was to have the same multiplicity in each point, i.e. the condition of equimultiplicity. To obtain such curves, Zariski used the following procedure: He showed that the set S of points of highest multiplicity on X is a subvariety, so every one dimensional component of S is an equimultiple curve.

In Hironaka's proof of resolution of singularities in characteristic 0 [H_1], the way of measuring and choosing the center of a monoidal transformation was based on the same idea. Hironaka generalized and refined the notion of equimultiplicity to normal flatness, and he replaced the multiplicity by a kind of "multiplicity sequence" ν^* to deal with the case of codimension > 1. Now ν^* was used as a measure for the singularity, and the corresponding condition on the centers of monoidal transformations was normal flatness, because it turned out that normal flatness could be expressed by using ν^* (see theorem H below). Then Hironaka showed that by blowing up the variety in a regular, normally flat center, the singularities don't become worse if measured by ν^*.

Now normal flatness can equally well be expressed by using Hilbert-Samuel functions. And again, blowing up X in a regular, normally flat center will not increase the Hilbert-Samuel functions of corresponding points (in an appropriate sense) ([H_2],[Si_1]). Furthermore, stability of ν^* in this process is equivalent to the Hilbert-Samuel function being unchanged ([H_2],[B]). Therefore, in Hironaka's desingularisation, the Hilbert-Samuel function may replace the character ν^*. Finally, by the semicontinuity of the Hilbert-Samuel function, the points of X with maximal Hilbert-Samuel function form again a subvariety S, and any regular subvariety of S is a natural candidate for a center of a permissible monoidal transformation, since it satisfies the condition of normal flatness.

From the algebraic point of view we have the following situation (locally): Let (Q,M_0) be a regular local ring, $P_0 \subset Q$ a prime ideal such that Q/P_0 is regular again. Let $A \subset P_0$ be an ideal and consider $R = Q/A$ and $P = P_0/A$. Then we have the following

<u>Theorem H.</u> (Hironaka) R is normally flat along P if and only if A can be generated by elements $f_1,\dots,f_r$ such that

(a) the initial forms of f_i generate the initial ideal $gr_M(A,Q)$ and

(b) $\nu_{M_0}(f_i) = \nu_{P_0}(f_i)$ for $i=1,\dots,r$.

Geometrically, condition (a) says that not only the variety X at the given point x is the intersection of the hypersurfaces $f_i = 0$, but also the tangent cone of X at x is the intersection of the corresponding hypersurfaces, defined by the initial forms of the f_i . As a special case of the above theorem we get the following easier result:

Proposition. The notations being the same as above, let $A = f \cdot Q$ define a hypersurface. Then the following conditions are equivalent:

(i) R is normally flat along P .

(ii) $e_0(R) = e_0(R_p)$.

(iii) $\nu_{M_0}(f) = \nu_{P_0}(f)$.

In the general case, normal flatness is stronger than eqimultiplicity, as can easily be seen. One of our aims was to get some idea of the gap between these two notions, and as a counterpart we tried to describe cases in which normal flatness is the same as equimultiplicity like in the hypersurface case above. In general, already the resolution of hypersurface singularities presents the main difficulties. Nevertheless one can ask if the coincidence of equimultiplicity and normal flatness will yield some simplification of resolution, in the sense that some additional complications can be avoided, which occur in the general case.

One of our main techniques in §2 was motivated by the question, under which conditions the blowing up of a local Cohen-Macaulay ring will be Cohen-Macaulay again. In this form the problem is still completely open, but some comments will be given in §2. In the main result of §2 we give conditions under which equimultiplicity implies normal flatness, generalizing the hypersurface case to some complete intersections. Examples will illustrate that these conditions are not too strong.

In §1, some conditions for equimultiplicity are given. This is an extension of our paper [HO_1], using a result by Schickhoff [Sch] (see also [Li_2]).

As a consequence of the result in [HO_1] we give an elementary proof that the blowing up of a surface in a regular curve is a finite morphism if and only if the surface is equimultiple along this curve. At the end of §1 we indicate how one of the conditions for equimultiplicity is related to flat, equimultiple families of ideals, a topic, which is related to the problem of simultaneous resolution. These results are mainly due to Teissier and can be found in [Li_2].

Some of our original questions are still unanswered, and some new questions arose as consequences of our partial results. These questions will occur at the proper place in the text, where they can be formulated more precisely.

§ 1 Conditions for equimultiplicity

In this section we study two conditions related to multiplicity. The first refers to the reduction of an ideal [NR] and the corresponding notion of analytic spread. The second uses a homomorphism constructed by Hironaka to characterize normal flatness; using this homomorphism one can also deduce the well-known criterion for normal flatness using Hilbert-Samuel functions.

If R is any local ring and P a prime ideal such that R/P is regular, then we say that R is equimultiple along P if $e_0(R) = e_0(R_P)$, e_0 being the multiplicity in the sense of Samuel. Now Dade [D] and R. Schmidt [R.S.] independently introduced a generalization of the definition of multiplicity to an arbitrary ideal $A \subset R$, using systems of parameters of R/A . This gives a more general definition of R being equimultiple along A .

Reductions of ideals are very closely related to multiplicity. If R is a quasi-unmixed local ring with maximal ideal M , then two M-primary ideals $A \subset B$ have the same multiplicity if and only if A

is a reduction of B [NR], [R]. We always assume R/M to be infinite. (A corresponding result holds for ideals of height less than dim R([Bö]). This is the reason why equimultiplicity of R along some ideal A can be related to the condition $ht(A) = l(A)$, where $l(A)$ denotes the analytic spread of A, i.e. the least number of generators of a minimal reduction of A. $l(A)$ has the following geometric interpretation (see [NR]): If $F : X \to Spec(R)$ is the blowing up of R with center A, then the dimension of the fibre $f^{-1}(M)$ is $l(A) - 1$. So the condition $ht(A) = l(A)$ means that all fibres of f have the same dimension.

If R, M, A are as above, for any elements $x_1, \dots, x_r \in R$ whose image in R/A form a system of parameters, we put $V = A + \sum_{i=1}^{r} x_1 R$, and we have a canonical graded epimorhism

$$\varphi : \bigoplus_{n \geq 0} A^n/MA^n[X_1, \dots, X_r] \to \bigoplus_{n \geq 0} V^n/MV^n$$

sending X_i to the class of x_i mod MV. If R/A is regular and if $x_1, \dots, x_r$ form a regular system of parameters in R/A, so that $V=M$, we know that R is normally flat along A if and only if φ is an isomorphism ([H_1],[HSV],[Si_2]). A weaker condition than φ being an isomorphism is the requirement that ker φ should be nilpotent.

So let us consider the following conditions:

(1) R is normally flat along A.

(2) For any system of parameters for R/A, φ has a nilpotent kernel.

(3) $ht(A) = l(A)$.

(4) R is equimultiple along A.

In (1.1), after giving the necessary definitions, we will show the easy implications (2) $\Rightarrow$ (3) $\Rightarrow$ (4), and if A is a prime ideal such that R/A is regular, (1) $\Rightarrow$ (2) for a regular system of parameters. So one could be tempted to believe that these conditions allow to pass from equimultiplicity to normal flatness step by step. Unfortunately (or fortunately?) it turned out that in a quasi-unmixed local ring (2), (3) and (4) are equivalent. The implications (4) ⇒ (3) and (3) ⇒ (2) will be shown in (1.2) and (1.3) respectively.

Our considerations lead us to state the following open problems:

Problem 1. Is there a reasonable condition properly between normal flatness and equimultiplicity? (Here reasonable could mean that such a condition can be related to the behaviour of singularities under monoidal transformations.)

Problem 2. If P is a prime ideal of R such that R/P is regular and $e(R) = e(R_P)$, and if R' is a monoidal transform of R with center P, when does it hold that $e(R') \leq e(R)$?

Problem 3. Does normal flatness of R along A imply that φ is an isomorphism for general A?

(1.1) Throughout §1, R denotes a local ring, M its maximal ideal, and we assume R/M to be infinite, although most of the results are true for a finite residue field too.

Let A be an ideal in R and $\underline{x} = \{x_1,\dots,x_r\}$ a multiplicity system for R/A (see [No] or [HSV]). Then we define a numerical function by

(1.1.1) $$H^{(o)}_{\underline{x},A,R}(n) = e(\underline{x}, gr^n_A(R)) ,$$

$e(\underline{x},-)$ denoting the multiplicity symbol introduced by Wright. As usual we put $H^{(i)}(n) = \sum_{k=o}^{n} H^{(i-1)}(k)$ for $i > 0$, and by an easy computation (see [HSV], p. 113) we get

(1.1.2) $$H^{(1)}_{\underline{x},A,R}(n) = \sum_P e(\underline{x},R/P)\cdot l(R_P/A^{n+1}R_P) ,$$

where the summation is taken over all minimal primes P of A. Let us remark that for such a prime we have $e(\underline{x},R/P) \neq 0$ if and only if $\dim R/P = \dim R/A$, so we could restrict the summation to this subset of $Ass(R/A)$, which will be denoted by $Assh(R/A)$.

By (1.1.2), the values of $H^{(1)}_{\underline{x},A,R}(n)$ for large n are given by a polynomial with rational coefficients. Let d and a be the degree and the highest coefficient of this polynomial respectively. Then we define

(1.1.3) $$e(\underline{x},A,R) = d!a .$$

If A is M-primary, then the empty set is the only non-trivial multiplicity system for R/A, and for this we recover the ordinary Hilbert-Samuel function $H^{(1)}_{A,R}(n)$ and the Samuel multiplicity $e_0(A)$.

If $ht(A)+\dim R/A = \dim R$, the relation to Samuel multiplicity is given by

(1.1.4) $$\underline{e}(\underline{x},A,R) = \sum_{P\in \mathrm{Assh}(R/A)} e(\underline{x},R/P)e_0(A\cdot R_P) .$$

All these facts can be found in [HSV]. The condition of R being equimultiple along A is now expressed in the following way: There exists a system of parameters $\underline{x} = \{x_1,\dots,x_r\}$ for R/A such that

(1.1.5) $$e(\underline{x},A,R) = e_0(\underline{x}\, R + A) .$$

Next let us recall some facts about analytic spread. If $B \subset A$ is an ideal, B will be called a reduction of A if $B A^n = A^{n+1}$ for some n. Every reduction of A (e.g. A itself) contains a minimal reduction of A, and if R/M is infinite (as we assume), every reduction of A has the same minimal number of generators, denoted by $l(A)$. It turns out that

(1.1.6) $$l(A) = \dim gr_A(R) \otimes_R R/M ,$$

where $gr_A(R)$ is the associated graded ring of R with respect to A and dim the Krull-dimension. If $\mu(A)$ denotes the minimal number of generators of A, then we have

(1.1.7) $$ht(A) \le l(A) \le \mu(A) .$$

If A is M- primary and B a reduction of A, then it is clear that $e_0(A) = e_0(B)$. Using (1.1.4) we see that also in the general case we have $e(\underline{x},A,R) = e(\underline{x},B,R)$ for every reduction B of A and every multiplicity system $\underline{x}$ of R/A, or more generally $e_0(A \cdot R_P) = e_0(B \cdot R_P)$ for every minimal prime P of A. This last observation has the following important converse, which will be used:

Theorem 1. ([Bö]) Let R be quasi-unmixed and let $B \subset A$ be ideals in R such that

a) A and B have the same radical and $ht(B) = l(B)$

b) $e_0(A \cdot R_P) = e_0(B \cdot R_P)$ for every minimal prime P of A (or B).

Then B is a reduction of A.

In the following we will also need the description of reductions by the notion of integral dependence on ideals. If $A \subset R$ is any ideal and $x \in R$, then x is said to be integral over A, if x satisfies an equation

(1.1.8) $$x^n + a_1 x^{n-1} + \ldots + a_n = 0 \ , \ a_i \in A^i \ .$$

All elements integral over A form again an ideal, the integral closure $\bar{A}$ of A. Now an ideal $B \subset A$ is a reduction of A if and only if A is integral over B, or $A \subset \bar{B}$ (see [NR] and [Li_3]). Therefore, if R is a domain, reductions may be described by using valuations.

Now let $A \subset R$ again be an ideal in R and let $x_1,\ldots,x_r$ be a system of parameters for R/A (by which we mean elements, whose images are a system of parameters in R/A). Let $V = A + \sum_{i=1}^{r} x_i R$. Then there is a natural homomorphism of graded rings

(1.1.9) $$\varphi : \bigoplus_{n \geq 0} A^n/M A^n[X_1,\ldots,X_r] \to \bigoplus_{n \geq 0} V^n/M V^n$$

sending X_i to $x_i + M V \in V/M V$. φ surjective by definition. If φ has a nilpotent kernel, then both rings in (1.1.9) have the same dimension. Now

(1.1.10) $$\dim\left(\bigoplus_{n\geq 0} A^n/M A^n[X_1,\ldots,X_r]\right) = l(A) + r ,$$

and

(1.1.11) $$\dim\left(\bigoplus_{n\geq 0} V^n/M V^n\right) = l(V) = \dim R$$

since V is M-primary (see [NR]). Therefore we get:

Lemma 1. Let $A \subset R$ be an ideal such that

a) $ht(A) + \dim R/A = \dim R$,

b) there exists a system of parameters of R/A , for which φ has a nilpotent kernel.

Then $ht(A) = l(A)$.

Lemma 2. Let $A \subset R$ be an ideal such that

a) $ht(A) + \dim R/A = \dim R$,

b) $ht(A) = l(A)$.

Then for any system of parameters $\underline{x} = \{x_1,\ldots,x_r\}$ for R/A we have

$$e(\underline{x},A,R) = e_0(A + \underline{x}\, R) .$$

Proof. Let $z_1,\ldots,z_s$ $(s = l(A) = ht(A))$ generate a minimal reduction B of A. Then B and A have the same minimal primes, and for each such prime P we have

(1.1.12) $$e_0(B\cdot R_P) = e_0(A\cdot R_P) .$$

Of course $\underline{x}\, R + B$ is a reduction of $\underline{x}\, R + A$, therefore $e_0(\underline{x}\, R+A) = e_0(\underline{x}\, R + B)$. By the associativity formula for multiplicities (see [No]), we have

(1.1.13) $$e_0(\underline{x}\, R + B) = \sum_P e(\underline{x},R/P)e_0(B\cdot R_P) ,$$

where P ranges over the minimal primes of B. Now the assertion follows by (1.1.4) and (1.1.12). (We remark that assumption a) is needed to assure that $x_1,\ldots,x_l, z_1,\ldots,z_s$ is a system of parameters for R.)

Later on we will apply these lemmas to quasi-unmixed local rings, where assumption a) of lemmas 1 and 2 holds for any ideal A.

(1.2) To prove the converse of lemma 2 we need the following

Lemma 3. Let $A \subset R$ be any ideal such that $r = \dim R/A > 0$. Let $y = \{y_1,\ldots,y_r\}$ be a system of parameters for R/A and put $V = A + \underline{y} R$. Then there is a sequence of superficial elements $x_1,\ldots,x_r$ for V with the following properties:

(1.2.1) $V = A + \underline{x} R$.

(1.2.2) $e(\underline{x},A,R) = e(\underline{y},A,R)$.

(1.2.3) $$e_o(V) = \begin{cases} e_o(\underline{x} R) & \text{if } ht(A) = 0 \\ e_o(V/\underline{x}R) & \text{if } ht(A) > 0 . \end{cases}$$

(1.2.4) $x_1,\ldots,x_r$ is part of a system of parameters for R.

Proof. Remember that R/M is assumed to be infinite. Therefore, by induction, we can choose x_i with the following properties:

- The image of x_i in $R/x_1R +\ldots+ x_{i-1} R$ is superficial of order 1 for $V/x_1R +\ldots+ x_{i-1} R$.

- $x_i \notin A + M V + x_1R +\ldots+ x_{i-1} R$.

- x_i is not contained in any minimal prime of

$$\begin{cases} x_1R +\ldots+ x_{i-1}R & \text{if } ht(A) = 0 \\ A+x_1 R+\ldots+ x_{i-1}R & \text{if } ht(A) > 0 . \end{cases}$$

Then (1.2.1) and (1.2.3) hold by construction and (1.2.4) follows from the fact that

$$\dim R/x_1 R+\ldots+ x_i R = \dim R-i .$$

Finally we get (1.2.2) from (1.1.4), since for every minimal prime P of A, the images of $x_1,\ldots,x_r$ and of $y_1,\ldots,y_r$ in R/P generate the same ideal.

Assume now that there is a system of parameters $\underline{x} = \{x_1,\ldots,x_r\}$ for R/A such that

(1.2.5) $$e(\underline{x},A,R) = e_0(\underline{x} R + A).$$

We want to prove finally that $ht(A) = l(A)$ (see lemma 5), which is trivial if $r = 0$. Suppose therefore that $r > 0$. Then we can assume in addition (1.2.3) and (1.2.4) by lemma 3.

Lemma 4. Under the assumptions (1.2.3), (1.2.4) and (1.2.5), suppose that $\dim R/A + ht(A) = \dim R$ and let $\underline{z} = \{z_1,\ldots,z_r\} \subset A$ be a system of parameters for $R/\underline{x} R$ such that

(1.2.6) $$e_0(V/\underline{x} R) = e_0(\underline{z} R + \underline{x} R/x R) .$$

Then

(1.2.7) $\operatorname{Assh}(R/A) \subset \operatorname{Assh}(R/\underline{z} R)$ and $e(\underline{x},A,R) = e(\underline{x},\underline{z} R, R)$.

(1.2.8) $\operatorname{Assh}(R/A) = \operatorname{Assh}(R/\underline{z} R)$ and $e_0(A R_P) = e_0(\underline{z} R_P)$ for every $P \in \operatorname{Assh}(R/A)$.

(Here $\underline{z} R = 0$ if $s = 0$. The existence of $z_1,\ldots,z_r$ follows from our assumption that R/M is infinite.)

Proof. From $r+s = \dim R$ we conclude $\dim R/\underline{z} R = \dim R/A$ and therefore $\operatorname{Assh}(R/A) \subset \operatorname{Assh}(R/\underline{z} R)$. Assume first that $s > 0$ and put $W = \underline{z} R + \underline{x} R$. The associativity law gives

(1.2.9) $$e_0(W) = e(\underline{x},\underline{z} R,R) \geq e(\underline{x},A,R) .$$

Since $\underline{x}$ is part of a system of parameters for R, we have $e_0(W) \leq e_0(W/\underline{x} R)$ and therefore

$$(1.2.10)\quad e(\underline{x},A,R) \le e(\underline{x},\underline{z}\,R,R) = e_0(W) \le e_0(W/x\,R) = e_0(V/\underline{x}\,R) = = e_0(V)\ .$$

Therefore assumption (1.2.5) implies (1.2.7) in the case $s > 0$. The case $s=0$ is similar.
To prove (1.2.8), consider the relations

$$e(\underline{x},A,R) = \sum_{P\in Assh(R/A)} e(\underline{x},R/P)e_0(A\cdot R_P)$$

$$\le \sum_{P\in Assh(R/A)} e(\underline{x},R/P)e_0(\underline{z}\,R_P)$$

$$= e(\underline{x},\underline{z}\,R\,,R) - \sum_{\substack{P\in Assh(R/z\,R)\\ \setminus Assh(R/A)}} e(\underline{x},R/P)e_0(\underline{z}\,R_P)$$

$$\le e(\underline{x},\underline{z}\,R\,,R)\ .$$

From (1.2.7) we get equality in every step, and (1.2.8) follows.

Lemma 5. Let $A,\underline{x},\underline{z}$ be as above and assume in addition that R is quasi-unmixed. Then $\underline{z}$ generates a reduction of A and consequently $ht(A) = l(A)$.

Proof. Using Böger's theorem, the result will follow from (1.2.8) if we can show that A and $\underline{z}\,R$ have the same radical. But if R is quasi-unmixed, $Assh(\underline{z}\,R)$ is just the set of minimal primes of $\underline{z}\,R$. Therefore (1.2.8) implies that also $Assh(R/A)$ is the set of minimal primes of A .

This lemma not only proves the desired result $ht(A) = l(A)$, but also gives some conditions on elements $z_1,\ldots,z_r$ to generate a reduction of A . If R is quasi-unmixed and $P \subset R$ is a prime ideal such that $ht(P) = l(P) = 1$, then lemma 5 allows to construct a system of generators $x_1,\ldots,x_m$ of P such that each x_i generates a minimal reduction of P . Using these generators it is easy to show that the morphism $f : Bl_P(R) \to Spec\ R$ obtained by blowing up P is affine. Since the dimension of the fibre over M is $l(P) - 1 = 0$, we get:

Proposition 1. Let R be a quasi-unmixed local ring and $P \subset R$ a regular prime of height 1, such that R is equimultiple along P. Then blowing up P gives a finite morphism $Bl_P(R) \to Spec(R)$.

For the details see [HO_1].

We remark that the condition $ht(A) = l(A)$ is independent of any system of parameters for R/A. Therefore, by lemmas 2 and 5, the equimultiplicity of R along A for some system of parameters for R/A implies the equimultiplicity for every system of parameters.

(1.3) We start with a quasi-unmixed ring R again and with an ideal A in R such that $ht(A) = l(A)$. Choosing any system of parameters $x_1,\ldots,x_r$ for R/A, we want to prove that the epimorphism (1.1.9) has a nilpotent kernel. We put $B = \underline{x}R$ and $V = A + B$. We have a canoncial epimorphism

$$(1.3.1)\quad \Psi : \bigoplus_{n\geq 0} A^n/V A^n \to \bigoplus_{n\geq 0} (V/B)^n/(V/B)^{n+1} = \bigoplus_{n\geq 0} A^n/A^{n+1}+B \cap A^n ,$$

and the first step will be to prove that Ψ has a nilpotent kernel. This is the content of lemma 6 and proposition 2 which are due to Lipman ([Li_2]). Let us note in advance that our assumption $ht(A) = l(A)$ implies

$$ht(V) = \dim R = \dim R/A + ht(A) = r + l(A) ,$$

and since $x_1,\ldots,x_r$ is part of system of parameters for R because R is quasi-unmixed, we get $l(B) = r$ and therefore

$$(1.3.2)\qquad ht(A+B) = l(A) + l(B) .$$

Lemma 6. Assume that $A^n \cap B \subset \overline{A^n B}$ for large n. (A bar denotes the integral closure as in (1.1).) Then Ψ (1.3.1) has a nilpotent kernel.

Proof. Let $x \in A^n$ and $\Psi(x+V A^n) = 0$, i.e. $x \in A^{n+1} + B \cap A^n$. By taking a power of $(x + V A^n)$ we may assume that $B \cap A^n$ is integral over $B A^n$. Then x satisfies an equation

$$(1.3.3)\qquad x^m + a_1 x^{m-1} + \ldots + a_m = 0$$

such that

(1.3.4) $a_i \in (A^{n+1} + B\,A^n)^i = A^{n-i}\,V^i$, $i=1,\ldots,m$.

It follows that for $i = 1,\ldots,m$ we have

(1.3.5) $a_i\,x^{m-i} \in A^{ni}\,V^i\,A^{n(m-i)} \in V\,A^{n\,m}$,

which means that $(x+V\,A^n)^m = 0$.

Proposition 2. Let R be a quasi-unmixed local ring and A,B ideals of R such that $ht(A+B) = l(A) + l(B)$. Then for all positive integers m,n we have that $A^m B^n$ is a reduction of $A^m \cap B^n$.

For the proof we refer to [Li_2], where the case $m=n=1$ is treated. (This is sufficient since $ht(A^m + B^n) = ht(A+B)$ and $l(A^m) = l(A)$.)

Remark. It is not difficult to show that the assumption $ht(A+B) = l(A) + l(B)$ implies that $Min(A\cdot B) = Min(A\cap B)$ is the disjoint union of $Min(A)$ and $Min(B)$, from which we conclude that

$$(AB)R_P = (A\cap B)R_P \quad \text{for all} \quad P \in Min(A\cdot B) .$$

Unfortunately Böger's theorem does not apply, since in general we have $ht(AB) < l(AB)$ (example: $A = xR$, $B = y\,R + z\,R$ in a regular local ring R with regular parameters x,y,z).

Corollary. Under the assumptions of the proposition we have

$$\overline{A\cdot B} = \bar{A} \cap \bar{B} .$$

Proof. The assumption implies $ht(\bar{A} + \bar{B}) = l(\bar{A}) + (\bar{B})$, therefore $\bar{A}\,\bar{B}$ is a reduction of $\bar{A} \cap \bar{B}$. But AB is a reduction of $\bar{A}\,\bar{B}$, which implies $\bar{A} \cap \bar{B} \subset \bar{A}\,\bar{B}$. The converse is obvious.

We now return to our assumption about R,A,B made at the beginning of (1.3). From (1.3.2), proposition 2 and lemma 6 we get that Ψ (1.3.1) has a nilpotent kernel, and we want to prove the same about the homomorphism φ (1.1.9). For this we embed Ψ in a commutative diagram

$$\begin{array}{ccc}
\bigoplus_{n\geq 0} A^n/V\,A^n & \xrightarrow{\ \Psi\ } & \bigoplus_{n\geq 0} (V/B)^n/(V/B)^{n+1} \\
\downarrow{\scriptstyle\alpha} & & \downarrow{\scriptstyle\beta} \\
\bigoplus_{n\geq 0} A^n/M\,A^n & \xrightarrow[\ \varphi_0\ ]{} & \bigoplus_{n\geq 0} (V/B)^n/M\cdot(V/B)^n
\end{array} ,$$

where α,β and φ_0 are the obvious epimorphism. Since V is M-primary, α and β have nilpotent kernels, so the same is true for φ_0. Let us introduce the following notation:

$$G(A) = \bigoplus_{n\geq 0} A^n/M\,A^n \quad , \quad G(V) = \bigoplus_{n\geq 0} V^n/M\,V^n \ ,$$

$$G(\tilde{V}) = \bigoplus_{n\geq 0} \tilde{V}^n/M\,\tilde{V}^n \quad , \quad \text{where } \tilde{V} = V/B \ .$$

Then we have a commutative diagram

$$\begin{array}{ccc}
H = G(A)[X_1,\dots,X_r] & \xrightarrow{\ \varphi\ } & G(V) \\
\downarrow{\scriptstyle\pi_1} & & \downarrow{\scriptstyle\pi_2} \\
G(A) & \xrightarrow{\ \varphi_0\ } & G(\tilde{V})
\end{array}$$

where $\pi_1(X_i) = 0$ and π_2 is induced by $R \to R/B$. Let P_1, P_2 be two minimal primes in $G(V)$. Our assumption that R is quasi-unmixed implies $\dim G(V)/P_1 = \dim G(V)/P_2$ (see [Ra]). Since

$$\begin{aligned}\dim H &= \dim G(A) + r = l(A) + r = ht(A) + r = \dim R \\ &= l(V) = \dim G(V)\end{aligned}$$

we conclude that <u>every</u> minimal prime of $\ker\varphi$ is a minimal prime in H, and these are of the form $P\cdot H$ for some minimal prime P in $G(A)$. To show that $\ker\varphi$ is nilpotent we have to show that $\ker\varphi \subset P\cdot H$ for every minimal prime P of $G(A)$. Now gives such a P we know that $\ker\varphi_0 \subset P$ and therefore

$$\ker\varphi \subset \ker\varphi_0\cdot H + (X_1,\dots,X_r) \subset P\cdot H + (X_1,\dots,X_r) \ .$$

The prime ideal $P H + (X_1,\dots,X_r)$ contains a minimal prime of $\ker\varphi$, which was shown to be of the form $P'\cdot H$ for some minimal prime P' of $G(A)$. Since

$$(P\cdot H + (X_1,\dots,X_r)) \cap G(A) = P ,$$

we conclude $P = P'$, and therefore $\ker\varphi \subset P\cdot H$. It follows that $\ker\varphi$ is nilpotent.

(1.4) We conclude §1 with some remarks how the condition $ht(A) = l(A)$ (or its equivalents) can be related to a suitable defined equimultiplicity of a flat family. For this purpose, let us consider a flat local homomorphism $h : (S,N) \to (R,M)$ of local rings. In R we consider an ideal A satisfying

(1.4.1) $$\begin{cases} h^{-1}(A) = 0 \\ R/A \text{ is a finite } S\text{-module .} \end{cases}$$

Then for any prime ideal P in S we can define a multiplicity $e_P(I)$ in the following manner. If $k(P)$ denotes the residue field at P, $R/I \otimes k(P)$ is finite dimensional over $k(P)$. Therefore for large n the function $\dim_{k(P)}(R/I^n \otimes k(P))$ is given by a rational polynomial. If d is the degree and a the highest coefficient of this polynomial, we define $e_P(I) = d!a$. The following theorem is Lipman's extension ([Li_2]) of a result by Teissier ([T]):

<u>Theorem 2.</u> Let $h: (S,N) \to (R,M)$ be a flat local homomorphism and $A \subset R$ an ideal with (1.4.1). If R is quasi-unmixed, the following conditions are equivalent:

(i) $ht(A) = l(A)$.

(ii) For every prime ideal P of S we have $e_P(I) = e_N(I)$.

The proof and the geometric interpretation are given in [Li_2].

The final result of all of this is

<u>Theorem 3.</u> Let R be a quasi-unmixed local ring and $A \subset R$ any

ideal. Then the following conditions are equivalent:

(i) There exists a system of parameters $\underline{x} = \{x_1,\dots,x_r\}$ for R/A such that $e(\underline{x},A,R) = e_0(\underline{x}\ R + A)$.

(i') For every system of parameters $\underline{x} = \{x_1,\dots,x_r\}$ for R/A we have $e(\underline{x},A,R) = e_0(\underline{x}\ R + A)$.

(ii) $ht(A) = l(A)$.

(iii) There exists a system of parameters $\underline{x} = \{x_1,\dots,x_r\}$ for R/A such that the corresponding epimorphism φ (1.1.9) has a nilpotent kernel.

(iii') For every system of parameters $\underline{x} = \{x_1,\dots,x_r\}$ for R/A, the corresponding epimorphism φ (1.1.9) has a nilpotent kernel.

Furthermore, if there exists a local subring (S,N) of (R,M) such that $A \cap S = 0$, $M \cap S = N$, $S \to R$ is flat and R/A a finite S-module, then these conditions are equivalent to

(iv) $$e_P(A) = e_N(A) \text{ for every prime ideal } P \text{ of } S .$$

The proof of (iv) ⇒ (ii) is similar to that of (i) ⇒ (ii), using Böger's theorem. But there is a different proof of (i) ⇒ (ii) in [AV] and in [D], which avoids Böger's result. Instead they use a "cutting lemma" by Dade [D], where a construction of an element $x \in A$ is given such that $l(A/x\ A) = l(A) - 1$. We conclude §1 by asking two questions:

Question 1. Is it possible to use Dade's method for a different proof of (iv) ⇒ (ii) or even of Böger's theorem?

Question 2. How can the equimultiplicity condition (iv) be expressed by using the multiplicity symbol $e(\underline{x},A,R)$?

§ 2 Normal flatness and equimultiplicity for complete intersections

It was known that equimultiplicity does not imply normal flatness, but in an example given by Hironaka, the ring was not even a Cohen-Macaulay ring. So we asked if under suitable assumptions on the ring we can deduce normal flatness from equimultiplicity. Using power series rings generated by monomials we gave a counterexample which is a Cohen-Macaulay ring, and Robbiano ([Ro]) independently found a similar example, which is a strict complete intersection (i.e. not only the local ring R, but also the graded ring $gr_M(R)$ with respect to the maximal ideal is a complete intersection).

The example is the following one: Let k be any field and $R = k[[t^4, t^{10}, u\,t^5, u^2]] \subset k\,[[t,u]]$. If we put $x = u^2$, $y = u\,t^5$, $z = t^{10}$ and $w = t^4$, then R can be written as

$$R = k[[X,Y,Z,W]] \;/\; (Z^2 - W^5,\ Y^2 - X\,Z).$$

Let $P = (y,z,w)\cdot R$. Then R/P is regular and $ht(P) = l(P) = 1$ (w generates a minimal reduction of P). Therefore $e_0(R) = e_0(R_P)$ by §1. But since $x\,z \in P^2$, we see that P/P^2 has torsion and consequently R is not normally flat along P. The initial forms of the defining equations with respect to $M_0 = (X,Y,Z,W)$ are a regular sequence, so by [VV_1] we get $gr_M(R) = k[X,Y,Z,W]/(Z^2, Y^2 - Z\,X)$, which shows that R is a strict complete intersection. But the initial forms with respect to $P_0 = (Y,Z,W)$ are $-\,X\,Z$ and Z^2 and therefore not a regular sequence.

To describe our result, let us make the following

<u>Definition</u>. Let (Q,M_0) be a local Cohen-Macaulay ring, $P_0 \subset Q$ a prime ideal such that Q/P_0 is regular, and let $A \subset P_0$ be any ideal. We call A (or $R = Q/A$) a strict complete intersection with respect to P_0, if A can be generated by elements $f_1,\dots,f_m \in Q$ such that

a) the initial forms of $f_1,\dots,f_m$ in $gr_{M_0}(Q)$ are a regular sequence, and

b) the initial forms of $f_1,\dots,f_m$ in $gr_{P_0}(Q)$ are a regular sequence.

(We remark that a) implies that $f_1,\dots,f_m$ is a regular sequence in Q, see proof of proposition 4.)

Then, as a special case of our main result in §2, we will get the following

Theorem 4. Let (Q,M_0) be a local Cohen-Macaulay ring and $A \subset P_0 \subset M_0$ ideals such that Q/P_0 is regular. Assume that Q is normally flat along P_0 and that A is a strict complete intersection with respect to P_0. Let $R = Q/A$ and $P = P_0/A$. Then the following conditions are equivalent:

(i) $e_0(R) = e_0(R_P)$.

(ii) R is normally flat along P.

If A is generated by $f_1,\dots,f_m$ with properties a) and b) above, then these conditions are equivalent to $\nu_{M_0}(f_i) = \nu_{P_0}(f_i)$, $i=1,\dots,m$. This will be used as a tool in the proof.

The generalization of this result is twofold. First, by modifying the definition of strict complete intersection with respect to an ideal, we can replace the prime ideal P_0 by any ideal $I_0 \supset A$. In this case we have to replace (ii) by the condition that $gr_I^n(R)$ is an R/I-Cohen-Macaulay module of depth equal to $\dim R/I$ for all $n (I=I_0/A)$. This condition can be expressed by using suitable Hilbert functions (see theorem 6). For the second generalization we replace the equality between Hilbert functions by an equality between the corresponding polynomials. For this result, in the assumptions a) and b) above, the condition "regular sequence" can be replaced by "weakly regular sequence", which will be defined in (2.1). This weaker condition still implies that $f_1,\dots,f_m$ is a regular sequence and that the blowing up of R with center P is again a complete intersection (see proposition 5). Especially these conditions assure that blowing up R with center P gives a Cohen-Macaulay ring again. But for this last property, these conditions seem to be much too strong.

Problem 4. a) Give (weaker) conditions under which a quadratic transformation of a local Cohen-Macaulay ring is Cohen-Macaulay again.

b) Give an example of a local Cohen-Macaulay ring having a quadratic transform which is not Cohen-Macaulay.

Problem 5. If A is a strict complete intersection with respect to P_0 in Q and if Q' is a monoidal transform of Q with center P_0, under which conditions is the strict transform of A in Q' again a strict complete intersection ?

We start in (2.1) by giving the necessary results about weakly regular sequences in graded rings of local rings. These are applied in (2.2) to prove the main results. They are followed by some applications and comments.

(2.1) We start with the following

Definition. Let $A = \bigoplus_{n \geq 0} A_n$ be a graded ring and $f_1,\ldots,f_m$ homogeneous elements of A. The sequence $f_1,\ldots,f_m$ is called weakly regular, if there exists an n_0 with the following property: For every $i \in \{1,\ldots,m\}$ and any $x \in A_n$ such that $n \geq n_0$, $x f_i \in f_1 A+\ldots+ f_{i-1}A$ implies $x \in f_1A+\ldots+ f_{i-1}A$. ($f_0 = 0$ of course.)

In many respects, weakly regular sequences behave like regular sequences (see [HO_2]), but there is one important exception: A permutation of a weakly regular sequence need not be weakly regular again. As an example take $A = k[[x]]\,[Z]/(x\,Z^2) = k[[x]]\,[z]$, graded with respect to z. Then the sequence (z^2, xz) is weakly regular, but xz is not a weakly regular element.

Weakly regular sequences are closely connected to superficial elements. To make this clear we need another definition:

Definition. Let (Q,M_0) be a local ring and $I \subset M_0$ any ideal, and let $f_1,\ldots,f_m \in I$.

(1) $(f_1,\ldots,f_m)$ is called a superficial sequence for I, if for all $i \in \{1,\ldots,m\}$, the image of f_i in $Q/f_1 Q+\ldots+ f_{i-1} Q$ is superficial for $I/f_1 Q+\ldots+f_{i-1} Q$.

(2) $(f_1,\ldots,f_m)$ is called a stable superficial sequence for I, if it is a superficial sequence for I and

$$f_i \notin I^{d_i+1} + f_1 Q + \ldots + f_{i-1} Q , \quad i=1,\ldots,r ,$$

where $d_i = \nu_I(f_i)$.

Lemma 7. Let (Q,M_0) be a local ring, $I,K \subset M_0$ ideals and $f \in I$. We put $J = K + f Q$, $\bar{Q} = Q/K$, $\bar{I} = I+K/K$, $d = \nu_I(f)$ and $\bar{f}$ denotes the image of f in $\bar{Q}$. Assume that

a) $\bar{f}$ is superficial for $\bar{I}$,

b) $\nu_{\bar{I}}(\bar{f}) = d$,

c) $\bar{f}$ is not a zero-divisor in $\bar{Q}$.

Then

(2.1.1) $$gr_I^n(J,Q) = gr_I^n(K,Q) + in_I f \cdot gr_I^{n-d}(Q) \quad \text{for large } n .$$

($gr_I(J,Q)$ denotes the initial ideal of J in $gr_I(Q)$ and $in_I f$ the inital form of f.)

Proof. By assumption there exists come c such that

(2.1.2) $$(\bar{I}^{n+d} : \bar{f}) \cap \bar{I}^c = \bar{I}^n \quad \text{for} \quad n \geq c .$$

Furthermore, since $\bar{f}$ is not a zero-divisor in $\bar{Q}$, there is some k such that

(2.1.3) $$\bar{I}^n : \bar{f} \subset \bar{I}^{n-k} \quad \text{for} \quad n \geq k .$$

Now take $n_0 = \max\{c + d, c + k\}$, let $n \geq n_0$ and let $x^* \in gr_I^n(J,Q)$ be the initial form of some element $x \in (I^n \cap J) \setminus I^{n+1}$. x can be written as $x = a + b f$, $a \in K$. Passing to $\bar{Q}$ we get $\bar{x} = \bar{b} \bar{f}$, and therefore $\bar{b} \in \bar{I}^{n-k} \subset \bar{I}^c$ (since $n-k \geq n_0 - k \geq c$). From this we get

(2.1.4) $$\bar{b} \in (\bar{I}^n : \bar{f}) \cap \bar{I}^c = \bar{I}^{n-d}$$

by (2.1.2), which means that $b \in I^{n-d} + K$. Therefore x can be written as

(2.1.5) $$x = g + h\,f\ ,\ g \in K,\ h \in I^{n-d}\ ,$$

and $g = x - h\,f \in I^n$. Let us denote by g^*, h^*, f^* the inital forms of g,h,f respectively in $gr_I(Q)$. If $h\,f \notin I^{n+1}$, we have $in_I(h\,f) = h^*\,f^*$, and therefore we get form (2.1.5)

$$x^* = \begin{cases} h^*\,f^* & \text{if } g \in I^{n+1} \\ g^* + h^*\,f^* & \text{if } g \notin I^{n+1} \end{cases}$$

Finally, if $h\,f \in I^{n+1}$, then $x^* = g^*$, which proves the non-trivial inclusion of (2.1.1).

Lemma 7 leads us to the definition of a weak standard base.

<u>Definition.</u> Let (Q,M_o) be a local ring and $I,K \subset M_o$ any ideals. The elements $f_1,\ldots,f_m \in K$ are called a weak standard base of K with respect to I, if there exists an n_o such that

(2.1.6) $$gr_I^n(K,Q) = \sum_{i=1}^{m} in_I f_i \ gr_I^{n-d_i}(Q) \quad \text{for} \quad n \geq n_o\ ,$$

where $d_i = \nu_I(f_i)$.

Applying lemma 7 inductively, we get the following

<u>Corollary.</u> Let Q and I be as above and assume that $f_1,\ldots,f_m$ is a regular sequence and a stable superficial sequence for I. Then for each $i \in \{1,\ldots,m\}$, $f_1,\ldots,f_i$ is a weak standard base of $f_1\,Q + \ldots + f_i\,Q$ with respect to I.

<u>Lemma 8</u>. Let (Q,M_o) be a local ring, $I,K \subset M_o$ ideals and $f \in M_o$. We put $J = K + f\,Q$ and $d = \nu_I(f)$. Assume that for $n \geq n_o$ we have

(2.1.7) $$gr_I^n(J,Q) = gr_I^n(K,Q) + in_I\ f\ gr_I^{n-d}(Q)\ .$$

Then

(2.1.8) $$(gr_I(K,Q) : in_I f)_n = gr_I^n(K{:}f,Q) \quad \text{for} \quad n \geq n_o\ .$$

Proof. First from our assumption (2.1) we get

$$I^n \cap J \subset I^n \cap K + I^{n-d} f + I^{n+1} , \quad n \geq n_0 .$$

Intersecting with J on both sides and making induction, this yields

$$I^n \cap J \subset I^n \cap K + I^{n-d} f + I^{n+t} , \quad n \geq n_0, \ t \geq 0$$

and therefore we get

(2.1.9) $$I^n \cap J = I^n \cap K + I^{n-d} f .$$

Now let $x^* \in (gr_I(K,Q) : in_I f)_n$ be the initial form of some $x \in I^n$, $n \geq n_0$. Then $x f = a + b$, $a \in K$, $b \in I^{n+d+1}$. By (2.1.9) we get

$$b = x f - a \in J \cap I^{n+d+1} = K \cap I^{n+d+1} + I^{n+1} f .$$

Therefore, if we write $x f - a = c + e f$, where $c \in K$ and $e \in I^{n+1}$, it follows that $(x-e) f = a + c \in K$. Hence

$$x^* = in_I(x-e) \in gr_I^n(K : f,Q) .$$

<u>Corollary.</u> Let Q,M_0,I be as above and let $f_1,\ldots,f_m \in M_0$ be a regular sequence. Assume that for each $i \in \{1,\ldots,m\}$ the elements $f_1,\ldots,f_i$ are a weak standard base of $f_1 Q+\ldots+ f_i Q$ with respect to I . Then $(in_I f_1,\ldots,in_I f_m)$ is a weakly regular sequence in $gr_I(Q)$.

The proof is made by induction again.

<u>Lemma 9.</u> If (Q,M_0) is a local ring, $I \subset M_0$ an ideal and $f \in I$ an element such that $in_I f$ is weakly regular in $gr_I(Q)$, then f is superficial for I .

This is well known. The next proposition gives the relationship between superficial sequences and weakly regular sequences.

<u>Proposition 3.</u> Let (Q,M_0) be a local ring, $I \subset M_0$ any ideal and let $f_1,\ldots,f_m \in I$ be a regular sequence. Then the following condi-

ions are equivalent:

i) $f_1,\ldots,f_m$ is a stable superficial sequence for I .

ii) $(in_I\, f_1,\ldots,in_I\, f_m)$ is a weakly regular sequence in $gr_I(Q)$.

roof. (i) ⇒ (ii) holds by the corollaries to lemmas 7 and 8. We rove (ii) ⇒ (i) by induction on m , the case $m = 1$ being treated n lemma 9. Assume therefore that $m > 1$ and that the conclusion olds for $f_1,\ldots,f_{m-1}$. From the corollary to lemma 7 we know that $_1,\ldots,f_{m-1}$ is a weak standard base of $f_1 Q+\ldots+ f_{m-1} Q$ with espect to I . Let us write $\bar{Q} = Q/f_1 Q+\ldots + f_{m-1} Q$, $= I/f_1 Q +\ldots+ f_{m-1} Q$ and $\bar{f}_r$ for the image of f_r in Q . Then he canonical epimorphism

$$\alpha : gr_I(Q)/ \sum_{i=1}^{m-1} in_I\, f_i gr_I(Q) \rightarrow gr_{\bar{I}}(\bar{Q})$$

s an isomorphism in large degrees. Let z denote the class of $n_I\, f_m$ $\mathrm{mod}(in_I\, f_1,\ldots,in_I\, f_{m-1})$. If $\alpha(z) = 0$ then - since $\alpha(z)$ s weakly regular - $gr^n_{\bar{I}}(\bar{Q}) = 0$ for large n, and therefore $\bar{I}$ is ilpotent. This is impossible since $\bar{I}$ contains the non-zero-diviso $_r$. It follows that $\alpha(z) \neq 0$ and therefore

$$\alpha(z) = in_{\bar{I}}(\bar{f}_m) .$$

y lemma 9 , $\bar{f}_m$ is superficial for $\bar{I}$, and $\alpha(z) \neq 0$ means that

$$f_m \notin I^{d+1} + f_1 Q +\ldots+ f_{m-1} Q , d = \nu_I(f_m) .$$

<u>roposition 4.</u> Let (Q,M_o) be a local Cohen-Macaulay ring, let I e an M_o- primary ideal and let $f_1,\ldots,f_m \in I$ be elements such hat $(in_I\, f_1,\ldots,in_I\, f_m)$ is a weakly regular sequence in $gr_I(Q)$. hen $f_1,\ldots,f_m$ is a stable superficial sequence for I .

The proof is by induction like (ii) ⇒ (i) of proposition 3. In he inductive step one has to show in addition that $\bar{f}_m$ is not a ero-divisor in $\bar{Q}$. This follows from the fact that $\bar{Q}$ is Cohen-

Macaulay and $\bar{f}_m$ is part of a system of parameters, since it is superficial for an open ideal.

Remark. In connection with proposition 4 we would be interested in an example of a local Cohen-Macaulay ring (Q,M_0) and a zero-divisor f such that f is superficial for some ideal I of height > 0. (Note that then I cannot be M_0- primary.)

Originally we considered weakly regular sequences to study properties of blowing ups. One result related to the next section is the following:

Proposition 5. Let Q be a regular local ring, P_0 a prime ideal such that Q/P_0 is regular and $f_1,\ldots,f_m \in P_0$ a regular sequence. We put $R = Q/f_1 Q+\ldots+ f_m Q$ and $P = P_0/f_1 Q+\ldots+ f_m Q$. Assume that the initial forms of $f_1,\ldots,f_m$ in $gr_{P_0}(Q)$ are a weakly regular sequence. Then the blowing up of R with center P is again a complete intersection.

Proof. We fix $x \in P_0$ and put $Q' = Q[P_0/x]$. Furthermore we use the notation $I_j = f_1 Q+\ldots+ f_j Q$, $d_j = \nu_{P_0}(f_j)$, $j=1,\ldots,m$. From the corollary to lemma 7 (see also (2.1.9)) we deduce

$$P_0^n \cap I_j = \sum_{i=1}^{j} P_0^{n-d_i} f_i \quad \text{for } n \geq n_0 .$$

If we fix some $s \geq n_0$, we get

$$P_0^s(P_0^n \cap I_j) \subset P_0^{s+n} \cap I_j = \sum_{i=1}^{j} P_0^{s+n-d_i} f_i \quad \text{for all } n .$$

By a result of Valabrega - Valla ([VV$_2$]) we conclude that $f_1/x^{d_1},\ldots,f_j/x^{d_j}$ generate the strict transform I_j' of I_j in Q'. Therefore $f_1/X^{d_1},\ldots,f_m/X^{t_m}$ is a regular sequence in Q' generating I_m' ([VV$_2$], Thm 3.1).

Remark. The conclusion that I'_m is generated by a regular sequence holds without the regularity assumptions about Q and Q/P_0. In particular, if the blowing up of Q with center P_0 is Cohen-Macaulay, then the blowing up of R with center P is Cohen-Macaulay too. We would like to know under which assumptions a quadratic transform of a local Cohen-Macaulay ring is Cohen-Macaulay again (see problem 4). The use of weakly regular sequences for this question is suggested by the following simple observation:

Lemma 10. Let $A = \bigoplus_{n \geq 0} A_n$ be a graded ring, $x \in A_1$ and $S = A_{(x)}$ (the elements of degree 0 in A_x). Let $f_i \in A$, $i = 1,\dots,m$, be homogeneous elements of degree d_i and let $g_i = f_i/x^{d_i} \in S$. If $f_1,\dots,f_m$ is a weakly regular sequence, then $g_1,\dots,g_m$ is a regular sequence in S.

The proof is easy and will be onmitted.

(2.2) In [HO_2] we proved theorem 4 as a special case of a more general result. We compared the Hilbert function of a local ring R with the Hilbert function of some localization R_P, and to conclude normal flatness we had to use Bennett's criterion for permissibility ([B]). Also we were faced with the following problem: If (Q,M_0) is local and P_0 any prime, it may happen that $P_n^{(n)} \not\subset M_0^n$ ($P_0^{(n)}$ is the n-th symbolic power of P_0). A corresponding example was shown to us by J. Giraud.

Using the generalized Hilbert functions $H_{\underline{x},A}$ and the multiplicity $e(\underline{x},A,R)$ introduced in (1.1), we can prove a similar result even for ideals which are not prime. To obtain theorem 4, Bennett's criterion for permissibility is replaced by a result of R. Schmidt ([R.S.], [HSV], see theorem 6).

We will keep the following notation: (Q,M_0) is a local ring, $I_0 \subset M_0$ an ideal, $f_1,\dots,f_m \in I_0$ and $A = f_1 Q + \dots + f_m Q$. $\underline{x} = \{x_1,\dots,x_r\} \subset Q$ denotes a system of parameters for Q/I_0. Then $V_0 = I_0 + \underline{x} Q$ is M_0- primary. We put $R = Q/A$, $I = I_0/A$ and $V = V_0/A = I + \underline{x} R$.

The image of x_i in R will be denoted by y_i, so that $\underline{y} = \{y_1,\ldots,y_r\}$ is a system of parameters for R/I.

Lemma 11. Let $f \in I_0$ and $s = \nu_I(f)$. If $\mathrm{in}_{I_0}(f)$ is weakly regular in $\mathrm{gr}_{I_0}(Q)$, and f a non-zerodivisor, then

(2.2.1) $\quad H^{(o)}_{\underline{x},I_0/f\,Q}(n+s) = H^{(o)}_{\underline{x},I_0}(n+s) - H^{(o)}_{\underline{x},I_0}(n)$ for large n,

and in particular

(2.2.2) $\quad e(\underline{x},I_0/f\,Q,Q/f\,Q) = s\cdot e(\underline{x},I_0,Q)$.

Proof. For large n we have an exact sequence

$$0 \to I_0^n/I_0^{n+1} \overset{\beta}{\to} I_0^{n+s}/I_0^{n+s+1} \to I_0^{n+s}/I_0^{n+s+1} + f\,Q\,I_0^{n+s} \to 0,$$

where β is induced by multiplication with f. Now $\underline{x}$ is a multiplicity system for all these modules, and since the multiplicity symbol $e(\underline{x},-)$ is additive on exact sequences, the assertion follows.

Proposition 6. Let Q be a Cohen-Macaulay ring and assume that $e_0(V_0) = e(\underline{x},I_0,Q)$. Assume furthermore that

a) $\mathrm{in}_{I_0} f_1,\ldots,\mathrm{in}_{I_0} f_m$ is a weakly regular sequence in $\mathrm{gr}_{I_0}(Q)$,

b) $\mathrm{in}_{V_0} f_1,\ldots,\mathrm{in}_{V_0} f_m$ is a weakly regular sequence in $\mathrm{gr}_{V_0}(Q)$.

Then the following conditions are equivalent:

(i) $\quad e_0(V) = e(\underline{x},I,R)$.

(ii) $\quad \nu_{I_0}(f_i) = \nu_{V_0}(f_i)$, $i=1,\ldots,m$.

Proof. First we note that $f_1,\ldots,f_m$ is a regular sequence by b). Therefore $f_1,\ldots,f_i$ is a weak standard base of $f_1Q+\ldots+f_i\,Q$ with respect to I_0 and V_0. This allows to apply lemma 11 inductively to get

$$e(\underline{y},I,R) = s_1\cdot\ldots\cdot s_m\, e(\underline{x},I_0,Q)$$

and

$$e_o(V) = t_1 \cdot \ldots \cdot t_m \; e_o(V_o) ,$$

where $s_i = \nu_{I_o}(f_i) \leq t_i = \nu_{V_o}(f_i)$. So clearly (i) and (ii) are equivalent.

Theorem 5. Assume that Q is a Cohen-Macaulay ring and that $H^{(o)}_{\underline{x},I_o}(n) = (\Delta^r H^{(o)}_{V_o,Q})(n)$ *) for large n. Assume furthermore that

a) $in_{I_o} f_1,\ldots,in_{I_o} f_m$ is a weakly regular sequence in $gr_{I_o}(Q)$,

b) $in_{V_o} f_1,\ldots,in_{V_o} f_m$ is a weakly regular sequence in $gr_{V_o}(Q)$.

Then the following conditions are equivalent:

(i) $e_o(V) = e(\underline{y},I,R)$.

(ii) $\nu_{I_o}(f_i) = \nu_{V_o}(f_i)$, $i=1,\ldots,m$.

(iii) $H^{(o)}_{\underline{y},I}(n) = (\Delta^r H^{(o)}_{V,R})(n)$ for large n.

Proof. The equivalence of (i) and (ii) was shown above, and (iii) $\Rightarrow$ (i) is trivial. To prove (ii) $\Rightarrow$ (iii), we note that $f_1,\ldots,f_m$ is a regular sequence, and $f_1,\ldots,f_i$ is a weak standard base of $f_1 Q + \ldots + f_i Q$ with respect to I_o and V_o. Therefore we can make induction on m, and the result follows from (2.2.1).

If we replace "weakly regular sequence" by "regular sequence" in a) and b) above, the same conclusions as before can be made for all n, i.e. for all degrees in the graded rings (see also [VV_1], of which our section (2.1) is a generalization). This gives the following result:

Theorem 5'. Assume that Q is a Cohen-Macaulay ring and that $H^{(r)}_{\underline{x},I_o}(n) = H^{(o)}_{V_o,Q}(n)$ for all n. Assume furthermore that

---

*) As usual, $\Delta H(n) = H(n+1) - H(n)$ and $\Delta^r = \Delta \cdot \Delta^{r-1}$.

a) $in_{I_0} f_1,\ldots,in_{I_0} f_m$ is a regular sequence in $gr_{I_0}(Q)$,

b) $in_{V_0} f_1,\ldots,in_{V_0} f_m$ is a regular sequence in $gr_{V_0}(Q)$.

Then the following conditions are equivalent:

(i) $e_0(V) = e(\underline{y},I,R)$.

(ii) $\nu_{I_0}(f_i) = \nu_{V_0}(f_i)$, $i=1,\ldots,m$.

(iii) $H^{(r)}_{\underline{y},I}(n) = H^{(o)}_{V,R}(n)$ for all n.

R. Schmidt has given an interpretation of condition (iii) (see [HSV], Satz 3.13, p. 121):

<u>Theorem 6.</u> The following conditions are equivalent:

(i) $H^{(r)}_{\underline{y},I}(n) = H^{(o)}_{V,R}(n)$ for all n.

(ii) For all n, $gr^n_I(R)$ is a Cohen-Macaulay module of depth r over R/I.

<u>Corollary.</u> Assume that Q is a Cohen-Macaulay ring and I_0 is a prime ideal such that Q/I_0 is regular and Q is normally flat along I_0 . If A is a strict complete intersection with respect to I_0 , then the following conditions are equivalent:

(i) $e_0(R) = e_0(R_I)$.

(ii) R is normally flat along I .

If in theorem 5' we assume Q/I_0 to be regular and $\underline{x}$ a regular system of parameters, a) could be replaced by the seemingly weaker assumption that the initial forms in $gr_{I_0}(Q)$ are a weakly regular sequence. This follows from ([H_1], theorem 2 in ch. II), since b) implies that $f_1,\ldots,f_m$ is a standard base of A with respect to M_0 , and therefore normal flatness follows from condition (ii). Using ([H_1], lemma 7, p. 190) it turns out that the weakly regular sequence $in_{I_0} f_1,\ldots,in_{I_0} f_m$ is in fact regular.

The next proposition is an application of theorem 5'.

Proposition 7. Assume that Q is a Cohen-Macaulay ring with infinite residue field, and I_o is a prime ideal such that Q/I_o is regular and Q normally flat along I_o. Assume furthermore that

a) $in_{M_o} f_1,\dots,in_{M_o} f_m$ are a regular sequence in $gr_{M_o}(Q)$,

b) $f_1,\dots,f_m$ are part of a minimal system of generators of a minimal reduction J of I_o such that $J I_o = I_o^2$.

Then the following conditions are equivalent:

(i) $e_o(R) = e_o(R_I)$.

(ii) $\nu_{I_o}(f_i) = \nu_{M_o}(f_i)$, $i=1,\dots,m$.

(iii) R is normally flat along I.

Proof. We only have to verify condition b) of theorem 5'. Let $f_1,\dots,f_s$ $(s \geq m)$ be a minimal system of generators for J. Then $s = ht(I_o)$ (see §1), and therefore $f_1,\dots,f_s$ is a regular sequence. On the other hand $J I_o = I_o^2$ implies that $f_1,\dots,f_s$ is a standard base of J with respect to I_o. By [VV1] we get that $in_{I_o} f_1,\dots,in_{I_o} f_s$ is a regular sequence in $gr_{I_o}(Q)$.

Remark. If I_o is prime, and if e and μ are the multiplicity and the embedding dimension of Q_{I_o} respectively, then it is known that $e \geq \mu - ht(I_o) + 1$. Condition b) of proposition 7 implies that $e = \mu - ht(I_o) + 1$. (See [Sa])

Example. Let $R = k[[t^2,t^3,t^2u^2,u^3]]$ and put $x = u^3$, $y = t^2u^2$, $z = t^3$ and $w = t^2$. Then R can be written as

$$R = k[[X,Y,Z,W]]/(Y^3 - Z^2X^2, Z^2 - W^3) .$$

Let $f_1 = Y^3 - Z^2 X^2$, $f_2 = Z^2 - W^3$ and $P_o = (Y,Z,W)$, $M_o = (X,Y,Z,W)$. R is equimultiple along $P = (y,z,w)$, and $\nu_{P_o}(f_1) \neq \nu_{M_o}(f_1)$.

Nevertheless R is normally flat along P . To see this we replace f_1 by $f_1' = Y^3 - W^3 X^2$ and remark that the initial forms of f_1', f_2 with respect to M_0 and P_0 are regular sequences.

Our result shows that in Robbiano's example we cannot improve the bad behaviour in $gr_{P_0}(Q)$ by choosing different equations to define R .

Added in proof:
The time between the conference and publication has been used to give some answers to the problems mentioned in the text:

Problem 1. A condition properly between normal flatness and equimultiplicity is " A^n/A^{n+1} is a flat R/A-module for large n " . This condition (for R/A regular) has been studied in: U. Orbanz - L. Robbiano, Projective normal flatness and Hilbert functions, to appear. In particular this condition gives a different approach to Theorem 5 in §2 with a slightly stronger result.

Problem 2. We have $e(R') \leq e(R)$ if R is quasi-unmixed, or more generally, if $ht(P) = l(P)$ and R' has the correct dimension. A proof of this and more is given in: U. Orbanz, Multiplicities and Hilbert functions under blowing up, Manuscripta math. 36(1981), 179-186.

Problem 3. φ is certainly an isomorphism if R/A is Cohen-Macaulay. This is contained in [HSV], p. 136, Satz 1.1.

Problem 4.a) There are certain papers (e.g. by Valla, Sally, Goto ...) concerned with the Cohen-Macaulay property of the rings $gr_I(R)$ and $\oplus I^n$. Our question is a little different, since the Cohen-Macaulayness of these rings is stronger than the Cohen-Macaulayness of $Bl_I(R)$.

Problem 4.b) If $R = k[[X,Y,Z,W]]/(Y^2W^3-Z^2, YZ-XW^3, XZ-Y^3)$ is blown up at the maximal ideal, the result is not Cohen-Macaulay, although R itself is Cohen-Macaulay. This example has been given by L. Robbiano in: On normal flatness and some related topics, to appear in "Commutative Algebra: Proceedings of the Trento Conference", Marcel Dekker, New York.

References

[AV] R. Achilles - W. Vogel, Über vollständige Durchschnitte in lokalen Ringen, Math. Nachr. 89 (1979), 285-298

[B] B.M. Bennett, On the characteristic functions of a local ring, Ann. of Math. 91 (1970), 25-87

[Bö] E. Böger, Eine Verallgemeinerung eines Multiplizitätensatzes von D. Rees, J. Algebra 12 (1969), 207-215

[D] E.C. Dade, Multiplicity and monoidal transformations, Thesis, Princeton University 1960, unpublished

[H_1] H. Hironaka, Resolution of singularities of an algebraic variety over a field of characteristic zero I-II, Ann. of Math. 79 (1964), 109-326

[H_2] H. Hironaka, Certain numerical characters of singularities, J. Math. Kyoto Univ. 10 (1970), 151-187

[HO_1] M. Herrmann - U. Orbanz, Faserdimensionen von Aufblasungen lokaler Ringe und Äquimultiplizität, J. Math. Kyoto Univ. 20 (1980), 651-659

[HO_2] M. Herrmann - U. Orbanz, Normale Flachheit und Äquimultiplizität für vollständige Durchschnitte, J. Algebra 70 (1981), 437-451

[HSV] M. Herrmann - R. Schmidt - W. Vogel, Theorie der normalen Flachheit, Teubner-Texte zur Mathematik, Leipzig 1977

[Li_1] J. Lipman, Desingularization of two-dimensional schemes, Ann. of Math. 107 (1978), 151-207

[Li_2] J. Lipman, Equimultiplicity, reduction and blowing up, in: "Commutative Algebra: Analytical methods", Marcel Dekker, New York

[Li_3] J. Lipman, Relative Lipschitz-saturation, Amer. J. Math. 93 (1975), 791-813

[No] D.G. Northcott, Lessons on rings, modules and multiplicities, Cambridge University Press 1968

[N-R] D.G. Norhcott - D. Rees, Reductions of ideals in local rings, Proc. Camb.Phil.Soc. 50 (1954), 145-158

[R] D. Rees, $\mathfrak{a}$-transforms of local rings and a theorem on multiplicities of ideals, Proc. Camb.Phil. Soc. 57 (1961), 8-17

[Ra] L.J. Ratliff Jr., On the prime divisors of zero in form rings, Pacific J. Math. 70 (1977), 489-517

[R.S.] R. Schmidt, Normale Flachheit als Spezialfall der Cohen-Macaulay-Eigenschaft von Graduierungen, Dissertation Humboldt-Universität Berlin 1976

[Sa] J.D. Sally, On the associated graded ring of a local Cohen-Macaulay ring, J. Math. Kyoto Univ. 17 (1977), 19-21

[Sch] W. Schickhoff, Whitneysche Tangentenkegel, Multiplizitätsverhalten, Normal-Pseudoflachheit und Äquisingularitätstheorie für Ramissche Räume, Schriftenreihe des Math. Instituts der Universität Münster, 2. Serie, Heft 12, 1977

[Si_1] B. Singh, Effect of a permissible blowing-up on the local Hilbert functions, Inventiones math. 26 (1974), 201-212

[Si_2] B. Singh, A numerical criterion for the permissibility of a blowing up, Comp. Math. 33 (1976), 15-28

[T] B. Teissier, Résolution simultanée et cycles évanescent, Lecture Notes in Math. no. 777, p. 82-146, Berlin - New York - Heidelberg 1980

[VV_1] P. Valabrega - G. Valla, Form rings and regular sequences, Nagoya Math. J. 72 (1978), 93-101

[VV_2] P. Valabrega - G. Valla, Standard bases and generators for the strict transform, preprint 1979

LIAISON ET RESIDU

par M. LEJEUNE-JALABERT

§ 1. - LIAISON.

On considère R un anneau local régulier (par exemple l'anneau des séries convergentes $\mathbb{C}\{x_1,\dots,x_n\}$ ou le localisé à l'origine de l'anneau de polynômes $\mathbb{C}[x_1,\dots,x_n]$ pour fixer les idées), deux idéaux propres I et J dans R et un idéal $K \subset I \cap J$ tel que R/K soit une intersection complète (i.e. K est engendré par des éléments qui forment une suite régulière de R). On dit que R/I et R/J sont liés par R/K si

1) R/I et R/J sont équidimensionnels (ceci signifie en particulier sans composantes immergées) .

2) $(K:I) = \{g \in R : gI \subset K\} = J$

   $(K:J) = \{f \in R : fJ \subset K\} = I$.

Cette définition généralise la situation géométrique suivante : la réunion de V_1 et V_2 supposées équidimensionnelles sans composantes irréductibles communes est une intersection compète X .

On trouve trace de cette notion dès le siècle dernier notamment chez M. Noether et G. Halphen. En 1916, chez F.S. Macaulay, dans le chapitre de "The Algebraic theory of modular systems" consacré au "système inverse" première apparition d'une théorie algébrique de la dualité, elle s'introduit assez naturellement sous la forme suivante : nous supposerons ici pour simplifier que $R = \mathbb{C}\{x_1,\dots,x_n\}$ (contrairement à Macaulay qui étudie le cas affine) et que R/I est un anneau de dim 0 . Il s'agit de savoir quelles conditions linéaires doivent satisfaire les $c_\alpha \in \mathbb{C}$ pour que $F = \sum_\alpha c_\alpha x^\alpha \in I$. Soit $M = (x_1,\dots,x_n)$ l'idéal maximal de R . Pour tout $n \in \mathbb{N}$, on considère l'injection de l'espace vectoriel de dimension finie IR/M^{n+1} dans R/M^{n+1} . Le sous-espace vectoriel $(R/I+M^{n+1})^*$ de $(R/M^{n+1})^*$

est l'espace vectoriel des conditions linéaires qui doivent être satisfaites par les c_α, $|\alpha| \leq n$ pour qu'il existe $F \in I$ tel que $F \equiv \sum c_\alpha x^\alpha \bmod M^{n+1}$. Ce sont les équations modulaires de I à l'ordre n. La réunion (ou limite inductive) des équations modulaires pour tous les ordres de I constitue le système inverse à I (cette terminologie rappelle qu'on a inversé les matrices représentant l'inclusion de IR/M^{n+1} dans R/M^{n+1}). Nous le désignerons par I^{-1}. C'est un sous-espace vectoriel de $\varinjlim (R/M^{n+1})^* = \hat{R}^*_{cont}$ ensemble des formes linéaires sur $\hat{R} = \mathbb{C}[[x_1,\dots,x_n]]$ annulant M^k dès que k est assez grand. La structure de R-module sur $\hat{R}^*_c$ définie par $P\cdot E(Q) = E(P\cdot Q)$ induit une structure de R-module sur I^{-1} de sorte que :

$$I^{-1} = \mathrm{Hom}_R(R/I, \hat{R}^*_c) .$$

On obtient facilement le

1.1. - THEOREME. - ([MAC] n°61) $F \in I \Leftrightarrow E(F) = 0$, $\forall E \in I^{-1}$.

Macaulay constate alors que le R-module des équations modulaires d'une intersection complète est monogène. Au sujet des liaisons, il remarque que connaissant les générateurs $F_1,\dots,F_n$ de I et l'équation modulaire de K, il obtient facilement non les générateurs de $J = (K:I)$ mais ses équations modulaires.

1.2. - PROPOSITION. - ([MAC] n°62) La suite
$0 \longrightarrow (K:I)^{-1} \longrightarrow K^{-1} \longrightarrow K^{-1} \otimes_R R/I \longrightarrow 0$ est exacte.

Démonstration. - $G \in (K:I) \Leftrightarrow \forall i = 1\dots n$, $GF_i \in K \Leftrightarrow \forall i = 1..n$, $\forall E \in K^{-1}$, $E(GF_i) = 0 \Leftrightarrow \forall i = 1..n$, $\forall E \in K^{-1}$, $F_i \cdot E(G) = 0 \Leftrightarrow \forall E' \in (K:I)^{-1}$ $E'(G) = 0$.

Autrement dit $(K:I)^{-1} = \sum F_i K^{-1}$.

Pour étudier les courbes gauches (d'abord lisses et irréductibles dans $\mathbb{P}^3$), travaillant une idée proposée par F. Sévéri, F. Gaeta [G] introduit la définition suivante : (1952).

1.3. - DEFINITION. - Un quotient R/I équidimensionnel de R local régulier de codimension 2 est à résiduel 0 si et seulement si I admet une base de 2 éléments.

R/I est à résiduel ρ , si R/I peut être lié par une intersection complète R/K à R/J de résiduel $\rho-1$ et s'il ne peut être lié à aucun R/J de résiduel moindre que $\rho-1$.

Apéry, Dubreil se sont intéressés aux propriétés qui se conservent par liaisont. Peskine et Szpiro [P.S.] ont repris ces questions (1974) disposant des notions de profondeur et des théorèmes de dualité, en particulier en vue d'étudier les déformations des variétés projectives de codim 2 à cône projetant de Cohen-Macaulay.

1.4. - THEOREME (Gaeta, Peskine, Szpiro). - Soit R/I un quotient équidimensionnel de codim 2 de R local régulier. Les conditions suivantes sont équivalentes :

1) R/I est Cohen-Macaulay ;

2) R/I est à résiduel fini.

Si I admet un système minimal de n générateurs, R/I peut être lié à R/J tel que J admette un système minimal de $n-1$ générateurs.

Watanabe, Buschbaum, Eisenbud $[B.E.]_1$ donnent un autre exemple d'anneau à résiduel fini cette fois-ci en codimension 3.

1.5. - THEOREME. - Si R/I est un anneau de Gorenstein de codim 3 (i.e. R/I est Cohen-Macaulay et son module dualisant est un R-module monogène), R/I est à résiduel fini. Il en est de même (toujours en codim 3) si R/I est presque une intersection complète (i.e. R/I est Cohen-Macaulay et I est engendré par 4 générateurs).

Ici R/I Gorenstein tel que I admette un système minimal de n générateurs (nécessairement n est impair) est lié à R/J presque intersection complète elle-même liée à R/I_1 Gorenstein tel que I_1 admette un système minimal de $n-2$ générateurs.

1.6. - <u>Exemple</u>. - Par contre, Buchweitz montre que $\mathbb{C}\{x_1,x_2,x_3\}/M^2$ où $M = (x_1,x_2,x_3)$ n'est pas à résiduel fini. (Ceci est néanmoins déterminantiel et on remarque que $\mathbb{C}\{x_1,x_2,x_3\}/M^2$ est lié à lui-même par $\mathbb{C}\{x_1,x_2,x_3\}/x_1^2,x_2^2,x_3^2)$.

§2. - RESIDU.

On connait bien la formule intégrale de Cauchy

$$\frac{1}{(2i\pi)^n}\int_{|z_i|=\epsilon}\frac{g(z_1,\dots,z_n)dz_1\wedge\dots\wedge dz_n}{z_1^{k_1+1}\dots z_n^{k_n+1}}=\frac{1}{(k_1+\dots+k_n)!}\frac{\partial^{k_1+\dots+k_n}g}{\partial z_1^{k_1}\dots\partial z_n^{k_n}}\quad(0)$$

et on sait par ailleurs que si $R=\mathbb{C}\{x_1,\dots,x_n\}$ et si $I=(f_1,\dots,f_n)$ où $f_1,\dots,f_n$ est une suite régulière

$$(2.1)\qquad rg_{\mathbb{C}}\, R/I=\frac{1}{(2i\pi)^n}\int_{|f_i|=\epsilon}\frac{df_1\wedge\dots\wedge df_n}{f_1\wedge\dots\wedge f_n}\ .$$

Nous avons cherché à obtenir de façon analogue par un calcul de résidu le $rg_{\mathbb{C}}\, R/I$ si I n'est plus engendré par une suite régulière mais reste primaire pour $M=(x_1,\dots,x_n)$. A partir de maintenant $R=\mathbb{C}\{x_1,\dots,x_n\}$, Ω^n est le R-module des n-formes différentielles et R/I est un anneau de dimension 0. Le théorème de dualité locale nous dit qu'il existe une application bilinéaire non dégénérée :

$$\mathrm{Ext}^n_R(R/I,\Omega^n)\times R/I\longrightarrow\mathbb{C}\ .$$

Elle se définit "assez facilement" de façon transcendante de la façon suivante : On considère la résolution de Dolbeault de Ω^n par les germes de courants (formes différentielles à coefficients distribution) qui est une résolution injective (à cause du théorème de division des distributions)

$$0\longrightarrow\Omega^n\longrightarrow{}'\mathcal{B}^{n,0}\longrightarrow{}'\mathcal{B}^{n,1}\longrightarrow\dots\longrightarrow{}'\mathcal{B}^{n,n}\longrightarrow 0$$

et $\mathrm{Ext}^n_R(R/I,\Omega^n)$ s'identifie au n-ième groupe de cohomologie de $\mathrm{Hom}_R(R/I,{}'\mathcal{B}^{n,\cdot})$. Si $\varphi : R/I\longrightarrow{}'\mathcal{B}^{n,n}$ est un représentant d'un de ces éléments et g un élément de R/I, l'accouplement $\langle\bar\varphi,g\rangle$ est donné par $\varphi(g)(\tilde 1)$ où $\tilde 1$ est une fonction C^∞ à support compact coïncidant avec la constante 1 sur un voisinage de 0. Par ailleurs, on peut calculer $\mathrm{Ext}^n_R(R/I,\Omega^n)$ avec une résolution libre de R/I. En particulier, si R/I est une intersection complète, le complexe de Koszul $\Lambda^{\cdot}R^n$ construit à partir de $f_1,\dots,f_n$ un système de générateurs de I est une résolution libre de R/I et $\mathrm{Ext}^n_R(R/I,\Omega^n)$ est le n-ième groupe de cohomogie de $\mathrm{Hom}(\Lambda^{\cdot}R^n,\Omega^n)$. Si $\omega\in\Omega^n$, $\begin{bmatrix}\omega\\ f_1,\dots,f_n\end{bmatrix}$ désigne la classe dans $\mathrm{Ext}^n_R(R/I,\Omega^n)$ de l'application R-linéaire envoyant $e_1\wedge\dots\wedge e_n$ sur ω et la formule (2.1) s'interprète alors sous la forme:

$$\mathrm{rg}_{\mathbb{C}}\, R/I = \langle \begin{bmatrix} df_1 \wedge \ldots \wedge df_n \\ f_1 \ldots f_n \end{bmatrix}, 1 \rangle .$$

Nous allons maintenant rappeler brièvement comment ces calculs de résidu se rattachent aux considérations précédentes : On peut montrer [S] que $\omega_{R/I} = \mathrm{Ext}^n_R(R/I, \Omega^n)$ s'identifie à $I^{-1} = \mathrm{Hom}_R(R/I, \hat{R}^*_{\mathrm{cont}})$ en interprétant $\hat{R}^*_{\mathrm{cont}}$ comme le n-ième groupe de cohomogie du complexe

$$0 \to {}'\mathcal{D}^{n,0}_{\{0\}} \to {}'\mathcal{D}^{n,1}_{\{0\}} \to \ldots \to {}'\mathcal{D}^{n,n}_{\{0\}} \to 0$$

des courants de support l'origine. Un élément de $\hat{R}^*_{\mathrm{cont}}$ est alors représenté par $\sum_{|\alpha| \leq k} c_\alpha D^\alpha \delta\, dz_1 \wedge \ldots \wedge dz_n \wedge \overline{dz_1} \wedge \ldots \wedge \overline{dz_n}$ où δ est la masse de Dirac. La proposition 1.2 de Macaulay devient alors :

$$0 \to \omega_{R/(K:I)} \to \omega_{R/K} \to \omega_{R/K} \otimes_R R/I \to 0$$

est exacte. Comme elle entraîne que $\mathrm{rg}_{\mathbb{C}} R/K = \mathrm{rg}_{\mathbb{C}} R/(K:I) + \mathrm{rg}_{\mathbb{C}} R/I$, il s'ensuit immédiatement que $K : (K:I) = I$ et que

$$0 \to \omega_{R/I} \to \omega_{R/K} \to \omega_{R/K} \otimes_R R/(K:I) \to 0 .$$

Si R/I n'est pas une intersection complète, le module dualisant $\omega_{R/I}$ (ou module des résidus du gros point R/I) s'interprète donc comme un sous-module du module des résidus de R/K l'intersection complète en terme de l'anneau R/J lié à R/I par R/K .

2.2. - DEFINITION. - Généralisant l'écriture des symboles de Grothendieck, si $\mathcal{R}_\bullet$
$0 \to R^{p_n} \xrightarrow{\Phi_n} R^{p_{n-1}} \to \ldots \to R^{p_1} \xrightarrow{\Phi_1} R \to 0$ est une résolution libre de type fini de longueur n de R/I, on désignera par $\begin{bmatrix} \omega_1, \ldots, \omega_{p_n} \\ t_{\Phi_n} \end{bmatrix}$ la classe dans $\omega_{R/I}$ de l'application R-linéaire $R^{p_n} \to \Omega^n$ envoyant e_i sur ω_i .

Soit $h_1, \ldots, h_n$ une suite régulière engendrant K, $\Lambda^\cdot R^n$ la résolution de Koszul construite à partir de $h_1, \ldots, h_n$. Soit $\alpha_\bullet : \Lambda^\cdot R^n \to \mathcal{R}_\bullet$ un morphisme de complexes déduit de l'inclusion $K \subset I$. Si $\alpha_n(\varepsilon_1 \wedge \ldots \wedge \varepsilon_n) = \sum_{i=1}^{p_n} g_i e_i$, $\begin{bmatrix} \omega_1, \ldots, \omega_{p_n} \\ t_{\Phi_n} \end{bmatrix}$ s'envoie sur $\begin{bmatrix} \sum g_i \omega_i \\ h_1, \ldots, h_n \end{bmatrix}$ et $(K:I) = J = (g_1, \ldots, g_{p_n} ; h_1, \ldots, h_n)$. En fait, connaissant $\mathcal{R}_\bullet$ et $\alpha_\bullet$, on peut non seulement trouver un système de générateurs de $(K:I)$ mais une syzygie.

2.3. - PROPOSITION (Ferrand). Le cône du morphisme dual $\alpha_\bullet^\vee : \mathcal{R}_\bullet^\vee \to (\Lambda^\bullet R^n)^\vee$ est une résolution libre de $R/(K:I)$.

Utilisant l'isomorphisme $(\Lambda^i R^n)^\vee \simeq \Lambda^{n-i} R^n$, on obtient une résolution de longueur n :

$$0 \to \mathcal{R}_1^\vee \to \mathcal{R}_2^\vee \oplus \Lambda^{n-1} R^n \to \mathcal{R}_3^\vee \oplus \Lambda^{n-2} R^n \to \ldots \to \mathcal{R}_n^\vee \oplus \Lambda^1 R^n \to R \to 0$$

où $d(f, \varepsilon_{i_1} \wedge \ldots \wedge \varepsilon_{i_k}) = (-d^\vee f, d(\varepsilon_{i_1} \wedge \ldots \wedge \varepsilon_{i_k}) + \alpha^\vee(f))$. Cependant même si $\mathcal{R}_\bullet$ est la résolution minimale de R/I , il se peut qu'on n'obtienne pas ainsi la résolution minimale de $R/(K:I)$ (cf. 1.4 et 1.5).

2.4. - THEOREME [LJ]. - Soit $R = \mathbb{C}\{x_1, \ldots, x_n\}$ et R/I un anneau de dimension 0 . Soit $\mathcal{R}_\bullet$ une résolution libre de type fini de longueur n de R/I comme dans 2.2, $h_1, \ldots, h_n$ une suite régulière engendrant $K \subset I$, $g_1, \ldots, g_{p_n}$; $h_1, \ldots, h_n$ le système de générateurs de $(K:I)$ déduit du morphisme $\alpha_n : \Lambda^n R^n \to \mathcal{R}_n$ comme ci-dessus. Soit

$$\eta_i = \sum_{i_{n-1}, \ldots, i_1} d^t\Phi_{n,i,i_{n-1}} \wedge \ldots \wedge d^t\Phi_{s,i_s,i_{s-1}} \wedge \ldots \wedge d^t\Phi_{1,i_1,1}$$

où $i_s = 1 \ldots p_s$, $s = 1 \ldots n-1$ et ${}^t\Phi_s : R^{p_{s-1}} \to R^{p_s}$ est l'application définie par la matrice transposée de Φ_s

$$\mathrm{rg}_{\mathbb{C}} R/I = \frac{1}{n!} \begin{bmatrix} \eta_1, \ldots, \eta_{p_n} \\ {}^t\Phi_n \end{bmatrix} (1) = \frac{1}{(2i\pi)^n n!} \int_{|h_j| = \varepsilon} \frac{\sum_{i=1}^{p_n} g_i \eta_i}{h_1 \ldots h_n} .$$

2.5. - Remarque. Le théorème 2.4 donne dans certains cas un moyen de calcul "effectif" pour tester si $F \in I$ idéal de $R = \mathbb{C}\{x_1, \ldots, x_n\}$ dont on sait a priori que $\mu = \mathrm{rg}_{\mathbb{C}} R/I < +\infty$.

Il s'agit en fait uniquement de calculer μ . En effet, on vérifie que $M^\mu \subset I$. Les équations modulaires de I (il y en a μ indépendantes d'après le théorème de dualité locale) coïncident donc avec ses équations modulaires à l'ordre $\mu-1$. Soit $(\omega_\alpha)_{\alpha \in A}$ les $\binom{\mu-1+n}{n}$ monômes de R dont les images dans R/M^μ constituent une $\mathbb{C}$-base de R/M^μ . Pour tout $\beta \in A$ tel que ω_β soit de degré inférieur ou égal à $\mu-2$, il existe $c_{\beta,i;\alpha} \in \mathbb{C}$ tel que

$$\omega_\beta f_i \equiv \sum_{\alpha\in A} c_{\beta,i;\alpha}\omega_\alpha \mod M^\mu .$$

I^{-1} est alors l'espace vectoriel des solutions du système linéaire

$$\sum_{\alpha\in A} c_{\beta,i;\alpha} X_\alpha = 0 .$$

La détermination de μ est donc ramené par 2.4 à un calcul de résidu.

Une syzygie explicite pour R/I fournit les n-formes η_i . Signalons qu'on trouve dans [E.N], $[B.E]_2$, [L] les syzygies des variétés déterminantielles. D'autre part, si $n = 2$ ou si $n = 3$ et R/I est de Gorenstein, on dispose de théorèmes indiquant la structure générale des syzygies. Dans le 1er cas [BUR] R/I est déterminantiel, dans le 2ème cas $[B.E]_1$ R/I est pfaffien.

On obtient une suite régulière $h_1,\ldots,h_n$ formés d'éléments de I en écrivant n combinaisons linéaires générales (à coefficients dans $\mathbb{C}$) des $f_1,\ldots,f_{p_1}$. Pour obtenir les g_i correspondants (décrivant l'anneau R/J lié à R/I par $R/h_1,\ldots,h_n$), il faut savoir déterminer $\alpha_\bullet : \Lambda^n R^n \to \mathcal{R}_\bullet$ étendant l'inclusion de $K = (h_1,\ldots,h_n)$ dans I . Ceci se fait explicitement dans les 2 cas précités ([P.S] §3 et $[B.E]_1$ Th. 5.3) et s'obtiendrait immédiatement si on savait munir $\mathcal{R}_\bullet$ d'une structure d'algèbre graduée différentielle associative et commutative.

Il reste donc à calculer un résidu relatif à une intersection complète $h_1,\ldots,h_n$. Pour ce faire, il suffit de déterminer des $r_1,\ldots,r_n \in \mathbb{N}$ tels que $x_i^{r_i} = \sum_j a_{ij}h_j$. On sait alors que :

$$\int_{|h_j|=\varepsilon} \frac{\omega}{h_1\cdots h_n} = \int_{|x_j|=\varepsilon} \frac{\det(a_{ij})\omega}{x_1^{r_1}\ldots x_n^{r_n}}$$

et on termine en appliquant la formule intégrale de Cauchy.

BIBLIOGRAPHIE

[A] B. ANGENIOL : Classes fondamentales et traces de différentielles (ce volume) (1981).

[B.E]$_1$ D. BUCHSBAUM et D. EISENBUD : Algebra structures for finite free resolutions, and some structure theorems for ideals of codim 3. Amer. J. Math. Vol. 99 n°3, pp. 447-485 (1977).

[B.E]$_2$ D. BUCHSBAUM et D. EISENBUD : Generic free resolutions and a family of generically perfect ideals. Advances in Math. 18, pp. 245-301 (1975).

[BUR] L. BURCH : On ideals of finite homological dimension in local rings. Proc. Cam. Phil. Soc. 64, pp. 941-946 (1968).

[E.N] J. EAGON et D. NORTHCOTT : Ideals defined by matrices and a certain complex associated to them. Proc. Royal Soc. of London, series A, t. 269, pp. 188-204 (1962).

[E.R.S] D. EISENBUD, O. RIEMENSCHNEIDER, F. SCHREYER : Projective resolutions of Cohen-Macaulay algebras. Preprint A paraitre (1980).

[G] F. GAETA : Quelques progrès récents dans la classification des variétés algébriques d'un espace projectif. Deuxième colloque de Géométrie algébrique Liège. C.B.R.M. 145-181 (1952).

[H] D. HILBERT : Über die Theorie der Algebraischen Formen. Math. Ann. 36 pp. 473-534 (1890).

[LA] A. LASCOUX : Syzygies des variétés déterminantales. Advances in Math. Vol n = 30 N°3, pp. 202-237 (1978).

[L.J] M. LEJEUNE-JALABERT : Remarque sur la classe fondamentale d'un cycle, Note au C.R.A.S. à paraître (1981).

[MAC] F.S. MACAULAY : The algebraic theory of modular systems Cambridge university press. (1916) ou New-York, London, Stechert-Hafner (1964) (Cambridge tracts in Mathematics and mathematical physics, 19).

[P.S] C. PESKINE et L. SZPIRO : Liaison des variétés algébriques. Inventiones Math. 26, pp. 271-302 (1974).

[S] J.P. SERRE : Sur les modules projectifs, Séminaire Dubreil (1960).

Elementary transformations in the theory of algebraic vector bundles

By

Masaki MARUYAMA

Introduction. About ten years ago I raised the following problem: Construct many vector bundles on a higher dimensional projective variety". While considering this question, I found an operation on algeraic vector bundles, an elementary transformation. By using this peration, many vector bundles were constructed on every projective noningular variety. On the other hand, it appears recently that the ransformation is a powerful tool in various directions of the theory of lgebraic vector bundles. Let me mention some examples:

(1) This played a key role in the deformation theory and the esingularization of the moduli spaces of vector bundles on curves Narasimhan and Ramanan [11], [12]).

(2) This has been used in essential way by R. Hartshorne to solve he following problem: Let E be a semi-stable vector bundle of rank on P^3. If m is an integer such that $\chi(E(m)) > 0$ and $m \geq 0$, hen $H^0(P^3, E(m)) \neq 0$.

(3) S. Langton exploited this operation to prove "valuative crierion" of the properness of moduli spaces of semi-stable sheaves ([5]).

(4) As I mentioned in the above, this is useful to construct many ector bundles (Maruyama [6] and §3 of this article).

(5) By using this we can determine the multiplicity at a point of he curve of jumping lines of a semi-stable vector bundle E on P^2 ith $r(E) = 2$ and $c_1(E) = 0$ (see §4 of this article).

It seems useful at this moment to give an expository account of lementary transformations. In §1 I will try to show what the elemenary transformation is. §2 is devoted to show several properties of the operation, the compatibility with base changes, the commutativity of elementary transformations, etc.. In the final part of this section the family of stable vector bundles E of rank 2 on P^3 with $c_1(E) = 0$ and $c_2(E) = 2$ is treated, based on a result of Hartshorne. I hope that the reader will realize how to use the operation through this example. To study the transformation we have two viewpoints, geometric and sheaf-theoretic. Though in [6] the geometric viewpoint was emphasized, in this note I will care about the interrelation between two interpretations as far as possible.

An existence theorem of stable vector bundles is stated in §3. As a proof of general cases was given in [9] Appendix, the case of rank 2 on threefolds will be proved here. The proof is slightly different from that of the general cases and depends on a fact which is peculiar to threefolds (Corollary 2.7). In §4 some results on (5) in the above are given. The proofs of the results are going to be published elsewhere because they are rather technical and too special for our purpose of this article. Instead I will construct several examples of the curves of jumping lines by applying our method to vector bundles on P^2.

§1. Elementary transformations.

Let us start with the elementary transformations of ruled surfaces. Let C be a non-singular projective curve over an algebraically closed field and $p : X \longrightarrow C$ a geometric ruled surface, that is, a P^1-bundle over C. Pick a point x of X and blow up X with the center x; $f : \tilde{X} \longrightarrow X$. The proper transform D of the fibre $p^{-1}(p(x))$ by f is an exceptional curve of the first kind; $D \cong P^1$ and $D^2 = -1$. By a theorem of Castelnuovo, we can contract D to a smooth point, $g : \tilde{X} \longrightarrow X'$. Then we obtain a new ruled surface $p' : X' \longrightarrow C$.

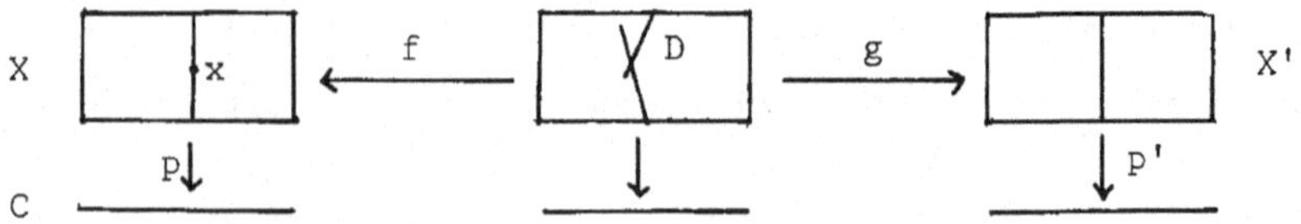

The birational map $gf^{-1} : X \longrightarrow X'$ is called the elementary transformation with the center x and denoted by elm_x. By the theorem of Tsen, $X \cong P(E)$ as P^1-bundles over C with E a vector bundle of rank 2 on C. Giving the point x is equivalent to doing a surjective homomorphism δ of E to $k(z)$, where $z = p(x)$:

$$0 \longrightarrow E' \longrightarrow E \xrightarrow{\delta} k(z) \longrightarrow 0.$$

Then $E' = \ker(\delta)$ is locally free and

Proposition 1.1. $X' \cong P(E')$.

Proof. By the theorem of Tsen $X' = P(E")$ with $E"$ a vector bundle of rank 2 on C. Let $O_{X'}(1)$ be the tautological line bundle of $E"$ on X'. Pick a rational section s of $O_{X'}(1)$ and set $G = (s)_0 - (s)_\infty$. The total transform of G to X defines a line bundle L on X such that $L|_{p^{-1}(t)} \cong O_{P^1}(1)$ for all $t \in C$. Then $p_*(L) \cong E \otimes M$ for a line bundle M on C. It is easy to see that $f^*(L) \otimes O(-F) \cong g^*(O_{X'}(1))$, where F is the exceptional divisor of f; $F = f^{-1}(x)$. Now $E" = p'_*(O_{X'}(1)) \cong p'_*(g_*g^*(O_{X'}(1))) \cong p_*f_*(f^*(L) \otimes O(-F)) \cong p_*(L \otimes I_x)$, where I_x is the ideal of x in X. On the other hand, from the exact sequence $0 \longrightarrow L \otimes I_x \longrightarrow L \longrightarrow k(x) \longrightarrow 0$, we obtain another exact sequence

$$0 \longrightarrow p_*(L \otimes I_x) \longrightarrow E \otimes M \xrightarrow{\delta'} k(z).$$

It is obvious that δ' is surjective and $\delta' = \delta \otimes M$. Thus we get an isomorphism $E" \otimes M \cong E'$ and hence $X' \cong P(E") \cong P(E')$. q. e. d.

We can generalize the above operation to the case of higher ranks and dimensions. Let S be a locally noetherian scheme and T a subscheme of S whose defining ideal I_T is a Cartier divisor on S. For a vector bundle E on S, assume that there is a surjective homomorphism δ of E to a vector bundle F on T with $r(F) < r(E)$. We denote this situation by the quadrulet (E, T, F, δ). A fact is

(1.2) $E' = \ker(\delta)$ is a vector bundle.

Our situation can be displayed in the following exact and commutative diagram:

$$(1.3)\qquad \begin{array}{ccccccccc} & & 0 & & 0 & & & & \\ & & \uparrow & & \uparrow & & & & \\ 0 & \longrightarrow & F' & \longrightarrow & E_{|T} & \longrightarrow & F & \longrightarrow & 0 \\ & & \uparrow & & \uparrow & & \| & & \\ 0 & \longrightarrow & E' & \longrightarrow & E & \longrightarrow & F & \longrightarrow & 0 \\ & & \uparrow & & \uparrow & & & & \\ & & E(-T) & = & E(-T) & & & & \\ & & \uparrow & & \uparrow & & & & \\ & & 0 & & 0 & & & & \end{array}$$

Let us give a geometric interpretation of the above. Set $X = P(E)$ and $Y = P(F)$. Then Y is a projective subbundle of $P(E)_T$ and a subscheme of $P(E)$. Blow up X along Y; $f : \tilde{X} \longrightarrow X$. Set $D = f^{-1}[P(E)_T]$, the proper transform of $P(E)_T$ and let G be the exceptional divisor of f on $\tilde{X}$.

Theorem 1.4. There exist a P^N-bundle $(r(E) = N + 1)$ $p' : X' \longrightarrow S$ which is the projective bundle $P(E'')$ associated with a vector bundle E'' on S and an S-morphism $g : \tilde{X} \longrightarrow X'$ such that

(1) the closed subscheme Y' defined by $g_*(I_D) \subset g_*(O_{\tilde{X}}) = O_{X'}$ is a projective subbundle $P(F'')$ of $P(E'')_T$, where I_D is the defining ideal of D in X,

(2) $g^*(O_{X'}(1)) \cong f^*(O_X(1)) \otimes O_X(-G)$, where $O_X(1)$ (or, $O_{X'}(1)$) is the tautological line bundle of E (or, E'', resp.),

(3) g is the monoidal transformation along Y'.

Moreover, P^N-bundle X' satisfying the above conditions is unique.

The rational map gf^{-1} is called the elementary transformation along Y and denoted by elm_Y. To indicate a proof of the theorem (for a complete proof, see [6] Chap. I, §2), let us mention the lemma ([6] Lemma 1.4) which is a local version of the theorem.

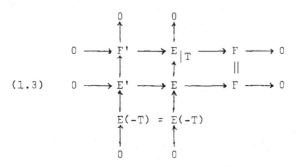

Lemma 1.5. Assume that $S = \mathrm{Spec}(A)$, $X = \mathrm{Proj}(A[x_0, \ldots, x_N])$ and the defining ideal of T (or, Y) is generated by $t \in A$ (or, t, $x_{n+1}, \ldots, x_N$, resp.). Then elm_Y exists and $\mathrm{elm}_Y(X) = \mathrm{Proj}(A[x'_0, \ldots, x'_N])$, where $x_i = x'_i$ $(0 \le i \le n)$ and $x_j = tx'_j$ $(n+1 \le j \le N)$. Moreover, the defining ideal $I_{Y'}$ of Y' is generated by $t, x'_0, \ldots, x'_n$.

This lemma enable us to compute explicitly the local properties of elementary transformations (for example, see [7] Appendix II). To combine Theorem 1.4 with (1.2) we need

Proposition 1.6. Under the same situation as in Theorem 1.4, we denote the defining ideal of Y in X by I_Y. Then $E' = p_*(I_Y \otimes O_X(1))$ is a locally free O_S-module, $P(E') \cong \mathrm{elm}_Y(X)$ and $R^i p_*(I_Y \otimes O_X(1)) = 0$ for $i > 0$, where $p : X \longrightarrow S$ is the natural projection.

Once one admits Theorem 1.4, the proof of the proposition is essentially the same as that of Proposition 1.1. On the other hand, the exact sequence

$$0 \longrightarrow I_Y \otimes O_X(1) \longrightarrow O_X(1) \longrightarrow O_X(1) \otimes O_Y \longrightarrow 0$$

provides us with another exact sequence

$$0 \longrightarrow p_*(I_Y \otimes O_X(1)) \longrightarrow E \xrightarrow{\delta'} F \longrightarrow R^1 p_*(I_Y \otimes O_X(1)) = 0$$

It is almost obvious that δ' is equal to the given surjective δ. Thus $X' = \mathrm{elm}_Y(X) \cong P(\ker(\delta))$. In the situation of Lemma 1.5 the F' in (1.3) is generated by $x'_0, \ldots, x'_n$ as a subsheaf of $E_{|T}$. This and Lemma 1.5 show that $P(F') = Y'$. Therefore, we have

(1.7) $\mathrm{elm}_Y(X) = P(\ker(\delta))$ and $\mathrm{elm}_Y^{-1} = \mathrm{elm}_{Y'}$ with $Y' = P(F')$, where F' is as in (1.3).

Example 1.8. To illustrate what happens to the fibre of X over a point t of T, let us pick the case of $r(E) = 3$. Then the fibre $X_t = p^{-1}(t) = P^2$. If $r(F) = 1$, then Y_t is a point of X_t. $\widetilde{X}_t$ is the union of F_1 and P^2 glued along the minimal section of F_1 and a line of P^2. By the morphism g, all the fibres of F_1 collaps to the line of P^2 and we get P^2 as X_t'. The Y_t' is the line to which the F_1 contracts.

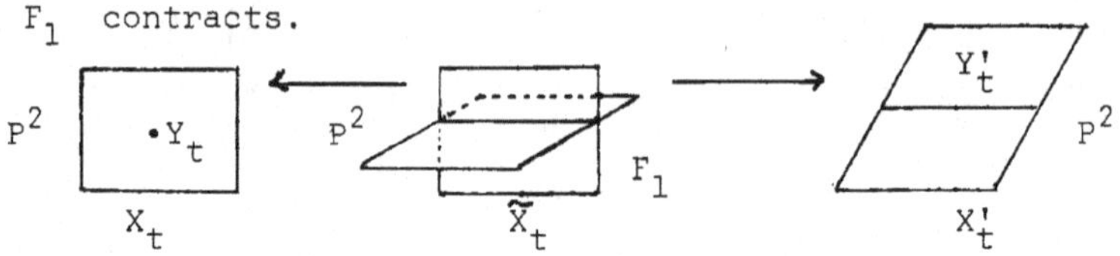

Assume $r(F) = 2$. In this case Y_t is a line in $X_t = P^2$ and hence $\widetilde{X}_t$ is the same as before. g induces the contraction of the minimal section of $F_1 \subset \widetilde{X}_t$ to a smooth point of P^2 while $P^2 \subset \widetilde{X}_t$ collapses to the point.

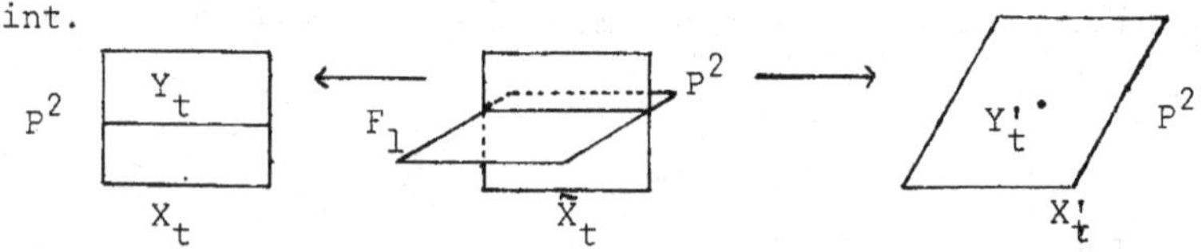

The second is the converse of the first and vice vasa.

§2. Some properties of elementary transformations.

In this section we shall show several properties of elementary transformations.

Proposition 2.1 (Compatibility with base changes). Under the same situation as in Theorem 1.4, suppose that we have a morphism $h : S' \longrightarrow S$ of locally noetherian schemes such that the defining ideal $I_{T'}$ of $T' = T \times_S S'$ is a Cartier divisor on S'. Then $\mathrm{elm}_Y(X) \times_S S' \cong \mathrm{elm}_{Y_{S'}}(X_{S'})$.

Proof. Consider the exact sequence

$$0 \longrightarrow E' \longrightarrow E \xrightarrow{\delta} F \longrightarrow 0.$$

By pulling back the sequence to S', we have

$$\underline{\mathrm{Tor}}_1^{O_{S'}}(F, O_{S'}) \longrightarrow h^*(E') \longrightarrow h^*(E) \xrightarrow{h^*(\delta)} h^*(F) \longrightarrow 0.$$

Our assertion is nothing but $h^*(E') \cong \ker(h^*(\delta))$ and hence $\underline{\mathrm{Tor}}_1^{O_{S'}}(F, O_{S'}) = 0$ is enough. Thus we may assume that S and S' are affine,

$S = \mathrm{Spec}(A)$ and $S' = \mathrm{Spec}(B)$, $I_T = tA$ and that F is a free O_T-module. By tensoring B to $0 \longrightarrow A \xrightarrow{t} A \longrightarrow A/tA \longrightarrow 0$, we have the exact sequence

$$0 \longrightarrow \mathrm{Tor}_1^A(A/tA, B) \longrightarrow B \xrightarrow{t} B \longrightarrow B/tB \longrightarrow 0.$$

Our assumption implies that t is a non-zero divisor of B. Thus $\mathrm{Tor}_1^A(A/tA, B) = 0$. q. e. d.

Assume that two quadruplets $(E_1, T_1, F_1, \delta_1)$ and $(E_2, T_2, F_2, \delta_2)$ on S are given. Let F_2' (or, F_1') be the image of $E_1 = \ker(\delta_1)$ (or, $E_2 = \ker(\delta_2)$, resp.) by δ_2 (or, δ_1, resp.). Then, by Snake lemma, the following is exact and commutative:

$$\begin{array}{ccccccccc}
 & & 0 & & 0 & & 0 & & \\
 & & \uparrow & & \uparrow & & \uparrow & & \\
0 & \longrightarrow & F_1' & \longrightarrow & F_1 & \longrightarrow & G & \longrightarrow & 0 \\
 & & \uparrow \psi_1 & & \delta_1 \uparrow & & \uparrow & & \\
0 & \longrightarrow & E_2 & \longrightarrow & E & \xrightarrow{\delta_2} & F_2 & \longrightarrow & 0 \\
 & & \uparrow & & \uparrow & & \uparrow & & \\
0 & \longrightarrow & E' & \longrightarrow & E_1 & \xrightarrow{\psi_2} & F_2' & \longrightarrow & 0 \\
 & & \uparrow & & \uparrow & & \uparrow & & \\
 & & 0 & & 0 & & 0 & &
\end{array}$$

where $G = F_1/F_1' \cong F_2/F_2'$ and $E' = \ker(\psi_2) \cong \ker(\psi_1)$. A geometric interpretation of this diagram is

Proposition 2.2 (Commutativety of elementary transformations). Put $Y_i = P(F_i)$. If the proper transform $[Y_1]$ (or, $[Y_2]$) of Y_1 (or, Y_2, resp.) by elm_{Y_2} (or, elm_{Y_1}, resp.) is a projective subbundle of $\mathrm{elm}_{Y_2}(X)_{T_1}$ (or, $\mathrm{elm}_{Y_1}(X)_{T_2}$, resp.), then $\mathrm{elm}_{[Y_1]}\mathrm{elm}_{Y_2} = \mathrm{elm}_{[Y_2]}\mathrm{elm}_{Y_1}$.

The simplest but trivial application of Proposition 2.2 is the case of $T_1 \cap T_2 = \phi$. We have another simple application.

Corollary 2.2.1. Assume that (1) $\dim S = 2$, (2) T_1 and T_2 has no common irreducible components and (3) every point x in $T_1 \cap T_2$

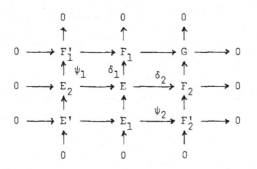

is smooth in T_1 and T_2. Then the commutativity of the elementary transformations holds.

Proof. Since $Supp(G) \subset T_1 \cap T_2$, $F_i^! = F_i$ outside $T_1 \cap T_2$. By the assumption (3) and the fact that $F_i^!$ is a torsion free O_{T_i}-module at every point x of $T_1 \cap T_2$, $F_i^!$ is a locally free O_{T_i}-module at the point x, too. q. e. d.

One of the main results on the elementary transformation is

Theorem 2.3. Let S be a non-singular quasi-projective variety over an algebraically closed field with $\dim S \leq 3$ and $p : P \longrightarrow S$ be a P^N-bundle (in Zariski topology). Then there exist a smooth subvariety T of codimension 1 in S and a P^{N-1}-subbundle Y of $P^N \times T$ ($\subset P^N \times S$) such that $elm_Y(P^N \times S) = P$ as P^N-bundle over S. Moreover, if $\dim S \geq 2$, then T can be chosen to be irreducible.

To show the idea of a proof of the theorem, let us give a proof in the case of $\dim S = 1$ and $N = 1$. For a complete proof, see [6] Theorem 1.12. Let E be a vector bundle of rank 2 on a non-singular curve S, $p : P = P(E) \longrightarrow S$ the projection and $O(1)$ the tautological line bundle of E on P. If M is an ample line bundle on S, $L = O(1) \otimes p^*(M^{\otimes m})$ is very ample for a sufficiently large m. When one chooses sufficiently general members D_1 and D_2 of the complete linear system $|L|$, they are sections of p and intersect transversally. Then $D_1 \cdot D_2 = x_1 + \ldots + x_r$ with $x_1, \ldots, x_r$ mutually distinct. Put $p(x_i) = z_i$ and $T = z_1 \cup \ldots \cup z_r$. T is smooth and Y' $x_1 \cup \ldots \cup x_r$ is a P^0-subbundle of P_T, For $P' = elm_{Y'}(P)$, the proper transform $D_i^!$ of D_i to P' by $elm_{Y'}$ is a section of $p' : P' \longrightarrow S$, $D_1^!$ is linearly equivalent to $D_2^!$ and $D_1^! \cap D_2^! = \phi$. Thus we can find another section $D_3^!$ in $|D_1^!| = |D_2^!|$ and $D_1^!$, $D_2^!$, $D_3^!$ form a system of global coordinates of the P^1-bundle $p' : P' \longrightarrow S$, that is, $P' = P^1 \times S$. If one takes the center Y of $elm_{Y'}^{-1}$, then the couple (T, Y) meets our requirement. q. e. d.

We shall give a sheaf-theoretic interpretation of Theorem 2.3. Let E be a vector bundle of rank r on S and $O_S(1)$ an ample line bundle on S. For a large integer m, $E(m) = F \otimes O_S(m)$ is very ample, that is, the tautological line bundle of $E(m)$ on $P(E)$ is very ample in the usual sense. For a suitable r-ple $(s_1, \ldots, s_r)$ of global sections of $E(m)$, we have an exact sequence

$$0 \longrightarrow O_S^{\oplus r} \xrightarrow[\times(s_1,\ldots,s_r)^t]{} E(m) \longrightarrow F \longrightarrow 0.$$

The proof of the theorem shows that under the assumption of Theorem 2.3 F is a line bundle on a smooth divisor T on S if $(s_1, \ldots, s_r)$ is sufficiently general. Our situation can be displayed in the following diagram:

$$(2.4)\qquad \begin{array}{ccccccccc}
 & & 0 & & 0 & & & & \\
 & & \uparrow & & \uparrow & & & & \\
0 & \longrightarrow & F' & \longrightarrow & E(m)|_T & \longrightarrow & F & \longrightarrow & 0 \\
 & & \uparrow & & \uparrow & & \| & & \\
0 & \longrightarrow & O_S^{\oplus r} & \longrightarrow & E(m) & \longrightarrow & F & \longrightarrow & 0 \\
 & & \uparrow & & \uparrow & & & & \\
 & & E(m)(-T) & = & E(m)(-T) & & & & \\
 & & \uparrow & & \uparrow & & & & \\
 & & 0 & & 0 & & & &
\end{array}$$

The Y in the theorem is nothing but $P(F') \subset P(O_S^{\oplus r}) = P^{r-1} \times S$. The above diagram shows that $E(m) = \ker(O_S(T)^{\oplus r} \longrightarrow F' \otimes O_S(T))$, which is the sheaf-theoretic interpretation of Theorem 2.3. As a special case, it is not difficult to see

(2.5) if $r(E) = 2$ and E is very ample on a non-singular projective variety S with $\dim S \leq 3$, then the first Chern class $c_1(E)$ of E is T, F is a line bundle on T, $F \cong O_T(T^2 - c_2(E))$ and $F' \cong O_T(c_2(E))$, where the second Chern class $c_2(E)$ of E can be regarded as a divisor on T in this case.

Corllary 2.6. Let S be a non-singular projective threefold over an algebraically closed field and E a very ample vector bundle of rank

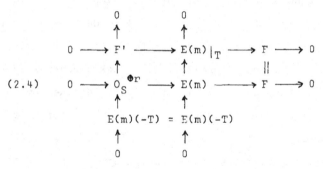

2 on S. If $s \in H^0(S, E) = H^0(P(E), O_{P(E)}(1))$ is a section which defines a smooth divisor on $P(E)$, then there is a <u>smooth</u> member in the complete linear system $|c_1(E)|$ such that (1) there is a surjective morphism $u : T \longrightarrow P^1$ and (2) for an $x \in P^1$, $u^{-1}(x) = (s)_0$.

<u>Corollary</u> 2.7. Let S be as in Corollary 2.6 and L_1, L_2 very ample line bundles on S. If $D_i \in |L_i|$ are so general that $D_1 \cdot D_2$ is a smooth curve, then there exists a <u>non-singular</u> member T in $|L_1 \oplus L_2|$ such that (1) there is a surjective morphism $u : T \longrightarrow P^1$ and (2) for a point x in P^1, $u^{-1}(x) = D_1 \cdot D_2$.

<u>Proof</u>. First of all we shall show that Corollary 2.7 is a special case of Corollary 2.6. Consider the vector bundle $E = L_1 \oplus L_2$. Then it is very ample and the assumption on D_1 and D_2 implies that for a section $s_i \in H^0(S, L_i)$ with $(s_i)_0 = D_i$, $s = s_1 \oplus s_2$ defines a smooth divisor on $P(E)$. Since $(s)_0 = (s_1)_0 \cdot (s_2)_0 = D_1 \cdot D_2$, the latter is deduced from the former. To prove Corollary 2.6, pick a general section t in $H^0(S, E)$. Then the T in Theorem 2.3 is defined by the ideal $\mathrm{Ann}(E/(s,t)O_S)$ (see the above observation). The Proof of the theorem ([6] p 108) shows that T is smooth. By (2.5), $T \in |c_1(E)|$. By using F' in the diagram (2.4), we have a morphism u of T to P^1; $u : T = P(F') \hookrightarrow P(O_T^{\oplus 2}) = P^1 \times T \longrightarrow P^1$. Since T is irreducible, u is not surjective if and only if $P(F') = y \times T$, that is, $F' \cong O_T$. Then $E = O_S \oplus O_S(T)$ which is not very ample. Thus u is surjective. For a point y of T, choose an open neighborhood V of y on which E is free. s and t can be written in the forms (s_1, s_2) and (t_1, t_2) respectively with $s_i, t_i \in \Gamma(V, O_S)$. The map u is given by $z \longrightarrow (s_1(z):t_1(z))$ or $(s_2(z):t_2(z))$ and T is defined by $s_1t_2 - s_2t_1 = 0$. Thus $u^{-1}(0) = V \cap (s)_0$. q. e. d.

<u>Example</u> 2.8. Let us give some examples for the corollaries in the case of $S = P^3$. Corollary 2.7 implies that for all $n \geq 2$, $|O_{P^3}(n)|$ contains a non-singular member with fibre structure. When $n = 2$ or 3

it is well-known that every non-singular member of $|O_{P^3}(n)|$ carries a fibre structure and the fibres are lines or conics in P^3 according as $n = 2$ or 3. $|O_{P^3}(4)|$ contains non-singular surfaces with fibre structure whose general fibres are elliptic curves of degree 3 (plane curves) or 4 (complete intersections of two non-singular quadrics). Next we shall make use of nullcorrelation bundles. A nullcorrelation bundle E fits in an exact sequence

$$0 \longrightarrow O_{P^3}(-1) \longrightarrow \Omega_{P^3}(1) \longrightarrow E \longrightarrow 0$$

and $c_1(E) = 0$, $c_2(E) = 1$. It is easy to see that $\Omega_{P^3}(2)$ and hence $E(1)$ are generated by their global sections. For general members s_1, s_2 in $H^0(P^3, E(1))$, we have the exact sequence

$$0 \longrightarrow O_{P^3}^{\oplus 2} \longrightarrow E(1) \longrightarrow F \longrightarrow 0,$$

where F is the line bundle of type (2, 0) on a non-singular quadric Q. $(s_i)_0$ is a union of two skew lines on Q. For a general member s in $H^0(P^3, E(2))$, $D = (s)_0$ is a smooth curve because $E(1)$ is generated by its global sections (see [4] 3.2 (iii), 2.4 (iii) and 3.5). Since D is a curve on a non-singular quartic and $D^2 = 0$, we see that $p_a(D) = 1$. Moreover, the degree of D in P^3 is 5. Then it is not difficult to see that D is a non-singular elliptic curve.

Let us study a special family of vector bundles on a non-singular projective variety of dimension ≥ 2. Fix a non-singular projective variety X with $\dim X \geq 2$. We shall consider a triple (T, F, δ) of a non-singular irreducible divisor T of X, a vector bundle F of rank $r-1$ on T and a surjective homomorphism $\delta : O_X(T)^{\oplus r} \longrightarrow F$. Then, by (1.2), $\ker(\delta) = E(T,F,\delta)$ is a vector bundle of rank r. Theorem 2.3 tells us that every vector bundle E is isomorphic to $E(T,F,\delta) \otimes L$ for a suitable (T, F, δ) and a line bundle L on X if $\dim X = 2$ or 3.

For a given triple (T, F, δ), the kernel of $\delta_{|T}$ is a line bundle

$O_T(D')$ on T. Taking the dual of $O_T(D') \longrightarrow O_T(T)^{\oplus r}$, we see that giving δ is equivalent to doing an r-ple $(s_1, \ldots, s_r)$ of global sections of $O_T(T^2-D')$ such that $\underset{i}{\cap}(s_i)_0 = \phi$, and to doing a morphism f of T to P^{r-1} such that $f^*(O_{P^{r-1}}(1)) \cong O_T(T^2-D')$. Putting $D = T^2-D'$, we can use the notation (T, D, δ) instead of (T, F, δ).

Theorem 2.9 ([6] Theorem 2.19). $ch(E(T,D,\delta)) = r(E(T,D,\delta)) + \sum_{\ell=1}^{\infty} T^{\ell}/\ell! + \sum_{m,n=1}^{\infty} \frac{(-1)^n T^{m-1} \cdot i_*(D^n)}{m!n!}$, where $ch(E)$ is the Chern character of E and $i : T \longrightarrow X$ is the closed immersion.

We shall close this section by studying the problem: When $E(T_1, D_1, \delta_1) \cong E(T_2, D_2, \delta_2)$? Unfortunately, we have no complete answer to this problem. A nice criterion for the isomorphism can, however, be given in a good case.

Theorem 2.10. Assume $H^0(T_1, O_{T_1}(T_1{}^2-D_1)) = 0$. Then $E(T_1, D_1, \delta_1) \cong E(T_2, D_2, \delta_2)$ if and only if (1) $T_1 = T_2$, (2) D_1 is linearly equivalent to D_2 and (3) $(s_{11}, \ldots, s_{1r}) = (s_{21}, \ldots, s_{2r})g$ with a g in $GL(r)$, where δ_i is given by r-ple $(s_{i1}, \ldots, s_{ir})$ of members of $H^0(T_i, O_{T_i}(D_i))$.

Proof. A geometric proof was given in [6] Proposition 2.12. We shall present here a sheaf-theoretic proof of the theorem. The "if" part is obvious. Let us prove the converse. First of all, put $E_i = E(T_i, D_i, \delta_i)$ and look at the display (1.3) for $\{E = O_X(T_i)^{\oplus r}, F' = O_{T_i}(T_i{}^2-D_i), E' = E_i\}$:

$$
\begin{array}{ccccccccc}
 & & 0 & & 0 & & & & \\
 & & \uparrow & & \uparrow & & & & \\
0 & \longrightarrow & O_{T_i}(T_i{}^2-D_i) & \longrightarrow & O_{T_i}(T_i)^{\oplus r} & \longrightarrow & F_i & \longrightarrow & 0 \\
 & & \uparrow & & \uparrow & & \| & & \\
0 & \longrightarrow & E_i & \xrightarrow{\psi_i} & O_X(T_i)^{\oplus r} & \xrightarrow{\delta_i} & F_i & \longrightarrow & 0 \\
 & & \alpha_i \uparrow & & \uparrow & & & & \\
 & & O_X{}^{\oplus r} & = & O_X{}^{\oplus r} & & & & \\
 & & \uparrow & & \uparrow & & & & \\
 & & 0 & & 0 & & & &
\end{array}
$$

Since $H^0(T_1, O_{T_1}(T_1{}^2-D_1)) = 0$, $k^{\oplus r} \cong H^0(X, O_X{}^{\oplus r}) \xrightarrow{H^0(\alpha_2)} H^0(X, E_2) \cong H^0(X, E_1) \xleftarrow[\sim]{H^0(\alpha_1)} H^0(X, O_X{}^{\oplus r}) \cong k^{\oplus r}$. Thus α_i is determined by global sections of E_i and hence the left columns of the displays are isomorphic with each other, in particular $T_1 = \mathrm{Supp}(O_{T_1}(T_1{}^2-D_1)) = \mathrm{Supp}(O_{T_2}(T_2{}^2-D_2)) = T_2$ and $O_{T_1}(T_1{}^2-D_1) \cong O_{T_2}(T_2{}^2-D_2)$ as line bundles on $T = T_1 = T_2$. These show (1) and (2). Now, identifying E_1 with E_2, put $E = E_1 = E_2$ and $D = D_1 \sim D_2$ and consider the display (1.3) for $\{E = E, F = O_T(T^2-D), E' = O_X{}^{\oplus r}\}$:

$$
\begin{array}{ccccccccc}
 & & 0 & & 0 & & & & \\
 & & \uparrow & & \uparrow & & & & \\
0 & \longrightarrow & F_i(-T) & \longrightarrow & E|_T & \longrightarrow & O_T(T^2-D) & \longrightarrow & 0 \\
 & & {\scriptstyle \delta_i(-T)}\uparrow & & \uparrow & & \| & & \\
0 & \longrightarrow & O_X{}^{\oplus r} & \xrightarrow{\eta_i} & E & \longrightarrow & O_T(T^2-D) & \longrightarrow & 0 \\
 & & {\scriptstyle \psi_i(-T)}\uparrow & & \uparrow & & & & \\
 & & E(-T) & = & E(-T) & & & & \\
 & & \uparrow & & \uparrow & & & & \\
 & & 0 & & 0 & & & &
\end{array}
$$

These displays define the inverses of the elementary transformations of the former displays. The homomorphisms $O_X{}^{\oplus r} \longrightarrow E$ in the middle row of the displays are determined by global sections of E because $k^{\oplus r} \cong H^0(X, O_X{}^{\oplus r}) \cong H^0(X, E)$. Hence $\eta_1 = \eta_2 h$ with an h in $GL(r) = \mathrm{Aut}(k^{\oplus r})$. Since $\eta_i\psi_i(-T)$ is the natural inclusion of $E(-T)$ to E, $\eta_2\psi_2(-T) = \eta_1\psi_1(-T) = \eta_2 h\psi_1(-T)$. The injectivity of η_2 forces that $\psi_2(-T) = h\psi_1(-T)$ and hence $\psi_2 = h\psi_1$. Thus our assertion (3) is obtained with $g = {}^t h$. q. e. d.

Remark 2.11. (1) In the above situation, $P(F)$ is a subscheme of $P^{r-1} \times T$ such that for each x in T, the fibre $P(F)_x$ is a hyperplane of $P^{r-1}_{k(x)}$. Thus we obtain a morphism f of T to the dual space $(P^{r-1})^* \cong P^{r-1}$. The morphism is given by $T \ni x \longrightarrow (s_1(x):\dots:s_r(x)) \in P^{r-1}$. Conversely, a morphism f of T to P^{r-1} gives a surjective homomorphism of $O_T(T)^{\oplus r}$ to a vector bundle F of rank $r-1$

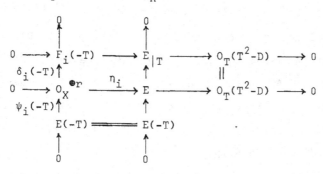

on T. It is easy to see that $f^*(O_{P^{r-1}}(1)) \cong O_T(D)$.

(2) The condition $H^0(T, O_T(T^2-D)) = 0$ in Theorem 2.10 is not so peculiar. In fact, we get a finite morphism f_m of P^{r-1} to itself by composing a Veronese morphism $P^{r-1} \longrightarrow P^N$ by m-forms with a general projection $P^N \longrightarrow P^{r-1}$. If a non-trivial morphism f of T to P^{r-1} is given, then $(f_m f)^*(O_{P^{r-1}}(1)) \cong O_T(mD)$, where $O_T(D) = f^*(O_{P^{r-1}}(1))$. When m is sufficiently large, $H^0(T, O_T(T^2-mD)) = 0$.

(3) A vector bundle E on a complete variety X over an algebraically closed field k is said to be simple if $\mathrm{End}_{O_X}(E) = \mathrm{Hom}_{O_X}(E, E) = k$ (= multiplication of constants). If $H^0(T, O_T(T^2-D)) = 0$ and δ corresponds to an r-ple $(s_1, \ldots, s_r)$ of elements of $H^0(T, O_T(D))$, then $E(T,D,\delta) = O_X^{\oplus r'} \oplus E$ with E a simple vector bundle, where r-r' is the dimension of the vector subspace of $H^0(T, O_T(D))$ generated by $\{s_1, \ldots, s_r\}$ ([6] Theorem 3.4). If r = 2 and $H^0(T, O_T(T^2-D)) = 0$, then every $E(T,D,\delta)$ is simple.

(4) In the case of rank 2, we have a good criterion for the simpleness of $E(T,D,\delta)$ ([6] Theorem 3.10). As a special case of the criterion, we know that $E(T,D,\delta)$ is simple if $H^0(T, O_T(T^2-2D)) = 0$ and $r(E(T,D,\delta)) = 2$.

Example 2.12. Let us study the family of stable vector bundles E (see Definition in §3) of rank 2 on P^3 with $c_1(E) = 0$ and $c_2(E) = 2$. In this case, the stability coincides with the simpleness ([7] Proposition A.1). As was shown in [3] p 268, we have the following exact sequence;

$$0 \longrightarrow O_{P^3}^{\oplus 2} \longrightarrow E(1) \longrightarrow O_Q(-1, 2) \longrightarrow 0,$$

where $O_Q(-1, 2)$ is the line bundle of type (-1, 2) on a non-singular quadric Q in P^3. This means that $O_{P^3}^{\oplus 2}$ is obtained from E(1) by an elementary transformation f along a section of $P(E)_Q$ over Q. Since the kernel of $E(1)_{|Q} \longrightarrow O_Q(-1, 2)$ is $O_Q(3, 0)$, the display of f^{-1} is as follows:

$$\begin{array}{ccccccccc}
 & & 0 & & 0 & & & & \\
 & & \uparrow & & \uparrow & & & & \\
0 & \longrightarrow & O_Q(-1,\ 2) & \longrightarrow & O_Q(2,\ 2)^{\oplus 2} & \longrightarrow & O_Q(5,\ 2) & \longrightarrow & 0 \\
 & & \uparrow & & \uparrow & & \| & & \\
0 & \text{---} & E(1) & \longrightarrow & O_{P^3}(2)^{\oplus 2} & \longrightarrow & O_Q(5,\ 2) & \longrightarrow & 0 \\
 & & \uparrow & & \uparrow & & & & \\
 & & O_{P^3}^{\oplus 2} & = & O_{P^3}^{\oplus 2} & & & & \\
 & & \uparrow & & \uparrow & & & & \\
 & & 0 & & 0 & & & &
\end{array}$$

Therefore, $E(1) \cong E(Q,O_Q(3,\ 0),\delta)$ with a suitable δ. Since the condition $H^0(T,\ O_T(T^2-D)) = 0$ holds for $T = Q$ and $O_T(D) = O_Q(3,\ 0)$, we can apply Theorem 2.10 to our case. Giving δ is equivalent to doing a surjective morphism g of Q to P^1 such that $g^{-1}(x)$ is three lines in the first family of lines on Q for all x in P^1 and the action of $GL(2)$ in the conclusion of the theorem is just changing coordinates of P^1. Moreover, every vector bundle of the form $E(Q,$ $O_Q(3,\ 0),\delta)$ is simple (Remark 2.11, (3)) and hence stable. Thus the set of isomorphism classes of stable vector bundles E of rank 2 on P^3 with $c_1(E) = 0$ and $c_2(E) = 2$ is in bijective correspondence with the set of couples $(Q,\ L)$ of a non-singular quadric Q and a linear pencil L in $|O_Q(3,\ 0)| \cong |O_{P^1}(3)| \cong P^3$ without base points (see [3] Theorem 9.7). The above observation provides us with an explanation of [3] Propositions 9.10 and 9.11 from our frame of reference. Indeed, if H is a hyperplane in P^3 which is transversal to Q, then $C = H\cdot Q$ is a non-singular conic in H and $O_Q(3,\ 0)\otimes O_H = O_C(D)$ with $\deg D = 3$. By virtue of proposition 2.1, $E(Q,O_Q(3,\ 0),\delta)_{|H} = E(C,D,$ $\delta_{|H})$. Remark 2.11, (4) tells us that $E(C,D,\delta_{|H})$ is simple and hence stable. Assume that H is tangent to Q. Then $C = H\cdot Q$ is the union of two lines ℓ_1 and ℓ_2 on Q such that ℓ_1 belongs to the first family of lines on Q and ℓ_2 does to the second. For the center Y of the elementary transformation $f : P^1\times P^3 \longrightarrow P(E(Q,O_Q(3,$ $0),\delta)$, $Y_{|\ell_i}$ is a section Y_i of $P^1\times\ell_i \longrightarrow \ell_i$ with $Y_1{}^2 = 0$ and $Y_2{}^2 = 6$. It is easy to see that $elm_{[Y_2]}elm_{Y_1} = elm_{Y_C}$, where $[Y_2]$

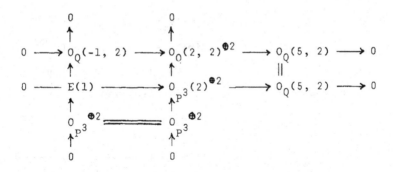

is the proper transform of Y_2 by elm_{Y_1}. elm_{Y_1} is defined by $O_H(2)^{\oplus 2} \xrightarrow{\delta_1} O_{\ell_1}(2)$ and the first direct summand of $O_H(2)^{\oplus 2}$ is sent to zero by δ_1. Thus $\ker(\delta_1) = O_H(2)\oplus O_H(1)$. Since $Y_1 \cap Y_2 \neq \phi$, $[Y_2]^2 = 5$. Then the situation of $elm_{[Y_2]}$ is displayed as follows:

$$\begin{array}{ccccccccc}
 & & 0 & & 0 & & & & \\
 & & \uparrow & & \uparrow & & & & \\
0 & \longrightarrow & O_{\ell_2}(-1) & \longrightarrow & O_{\ell_2}(2)\oplus O_{\ell_2}(1) & \longrightarrow & O_{\ell_2}(4) & \longrightarrow & 0 \\
 & & \uparrow & & \uparrow & & \| & & \\
0 & \longrightarrow & E' & \longrightarrow & O_H(2)\oplus O_H(1) & \longrightarrow & O_{\ell_2}(4) & \longrightarrow & 0 \\
 & & \uparrow & & \uparrow & & & & \\
 & & O_H(1)\oplus O_H & = & O_H(1)\oplus O_H & & & & \\
 & & \uparrow & & \uparrow & & & & \\
 & & 0 & & 0 & & & &
\end{array}$$

where $E' = E(Q, O_Q(3, 0), \delta)_{|H}$. Thus $H^0(H, E'(-1)) = k$ and $H^0(H, E(-2)) = 0$. Since $c_1(E') = 2$, E' is μ-semi-stable but not stable.

To see [3] Proposition 9.11 which determines $E(Q, O_Q(3, 0), \delta)_{|\ell}$ for all the lines ℓ in P^3, we need the following general lemma.

<u>Lemma</u> 2.13. For a quadruplet (E, T, F, δ) in (1.2), if $r(E) = 2$, then $E'_{|T} = \gamma_*(N^\vee_{P(F)/P(E)}) \otimes F$, where $E' = \ker(\delta)$, γ is the isomorphism of $P(F)$ to T and $N_{P(F)/P(E)}$ is the normal bundle of $P(F)$ in $P(E)$.

<u>Proof</u>. By the geometric meaning of the elementary transformation it is clear that $P(E')_T \cong P(\gamma_*(N^\vee_{P(F)/P(E)}))$. Tensoring O_T to the middle row of the display (1.3), we have the exact sequence

$$0 \longrightarrow \underline{Tor}_1^{O_S}(F, O_T) \longrightarrow E'_{|T} \longrightarrow F' \longrightarrow 0.$$

Since T is a Cartier divisor, $\underline{Tor}_1^{O_S}(F, O_T) \cong F \otimes I/I^2 = F \otimes N^\vee_{T/S}$, where I is the defining ideal of T in S. It is easy to see that $F' \cong \gamma_*(N^\vee_{P(F)/P(E)_T}) \otimes F$ and the above exact sequence is obtained from the canonical one of conormal bundles corresponding to the inclusions $P(F) \subset P(E)_T \subset P(E)$. q. e. d.

Now let us go back to $E = E(Q, O_Q(3, 0), \delta)(-1)$. Let ℓ_1 (or, ℓ_2)

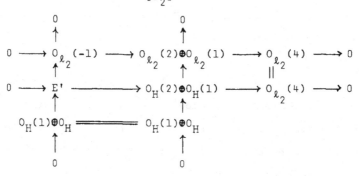

be an line on Q which belongs to the first (or, the second, resp.) family of lines on Q. If H is the hyperplane in P^3 spanned by ℓ_1 and ℓ_2, then $H \cdot Q = \ell_1 + \ell_2$. Let Y_i be as before. Then the minimal section Z of $\mathrm{elm}_{Y_1}(P^1 \times H)|_{\ell_1}$ meets $[Y_2]$ if and only if Y_2 is tangent to A at $y = Y_1 \cap Y_2$ (as we can see easily by applying Lemma 1.5 to our situation), where $A = y \times H \subset P^1 \times H$. This occurs if and only if ℓ_1 is contained in the double or triple line of a member of the linear pencil in $|O_Q(3, 0)|$ which defines δ. On the other hand, $(\mathrm{elm}_{[Y_2]}\mathrm{elm}_{Y_1}(P^1 \times H))|_{\ell_1} = F_2$ or F_1 according as $[Y_2]$ meets Z or not. Thus we obtain (b) and (d) of [3] Proposition 9.11. Since $[Y_2]^2 = 5$ and $\ell_2{}^2 = 1$, $(\mathrm{elm}_{[Y_2]}\mathrm{elm}_{Y_1}(P^1 \times H))|_{\ell_2} = F_4$ by Lemma 2.13 which is equivalent to (a) of the proposition. The proof of the other cases is similar to the above and much easier.

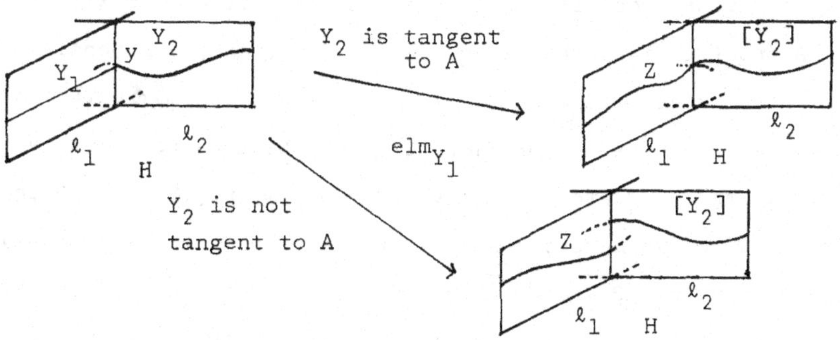

§3. Construction of stable bundles.

Theorem 2.3 shows that on a non-singular quasi-projective variety of dimension ≤ 3 every vector bundle can be constructed, in principle, by using elementary transformations. Furthermore, many vector bundles can be constructed on every non-singular projective variety through this technique. To show the fact let us recall the notion of stable vector bundles.

Definition. Let $(X, O_X(1))$ be a couple of a non-singular projective variety X over an algebraically closed field k and an ample invertible sheaf $O_X(1)$ on X. For a coherent sheaf F on X, $d(F, O_X(1))$ is the degree of the first Cheren class $c_1(F)$ of F with respect to $O_X(1)$. When the rank $r(F)$ of F is not zero, we put $\mu(F) = d(F, O_X(1))/r(F)$. A vector bundle E on $(X, O_X(1))$ is said to be μ-stable if for every coherent subsheaf F of E with $1 \leq r(F) < r(E)$, $\mu(F) < \mu(E)$.

Our theorem on existence of vector bundles is stated as follows.

<u>Theorem</u> 3.1. Let $(X, O_X(1))$ be as in the definition and assume $\dim X \geq 2$. For every divisor class D on X and integers r, s with $r \geq \dim X$, there is a μ-stable vector bundle E on $(X, O_X(1))$ with $r(E) = r$, $c_1(E) = D$ and $d(c_2(E), O_X(1)) \geq s$. Moreover, if $\dim X = 3$, the above holds for $r = 2$, too.

Let us give a proof of the theorem in the case of $\dim X = 3$ and $r = 2$. A proof of the other cases can be found in [9] Appendix. For a sufficiently large integer m, both $O_X(m)$ and $O_X(m) \otimes O_X(D)$ are very ample. By virtue of Corollary 2.7, we obtain a non-singular member T of the complete linear system $|O_X(2m) \otimes O_X(D)|$ such that there is a surjective morphism $u : T \longrightarrow P^1$ such that $A = u^{-1}(x) = T_1 \cdot T_2$ for an x in P^1, a T_1 in $|O_X(m)|$ and for a T_2 in $|O_X(m) \otimes O_X(D)|$. We may assume that A is a non-singular curve on T. Setting $F = O_X^{\oplus 2}$, let $B_0 = \{G \mid G$ is a quotient coherent sheaf of F with the following three properties (a), (b), (c)$\}$: (a) G is torsion free, (b) $r(G) = 1$ and (c) $d(G, O_X(1)) \leq d(T, O_X(1))/2$. Then B_0 is a bounded family by Corollary 1.2.1 of [8] and hence $B = \{\ker(F \otimes O_T \longrightarrow G \otimes O_T) \mid G \in B_0\}$ is bounded, too. We claim

(3.1.1) There exists an integer n_0 such that for all integers $n \geq n_0$, $F_T = F \otimes O_T$ contains $O_T(-nA)$ as a subsheaf with the following properties; (i) $F_T/O_T(-nA)$ is a line bundle on T and (ii) $O_T(-nA) \cap H = 0$ for all $H \in B$.

Proof of (3.1.1). Since B is bounded, $\{d(H, O_T(1)) \mid H \in B\}$ is a bounded set. Thus $\{d(\bar{H}, O_T(1)) \mid H \in B\}$ is bounded below, where $\bar{H}$ is the inverse image of the torsion part of F_T/H by $F_T \longrightarrow F_T/H$. Hence, by [8] Corollary 1.2.1 again, $\bar{B} = \{\bar{H} \mid H \in B\}$ is bounded. Replacing B by $\bar{B}$, we may assume that for every member H of B, F_T/H is torsion free and then H is an invertible sheaf on T. After this replacement it is enough for (ii) to show (iii) $O_T(-nA) \not\subset H$ for all $H \in B$. In fact, if $I = H \cap O_T(-nA) \neq 0$, then $r(I) = 1$, $I \neq O_T(-nA)$ and hence $O_T(-nA)/I$ is a non-zero torsion sheaf. On the other hand, $O_T(-nA)/I$ is a subsheaf of the torsion free F_T/H. This is a contradiction. For every $H \in B$, since $|c_1(F_T/H)| \neq \phi$, $d(H, O_T(1)) \leq 0$. This implies that $u_*(H) = O_{P^1}(a_H)$ or 0 with $a_H \leq 0$. By the projection formula, $u_*(H \otimes O_T(nA)) \cong u_*(H \otimes u^*(O_{P^1}(n))) \cong u_*(H) \otimes O_{P^1}(n)$ which implies that $\dim H^0(T, H \otimes O_T(nA)) \leq n+1$ for all $H \in B$. Similarly, $\dim H^0(T, F_T(nA)) = \dim H^0(P^1, O_{P^1}(n)^{\oplus 2}) = 2n+2$. Since B is bounded, there exist an algebraic k-scheme Z and a coherent subsheaf E of $F' = F_T \otimes_k O_Z$ such that F'/E is flat over Z and for all $H \in B$, we can find a k-valued point z of Z with $E \otimes k(z) \cong H$ as subsheaves of F_T. Now if n is large enough, then $\dim H^0(T, F_T \otimes O_T(nA)) > \dim H^0(T, H(nA)) + \dim Z$ for all $H \in B$. By shrinking Z and breaking it into a union of subschemes, we may assume that (1) T is reduced, (2) $\dim H^0(T, E \otimes k(z))$ is constant on each connected component of Z and (3) $\dim H^0(T, F_T(nA)) > \dim H^0(T, (E\ k(z))(nA)) + \dim Z$ for all $z \in Z$. By (1) and (2), $\tilde{E} = q_*(E \otimes p^*(O_T(nA)))$ is locally free and for all $z \in Z(k)$, $\tilde{E} \otimes k(z) \cong H^0(T, (E \otimes k(z))(nA))$, where $p : T \times Z \longrightarrow T$ and $q : T \times Z \longrightarrow Z$ are the projections. Then $\tilde{E}$ is a vector subbundle of $\tilde{F} = q_*(F' \otimes p^*(O_T(nA))) = H^0(T, F_T(nA)) \otimes_k O_Z$ on each connected component of Z. For $P = P(H^0(T, F_T(nA))^{\vee})$, $P(\tilde{E}^{\vee})$ is a closed subscheme of $P \times_k Z = P(\tilde{F}^{\vee})$. By (3) above, $\dim P > \dim P(\tilde{E}^{\vee})$. Therefore, for the projection g of $P \times_k Z \longrightarrow P$, the closure P_0 of $g(P(\tilde{E}^{\vee}))$ in P is a proper closed subset of P. Then, for a k-valued point y of $P - P_0$, a corresponding section t_y in $H^0(T, F_T(nA))$ is not contained in $\bigcup_{H \in B} H^0(T, H(nA))$. On the other hand, for a general section $t = (t_1, t_2) \in H^0(T, O_T(nA))^{\oplus 2} \cong H^0(T, F_T(nA))$, $(t_1)_0 \cap (t_2)_0 = \phi$, that is, the cokernel of $O_T(-nA) \xrightarrow{\times t} F_T$ is a line bundle. Therefore, if t is a sufficiently general section in $H^0(T, F_T(nA))$, $O_T(-nA)$ embedded into F_T by the multiplication of t meets our requirement.

Now let us go back to the proof of the theorem. We are going to construct a μ-stable vector bundle E of rank 2 on X such that $c_1(E) = T$ and $d(c_2(E), O_X(1)) \geq s'$ for a given integer s'. $E(-m)$ is one of the required vector bundles because $c_1(E(-m)) = D$ and $d(c_2(E), O_X(1)) \geq s$ if s' is large enough. Fix an integer n such that n is greater than n_0 in (3.1.1) and $d(O_T(nA), O_X(1)) \geq s'$. $O_T(-nA)$ is contained in F_T so that the properties (i) and (ii) of (3.1.1) are enjoyed. Set $F_0 = F_T/O_T(-nA)$ and $E = \ker(F(T) \longrightarrow F_0(T))$. We shall show that E has the required propertis. First of all, by Theorem 2.9, $c_1(E) = T$ and $c_2(E) = nA$, whence $d(c_2(E), O_X(1)) \geq s'$. To prove the stability, pick a coherent subsheaf L of E with $r(E) = 1$. What we have to show is the inequality $d(T, O_X(1))/2 > d(L, O_X(1))$. Let G be the torsion part of $F(T)/L$ and set $(F(T)/L)/G = M$. On the one hand, if $M(-T) \notin B_0$, then $d(M, O_X(1)) = d(T, O_X(1)) + d(M(-T), O_X(1)) > 3d(T, O_X(1))/2$ and we have that $d(L, O_X(1)) \leq d(F(T), O_X(1)) - d(M, O_X(1)) < d(2T, O_X(1)) - 3d(T, O_X(1))/2 = d(T, O_X(1))/2$. On the other hand, if $M(-T) \in B_0$, then for $M' = \ker(F_T \rightarrow M(-T))$, $M' \cap O_T(-nA) = 0$ because of the choice of $O_T(-nA) \longrightarrow F_T$. Since the image of $L(-T) \otimes O_T$ in F_T is contained in M', L goes to zero by $\psi : E \longrightarrow O(-nA)(T)$, that is, $L \subset \ker(\psi) \cong F$. This and the fact that F is semi-stable imply that $d(L, O_X(1)) \leq d(F, O_X(1))/2 = 0$.

q. e. d.

By a similar method we have

Theorem 3.2 ([10] Appendix). Let X be a non-singular projective surface, D a divisor on X and c an integer. Put $F(D, c) = \{E \mid E$ is an indecomposable vector bundle of rank 2 on X with $c_1(E) = D$ (rational equivalence) and $c_2(E) = c$ (algebraic equivalence)$\}$. Then $F(D, c)$ is *not* bounded.

§4. Singularities of the curve of jumping lines.

Let k be an algebraically closed field and E a vector bundle of rank 2 on $P = P_k^2$ with the following properties:

(4.1.1) $c_1(E) = 0$ and $c_2(E) = n$.

(4.1.2) For a line ℓ in P, $E_{|\ell} \cong O_\ell^{\oplus 2}$.

When the characteristic of k is zero, the property (4.1.2) is equivalent to the μ-semi-stability of E by the theorem of Grauert-Mülich.

$E(-1)$ fits in the following exact sequence;

$$(4.2) \qquad 0 \longrightarrow U \longrightarrow \bigoplus_{\text{finite}} O_P(m_i) \longrightarrow E(-1) \longrightarrow 0$$

with m_i negative integers. Let us consider the diagram

$$\begin{array}{ccc} F & \xrightarrow{\ q\ } & P^* \\ {\scriptstyle p}\big\downarrow & & \\ P & & \end{array}$$

where P^* is the dual space of $P = P_k^2$ and F is the flag manifold which defines the incidence correspondence between P and P^*. (4.1.2) implies that $q_*p^*(E(-1))$ is a torsion sheaf on P^*. On the other hand, it is torsion free because so is $p^*(E(-1))$. Thus $q_*p^*(E(-1)) = 0$. For every line ℓ in P, $H^0(\ell, U_{|\ell}) = H^0(\ell, \oplus O_P(m_i)_{|\ell}) = 0$. It is deduced from this that both $M = R^1q_*p^*(\oplus O_P(m_i))$ and $N = R^1q_*p^*(U)$ are locally free on P^*. Putting $L(E) = R^1q_*p^*(E(-1))$, we obtain an exact sequence;

$$0 \longrightarrow N \xrightarrow{\lambda} M \longrightarrow L(E) \longrightarrow 0$$

By (4.1.2) again, $L(E)$ is a torsion sheaf on P^*, whence $r = r(N) = r(M)$. We can, therefore, define $\det(\lambda)$ and it is contained in $H^0(P^*, \overset{r}{\Lambda} M \otimes (\overset{r}{\Lambda} N)^\vee)$. We see

(4.3.1) $\det(\lambda)$ is independent of the choice of (4.2),

(4.3.2) $(\det(\lambda))_0 = \mathrm{Supp}(L(E)) = \{\ell \in P^* \mid H^1(\ell, E(-1)_{|\ell}) \neq 0\} = \{\ell \in P^* \mid E_{|\ell} \not\cong O_\ell^{\oplus 2}\}$.

Thus the following definition seems to be adequate.

<u>Definition</u>. The curve in P^* defined by $\det(\lambda)$ is called the <u>curve of jumping lines</u> of E and denoted by $S(E)$.

The following is due to W. Barth.

<u>Proposition</u> 4.4 ([1] Theorem 2).

(1) $\deg S(E) = n$.

(2) $S(E)$ depends algebraically on E, that is, if $\tilde{E}$ is a vector bundle on $P_S = P \times_k S$ with S locally noetherian such that for all $s \in S(k)$, $\tilde{E}_s = \tilde{E} \otimes k(s)$ has the properties (4.1.1) and (4.1.2), then there is a relative Cartier divisor $S(\tilde{E})$ on $P^* \times_k S$ such that for all $s \in S(k)$, $S(\tilde{E})_s = S(\tilde{E}_s)$.

Let $\ell \in S(E)$. Then, by the definition of $S(E)$, $E_{|\ell} = O_\ell(a) \oplus O_\ell(-a)$ with $a > 0$. W. Barth proved in [1] Theorem 2 that the multiplicity $\mathrm{mult}_\ell(S(E))$ of $S(E)$ at ℓ is not less than a. Several examples show, however, the inequality can be strict (see [2]). On the one hand, it is natural to suspect that the infinitesimal neighborhood of ℓ, more precisely, of the minimal section of the ruled surface $P(E)_\ell$ must determine $\mathrm{mult}_\ell(S(E))$. On the other hand, the neighborhood can be developed by elementary transformation. Indeed, this idea works quite well.

Let us define a sequence of positive integers relating to the infinitesimal neighborhood of the minimal section of $P(E)_\ell$. Set $a_0 = a$, that is, $E|_\ell \cong O_\ell(a_0) \oplus O_\ell(-a_0)$ with $a_0 > 0$. Let E_1 be $\ker(E \longrightarrow O_\ell(-a_0))$. E_1 is the elementary transform of E_1 along the minimal section of $P(E)_\ell$. Since $c_1(E_1) = -1$, $E_1|_\ell = O_\ell(a_1) \oplus O_\ell(a_1')$ with $a_1 + a_1' = -1$ and $a_1 \geq a_1'$. If $a_1 > 0$, we shall define E_2 to be the elementary transform of E_1 along the minimal section of $P(E)_\ell$; $E_2 = \ker(E_1 \longrightarrow O_\ell(a_1'))$. Continuing this process until we reach such an E_{i+1} that $E_{i+1}|_\ell = O_\ell(a_{i+1}) \oplus O_\ell(a_{i+1}')$ with $a_{i+1} + a_{i+1}' = -i-1$, $a_{i+1} \geq a_{i+1}'$ and $a_{i+1} \leq 0$, we obtain a sequence of positive integers $(a_0, \ldots, a_i)$. Geometrically this means $P(E)_\ell \cong F_{2a_0}$, $P(E_j)_\ell \cong F_{j+2a_j}$ $(0 < j \leq i,\ a_j > 0)$ and $P(E_{i+1})_\ell \cong F_t$ with $t \leq i+1$.

<u>Definition</u>. $M(E, \ell) = (a_0, \ldots, a_i)$ and $|M(E, \ell)| = \Sigma a_j$. The length of $M(E, \ell)$ is i.

Our result on $\mathrm{mult}_\ell(S(E))$ is stated as follows.

<u>Theorem</u> 4.5. Let E is a vector bundle of rank 2 on P_k^2. Suppose that E has the property (4.1.1) and (4.1.2) and the characteristic of k is zero. Let $\ell \in S(E)$.

(1) $M(E, \ell)$ is a decreasing sequence and its length is finite.

(2) $\mathrm{mult}_\ell(S(E)) = |M(E, \ell)|$.

Instead of giving a proof of the theorem, we shall explain the meaning of it through examples because we have to prepare many technical results before proving it.

<u>Example</u> 4.6. Let C be a non-singular conic in $P = P_k^2$ and M_n a line bundle of degree n on C. For a surjective homomorphism δ of $O_P^{\oplus 2}$ to M_n, put $E = E(C, M_n, \delta)(-1)$. Then

(4.6.1) $c_1(E) = 0$ and $c_2(E) = n - 1$ (Theorem 2.9),

(4.6.2) if $n \geq 3$, E is simple (Remark 2.11, (4)) and then it is stable in this case.

δ defines a section $C \xrightarrow{\sim} D = P(M_n) \subset P(O_C^{\oplus 2}) = P^1 \times C$ and $P(E) \cong elm_D(P_P^1)$. Let ℓ be a line in P, $\ell \cap C = \{z_1, z_2\}$ and $y_i = D \cap P^1_{z_i}$. G_0 denotes the section $y \times P$ of $P_P^1 \longrightarrow P$ which passes through y_1. Let us define inductively Y_i to be $G_i \cdot P_\ell^1$, H_i to be $G_i \cdot P_C^1$, G_{i+1} to be the proper transform of G_i by elm_{Y_i} and X_{i+1} to be $elm_{Y_i}(X_i)$ $(X_0 = P_P^1)$. Then $X_i \cong P(O_P(i) \oplus O_P)$ and Y_i is the minimal section of $(X_i)_\ell \cong F_i$. If $[D_{i+1}]$ (or, $[Y_i]$) is the proper transform of D (or, Y_i, resp) by $elm_{Y_i} \ldots elm_{Y_0}$ (or, $elm_{[D_i]}$, resp.), then we have $elm_{[D_{i+1}]} elm_{Y_i} \ldots elm_{Y_0} \cong elm_{[Y_i]} \ldots elm_{[Y_0]} elm_D$ by Corollary 2.2.1. Set $M(E, \ell) = (a_0, a_1, \ldots)$.

(I) The case of $z_1 \neq z_2$. Then $H_0 \cdot D = b_1 y_1 + b_2 y_2 + A$ in P_C^1 with $b_1 > 0$, $b_2 \geq 0$ and $y_1, y_2 \notin A$. We may assume that $b_1 \geq b_2$. Since $P(E)_\ell \cong elm_{\{y_1, y_2\}}(P_\ell^1)$ by Proposition 2.1, $P(E)_\ell \cong F_0$ or F_2 according as $b_2 = 0$ or $b_2 > 0$. Thus $\ell \in S(E)$ if and only if $b_2 > 0$, and then $a_0 = 1$. Assume that $a_0 = 1$. By Theorem 4.5, every integer appearing in $M(E, \ell)$ is 1. To compute a_1, we have to look at the minimal section of $(elm_{[Y_0]}(P(E)))_\ell \cong (elm_{[D_1]}(X_1))_\ell$. Put $y_i' = (X_1)_{z_i} \cap [D_1]$. By using Lemma 1.5, we see that $H_1 \cdot [D_1] = b_1' y_1' + b_2' y_2' + A'$ in P_C^1 with $b_i' = \max(b_i - 1, 0)$ and $y_i' \notin A'$. Since $(elm_{[D_1]}(X_1))_\ell \cong elm_{\{y_1', y_2'\}}(X_1)_\ell \cong F_3$ or F_1 according as both b_1' and b_2' are positive or not. Therefore, $a_1 = 1$ if and only if $b_1 \geq b_2 \geq 2$. If $a_1 = 1$, then a similar observation shows that $(elm_{[Y_1]} elm_{[Y_0]}(P(E)))_\ell \cong (elm_{[D_2]}(X_2))_\ell \cong F_4$, F_2 or F_0 according as $b_1 \geq b_2 \geq 3$, $b_1 \geq 3$ and $b_2 = 2$ or $b_1 = b_2 = 2$. By the definition of $M(E, \ell)$, $a_2 = 1$ if and only if $b_1 \geq b_2 \geq 3$. Continuing this process, we see that $M(E, \ell) = (1, 1, \ldots, 1)$ and the length of $M(E, \ell)$ is equal to b_2.

(II) The case of $z_1 = z_2$. Then $H_0 \cdot D = b_1 y_1 + A$ in P_C^1 with $b_1 > 0$ and $y_1 \notin A$. Similarly to the above we see that $\ell \in S(E)$ if and only if $b_1 \geq 2$. Moreover, $M(E, \ell) = (1, \ldots, 1)$ and the length of $M(E, \ell)$ is equal to $[b_1/2]$.

Let (x_0, x_1) be a system of homogeneous coordinates of $C = P^1$. Define a homomorphism of $O_C^{\oplus 2} \longrightarrow M_n$ by taking the global sections x_0^n and x_1^n of M_n. Let z_0 and z_1 be the points $(0, 1)$ and $(1, 0)$ on C, respectively and ℓ_i be the line which is tangent to C at z_i. Then, in the similar notation to the above for $\ell = \ell_i$, $H_0 \cdot D$ $ny_i + A$ with $y_i \notin A$, where $y_i = D \cap P^1_{z_i}$. Thus $M(E, \ell_i) = (1, \ldots, 1)$ and the length of the $M(E, \ell_i)$ is $[n/2]$. Since $\deg S(E) = n - 1$ by Proposition 4.4 and (4.6.1), the line spanned by z_0 and z_1 is contained in $S(E)$ if n is even.

Example 4.7. Let ℓ be a line in $P = P_k^2$ and N_i be a line bundle of degree n_i on ℓ. For a surjective homomorphism δ_1 of $O_P^{\oplus 2}$ to N_1, put $E_1 = \ker(\delta_1) \otimes O_P(1)$. By Lemma 2.13, $E_1|_\ell \cong O_\ell(n_1) \oplus O_\ell(1-n_1)$. Thus if $n_2 \geq n_1 > 1$ or if $n_2 > n_1 = 1$, we have a surjective $\delta_2 : E_1 \longrightarrow N_2$. Set $\ker(\delta_2) = E$. Then $E|_\ell \cong O_\ell(n_2-1) \oplus O_\ell(1-n_2)$ and

(4.7.1) $c_1(E) = 0$ and $c_2(E) = n_1 + n_2 - 1$.

The section $P(N_2) \subset P(E_1)_\ell$ is a divisor of self-intersection number $2n_2 - 1$. On the other hand, every member D of $|O_{P(E_1)}(1)|$ cuts a divisor D' on $P(E_1)_\ell$ such that $D'^2 = 1$ and D' is a union of a section and fibres (note that $H^0(P, E_1(-1)) = 0$ and hence $D \not\supset P(E_1)_\ell$ if $n_1 \geq 1$). Thus $D \not\supset P(N_2)$ if $n_2 \geq 2$. This means that if $n_2 \geq 2$, then $H^0(P, E) = 0$ and hence E is stable by (4.7.1). Moreover, we see by the definition of $M(E, \ell)$

(4.7.2) $M(E, \ell) = (n_2-1, n_1-1)$ (when $n_1 - 1 = 0$, we omit it).

By a similar method to the above, we can produce examples of various types of $M(E, \ell)$.

References

[1] W. Barth, Some properties of stable rank-2 vector bundles on P^n, Math. Ann., 226, 1977, 125-150.

[2] W. Barth, Moduli of vector bundles on the projective plane, Invent. Math., 42, 1977, 63-91.

[3] R. Hartshorne, Stable vector bundles of rank 2 on P^3, Math. Ann., 238, 1978, 229-280.

[4] S. L. Kleiman, Geometry on Grassmannians and applications to splitting bundles and smoothing cycles, Publ. Math. I.H.E.S., 36, 1969, 281-298.

[5] S. Langton, Valuative criteria for families of vector bundles on algebraic varieties, Ann. of Math., 101, 1975, 88-110.

[6] M. Maruyama, On a family of algebraic vector bundles, Number Theory, Algebraic Geometry and Commutative Algebra, in honor of Y. Akizuki, Kinokuniya, Tokyo, 1973, 95-146.

[7] M. Maruyama, Stable vector bundles on an algebraic surface, Nagoya Math. J., 58, 1975,25-68.

[8] M. Maruyama, Openness of a family of torsion free sheaves, J. Math. Kyoto Univ., 16, 1976, 627-637.

[9] M. Maruyama, Moduli of stable sheaves, II, J. Math. Kyoto Univ., 18, 1978, 557-614.

[10] M. Maruyama, On boundedness of families of torsion free sheaves, to appear in J. Math. Kyoto Univ..

[11] M. S. Narasimhan and S. Ramanan, Deformations of the moduli space of vector bundles on curves, Ann. of Math., 101, 1975, 391-417.

[12] M. S. Narasimhan and S. Ramanan, Geometries of Hecke cycles - I, C. P. Ramanujam - A Tribute, Springer, Berlin-Heidelberg-New York, 1978, 291-345.

Department of Mathematics
Faculty of Science
Kyoto University
Kyoto 606
Japan

Déformations semiuniverselles et germes d'espaces analytiques $\mathbb{C}^*$-équivariantes

F. Puerta Sales

0. Dans (10) et (11) H. Pinkham démontre l'existence d'une déformation semiuniverselle $\mathbb{C}^*$ - équivariante (équivariante, dans ce qui suit) dans le cas formel. Il suggere dans (11) l'intérêt de démontrer des resultats analogues dans le cas analytique. Le cas analytique sans $\mathbb{C}^*$ action a déja été résolu par Donin (2) et Grauert (7) (voir aussi (6)) Ce dernier utilise la méthode de Schl esinger et un théoreme d'approximation, tandis que le premier emploie la théorie de Douady des espaces analytiques de Banach. Nous suivons la méthode de Donin a fin de résoudre le cas analytique équivariant. En outre on utilise le théoreme de preparation de Weierstrass dans la version de H. Hironaka pour démontrer une proposition qui est essentielle.

1. Generalités sur les $\mathbb{C}^*$ actions

1.1 Une $\mathbb{C}^*$ action dans $\mathbb{C}^n$ est une operation continue du groupe multiplicatif $\mathbb{C}^* = \mathbb{C} - \{0\}$ dans $\mathbb{C}^n$. On démontre (1) que, dans un systeme convenable de coordonées, elle est de la forme

$$(\lambda, x) = (\lambda, (x_1,\dots,x_n)) \longrightarrow \lambda . x = (\lambda^{q_1}x_1,\dots,\lambda^{q_n}x_n)$$

ou $q_1, \dots, q_n$ sont des entiers que l'on appelle <u>poids</u> de la $\mathbb{C}^*$ action. Dans ce qui suit nous supposerons fixée une $\mathbb{C}^*$ action dans $\mathbb{C}^n$ de poids $q_1,\dots,q_n$.

1.2 Toute $\mathbb{C}^*$ action induit dans l'algèbre $\mathcal{O}_n$, ainsi que dans les quotients $\mathcal{O}_n/I$ par idéaux stables par l'action, une structure d'algèbre topologiquement graduée. Cette structure et, en general celle de module topologiquement gradué jouent un role important dans tout ce qui suit. On peut consulter (5) pour une étude détaillée des graduations topologiques. Pour la commodité du lecteur, on énonce les principales propietés.

1.3 Par dualité $\mathbb{C}^*$ opère dans $\mathcal{O}_n$. On note $\mathcal{O}_n^{(k)}$ le sous-espace propre de $\mathcal{O}_n$ de valeur propre λ^k sous l'action de λ. On appelle q-homogenes de degré k les éléments non nuls de $\mathcal{O}_n^{(k)}$. Alors tout f de $\mathcal{O}_n$ peut s'écrire de manière unique comme

$$f = \sum_k f_k$$

où $f_k \in \mathcal{O}_n^{(k)}$ et la série est convergente dans $\mathcal{O}_n$ muni de la topologie séquentielle ou analytique (7). On dit que f_k est la k-ième composante q-homogène de f. On notera π_k la projection de $\mathcal{O}_n$ dans $\mathcal{O}_n^{(k)}$ définie par $\pi_k(f) = f_k$, et on dit alors que la famille$(\mathcal{O}_n^{(k)}, \pi_k)$ munit $\mathcal{O}_n$ d'une structure d'algèbre topologiquement graduée.

1.4 Un idéal I de $\mathcal{O}_n$ est équivariant si $\pi_k(I) = I$ pour tout k de $\mathbb{Z}$. On démontre que I est équivariant si et seulement s' il est engendré par un nombre fini d'éléments q-homogenes. Alors $\mathbb{C}^*. I \subset I$; donc, $\mathbb{C}^*$ opere dans $\mathcal{O}_n/I$ et ceci permet (comme dans 1.3) munir $\mathcal{O}_n/I$ d'une structure d'algèbre topologiquement graduée.

1.5 Plus généralement, soit $A = \mathcal{O}_n/I$ avec I équivariant et $M = A^p$. Alors chaque $b = (b_1, \ldots, b_n) \in \mathbb{Z}^p$ permet de définir dans M une structure de A-module topologiquement gradué, M_b, en prenant comme ensemble d'éléments k-homogenes de M_b

$$M_b^{(k)} = A^{(k+b_1)} x \overset{p)}{\ldots} x A^{(k+b_p)} \quad , k \in \mathbb{Z}$$

Les projections π_k sont définies de maniere analogue au cas des algèbres, ainsi que la notion de sous-module équivariant.

1.6 Un morphisme entre modules topologiquement gradués (en particulier, algèbres) est équivariant s'il envoie les éléments k-homogenes dans des éléments k-homogenes. Si φ est un morphisme de A_b^p dans A_d^q, on démontre qu'il est équivariant si et suelement si les colonnes de sa matrice sont formées par des éléments q-homogenes de A_d^q de dégrés $-b_1, \ldots, -b_p$.

2. Déformations équivariantes.

2.1 Rappelons qu'on a fixée une $\mathbb{C}^*$ action dans $\mathbb{C}^n$ de poids $q_1, \ldots, q_n$. Soit I un idéal de $\mathcal{O}_n$ et (X,0) le germe d'espace analytique défini par I. Si I est équivariant on dit que (X, 0) est un germe analytique équivariant.

2.2 Soit

$$\begin{array}{ccc} \mathcal{O}_{X,o} = \mathcal{O}_n/I & \longleftarrow & A = \mathcal{O}_n\{t\}/\mathcal{J} \\ \uparrow & & \uparrow \\ \mathbb{C} & \longleftarrow & B = \mathbb{C}\{t\}/\mathcal{I} \end{array}$$

une déformation de (X, 0), avec $t = (t_1, \ldots, t_s)$. On dit que cette déformation est <u>équivariante</u> s'il existe une $\mathbb{C}^*$ action dans $\mathbb{C}^s$ telle que les idéaux $\mathcal{J}$ et $\mathcal{I}$ soient équivariants, ainsi que tous les morphismes du carré ci-dessus.

On va caractériser les déformations équivariantes a travers d'extensions convenables d'une résolution de l'anneau $\mathcal{O}_{X,0}$.

2.3 <u>Lemme</u>.- Soit (X, 0) un germe analytique équivariant. Alors, il existe une résolution, $\mathcal{R}^\cdot$, de $\mathcal{O}_{X,o}$

$$0 \to \mathcal{O}_n^{p_r} \to \ldots \to \mathcal{O}_n^{p_i} \xrightarrow{a_i} \mathcal{O}_n^{p_{i-1}} \to \ldots \to \mathcal{O}_n$$

telle que tous les morphismes a^i sont équivariants par rapport à des graduations topologiques convenables des modules libres correspondents. On dit que $\mathcal{R}^\cdot$ est <u>équivariante</u>.

Nous fixons d'une fois pour toutes la résolution équivariante du germe équivariante (X, 0).

2.4 Avec les notations de 2.2 posons $\mathbb{C} = \mathcal{O}_n\{t\}/\mathcal{I}$. Si $\mathcal{A}^\cdot$ est un complexe de c-modules on pose

$$\mathcal{A}^\cdot(0) = \mathcal{A}^\cdot \otimes_B C$$

On a alors,

2.5 <u>Proposition</u>.- Soit (X, 0) un germe analytique équivariante.Les assertions suivantes sont équivalentes

1) Donner une déformation équivariante de (X, 0)
2) Donner une $\mathbb{C}^*$ action dans $\mathbb{C}^s$ et un complexe libre fini de C-modules ${}^*\mathcal{R}^\cdot$ tel que
   a) $\mathcal{I}$ est équivariant
   b) ${}^*\mathcal{R}^\cdot(0) = \mathcal{R}^\cdot$
   c) Le complexe ${}^*\mathcal{R}^\cdot$ est équivariant, et les morphismes $A^i = (A^i_{jh}) : C^{p_i} \longrightarrow C^{p_{i-1}}$ sont tels que $\deg A^i_{jh} = \deg a^i_{jh}$.

3. Déformations infinitésimales équivariantes.

3.1 Soient U un voisinage de 0 dans $\mathbb{C}^n$ tel que les matrices a^i qui définissent la résolution $\mathcal{R}^\bullet$ soient définies dans U, et K un polycylindre compact qui est un voisinage de 0 dans $\mathbb{C}^n$ contenu dans U. D'après Donin on considère les espaces $F^i(K)$ définis de la manière suivante.

3.2 $F^1(K)$ est l'espace des r-tuples $b = (b^1, \ldots, b^r)$ ou $b^i = (b^i_{jh})$ sont des matrices dont les coéficients b^i_{jh} appartiennent a H(K), espace des fonctions holomorphes dans $\mathring{K}$ et continues dans K, $1 \leqslant i \leqslant r$, $1 \leqslant j \leqslant p_{i-1}$, $1 \leqslant h \leqslant p_i$.

3.3 $F^2(K)$ est l'espace des r-tuples de matrices $c = (c^2, \ldots, c^r)$ ou $c^i = (c^i_{jh})$ avec $c^i_{jh} \in H(K)$, $2 \leqslant i \leqslant r$, $1 \leqslant j \leqslant p_{i-2}$, $1 \leqslant h \leqslant p_i$.

3.4 Finalement, $F^o(K)$ est l'espace des paires $(e, \varphi) = (e^o, \ldots, e^r; \varphi_1, \ldots, \varphi_n)$ où les matrices $e^i = (e^i_{jh})$ ont comme les precedentes les coéficients dans H(K), $0 \leqslant i \leqslant r$, $1 \leqslant j,h \leqslant p_i$ et $\varphi = (\varphi_1, \ldots, \varphi_n)$ est une fonction de K dans $\mathbb{C}^n$ dont les composantes appartiennent a H(K).

Avec la norme du supremum les $F^i(K)$ sont des espaces de Banach.
Dans ces espaces on considere les points distingués suivants

$$a \in F^1(K) \;, \qquad 0 \in F^2(K), \qquad (1, z) \in F^o(K)$$

où $a = (a^1, \ldots, a^r)$ est la r-tuple de 2.3 et (1;z) est la paire formée en prennant les matrices identités et les fonctions coordonées de $\mathbb{C}^n$.

On note $F^i(K)$ soit l'espace de Banach correspondant soit un voisinage convenable du point distingué.

On considere maintenant la suite

$$F^o(K) \xrightarrow{\delta^o_K} F^1(K) \xrightarrow{\delta^1_K} F^2(K)$$

où

$$\delta^o_K(e; \varphi) = (e^{i-1}\, a^i(\varphi)(e^i)^{-1})_{1 \leqslant i \leqslant r}$$

$$\delta^1_K(b) = (b^{i-1} b^i)_{2 \leqslant i \leqslant r}$$

Il est évident que les points distingués s'appliquent dans les points distingués.

3.5 Soit $Z^1(K)$ l'espace analytique de Banach défini par

$$Z^1(K) = (\delta^1_K)^{-1}(0)$$

Les déformations de base S s'identifient alors aux morphismes analytiques de S dans $Z^1(K)$.

3.6 Posons $\dot{\delta}^o_K = d\,\delta^o_{K,\,(1;z)}$ et $\dot{\delta}^1_K = d\,\delta^1_{K,\,a}$. On a la suit d'espaces de Banach

$$F^0(K) \xrightarrow{\dot{\delta}^o_K} F^1(K) \xrightarrow{\dot{\delta}^1_K} F^2(K)$$

Par passage à la limite inductive, les K parcourant un systeme de voisinages de 0 dans $\mathbb{C}^n$, on obtient la suite

$$F^o \xrightarrow{\dot{\delta}^o} F^1 \xrightarrow{\dot{\delta}^1} F^2$$

ou $F^o = \lim F^0(K)$, etc.

3.7Il en résulte que $T^1(X) = \ker \dot{\delta}^1 / \operatorname{im} \dot{\delta}^o$, où $T^1(X)$ est l'espace des déformations infinitésimales (c'est à dire, sur le point double).

Chaque F^i est un $\mathcal{O}_n$-module libre de rang convenable. Il est possible de munir chaque F^i d'une structure de $\mathcal{O}_n$-module topologiquement gradué de sorte que les morphismes $\dot{\delta}^i$ soient équivariants. On a alors,

3.8 <u>Proposition</u>.- L'espace $T^1(X)$ est un $\mathcal{O}_n$-module (en fait, un $O_{X,o}$-module) topologiquement gradué.
Si π est la projection naturelle de $\ker \dot{\delta}^1$ dans $T^1(X)$ on a

$$T^1(\nu) =: T^1(X)^{(\nu)} = \pi(\ker \dot{\delta}^1)^{(\nu)}$$

c'est à dire, les déformations infinitésimales de degrée ν sont celles qui peuvent être représentées par des éléments $(\alpha^1, \ldots, \alpha^r)$ appartenant a $F^{1(\nu)}$. Si $\alpha^i = (\alpha^i_{jh})$ ceci est équivalent à dire que

a) $\alpha^i_{jh} \in \mathcal{O}_n^{(d^i_{jh}+\nu)}$

b) $\alpha^{i-1}a^i + a^{i-1}\alpha^i = 0$.

4. <u>Équivariance de la déformation semiuniverselle</u>.

4.1 Dans ce qui reste on suppose que $T^1(X)$ est un $\mathbb{C}$-espace vectoriel de dimension finie ($=\tau$).
Soit $(\alpha_1, \ldots, \alpha_\tau)$ une base de $T^1(X)$ formée par déformations

q-homogenes de degrées $\nu_1, \ldots, \nu_\tau$. On identifie $T^1(X)$ avec $\mathbb{C}^\tau$ à travers cette base. On note $t_1, \ldots, t_\tau$ les coordonées de $\mathbb{C}^\tau$. Dans $\mathbb{C}^\tau$ on définit une $\mathbb{C}^*$action moyennant

$$\lambda.t = (\lambda^{e_1} t_1, \ldots, \lambda^{e_\tau} t_\tau)$$

avec $e_i = -\nu_i$, $1 \leqslant i \leqslant \tau$.

La construction de la déformation semiuniverselle cherchée est faite de la manière suivante

4.2 La suite 3.5 permet de considérer une suite de faisceaux libres sur U

$$\mathcal{N}_o \xrightarrow{\dot{\delta}^o} \mathcal{N}_1 \xrightarrow{\dot{\delta}^1} \mathcal{N}_2$$

telle que la fibre au-dessus de 0 du faisceau $Ex^1(X) = \ker \dot{\delta}^1 / \mathrm{im}\ \dot{\delta}^0$ est $T^1(X)$.

4.3 Soit (K) une base de voisinages de 0 privilegiés par rapport aux morphismes $\dot{\delta}^o$ et $\dot{\delta}^1$ (4). Puisque dim $T^1(X)$ est finie, il existe un voisinage K de cette base tel que $\mathrm{sop}_K Ex^1(X) = \{0\}$. Fixons ce voisinage K. Alors, compte tenu du fait que K est privilegié et de la version de Hironaka du theoreme de préparation de Weierstrass (8) on a

4.4 <u>Proposition</u>.- Il existe un sous-espace E_K' de $F^1(K)$ et une varieté N_K tels que

1) $\ker \dot{\delta}^1_K = \mathrm{im}\ \dot{\delta}^o_K \oplus E_K'$

2) E_K' est invariant par $\mathbb{C}^*$

3) $T_a N_K = E_K'$

4) $(N_K \cap Z^1(K)\ , 0)$ est un germe analytique équivariant.

De (2) on conclue le theoreme suivant

4.5 <u>Theoreme</u>.- Soit (X, 0) un germe analytique équivariant. Si $\dim_{\mathbb{C}} T^1(X)$ est finie, il existe une déformation semiuniverselle équivariant de (X, 0).

5 <u>Un exemple</u>.

Dans (5) on démontre que la base S de la déformation semiuniverselle du germe à l'origine des axes coordonés de $\mathbb{C}^n$, X_n, est le germe dans 0 de $\mathbb{C}^{n(n-2)}$ défini par

$$p_k^{ij} = p_h^{ij}$$

avec

$$p_k^{ij} = u_j^i u_k^j + u_i^j u_k^i - u_k^i u_k^j$$

où u_j^i **sont** les fonctions coordonées de $\mathbb{C}^{n(n-2)}$ (on suppose $1 \leqslant i, j \leqslant n$, $i \neq j$, $i \neq j-1$).

L'espace total est défini dans $\mathbb{C}^n \times S$ par

$$x_i x_j + u_i^j x_i + u_j^i x_j + p_k^{ij} = 0$$

On remarquera que X_n est équivariant par rapport a n'importe quelle $\mathbb{C}^*$action dans $\mathbb{C}^n$ et que , par consequent, la déformation semiuniverselle ci-dessus est également équivariante par rapport a la $\mathbb{C}^*$ actio: dans $\mathbb{C}^{n(n-2)}$

$$\lambda . u_j^i = \lambda^{q_i} u_j^i$$

ou q_1, ..., q_n sont les poids de l'action.

Bibliographie

(1) Chevalley, C; Séminaire Chevalley 1956/58, exposé 4, séminaire 1.

(2) Donin, I.F.; Complete families of deformations of germs of complex spaces, Math. USSR Sbornik, vol 18 (1972) No 3, 397-406

(3) Douady, A; Séminaire Bourbaki, 1964/65, exposé 277

(4) Douady, A; Le problème des modules pour les sous-espaces analytiques compacts d' un espace analytique, Ann Inst. Fourier, 16, (1966), fasc. 1, 1-95

(5) Ferrer, J., Puerta, F.; Deformaciones de gérmenes analíticos equivariantes (a paraître)

(6) Galligo, A., Houzel, C.; Déformations semiuniverselles de germes d'espaces analytiques (d'apres Verdier et Grauert) Astérisque 7-8, 139-163 (1973)

(7) Grauert, H.; Über die Deformation isolierter Singularitäten analytischer Mengen, Inv. Math., 15 (1972), 171-198

(8) Grauert, H., Remmert, R.; Analytische Stellenalgebren, G.M. 176, Springer-Verlag.

(9) Hironaka, H.; Stratification and flatness, Real and complex singularities, Oslo 1976

(10) Pinkham, H.; Deformations of algebraic varieties with G_m actio
Astérisque 20, (1974), 1-131

(11) Pinkham, H.; Deformations of normal surface singularities with
C action, Math. Ann. 232, (1978) 65-84.

F. Puerta Sales
Departamento de Matemáticas de la E.T.S.I.I.B.
Universidad Politécnica de Barcelona
Diagonal 647. Barcelona (28) España.

A VANISHING THEOREM FOR BIRATIONAL MORPHISMS

Juan B. Sancho de Salas

The purpose of this paper is to prove the following vanishing theorem and provide some applications.

Theorem: Let X *be a local complete intersection, embeddable in a smooth variety, and*

$$\pi : X' \longrightarrow X$$

be the birational blowing up of X *along a smooth subvariety* Y . *Suppose also, that* X' *is a local complete intersection in a smooth ambient variety. Then if* $\omega_{X'}$ *is the dualizing sheaf of* X' , *one has*

$$R^i \pi_* \omega_{X'} = 0 \quad \text{for all} \quad i > 0$$

This result allows one to compute the direct images of the sheaf of local rings and the change in the Euler-Poincaré characteristic when passing from X to X' .

Other results of this kind have been used for different purposes. For example to desingularize two-dimensional schemes Lipman [2] uses, in a very essential way, the part that the relative dualizing sheaf has zero high direct images by a birational map between normal Gorenstein surfaces. Other strong vanishing theorems have been proven by Grauert-Riemenschneider [1] (and used to generalize the Kodaira vanishing theorem to singular varieties) and Wahl [3] .

No further mention will be made of the ground field which is assumed to be of arbitrary characteristic.

One starts with the following

Lemma 1: Let X *be an hypersurface of a smooth ambient variety* Z , *and let* π , π' *be the respective blowing ups of* X , Z *along a smooth integral subvariety* Y *of* X , *so one has a diagram*

$$\begin{array}{ccc} X' & \hookrightarrow & Z' \\ \downarrow{\scriptstyle \pi} & & \downarrow{\scriptstyle \pi'} \\ X & \hookrightarrow & Z \end{array}$$

If d *is de codimension of* Y *in* Z , *and* m *is the multiplicity of* X *at the generic point of* Y , *then for the dualizing sheaf* $\omega_{Z'}$ *of* Z' *one has*

a) $\omega_{Z'} = \pi'^{*}\mathcal{O}_Z \otimes \mathcal{O}_{Z'}(-d+1)$.

b) $\mathcal{O}_{Z'}(X') = \pi'^{*}\mathcal{O}_Z(X) \otimes \mathcal{O}_{Z'}(m)$.

Proof of a): One certainly has

$$\omega_{Z'} = \pi'^{*}\ {}_Z \otimes \mathcal{O}_{Z'}(n) ,$$

for some $n \in \mathbb{Z}$. Let $E = \pi'^{-1}(Y)$. By the adjunction formula one gets

$$\omega_E = (\omega_{Z'} \otimes \mathcal{O}_{Z'}(-1)) \otimes \mathcal{O}_E = \pi'^{*}\omega_Z \otimes \mathcal{O}_{Z'}(n-1) \otimes \mathcal{O}_E .$$

On the other hand, E is a projective bundle over Y , so ω_E and $\mathcal{O}_E(-d)$ are locally isomorphic along Y . Comparing both results one concludes that $m = -d+1$.

Proof of b): Start with

$$\mathcal{O}_{Z'}(X') = \pi'^{*}\mathcal{O}_Z(X) \otimes \mathcal{O}_{Z'}(n) ,$$

for some integer n . To see that $n = m$, restrict $\mathcal{O}_{Z'}(-X')$ to the fiber of π' over the generic point of Y . Then consider the following fact: the exceptional fiber of the blowing up of a point is a projective variety of degree equal to the multiplicity of the point. One concludes from this that the restriction of $\mathcal{O}_{Z'}(-X')$ is the sheaf of ideals of a hypersurface of degree m in $\mathbb{P}^{d-1}$. This implies easily that $n = m$.

Theorem 2: *Let* Z *be a smooth variety and*

$$\pi : Z' \longrightarrow Z$$

be the blowing up of Z *along a smooth integral subvariety* Y *of codimension* d . *One has*:

a) $\pi_{*}\mathcal{O}_{Z'} = \mathcal{O}_Z$.

b) $R^i\pi_{*}\mathcal{O}_{Z'}(n) = 0$ *for all* $i > 0$ *and all* $n > -d$.

Proof: a) Follows from the fact that $\pi_{*}\mathcal{O}_{Z'}$ is coherent and Z is normal.
b) Call E the exceptional fiber of π . One has an exact sequence

$$0 \longrightarrow \mathcal{O}_{Z'}(n+1) \longrightarrow \mathcal{O}_{Z'}(n) \longrightarrow \mathcal{O}_E(n) \longrightarrow 0 .$$

By Serre's theorem $R^i\pi_{*}\mathcal{O}_{Z'}(n) = 0$ for $i > 0$ and $n >> 0$. The theorem follows by descending induction on n , observing that $R^i\pi_{*}\mathcal{O}_E(n) = 0$ for $i > 0$ and $n > -d$ (because $\pi : E \longrightarrow Y$ is a projective bundle).

And here comes the main theorem

Vanishing theorem 3: Let X be a local complete intersection in a smooth ambient variety Z. Let $\pi: X' \longrightarrow X$, $\pi': Z' \longrightarrow Z$ the respective blowing ups of X, Z along a smooth irreducible subvariety Y of X. Suppose X' is also a local complete intersection in Z'. Then if $\omega_{X'}$ denotes the dualizing sheaf of X' one has

$$R^i\pi_*\omega_{X'}(n) = 0 \quad \text{for all} \quad i > 0 \quad \text{and all} \quad n \geq 0 .$$

Proof: The assertion being local on X, one can suppose that X is a complete intersection: $X = H_1 \cap \ldots \cap H_r$. Suppose for a moment that the proper transform X' of X is the intersection of the proper transforms H_i' of H_i. The theorem goes forward now by induction on r. Let m_i be the multiplicity of H_i at the generic point of Y, and write $X_i = H_1 \cap \ldots \cap H_i$. For $r = 0$, one has $X_0 = Z$ and $X_0' = Z'$. Then by lemma 1 and theorem 2 it follows

$$R^i\pi_*\omega_{X_0'}(n) = R^i\pi_*^!\omega_{Z'}(n) = \omega_Z \otimes R^i\pi_*^! \mathcal{O}_{Z'}(n-d+1) = 0 \quad \text{for} \quad i > 0 \quad \text{and} \quad n \geq 0 .$$

Now for general r write $\mathcal{O}_r = \mathcal{O}_{X_r'}$, $\omega_r = \omega_{X_r'}$; there is an exact sequence of sheaves

$$0 \longrightarrow \mathcal{O}_{r-1}(-H_r') \longrightarrow \mathcal{O}_{r-1} \longrightarrow \mathcal{O}_r \longrightarrow 0 .$$

Tensoring by $\omega_{r-1}(n) \otimes \mathcal{O}_{r-1}(H_r')$ and using the adjunction formula one gets another exact sequence

$$0 \longrightarrow \omega_{r-1}(n) \longrightarrow \omega_{r-1}(n) \otimes \mathcal{O}_{r-1}(H_r') \longrightarrow \omega_r(n) \longrightarrow 0 .$$

Now, the assertion is local on X_{r-1} so one can suppose by lemma 1 that the middle sheaf is isomorphic to $\omega_{r-1}(n + m_r)$, and by the induction hyphotesis:

$$R^i\pi_*\omega_r(n) = 0 \quad \text{for} \quad i > 0 \quad \text{and} \quad n \geq 0 .$$

To conclude, it is enough to see that given a closed point $x \in Z$ one can choose $H_1, \ldots, H_r$ locally near x in such a way that X' is a connected component of the intersection of the H_i'. Let p_X, p_Y be the ideals defining X, Y in the local ring $\mathcal{O}_{Z,x}$. The graded module $\bigoplus_{n \geq 0} p_X \cap p_Y^n$, which is contained in $\bigoplus_{n \geq 0} p_Y^n$, defines by homogeneous localization the ideal of X'. Take then a minimal system of generators $f_1, \ldots, f_r$ of p_X in such a way that they form a base of $p_X \otimes_{\mathcal{O}_{Z,x}} \mathcal{O}_{Z,x}/m_{Z,x}$ and that all vector spaces defined by the images of the maps

$$p_X \cap p_Y^n \longrightarrow p_X \otimes_{\mathcal{O}_{Z,x}} \mathcal{O}_{Z,x}/m_{Z,x}$$

have a base consisting of elements of $\{f_1, \ldots, f_r\}$. The hypersurfaces given by $f_i = 0$ are those required. Indeed, given $x' \in \pi^{-1}(x)$, let $\bar{f}_1 = 0, \ldots, \bar{f}_r = 0$ be hypersurfaces such that $p_X = (\bar{f}_1,\dots,\bar{f}_r)$ and the intersection of their proper transforms is X' locally near x'. (These exists because X' is local complete intersection). Now, obviously, from the choice of the f_i it follows that the homogeneous elements of $\bigoplus_{n\geq 0} p_Y^n$ defined by the $\bar{f}_i$ are linear combinations of the homogeneous elements defined by the f_i. This implies that the intersection of the proper transforms of the $\bar{f}_i$ contains the intersection of the proper transforms of the $f_i = 0$.

Note 4: As $\omega_{X'} = \pi^*\omega_X \otimes \omega_{X'/X}$ one gets a vanishing theorem for the relative dualizing sheaf $\omega_{X'/X}$:

$$R^i\pi_*\omega_{X'/X}(n) = 0 \quad \text{for} \quad i>0 \quad \text{and all} \quad n\geq 0 .$$

Corollary 5: For every closed point x of Y one has a formal duality isomorphism:

$$(R^i\pi_*\mathcal{O}_{X'})^{\wedge}_x = H^{n-i}_{\{x\}}(X,\pi_*\omega_{X'})^*$$

where the dual module is taken with respect to the injective envelope of the residual field of the point x, n is the dimension of X and the sheaf $\pi_*\omega_{X'}$ can be computed by theorem 8 below.

Proof: See [2] to get the duality theorem:

$$(R^i\pi_*\mathcal{O}_{X'})^{\wedge}_x = H^{n-i}_E(X',\omega_{X'/X})^* ,$$

where $E = \pi^{-1}(x)$. Then by the vanishing theorem one gets

$$H^{n-i}_E(X',\omega_{X'/X}) = H^{n-i}_{\{x\}}(X,\pi_*\omega_{X'/X}) .$$

To conclude, use the fact that

$$\pi_*\omega_{X'} = \omega_X \otimes \pi_*\omega_{X'/X} .$$

To compute the change in the dualizing sheaf, let Z be a smooth variety, X a complete intersection in Z of r hypersurfaces $H_1,\dots,H_r$ and let $\pi: X' \longrightarrow X$, $\pi': Z' \longrightarrow Z$ be the respective blowing ups of X, Z along a smooth integral subvariety Y of X. Suppose that the proper transform X' is also the complete intersection of the proper transforms H_i' of the hypersurfaces H_i. And let d be the codimension of Y in Z, and m_i the multiplicity of H_i at the generic point of Y. One the has

Theorem 6: The dualizing sheaf $\omega_{X'}$ of X' is the sheaf

$$\omega_{X'} = \pi^*\omega_X \otimes \mathcal{O}_{X'}(m_1 + \dots + m_r - d + 1) .$$

Proof: By induction on r. Write $X_i = H_1 \cap \dots \cap H_i$, $\omega'_r = \omega_{X'_r}$ and $\omega_r = \omega_{X_r}$. The case $r = 0$ is done in lemma 1 a).

For the general r case one has equalities (adjunction formula)

$$\omega'_r = (\omega'_{r-1} \otimes \mathcal{O}'_{r-1}(H'_r)) \otimes_{\mathcal{O}'_{r-1}} \mathcal{O}'_r \ ,$$

$$\omega_r = (\omega_{r-1} \otimes \mathcal{O}_{r-1}(H_r)) \otimes_{\mathcal{O}_{r-1}} \mathcal{O}_r \ .$$

Comparing these and recalling that

a) $\omega'_{r-1} = \pi^*\omega_{r-1} \otimes \mathcal{O}'_{r-1}(m_1 + \dots + m_{r-1} - d + 1)$ (by induction)

b) $\mathcal{O}'_{r-1}(H'_r) = \pi^*\mathcal{O}_{r-1}(H_r) \otimes \mathcal{O}'_{r-1}(m_r)$ (by lemma 1 b)

one concludes the theorem.

Corollary 7: $R^i\pi_*\mathcal{O}_{X'}(n) = 0$ for $i > 0$ and all $n \geq m_1 + \dots + m_r - d + 1$.

Proof: Follows from theorems 3 and 6

Finally it is interesting to observe that the vanishing theorem is false for the composition of two monoidal transformations with smooth centers. To get a counter-example, let X be the $\mathrm{Spec}(k[x,y,z,t]/(x^3 - y^7))$. Let

$$f : X' \longrightarrow X$$

be the blowing up of the origin, and let

$$g : X'' \longrightarrow X'$$

the blowing up of X' along a smooth curve C of positive genus, contained in the exceptional fiber of f. It is easily seen that the exceptional fiber of f is a projective plane and that X' has multiplicity 3 at the generic point of C.

Denote by p_C the sheaf of ideals of C in X'. Tensoring by $\omega_{X'}$ the exact sequence

$$0 \longrightarrow p_C \longrightarrow \mathcal{O}_{X'} \longrightarrow \mathcal{O}_C \longrightarrow 0 \ ,$$

one obtains the also exact sequence

$$0 \longrightarrow \omega_{X'} \otimes p_C \longrightarrow \omega_{X'} \longrightarrow \omega_{X'} \otimes \mathcal{O}_C \longrightarrow 0 \ .$$

Substituting the equalities

$$\omega_{X'} \otimes p_C = g_*\omega_{X''} \quad \text{(by theorem 8)} \ ,$$

$$\omega_{X'} = f^*\omega_X \quad \text{(by theorem 6)} \ ,$$

one has

$$0 \longrightarrow g_*\omega_{X'} \longrightarrow \omega_{X'} \longrightarrow \mathcal{O}_C \longrightarrow 0 \ .$$

Now as $R^i f_* \omega_{X'} = 0$ for $i > 0$, one has

$$R^2 f_*(g_*\omega_{X''}) = R^1 f_* \mathcal{O}_C = H^1(C,\mathcal{O}_C) \neq 0 \ .$$

Besides $R^i g_* \omega_{X''} = 0$ for $i > 0$, so

$$R^2(f \circ g)_* \omega_{X''} = R^2 f_*(g_*\omega_{X''}) \neq 0 \ .$$

To compute the change in the Euler-Poincaré characteristic, observe that the vanishing theorem implies the degeneration of the Leray spectral sequence

$$E_2^{pq} = H^p(X, R^q\pi_*\omega_{X'}) \Rightarrow H^{p+q}(X', \omega_{X'}) \ ,$$

so one gets the usual isomorphisms

$$H^p(X, \pi_*\omega_{X'}) = H^p(X', \omega_{X'}) \ .$$

Even more:

Theorem 8: With the hypothesis of theorem 6, let $\mathcal{I}$ be the sheaf of ideals on X of the center Y of the blowing up. And write $m = m_1 + \dots + m_r$. One then has

$$\pi_*\omega_{X'} = \begin{cases} \omega_X \otimes \mathcal{I}^{m-d+1} & \text{if } m-d+1 \geqslant 0 \ , \\ \omega_X & \text{if } m-d+1 \leqslant 0 \ . \end{cases}$$

Proof: By theorem 6,

$$\omega_{X'} = \pi^*\omega_X \otimes \mathcal{O}_{X'}(m-d+1) \ ,$$

so it is enough to see for $n \geqslant m-d+1$ that

$$\pi_*\mathcal{O}_{X'}(n) = \begin{cases} \mathcal{I}^n & \text{if } n \geqslant 0 \ , \\ \mathcal{O}_X & \text{if } n \leqslant 0 \ . \end{cases}$$

We will show this by induction on r . Let X_i be the intersection of $H_1, \dots, H_i$ and $\mathcal{I}_i$ be the sheaf of ideals de Y in X_i . The $r = 0$ case is easy as X is smooth. For the general case, consider the exact sequence of sehaves:

$$0 \longrightarrow \mathcal{O}_{X'_{r-1}}(-H'_r + n) \longrightarrow \mathcal{O}_{X'_{r-1}}(n) \longrightarrow \mathcal{O}_{X'_r}(n) \longrightarrow 0 \ .$$

By induction, it follows that

$$(*) \qquad \pi_*\mathcal{O}_{X'_{r-1}}(n) = \begin{cases} \mathcal{I}^n_{r-1} & \text{if } n \geq 0 , \\ \mathcal{O}_{X_{r-1}} & \text{if } n \leq 0 . \end{cases}$$

On the other hand, the restriction map

$$(**) \qquad \pi_*\mathcal{O}_{X'_{r-1}}(n) \longrightarrow \pi_*\mathcal{O}_{X'_r}(n)$$

is surjective, as

$$R^1\pi_*\mathcal{O}_{X'_{r-1}}(-H'_r + n) = 0$$

To see this, the assertion being local on X_{r-1} , one can suppose that the sheaf $\mathcal{O}_{X'_{r-1}}(-H'_r + n)$ is isomorphic to $\mathcal{O}_{X'}(n - m_r)$ where $n - m_r \geq m_1 + \ldots + m_{r-1} - d+1$ But $\mathcal{O}_{X'_{r-1}}(n - m_r)$ has high direct images equal to 0 by corollary 7. Combining (*) and (**), one concludes.

Tensoring by ω_X the exact sequence

$$0 \longrightarrow \mathcal{I}^{m-d+1} \longrightarrow \mathcal{O}_X \longrightarrow \mathcal{O}_X/\mathcal{I}^{m-d+1} \longrightarrow 0 ,$$

(with $m - d + 1 \geq 0$) one gets the also exact sequence

$$0 \longrightarrow \omega_X \otimes \mathcal{I}^{m-d+1} \longrightarrow \omega_X \longrightarrow \omega_X \otimes \mathcal{O}_X/\mathcal{I}^{m-d+1} \longrightarrow 0 .$$

Now the first sheaf on the left is $\pi_*\omega_{X'}$, so taking the characteristic one obtains

$$\chi(X,\omega_X) - \chi(X',\omega_{X'}) = \chi(X,\omega_X \otimes \mathcal{O}_X/\mathcal{I}^{m-d+1}) .$$

If X is an hypersurface of say Z , and $\mathcal{J}$ is the sheaf of ideals of Y in Z then obviously

$$\mathcal{O}_Z/\mathcal{J}^{m-d+1} = \mathcal{O}_X/\mathcal{I}^{m-d+1} ,$$

so in this case it follows that

$$\chi(X,\omega_X) - \chi(X',\omega_{X'}) = \chi(X,\omega_X \otimes \mathcal{O}_Z/\mathcal{J}^{m-d+1}) = \sum_{i=0}^{m-d} \chi(Y,\omega_X \otimes \mathcal{J}^i/\mathcal{J}^{i+1}) .$$

Now using the fact that for a smooth subvariety Y of a smooth variety Z , there are isomorphisms

$$\mathcal{J}^i/\mathcal{J}^{i+1} = S^i N^*_{Y/Z} ,$$

where $N^*_{Y/Z}$ denotes the conormal sheaf to Y in Z , one concludes finally that

$$\chi(X,\omega_X) - \chi(X',\omega_{X'}) = \sum_{i=0}^{m-d} \chi(Y,\omega_X \otimes S^i N^*_{Y/Z}) .$$

We record it as theorem:

Theorem 9: *Let X be a complete intersection subvariety of a smooth variety Z, and let*

$$\pi : X' \longrightarrow X$$

be the blowing up of X along a smooth integral subvariety Y given by the sheaf of ideals I. If $H_1,\dots,H_r$ are like in theorem 6, $m = m_1 + \dots + m_r$ is the sum of the multiplicities of $H_1,\dots,H_r$ at the generic point of Y and d is the codimension of Y in Z, one has:

a) $H^i(X,\omega_X) = H^i(X',\omega_{X'})$ *if* $m-d+1 \leq 0$.

b) $\chi(X,\omega_X) - \chi(X',\omega_{X'}) = \chi(X,\omega_X \otimes \mathcal{O}_X/I^{m-d+1})$ *if* $m-d+1 \geq 0$.

c) *If X is an hypersurface of Z then for $m-d+1 > 0$ one has*

$$\chi(X,\omega_X) - \chi(X',\omega_{X'}) = \sum_{i=o}^{m-d} \chi(Y,\omega_X \otimes S^i N^*_{Y/Z}) \quad .$$

Examples: 1) Let X be an hypersurface of a smooth n-dimensional variety Z and let

$$\pi : X' \longrightarrow X$$

be the blowing up of X at a point x of multiplicity m.

If $m \geq n$, the change in the characteristic is

$$\chi(X,\omega_X) - \chi(X',\omega_{X'}) = \sum_{i=o}^{m-n} \dim m^i_{Z,x}/m^{i+1}_{Z,x} = \binom{m}{n} \quad .$$

Note that this formula permits one to compute the geometric genus of a plane curve as is well known.

If $m < n$, then $H^i(X,\mathcal{O}_X) = H^i(X',\mathcal{O}_{X'})$ for all i.

2) Let S be a surface of $\mathbb{P}_3$ and n its degree. Suppose S is singular along a curve C whose points have multiplicity m. Let $\pi : S' \longrightarrow S$ the blowing up of S with center C. To compute the change in the characteristic suppose C irreducible and smooth of genus g and degree d. Then:

$$\chi(S,\omega_S) - \chi(S',\omega_{S'}) = \binom{m}{2} (1-g+(n-4)d) - \binom{m}{3} (2g-2+4d) \quad .$$

To see this observe that

$$\chi(S,\omega_S) - \chi(S',\omega_{S'}) = \sum_{i=o}^{m-2} \chi(C, S^i N^*_{C/\mathbb{P}_3} \otimes \mathcal{O}_C(n-4)) \quad .$$

But by the Riemann-Roch theorem, it follows that

$$\chi(C,S^iN^*_{C/\mathbb{P}_3}\otimes\mathcal{O}_C(n-4)) = (i+1)(1-g)+\deg c_1(S^iN^*_{C/\mathbb{P}_3}\otimes\mathcal{O}_C(n-4)) .$$

A simple computation gives

a) $c_1(C,S^iN^*_{C/\mathbb{P}_3}\otimes\mathcal{O}_C(n-4)) = c_1(S^iN^*_{C/\mathbb{P}_3})+(i+1)c_1(\mathcal{O}_C(n-4))$.

b) $c_1(S^iN^*_{C/\mathbb{P}_3}) = \binom{i+1}{2}\, c_1(N^*_{C/\mathbb{P}_3})$.

Finally, $c_1(N^*_{C/\mathbb{P}_3})$ can be computed from the exact sequence

$$0\longrightarrow N^*_{C/\mathbb{P}_3}\longrightarrow i^*\Omega^1_{\mathbb{P}_3}\longrightarrow\Omega^1_C\longrightarrow 0 .$$

One has:

$$c_1(N^*_{C/\mathbb{P}_3}) = i^*c_1(\Omega^1_{\mathbb{P}_3})-c_1(\Omega^1_C) ,$$

from where

$$\deg c_1(N^*_{C/\mathbb{P}_3}) = -4d-(2g-2) .$$

Combining everything one concludes.

3) Let S be a surface in the projective space $\mathbb{P}_3$ with ordinary singularities. This is, S may be singular along a curve C whose points have multiplicity 2 except at a finite set of them which may have multiplicity 3 (in both C and S) . Also a desingularization S'' of S can be obtained by blowing up first all the triple points and blowing up after the double curve:

$$S''\longrightarrow S'\longrightarrow S .$$

To compute the change in the characteristic, let n be the degree of S , d the degree of C , g the geometric genus of C , c the number of irreducible components of C and t the number of triple points.

The blowing up of the triple points gives

$$\chi(S,\omega_S)-\chi(S',\omega_{S'}) = \sum_{x_1,\dots,x_t}\dim\mathcal{O}_S/m_{x_i} = t .$$

The blowing up of the double curve C' (which is supposed to be smooth) gives

$$\chi(S',\omega_{S'})-\chi(S'',\omega_{S''}) = \chi(C',\omega_{S'}\otimes_{\mathcal{O}_{S'}}\mathcal{O}_{C'}) = c-g+(n-4)d-3t$$

where $\omega_{S'}$ is calculated by theorem 6.

Putting them together one gets

$$\chi(S,\omega_S)-\chi(S'',\omega_{S''}) = c-g+(n-4)d-2t .$$

REFERENCES:

[1] H. Grauert and O. Riemenschneider - Verschwinchungssätze für analytische Komologiegruppen auf komplesen Raumen, Inv. Math., 11 (1970), 263-292.

[2] J. Lipman - Desingularization of two dimensional schemes, Annals of Math., 107 (1978), 151-207.

[3] J. Wahl - Vanishing theorems for resolutions of surface singularities, Inv. Math., 31 (1975), 17-41.

Juan B. Sancho
Dept. de Matemáticas
Universidad de Salamanca (Spain)

CHEVALLEY GROUPS OVER $\mathbb{C}((t))$ AND DEFORMATIONS OF SIMPLY ELLIPTIC SINGULARITIES

by

PETER SLODOWY

In these notes we are going to relate the deformation theory of the so called simply elliptic singularities to the corresponding Chevalley groups over the formal series field $\mathbb{C}((t))$ in a similar, however less complete way as Brieskorn [2] has done for the simple singularities and the associated simple complex Lie groups. We give only a survey of the main results and the basic concepts involved. Complete details will be found in a forthcoming work on adjoint quotients for certain groups attached to arbitrary Kac Moody Lie algebras [22]. These general results pertain to a much wider class of singularities which in addition includes at least the cusp singularities of degree ≤ 5 whose deformation theory has recently been studied by Looijenga [13]. Here we restrict to the special situation given by the simply elliptic singularities where it is possible to avoid the technical machinery needed for the general case.

I. Simple Singularities and Simple Lie Groups.

In this part we quickly recall the relation between the simple singularities (equivalently: Kleinian singularities or rational double points) and certain simple Lie groups. For complete details we refer to [21].

1.- Simple singularities are normal surface singularities with a very special minimal resolution. The dual graph of the exceptional divisor of such a resolution is a Dynkin diagram of type A_r, $r \geq 1$, D_r, $r \geq 4$, E_6, E_7, E_8. Up to analytic isomorphism these diagrams classify the corresponding singularities.

2. Let G be a semisimple, simply connected, complex algebraic group and $T \subset G$ a maximal torus with corresponding Weyl group $W = N_G(T)/T$. We have $r = \operatorname{rank} G = \dim T$. Denote by $X^*(T)$ the group $\operatorname{Hom}(T, G_m) \cong \mathbb{Z}^r$ of algebraic characters of T and by $X_*(T)$ the dual group $\operatorname{Hom}(G_m, T) \cong \mathbb{Z}^r$ of multiplicative one parameter subgroups. Let $\Sigma \subset X^*(T)$ be the system of roots of T in G. For each $\alpha \in \Sigma$ we fix an isomorphism

$$u_\alpha : G_a \xrightarrow{\sim} U_\alpha \subset G$$

from the additive group G_a onto the root subgroup U_α. For all $s \in T$ we have

$$s\, u_\alpha(c) s^{-1} = u_\alpha(\alpha(s)c), \quad c \in \mathbb{C}.$$

Let $\Delta = \{\alpha_1, \ldots, \alpha_r\}$ be a system of simple roots of Σ corresponding to the choice of a Borel subgroup $B \supset T$, and let $\Delta^\vee = \{\alpha_1^\vee, \ldots, \alpha_r^\vee\}$ be the simple coroots in $X_*(T)$. Since G is simply connected $X^*(T)$ is spanned freely (over $\mathbb{Z}$) by the fundamental dominant weights $w_1, \ldots w_r$ which are determined by the condition $\langle w_i, \alpha_j^\vee \rangle = \delta_{ij}$. To each w_i there corresponds a fundamental irreducible representation

$$\rho_i : G \text{ -------- } GL(V_i)$$

of G on a finite dimensional vector space V_i. Let

$$X_i : G \text{ -------- } \mathbb{C}$$

$$X_i(g) = \operatorname{trace} \rho_i(g)$$

be the corresponding fundamental character. Then the <u>adjoint quotient</u> of G is given by the morphism

$$X : G \text{ -------- } \mathbb{C}^r$$

$$X(g) = (X_1(g), \ldots, X_r(g)).$$

The morphism X is the algebraic quotient of the adjoint action of G on itself. Any fibre of X is the union of finitely many conjugacy classes and its dimension is $\dim G - r$. The restriction of X to T coincides with the natural quotient $T \longrightarrow T/W$ and T/W can be identified with $\mathbb{C}^r$.

3. Now we will look at the fibres of X more closely. Any fibre of X can be written in the form $X^{-1}(X(s))$ for a suitable $s \in T$. Let us first look at s =1. The corresponding fibre consists of the unipotent elements in G, i.e. those which are represented by unipotent matrices in all rational representations of G. It is called the <u>unipotent variety</u> Uni(G) of G. For arbitrary $s \in T$ there is a reduction to the centralizer Z(s) of s in G which is a reductive subgroup. It is generated by T and the root subgroups U_α for which $\alpha(s) = 1$. The unipotent variety Uni(s) of Z(s) (i.e. that of its semisimple part) is the product of the unipotent varieties of its simple (almost)-factors. The fibre $X^{-1}(X(s))$ is G-isomorphic to the homogeneous bundle $G \times^{Z(s)} \mathrm{Uni}(s)$ associated to the principal fibration $G \longrightarrow G/Z(s)$ and the adjoint action of Z(s) on Uni(s).

An element $x \in G$ is called <u>regular</u> (resp. <u>subregular</u>) exactly when $\dim Z_G(x) = r$ (resp. $r + 2$) which is the same as the condition dim(conjugate class of x) = dim G − r (resp. dim G − r − 2). There is exactly one regular orbit in the unipotent variety and hence in any fibre of X. If G is simple there is exactly one subregular unipotent orbit, and this is the orbit of greatest dimension among the nonregular orbits in Uni(G). If G is semisimple there are as many subregular unipotent orbits as there are simple factors.

4. Now let G be simple of type $\Delta = A_r, D_r, E_r$ and choose a sufficiently small normal slice $S \subset G$ to the subregular unipotent orbit of G. We may assume that S is transversal to all orbits and that is meets the subregular unipotent orbit exactly once.

<u>Theorem</u> (Brieskorn, [2]):

i) $S \cap \mathrm{Uni}(G)$ <u>is a simple singularity of type</u> Δ.

ii) <u>The restriction</u> $X|_S : S \longrightarrow T/W$ <u>of the adjoint quotient realizes a semiuniversal deformation of the simple singularity</u> $S \cap \mathrm{Uni}(G)$.

From this theorem we can derive many useful informations concerning simple singularities and their deformations. To determine the singularities in the fibres of a semiuniversal deformation we have to look at the singularities in $S \cap X^{-1}$ (X(s)) for $s \in T$ sufficiently close to 1. In this case a basis $\Delta(s)$ of the root system $\Sigma(s) = \{\alpha \in \Sigma \mid \alpha(s) = 1\}$ of Z(s) may be embedded into a base Δ of Σ, and for each connected component of $\Delta(s)$ there exists a simple singularity in $S \cap X^{-1}$ (X(s)) of the corresponding type. This description also shows that the <u>discriminant</u> (i.e. the critical set) of $X|_S$ coincides with the discriminant of the ramified covering $T \longrightarrow T/W$ (near X(1)).

5. All deformations of a simple singularity admit a simultaneous resolution. This fact can be derived from the following construction. Let $B \supset T$ be a Borel subgroup of G. Then B can be written as a semidirect product $B = T \ltimes U$ where U is the unipotent radical of B. Let $G \times^B (B)$ be the bundle associated to the principal fibration $G \longrightarrow G/B$ and the adjoint action of B on itself (B). We obtain a commutative diagram

$$\begin{array}{ccc} G \times^B (B) & \xrightarrow{\emptyset} & G \\ \Big\downarrow \Theta & & \Big\downarrow X \\ T & \xrightarrow{\Psi} & T/W \end{array}$$

where $\emptyset(g * b) = gbg^{-1}$, $\Theta(g * b) = \Theta(g * tu) = t$ and Ψ is the natural quotient map (we denote the class of (g,b) in $G \times^B (B)$ by $g * b$).

<u>Theorem</u> (Grothendieck, Springer):

<u>The diagram above is simultaneous resolution of</u> X, i.e. Θ <u>is smooth</u>, $\emptyset$ <u>is proper and for all</u> $s \in T$ <u>the restriction</u> $\emptyset_s : \Theta^{-1}(s) \longrightarrow X^{-1}(\Psi(s))$ <u>is a resolution of singularities</u>.

II. Simply Elliptic Singularities

We now review some properties of simply elliptic singularities and their semiuniversal deformations. Details can be found in the references [9] , [10], [11], [12], [13], [14], [18], [19], [20].

6. A normal surface singularity (X_o, x) with isolated singular point x is called simply elliptic exactly when the exceptional divisor $E = \pi^{-1}(x)$ in the minimal resolution $\pi : Y \longrightarrow X_o$ consists of a single elliptic curve. The selfintersection number of E is necessarily negative, $E \circ E = -d$ for some integer $d \geq 1$. We call d the degree of the singularity. Up to analytic isomorphism (X_o, x) is determined by its degree d and the analytic structure of E. Hence any simply elliptic singularity can be obtained as the contraction of the zero section in some negative line bundle over a suitable elliptic curve E.

The embedding dimension of (X_o, x) is max(3,d). For d = 1,2,3 we obtain the "parabolic" hypersurfaces in the sense of Arnol'd [1] which were studied by Saito [20].

$$X^6 + Y^3 + Z^2 + \tau XYZ = 0 \qquad d = 1$$

$$X^4 + Y^4 + Z^2 + \tau XYZ = 0 \qquad d = 2$$

$$X^3 + Y^3 + Z^3 + \tau XYZ = 0 \qquad d = 3$$

(Here the parameter τ is related to the j-invariant of the elliptic curve E, cf. [20]). For d = 4 we obtain the complete intersection of two quadrics in $\mathbb{C}^4$.

To the first six simply elliptic singularities there is associated an affine Dynkin diagram $\tilde{\Delta}$:

d	1	2	3	4	5	6
$\tilde{\Delta}$	$\tilde{E}_8$	$\tilde{E}_7$	$\tilde{E}_6$	$\tilde{D}_5$	$\tilde{A}_4$	$A_2 \tilde{\times} A_1$

The deformation theory of these singularities can be described completely in terms of the corresponding diagrams. This was suggested already in the work of Saito

[20] and established precisely in the works of Knörrer [9], Looijenga [10], [11], [12], Pinkham [19], and Merindol [14]. The basic tool in Pinkham's approach is the theory of the corresponding Del Pezzo surfaces.

7. We first give a rough picture of the semiuniversal deformation $\Phi : X \to V$ of a simply elliptic singularity X_o, cf. [14], [18]. We may choose Φ to be equivariant with respect to natural G_m-actions on X and V and we may assume that there is a projection
$p: V \to \Omega^+ = \{\lambda \in \mathbb{C} \mid |\lambda| > 1\}$ as well as a section $s : \Omega^+ \to V$ of p mapping Ω^+ onto the fixed points of G_m in V with the following properties. Decompose $V = V_e \cup V_f$, where $V_e = s(\Omega^+)$, $V_f = V - V_e$. Then a fibre $\Phi^{-1}(s(\lambda))$ has a simply elliptic singularity of the same degree d as X_o and the exceptional elliptic curve E of its minimal resolution is isomorphic to $\mathbb{C}^*/\langle \lambda^i | i \in \mathbb{Z}\rangle$. A fibre over V_f is either smooth or has at most simple singularities.

The dimension of V is max(11-d,1) and it is smooth exactly when $d \leq 5$. Then $V \cong \Omega^+ \times \mathbb{C}^{r+1}$, where $r = 9-d$. For d = 6 we obtain $V \cong \Omega^+ \times C(\mathbb{P}^1 \times \mathbb{P}^2)$ where $C(\mathbb{P}^1 \times \mathbb{P}^2)$ is the affine cone over the Segré embedding of $\mathbb{P}^1 \times \mathbb{P}^2$ into $\mathbb{P}^5$. For d = 7 each slice $V_\lambda = p^{-1}(\lambda)$ is a cone over a surface of degree 7 in $\mathbb{P}^6$ (depending on λ). For d = 8 there are two components $V = V_1 \cup V_2$ which intersect along V_e. Each slice $V_{1,\lambda}$ is a cone over an embedding of the elliptic curve $\mathbb{C}^*/\langle \lambda^i | i \in \mathbb{Z}\rangle$ into $\mathbb{P}^7$ and $V_2 \cong \Omega^+ \times \mathbb{C}^2$. For d = 9 we have $V_{red} \cong \Omega^+ \times \mathbb{C}$, however V has embedded components along V_e. For $d \geq 10$ we have $V_{red} = V_e \cong \Omega^+$, but again V is not reduced. (Under the isomorphisms given above p and s will always have the canonical form.)

A simply elliptic singularity can be smoothed by deformation if and only if $d \leq 9$. If d = 9 there are no singular fibres over V_f, and if d = 8 there are none above $V_1 \cap V_f$. In the other cases ($d \leq 8$) the discriminant of Φ is of particular interest. It was described in a uniform way by Pinkham and Looijenga ($d \leq 3$).

8. We now recall this construction. It suffices to consider the discriminant $D_\lambda \subset V_{f,\lambda} = V_f \cap V_\lambda$ of the restriction $\Phi_\lambda : \Phi^{-1}(V_{f,\lambda}) \to V_{f,\lambda}$.

The exceptional elliptic curve E in the resolution of $\Phi^{-1}(s(\lambda))$ is then ismorphic to $\mathbb{C}^*/\langle\lambda^i\rangle$.

Let $X_*(T)$ denote the lattice generated by the coroots of some root system Σ and let $T = X_*(T) \otimes \mathbb{C}^*$ be a maximal torus of the corresponding simply connected complex Lie group. By A we denote the abelian variety $X_*(T) \otimes E = T/X_*(T) \otimes \langle\lambda^i\rangle$. The Weyl group W of Σ acts naturally on $X_*(T)$ and A. There is an essentially unique $\mathbb{C}^*$-bundle L over A endowed with a W-action and such that its first Chern class $c_1(L)$ equals the negative normalized Killing form on $X_*(T)$ (value 2 on short coroots). Here we use the Apell-Humbert identification $c_1(\mathrm{Pic}\, A) \cong S^2X_*(T) \subset H^2(A,\mathbb{C})$. If Σ is irreducible or if Σ contains only components of type A_n or C_n then the isotropy groups of W on L are generated by reflections. Therefore L/W is a smooth space. Let $D \subset L/W$ denote the discriminant of the ramified covering $L \longrightarrow L/W$.

Theorem (Looijenga, Pinkham):

Let Σ be a root system of type E_8, E_7, E_6, D_5, A_4, $A_2 \times A_1$. Then the pair $(L/W,D)$ is isomorphic to the pair $(V_{f,\lambda}, D_\lambda)$ for the corresponding simply elliptic singularity. Let $\bar{s} \in V_{f,\lambda} = L/W$ be the image of a point $s \in L$ and W_s the stabilizer of s in W. Then there is a type preserving bijection between the irreducible factors of W_s and the (simple) singularities in the fibre $\Phi^{-1}(\bar{s})$ of the deformation Φ.

Pinkham actually gives a construction of the total space of Φ, too. His method also extends to the cases $d = 7$ and 8. If $d = 8$ the pair $(V_2 \cap V_{f,\lambda}, D_\lambda)$ is obtained by putting $\Sigma = A_1$. The case $d = 7$ can be described similarly by using a rank two lattice $X_*(T) = \mathbb{Z} \oplus \mathbb{Z}$ with " Killing form" $\begin{pmatrix} 4 & 3 \\ 3 & 4 \end{pmatrix}$ which contains an A_1-lattice, i.e. $\mathbb{Z}.(1,-1)$.

For later use we note the following. We may pull back L to a trivial $\mathbb{C}^*$-bundle $T \times \mathbb{C}^*$ over T equipped with an action of the affine Weyl group $\widetilde{W} = W \ltimes X_*(T)$. The translations $X_*(T)$ will then operate on T as the subgroup $X_*(T) \otimes \langle\lambda^i\rangle \cong X_*(T)$ of T, and the action on $T \times \mathbb{C}^*$ will be determined by an automorphy factor $e : T \times X_*(T) \longrightarrow \mathbb{C}^*$. Since $X_*(T)$ acts freely on $T \times \mathbb{C}^*$ we obtain the same isotropy groups for $\tilde{W}$ on $T \times \mathbb{C}^*$ as for the action of W on L. By the same reason the discriminants of the ramified coverings $T \times \mathbb{C}^* \longrightarrow (T \times \mathbb{C}^*)/\tilde{W} = L/W$ and $L \to L/W$ coincide.

III. Chevalley Groups over $\mathbb{C}((t))$

Extended Dynkin diagrams, affine root systems and affine Weyl groups arise in the study of algebraic groups over local fields [3], [5]. It has been natural to ask (cf. for example [19]) whether there would be a similar relation between simply elliptic singularities and the corresponding Chevalley groups over $\mathbb{C}((t))$ as there was between simple singularities and simple complex Lie groups.

A first attemp is to repeat the construction of the morphism $\chi : G \to T/W$ over the base field $K = \mathbb{C}((t))$. However, this will lead only to forms of simple singularities over K (cf. [21] Appendix 1). In particular one does not end up with finite dimensional objects over $\mathbb{C}$. To remedy these defects one has to modify a Chevalley group over K in a way suggested by the theory of the closely related Euclidean Kac Moody-Lie Algebras [6], [15], [16].

9. Let G be a semisimple simply connected algebraic group over $\mathbb{C}$. Let $K = \mathbb{C}((t)) = \left\{ \sum_{i \geq i_o} a_i t^i \mid a_i \in \mathbb{C}, i_o \in \mathbb{Z} \right\}$ be the field of power series over $\mathbb{C}$ and $G(K)$ the group of points of G over K. The most important modification of $G(K)$ will be the following semidirect product. By $\Omega \cong \mathbb{C}^*$ we denote the group of $\mathbb{C}$-automorphisms of K given by

$$\lambda_{p(t)} = p(\lambda t)$$

where $\lambda \in \mathbb{C}^*$ and $p(t)$ is a power series in t. This group acts naturally on $G(K)$ and we may form the semidirect product $G(K) \rtimes \Omega$. The projection $p : G(K) \rtimes \Omega \longrightarrow \Omega$ is invariant under conjugation by $G(K)$. If (g, λ) is an element of the fibre $p^{-1}(\lambda)$ conjugation by an element $x \in G(K)$

will look like

$$x(g, \lambda)\, x^{-1} = (xg^{\lambda} x^{-1}, \lambda)$$

where $^{\lambda}x^{-1} = \lambda x^{-1} \lambda^{-1} = x(\lambda t)^{-1}$ if we write x as a power series in t. It will turn out later that for $|\lambda| \neq 1$ the corresponding conjugacy classes will have finite $\mathbb{C}$-codimension in $G(K) \rtimes \Omega$ and that it is possible to define a quotient of (a part of) $G(K) \rtimes \Omega$ with respect to conjugation. However, to obtain a complete picture we need a further modification of the group $G(K)$.

10. The Kac Moody Lie algebra $\hat{g}$ corresponding to $G(K)$ is given as the following one-dimensional central extension of the points $\underline{g} \otimes \mathbb{C}[t,t^{-1}]$ of the Lie algebra $\underline{g}$ of G over the Laurent polynomial ring $\mathbb{C}[t,t^{-1}]$. We have $\hat{\underline{g}} = (\underline{g} \otimes \mathbb{C}[t,t^{-1}]) \oplus \mathbb{C}.c$ as a $\mathbb{C}$-vector space and the Lie bracket for elements $x \otimes u$, $y \otimes v$, $x,y \in \underline{g}$, $u,v \in \mathbb{C}[t,t^{-1}]$ is defined by

$$[x \otimes u, y \otimes v] = [x,y] \otimes uv - \mathrm{Res}\,(udv)\,(x,y).c$$

where $(\ ,\ )$ is the Killing form on $\underline{g}$ and $\mathrm{Res}(udv)$ means the residue of the differential form udv (cf. [4]). Kac has developed a theory of highest weight representations for $\hat{\underline{g}}$ including an analogue of the Weyl character formula [7]. Let G be simple of rank r, then there are $r+1$ fundamental representations corresponding to fundamental dominant weights similarly as in the classical theory. All these representations are of infinite dimension over $\mathbb{C}$. In [4] Garland has shown how to lift these representations to representations of a central extension $\tilde{G}$ of $G(K) \rtimes \Omega$

$$1 \longrightarrow \mathbb{C}^* \longrightarrow \tilde{G} \xrightarrow{\ \varepsilon\ } G(K) \rtimes \Omega \longrightarrow 1 .$$

To describe this extension it is, according to the theory of Moore [17], sufficient to know the restriction of this extension to the maximal torus K^* of some $SL_2(K)$ subgroup of $G(K)$ associated to a long root. Garland shows that the extension ε is defined by the inverse of the tame symbol , i.e. two elements u,v of the torus K^* lifted to $\tilde{G}$ in a special way multiply according to

$$u.v = uv\,(-1)^{\nu(u)\nu(v)}\; c(v^{\nu(u)} u^{-\nu(v)})$$

where $u.v$ is the product in $\tilde{G}$, uv is the product in K, $\nu : K \longrightarrow \mathbb{Z}$ is the t-valuation on $K = \mathbb{C}((t))$ and $c : K \longrightarrow \mathbb{C}$ is the constant term of power series. In the formula above the value of c lies in $\mathbb{C}^*$ and is regarded as an element of the center of $\tilde{G}$.

For a reductive group G there is a similarly defined central extension $1 \rightarrow \mathbb{C}^* \rightarrow \tilde{G} \rightarrow G(K) \rtimes \Omega \rightarrow 1$ which is uniquely determined once a Killing form for G, i.e. one on $X_*(T)$, is chosen.

11. In the following we will keep the notations of section 2. The composition $\tilde{G} \rightarrow G(K) \rtimes \Omega \longrightarrow \Omega$ is also denoted by p. By $T = X_*(T) \otimes \mathbb{C}^*$ we denote the complex points of a maximal torus of G. We regard T as a subgroup of $T(K) \subset G(K)$ which are the K-valued points of T and G. Let $\tilde{T} = \varepsilon^{-1}(T \times \Omega)$. Then $\tilde{T}$ is a maximal $\mathbb{C}$-torus in $\tilde{G}$ of dimension $r + 2$ which, using a section of T in $\tilde{G}$, can be written as a product $T \times \mathbb{C}^* \times \Omega$. Using the ordinary Bruhat decomposition of $G(K)$ one proves:

<u>Proposition 1:</u> <u>Let</u> $\tilde{N}$ <u>be the normalizer of</u> $\tilde{T}$ <u>in</u> $\tilde{G}$. <u>Then there is an exact sequence</u>

$$1 \longrightarrow \tilde{T} \longrightarrow \tilde{N} \longrightarrow \tilde{W} \longrightarrow 1$$

<u>where</u> $\tilde{W}$ <u>is the affine Weyl group</u> $\tilde{W} = W \ltimes X_*(T)$

There is a particularly nice section of $X_*(T)$ into the image of $\tilde{N}$ in $G(K) \rtimes \Omega$ given by the subgroup

$$X_*(T) \otimes \langle t^i \mid i \in \mathbb{Z} \rangle \subset X_*(T) \otimes K^* = T(K)$$

in $T(K)$.

Let $\tilde{T}_\lambda = p^{-1}(\lambda) \cap \tilde{T}$. Then $\tilde{T}_\lambda$ is a $\mathbb{C}^*$-bundle over $T_\lambda = \varepsilon(\tilde{T}_\lambda) \subset G(K) \rtimes \Omega$ equipped with the natural action of $\tilde{N}/\tilde{T} \cong \tilde{W}$. The subgroup $X_*(T)$ of $\tilde{W}$ operates on T_λ by translations through the subgroup

$X_*(T) \otimes \langle \lambda^i \mid i \in \mathbb{Z} \rangle$ of T. Comparing the automorphy factor for the action of $X_*(T)$ on $\tilde{T}_\lambda$ with the automorphy factor of the construction at the end of section 8 one obtains:

Proposition 2: The $\tilde{W}$-actions on the $\mathbb{C}^*$-bundle $\tilde{T}_\lambda$ and the $\mathbb{C}^*$-bundle $T \times \mathbb{C}^*$ defined in section 8 for the same λ coincide with respect to the natural identification

$$\begin{array}{ccccc} \tilde{T}_\lambda & = & T \times \mathbb{C}^* \times \{\lambda\} & \simeq & T \times \mathbb{C}^* \\ \downarrow & & \downarrow & & \downarrow \\ T_\lambda & = & T \times \{\lambda\} & \simeq & T \end{array}$$

12. Consider the points $G(\mathbb{C}[[t]])$ of G over the formal power series ring and the natural reduction homomorphism "mod t":

$$r : G(\mathbb{C}[[t]]) \longrightarrow G = G(\mathbb{C})$$

Let $B \subset G$ be a Borel subgroup containing T and $B' = r^{-1}(B)$ its preimage under r. We consider B' as a subgroup of $G(K)$ and call $\tilde{B} := \mathcal{E}^{-1}(B' \rtimes \Omega)$ an Iwahori subgroup of $\tilde{G}$. According to Moore's theory [17] we have an isomorphism of groups $\tilde{B} \cong (B' \rtimes \Omega) \times \mathbb{C}^*$ (with $\mathbb{C}^*$ the center of $\tilde{G}$) giving rise to a semidirect product decomposition $\tilde{B} = \tilde{T} \ltimes \tilde{U}$ where $\tilde{U}$ is the kernel of the obvious projection $\tilde{B} \longrightarrow \tilde{T}$. We then have the affine Bruhat decomposition originally due to Iwahori and Matsumoto [5] and adapted to our context by Garland [4]:

Theorem 1: The group $\tilde{G}$ is the disjoint union of the distinct double cosets $\tilde{U} w \tilde{B}$, $w \in \tilde{N}/\tilde{T}$.

Let $\Sigma \subset X^*(T)$ be the root system of T in G and $u_\alpha : G_a \xrightarrow{\sim} U_\alpha \subset G$ the fixed additive one parameter subgroup corresponding to an $\alpha \in \Sigma$ (cf: section 2). Through the projection $\tilde{T} \rightarrow T$ we consider Σ as a subset of

$X^*(\tilde{T})$. Let $\delta \in X^*(\tilde{T})$ be the character defined by the composition $\tilde{T} \hookrightarrow \tilde{G} \xrightarrow{p} \Omega \cong \mathbb{C}^*$. The affine root system of $\tilde{T}$ in $\tilde{G}$ is now defined as $\tilde{\Sigma} = \{\alpha + i\delta \in X^*(\tilde{T}) \mid \alpha \in \Sigma,\ i \in \mathbb{Z}\}$. For any affine root $a = \alpha + i\delta \in \tilde{\Sigma}$ we obtain a complex one parameter group

$$u_a : \mathbb{C} \xrightarrow{\sim} \tilde{U}_a \subset \tilde{G}$$

with the property

$$s\, u_a(c) s^{-1} = u_a(a(s)c)$$

for all $s \in \tilde{T}$, $c \in \mathbb{C}$, by composing

$$\begin{array}{ccccccc} \mathbb{C} & \longrightarrow & \mathbb{C}t^i & \hookrightarrow & K & \longrightarrow & U_\alpha(K) \\ c & \longmapsto & ct^i & & \longmapsto & & u_\alpha(ct^i) \end{array}$$

with a fixed grouptheoretic section $U_\alpha(K) \longrightarrow \tilde{G}$.

Using either the ordinary Bruhat decomposition for $G(K)$ or the affine one for $\tilde{G}$ one can investigate the structure of the centralizers $Z(s)$ of elements $s \in \tilde{T}$.

Theorem: Let $s \in \tilde{T}$ such that $|p(s)| = |\delta(s)| \neq 1$ and $\tilde{\Sigma}(s) = \{a \in \tilde{\Sigma} \mid a(s) = 1\}$. Then $\tilde{\Sigma}(s)$ is a finite subroot system of $\tilde{\Sigma}$ and $Z(s)$ is a finite dimensional complex reductive group with root system $\tilde{\Sigma}(s)$ generated by the subgroups $\tilde{T}$ and $\tilde{U}_a$, $a \in \tilde{\Sigma}(s)$.

13. Let $\overset{\sim o}{G}$ consist of the elements in $\tilde{G}$ which are conjugate into a fixed Iwahori subgroup $\tilde{B} = \tilde{T} \ltimes U$. From a refinement of the ordinary or affine Bruhat decomposition one deduces:

Proposition 1: Let s, s' be elements in $\tilde{T}$ conjugate under $\tilde{G}$. Then s and s' are conjugate by an element in $\tilde{N}$.

As a corollary we obtain a set-theoretic map

$$\tau : \tilde{G}^o \longrightarrow \tilde{T}/\tilde{W}$$

(here $\tilde{T}/\tilde{W}$ is the set-theoretic quotient!) defined uniquely in the following way. An element $g \in \tilde{G}^o$ is conjugate to some $b \in \tilde{B}$. Let s be the projection of b onto $\tilde{T}$. Then $\tau(g)$ is the class of s in $\tilde{T}/\tilde{W}$.

To form an analytic quotient $\tilde{T}/\tilde{W}$ we have to delete the points $s \in \tilde{T}$ with $|p(s)| = 1$. Now let G be simple and $\rho_i : \tilde{G} \longrightarrow GL(V_i)$, $i = 0, \ldots, r$ the fundamental irreducible representations of $\tilde{G}$ introduced in section 10. By X_i we denote the formal character of $\tilde{T}$ on V_i given by the Kac-Weyl character formula [7]. Let $\tilde{T}_{>1} := \{ s \in \tilde{T} \mid |p(s)| > 1 \}$.

Proposition 2: The characters X_i are $\tilde{W}$-invariant holomorphic functions on $\tilde{T}_{>1}$. The map $\tilde{T}_\lambda/\tilde{W} \longrightarrow \mathbb{C}^{r+1}$ sending the $\tilde{W}$-orbit of an element $s \in \tilde{T}_\lambda$ to $(X_o(s), \ldots, X_r(s))$ is an analytic isomorphism onto $\mathbb{C}^{r+1} \setminus \{0\}$ for all $\lambda \in \Omega$, $|\lambda| > 1$, with the possible exception of a discrete subset of Ω bounded from above.

Let $\tilde{G}^o_{>1} = \{ g \in \tilde{G}^o \mid |p(g)| > 1 \}$. The map $\tau : \tilde{G}^o_{>1} \longrightarrow \tilde{T}_{>1}/\tilde{W}$ may be composed with the morphism $\tilde{T}_{>1}/\tilde{W} \longrightarrow \Omega \times \mathbb{C}^{r+1}$, $(s \bmod \tilde{W}) \longmapsto (p(s), X_o(s), \ldots, X_r(s))$, to give the algebraic trace map

$$X : \tilde{G}^o_{>1} \longrightarrow \Omega \times \mathbb{C}^{r+1}.$$

If $g \in \tilde{G}^o_{>1}$ maps to a convergent power series in $G(\mathbb{C}((t))) \rtimes \Omega$ then $\rho_i(g)$ may be considered as an operator of trace class on V_i with respect to the hermitian product introduced by Garland [4]. Then $X_i(g) = X_i(\tau(g))$ will coincide with the analytic trace of $\rho_i(g)$.

If $G = G_1 \times \ldots \times G_k$ is semisimple with simple factors G_i the extension $\tilde{G}$ is only a quotient of the product $\tilde{G}_1 \times \ldots \times \tilde{G}_k$ by a $(k-1)$-dimensional central torus. Accordingly fundamental characters of $\tilde{G}$ are given by products of fundamental characters of the $\tilde{G}_i$ satisfying certain conditions. These characters will in general not be algebraically independent as is the case for simple G.

14. Let $s \in \tilde{T}$, $|p(s)| > 1$. We know from section 12 that the centralizer $Z(s)$ of s in $\tilde{G}$ is a finite dimensional reductive group. We denote its unipotent variey by $\mathrm{Uni}(s)$.

Theorem 1: <u>The fibre $\tau^{-1}(\tau(s))$ is $\tilde{G}$-isomorphic to the associated bundle $\tilde{G} \times^{Z(s)} \mathrm{Uni}(s)$.</u>

Corollary: <u>The fibre $\tau^{-1}(\tau(s))$ contains only finitely many conjugacy classes which are all of finite $\mathbb{C}$-codimension.</u>

Let $\tilde{B}_{>1} = \tilde{B} \cap \tilde{G}^{o}_{>1}$. Then there is a commutative diagram

$$\begin{array}{ccc} \tilde{G} \times^{\tilde{B}} (\tilde{B}_{>1}) & \xrightarrow{\;\emptyset\;} & \tilde{G}^{o}_{>1} \\ \Theta \big\downarrow & & \big\downarrow \tau \\ \tilde{T}_{>1} & \xrightarrow{\;\Psi\;} & \tilde{T}_{>1}/\tilde{W} \end{array}$$

defined in the same way as in section 5.

Theorem 2: <u>The diagram above is a simultaneous resolution for τ.</u>

The proof of these theorems is analogous to the proof in the classical case (cf. [21]), the crucial starting point being a theory of Jordan normal form for elements in $\tilde{G}^{o}$.

IV. Conclusion.

15. Combining the descriptions of the semiuniversal deformation $\emptyset : X \to V$ of a simply elliptic singularity and the fibres of $\tau : \tilde{G}^{o}_{>1} \to \tilde{T}_{>1}/\tilde{W}$ for the corresponding group $\tilde{G}$ we obtain the following result. Let $X_f := \emptyset^{-1}(V_f)$.

Theorem: For simply elliptic singularities of degree $d \leq 6$ there is an identification $V_f \cong \tilde{T}_{>1}/\tilde{W}$ such that for all $x \in X_f$ there is a neighborhood $X_f(x)$ of x and an inclusion $X_f(x) \hookrightarrow \tilde{G}^o_{>1}$ making the following diagram commute.

$$\begin{array}{ccccccc} X & \supset & X_f & \supset & X_f(x) & \subset & \tilde{G}^o_{>1} \\ \Big\downarrow \emptyset & & \Big\downarrow & & & & \Big\downarrow \tau \\ V & \supset & V_f & & \overset{\sim}{=\!=\!=\!=\!=} & & \tilde{T}_{>1}/\tilde{W} \end{array}$$

A similar statement holds for degree 7 and 8. If $d = 7$ the corresponding group $\tilde{G}$ is the central extension of $GL_2(K)$ defined by the "unusual" Killing form $\begin{pmatrix} 4 & 3 \\ 3 & 4 \end{pmatrix}$ mentioned in section 8. If $d = 8$ there are two components V_1 and V_2 in the semiuniversal deformation. The group corresponding to V_1 is the central extension of $G_m(K)$ defined by the Killing form (8) on $\mathbb{Z} = X_*(G_m)$ and the group corresponding to V_2 is the ordinary central extension of $SL_2(K)$ defined in section 10.

The theorem above leaves open an important problem. How can one realize the whole semiuniversal deformation, not only a part of it, in a group theoretic way. In particular this raises the question of constructing a simply elliptic singularity in Lie group theoretic terms. Wat is desired is a partial $\tilde{G}$-equivariant completion of $\tilde{G}^o$ which fills the hole corresponding to the value zero for the fundamental characters. One idea would be to use the conjugacy classes of $\tilde{G}$ in the complement of $\tilde{G}^o$. However, the corresponding representation operators are far away from being of trace class. Also the following aspects should be regarded. The completion may not be smooth, since, for example, the total space of a singularity of type $\tilde{A}_4$ is not smooth, and it may not be unique, since the morphism τ for $\tilde{SL}_2$ realizes a subdeformation for the singularity $\tilde{D}_5$ of degree 4 as well as for a singularity of degree 8. In addition, the root systems attached to simply elliptic singularities form only a small part of all root systems. Some further root systems can be attached to simply elliptic singularities equipped with a group of symmetries (cf. [8]). An analogue of the theorem above then holds for the deformations conserving symmetries. Here groups over $\mathbb{C}((t))$ come into play which are not of Chevalley type. They will be dealt with in the work on general Kac Moody algebras [22].

BIBLIOGRAPHY.

[1] V. I. Arnol'd: Critical points of smooth functions, Proc. Intern. Congress of Math., Vanoouver 1.974.

[2] E. Brieskorn: Singular elements of semisimple algebraic groups, Actes Congrès Intern. Math. 1.970, t. 2 , 279-284.

[3] F. Bruhat, J. Tits: Groupes réductifs sur un corps local, Publ. Math. I.H.E.S. 41 (1972), 1 - 251.

[4] H. Garland: The aritmetic theory of loop groups, Publ. Math. I.H.E.S. 52, to appear.

[5] N. Iwahori, H. Matsumoto: On some Bruhat decomposition and the structure of the Hecke rings of p-adic Chevalley groups, Publ. Math. I.H.E.S. 25 (1.965), 5-48.

[6] V.G. Kac: Simple irreducible graded Lie algebras of finite growth, Math. USSR Izvestija 2 (1968), 1271-1311.

[7] V.G. Kac: Infinite dimensional Lie algebras and Dedekind's η-function, Functional Analysis and its Applications 8 (1974), 68-70.

[8] M. Kato, T. Yano: Free deformations of simply elliptic singularities, Science Reports of the Saitama University, Series A, Vol. IX, No. 3, 1.980, 71-79.

[9] H. Knörrer: Die Singularitäten vom Typ $\tilde{D}$, Math. ann. 251 (1980), 135-150.

[10] E.J. Looijenga: Root systems and elliptic curves, Invent. math. 38 (1976), 17-32

[11] E. J. Looijenga: On the semiuniversal deformation of a simple elliptic singularity II, Topology 17 (1978), 23-40.

[12] E.J. Looijenga: Invariant theory for generalized root systems, Invent. math. 61 (1980), 1-32.

[13] E.J. Looijenga: Rational surfaces with an anti-canonical cycle, Annals of Math., to appear.

[14] Y. Merindol: Déformations des surfaces de Del Pezzo, de points doubles rationnels et des cônes sur une courbe elliptique.
Thèse 3ème cycle, Université Paris VII, 1980.

[15] R.V. Moody: A new class of Lie algebras, J. Algebra 10 (1968), 211-230.

[16] R.V. Moody: Euclidean Lie algebras, Can. J. Math. 21 (1969), 1432-1454.

[17] C.C. Moore: Group extensions of p-adic and adelic linear groups, Publ. Math. I.H.E.S. 35 (1968), 157-222.

[18] H. Pinkham: Deformations of algebraic varieties with G_m-action, Asterisque 20, Soc. Math. France, 1974.

[19] H. Pinkham: Simple elliptic singularities, Del Pezzo surfaces and Cremona transformations, Proc. of Symp. in Pure Math. 30 (1977), 69-71.

[20] K. Saito: Einfach elliptische singularitäten, Invent. math. 23 (1974), 289-325.

[21] P. Slodowy: Simple singularities and simple algebraic groups, Lecture Notes in Math. 815, Springer, Berlin-Heidelberg-New York, 1980.

[22] P. Slodowy: Kac Moody Lie algebras, associated groups and special singularities, in preparation.

Address: Peter Slodowy
Mathematisches Institut der Universität Bonn
Wegelerstr. 10

Sonderforschungsbereich Theoretische Mathematik
Universität Bonn
Beringstr. 4

5300 BONN 1

Federal Republic Germany

ON THE PICARD GROUP OF CERTAIN SMOOTH SURFACES IN WEIGHTED PROJECTIVE SPACES

by Joseph Steenbrink

Abstract. We consider a general member of a Lefschetz pencil of surfaces in weighted projective 3-spaces of type $(1,1,a,b)$ where $\gcd(a,b) = 1$. We show that such a surface either has Picard number equal to 1 or all of its 2-cohomolgy is algebraic.

1. Introduction.

This paper originated from the following problem. In $\mathbb{P}^6$, consider the cone X over the Veronese surface in $\mathbb{P}^5$. Let S be a general hypersurface section of X of degree 3. This is a type of Horikawa surface. Is it true that $\mathrm{Pic}(S) = \mathbb{Z}$?

The space X appears to be the weighted projective 3-space of type $(1,1,1,2)$. Thus one is led to the problem: determine the Picard group of a general complete intersection in weighted projective space, which is smooth of dimension two. In this paper we investigate the classical method of Lefschetz pencils.

The condition that our surface has to be smooth forces us to assume that the singular locus of the ambient space has codimension at least 3. If the codimension is at least 4, one can consider the surface as a member of a Lefschetz pencil on a smooth threedimensional complete intersection, and the classical methods apply immediately. If the singular locus of the ambient space has codimension equal to 3, we must consider also pencils with members passing through the singular points. The problems one meets then are of two kinds: the number of singularities can be too big (e.g. more than 2) or the singularities are so complicated, that they interfere with the ordinary Morse singularities of the pencil. We can deal with these problems if we take the case of two singularities such that their local Picard-Lefschetz transformations are unipotent. This occurs in projective 3-space of type $(1,1,a,b)$ when $\gcd(a,b) = 1$.

Our method also applies in some other cases, e.g. complete intersections of type $(2a,ma)$ in $\mathbb{P}(1,1,1,a,a)$, or certain branched coverings of surfaces.

We thank Arie Toet for suggesting the problem, and Horst Knörrer and Chris Peters for stimulating conversations.

2. Lefschetz pencils on threefolds with isolated singularities.

Let $X \subset \mathbb{P}^n(\mathbb{C})$ be a threedimensional projective variety with isolated singular points $x_1,\ldots,x_s$. Assume that X is not contained in any hyperplane. Let $\check{X} \subset \check{\mathbb{P}}^n$ denote the set of hyperplanes $H \subset \mathbb{P}^n$ such that $H \cap X$ is singular. Then $\check{X}$ consists of $s+1$ irreducible components:

$$\check{X} = \check{X}_0 \cup \ldots \cup \check{X}_s$$

where the generic point of $\check{X}_0$ corresponds to a hyperplane, tangent to X at a smooth point such that the intersection has one ordinary double point, and $\check{X}_i$ is the set of hyperplanes containing x_i for $i = 1,\ldots,s$.

A <u>Lefschetz pencil</u> on X is a family of hyperplane sections of X, parametrized by a line $L \subset \check{\mathbb{P}}^n$ which is transverse to $\check{X}$. This means that L is transverse to all $\check{X}_i$ and does not pass through their intersections.

For $t \in L$ let $X_t = H_t \cap X$ and define $\tilde{X} = \{(x,t) \mid t \in L \text{ and } x \in X_t\}$. Let $f: \tilde{X} \to L$ be the projection. Then f has the critical values

$$t_1,\ldots,t_r,\tau_1,\ldots,\tau_s$$

where $t_1,\ldots,t_r$ are the intersection points of L with $\check{X}_0$ and τ_i is the intersection point of L with $\check{X}_i$ for $i = 1,\ldots,s$. Let $U = L-\check{X}$; then f is a C^∞ fibration over U, so the direct image sheaf

$$R^2 f_* \mathbb{Q}_{|U}$$

is a local system on U, even a variation of Hodge structures, giving rise to a representation

$$\rho : \pi_1(U,t) \to \mathrm{Aut}\, H^2(X_t,\mathbb{Q})$$

where $t \in U$.

To describe this representation, we choose generators for $\pi_1(U,t)$ as follows. Take points $t_1',\ldots,t_r'$ near $t_1,\ldots,t_r$, small loops c_i from t_i' winding once around t_i and paths p_i from t to t_i'. Let $\gamma_i \in \pi_1(U,t)$ denote the class of the loop $p_i c_i p_i^{-1}$. Taking the points τ_j instead of the t_i one obtains elements β_j of $\pi_1(U,t)$; it is clear that $\pi_1(U,t)$ is generated by $\gamma_1,\ldots,\gamma_r$,

$\beta_1,\ldots,\beta_s$.

Lemma (2.1). For all i,j the elements $\rho(\gamma_i)$ and $\rho(\gamma_j)$ are conjugate in the group $\rho(\pi_1(U,t))$.

Proof. (Cf [4], Exp. XVIII 6.6.2)

Let $q: U \to \check{\mathbb{P}}^n - \check{X}$ be the inclusion. Because $f|_{f^{-1}(U)}$ is in fact the restriction to U of a fibre bundle over $\check{\mathbb{P}}^n - \check{X}$, we have a commutative diagram

$$\begin{array}{ccccc} & & \pi_1(\check{\mathbb{P}}^n-\check{X},t) & & \\ & \overset{q_*}{\nearrow} & & \searrow^{\bar{\rho}} & \\ \pi_1(U,t) & & \xrightarrow{\quad\rho\quad} & & \mathrm{Aut}\, H^2(X_t,\mathbb{Q}) \end{array}$$

Moreover the map q_* is surjective (see [8] and [10]). Hence it suffices to show that $q_*(\gamma_i)$ and $q_*(\gamma_j)$ are conjugate.

As $\check{X}_0$ is irreducible, there exists a path $\bar{v}$ on the regular locus of $\check{X}_0$ connecting t_i with t_j . Hence we can choose a path v on the boundary of a tubular neighbourhood of $\check{X}_0^{reg}$ connecting t_i' with t_j' such that the loops c_i and vc_jv^{-1} are homotopic. Then one obtains obtains

$$q_*(\gamma_j) = \sigma^{-1} q_*(\gamma_i)\sigma$$

where σ is the homotopy class of the loop $p_i v p_j^{-1}$ in $\pi_1(\check{\mathbb{P}}^n-\check{X},t)$.

QED

We choose the point $t \in U$ such that it has the following property: if σ is a horizontal section of $R^2 f_* \mathbb{Q}$ on a neighbourhood V of t in U such that $\sigma(t) \in H^2(X_t, \mathbb{Q})$ is the cohomology class of an algebraic cycle on X_t, then for all $t' \in V$ the class $\sigma(t')$ is algebraic. This property excludes at most a countable subset of U (cf. [9]).

The Picard-Lefschetz formula tells us, that there exist elements $\delta_1, \ldots, \delta_r \in H^2(X_t, \mathbb{Q})$ such that for all $z \in H^2(X_t, \mathbb{Q})$ one has

$$\rho(\gamma_i)(z) = z + \langle z, \delta_i \rangle \delta_i \qquad i = 1, \ldots, r ;$$

here $\langle .,. \rangle$ is the intersection form on $H^2(X_t, \mathbb{Q})$.

Let E be the subspace of $H^2(X_t, \mathbb{Q})$ generated by these vanishing cycles and $\bar{E}$ the smallest invariant subspace containing E. Recall that $\langle \delta_i, \delta_i \rangle = -2$ so $\rho(\gamma_i)$ is the reflection with respect to the hyperplane in $H^2(X_t, \mathbb{Q})$ orthogonal to δ_i. Moreover if $g \in \pi_1(\check{\mathbb{P}}^n - \check{X}, t)$ is such that $g_*(\gamma_j) = g^{-1} g_*(\gamma_i) g$, then the Picard-Lefschetz formula implies, that $\bar{\rho}(g)(\delta_j) = \pm \delta_i$. Hence the restriction of ρ to $\bar{E}$ is irreducible.

We recall the following facts from [3]. If $T \subset H^2(X_t, \mathbb{Q})$ is a π_1-stable subspace, then also its orthogonal complement $T^{\perp}$ is π_1-stable. Let V denote the space of algebraic cycle classes in $H^2(X_t, \mathbb{Q})$. Then the assumption on t implies that V is stable under the action of $\pi_1(U,t)$.

Let $\bar{X}$ denote a resolution of singularities of X. Then the image I of the restriction map $H^2(\bar{X},\mathbb{Q}) \to H^2(X_t,\mathbb{Q})$ coincides with the subspace of π_1-invariant elements.

<u>Theorem (2.2)</u>. Keeping the above notations, suppose that no three of the points $x_1,\ldots,x_s$ are on a line and that the restriction of $\rho(\beta_i)$ to $\bar{E}$ is unipotent for $i = 1,\ldots,s-1$. Then

$$H^2(X_t,\mathbb{Q}) = \bar{E} \oplus I .$$

<u>Proof</u>. Remark that the conditions of the theorem are fulfilled in the case $s = 1$, for in that case $E = \bar{E}$, $\pi_1(U,t)$ is generated by $\gamma_1,\ldots,\gamma_r$ and these act trivially on $E^{\perp}$.

Let $U' = \check{\mathbb{P}}^n - (\check{X}_1 \cup \ldots \cup \check{X}_s)$. Because $\rho(\gamma_i)$ acts trivially on $\bar{E}^{\perp}$ for $i = 1,\ldots,r$, the restriction of ρ to $\bar{E}^{\perp}$ factorizes through $\pi_1(U',t)$. The condition that no three of $x_1,\ldots,x_s$ are on a line, implies that $\pi_1(U',t)$ is abelian. For a general plane intersects U' in the complement of a curve with only nodes, and this intersection hence has abelian fundamental group [5], which is isomorphic to the fundamental group of U' [8].

Let Γ be the image of $\pi_1(U,t)$ in $\mathrm{Aut}(\bar{E}^{\perp})$. Then Γ is generated by the unipotent elements $\rho(\beta_j)|_{\bar{E}^{\perp}}$ for $j = 1,\ldots,s-1$. Moreover $\bar{E}$ is a semisimple Γ-module and Γ is commutative. This implies that $\Gamma = \{Id\}$. Consequently $\bar{E}^{\perp} = I$.

<u>Corollary (2.3)</u>. Suppose that all of $H^2(\bar{X},\mathbb{Q})$ is algebraic. Then under the assumptions of Theorem (2.2) either $V = H^2(X,\mathbb{Q})$ or $V = I$.

Proof. As all of $H^2(\bar{X},\mathbb{Q})$ is algebraic, one has $I \subset V$. Because V is π_1-stable and $\bar{E}$ is irreducible, either $\bar{E} \subset V$ or $V \subset I = \bar{E}^{\perp}$.

Remark. One would like to know if I equals the image of the map $H^2(X,\mathbb{Q}) \to H^2(X_t,\mathbb{Q})$. Because $\tilde{X}$ is the blowing-up of X along a smooth curve contained in X_t , it is easy to show, that $H^2(X,\mathbb{Q})$ and $H^2(\tilde{X},\mathbb{Q})$ have the same image in $H^2(X_t,\mathbb{Q})$.

Lemma (2.4). Suppose that $\tilde{X}$ (or X) is a rational homology manifold Then $H^2(\bar{X},\mathbb{Q})$ and $H^2(\tilde{X},\mathbb{Q})$ have the same image in $H^2(X_t,\mathbb{Q})$.

Proof. Consider the chain of mappings

$$H^2(\tilde{X},\mathbb{Q}) \to H^2(\bar{X},\mathbb{Q}) \to H^2(\tilde{X}-\Sigma,\mathbb{Q}) \to H^2(X_t,\mathbb{Q})$$

where Σ is the singular locus of $\tilde{X}$. Because $\tilde{X}$ is a rational homology manifold, $H^k_{\Sigma}(\tilde{X},\mathbb{Q}) = 0$ for $k = 2,3$ so the restriction map from $H^2(\tilde{X},Q)$ to $H^2(\tilde{X}-\Sigma,\mathbb{Q})$ is an isomorphism.

QED

This lemma applies in particular if X has only isolated quotient singularities.

3. Weighted projective spaces.

We first mention some facts concerning weighted projective spaces. For proofs see [6] and [7].

Let $Q = (q_0,\ldots,q_n)$ be a sequence of positive integers. Let $S(Q)$

denote the polynomial algebra $\mathbb{C}[z_0,\dots,z_n]$ with the grading such that $\deg(z_i) = q_i$. Then $\mathbb{P}(Q) = \operatorname{Proj}(S(Q))$ is the weighted projective space of type Q.

Without loss of generality one may assume that $\gcd(q_0,\dots,q_{i-1},q_{i+1},\dots,q_n)$ is equal to 1 for all i.

Let $\mu(m)$ denote the group of mth roots of unity. Then one can consider $\mathbb{P}(Q)$ as the quotient of $\mathbb{P}^n$ under the diagonal action of $\mu(q_0) \times \dots \times \mu(q_n)$.

We are mainly interested in the case $n = 3$. Then $\mathbb{P}(Q)$ has only isolated singularities if **and** only if $\gcd(q_i,q_j) = 1$ for $i \neq j$. In that case the number of singularities is equal to the number of the q_i which are different from 1. We first assume that $Q = (1,1,a,b)$ with $\gcd(a,b) = 1$.

If both a and b are equal to 1, we are in the smooth classical case. Then Noether's theorem tells us that a general surface of degree at least 4 has Picard group $\mathbb{Z}$, and for surfaces of lower degree all cohomology is algebraic.

Suppose that $a = 1$ but $b > 1$. A basis of $S(Q)_b$ embeds $\mathbb{P}(Q)$ in $\mathbb{P}^N$, $N = (b+1)(b+2)/2$. The image is a cone over the b-fold Veronese embedding of $\mathbb{P}^2$.

Let S be the intersection of $\mathbb{P}(Q)$ with a general hypersurface of degree d in $\mathbb{P}^N$. Projection **from** the vertex P of the cone shows, that S is a covering of $\mathbb{P}^2$ of degree d, with a ramification curve of degree $bd(d-1)$.

It is shown in [7] that $p_g(S) = \dim S(Q)_{b(d-1)-3}$, and all of $H^2(S,\mathbb{Q})$ is **algebraic** if and only if $p_g(S) = 0$, i.e. $b(d-1) < 3$.

Thus one obtains from Theorem (2.2) and Lemma (2.4), together with the fact that S is simply-connected, that $\mathrm{Pic}(S) \cong \mathbb{Z}$, except in the case

$d=2$, $b=2$, S is a double covering of $\mathbb{P}^2$ ramified along a smooth quartic curve, so S is isomorphic to the blowing up of $\mathbb{P}^2$ in 7 points.

Remark that $d = 1$ implies that S is isomorphic to $\mathbb{P}^2$.

Next suppose that $a,b > 1$ but $\gcd(a,b) = 1$. Then $\mathbb{P}(Q) = X$ has the singular points x_1 and x_2, images of the points $(0,0,0,1)$ resp. $(0,0,1,0)$ under the quotient mapping $\mathbb{P}^3 \to \mathbb{P}(Q)$.

It is sufficient to show, that the map $\rho(\beta_1)$ is unipotent. We keep the notations of section 2.

An affine neighbourhood of x_1 is isomorphic to $\mathbb{C}^3/\mu_b$ where μ_b acts on $\mathbb{C}^3$ by $\zeta.(x,y,z) = (\zeta x, \zeta y, \zeta^a z)$. Choose a local parameter at τ_1 on L. A Lefschetz pencil then determines a germ of function $f: (X,x_1) \to (\mathbb{C},0)$ and by composition with the quotient mapping a germ $\bar{f}: (\mathbb{C}^3,0) \to (\mathbb{C},0)$.

Let $M = \{(u,v,w) \in \mathbb{Z}^3 | u,v,w \geq 0, u+v+w > 0$ and b divides $u+v+aw\}$.

Let Δ denote the convex hull of $\bigcup_{m \in M} m + \mathbb{R}^3_+$. Because $\bar{f}$ is an invariant function on $\mathbb{C}^3$ with $\bar{f}(0) = 0$, the Taylor expansion of $\bar{f}$ at 0 involves only monomials in x,y,z such that the exponents are in M. We impose the condition on the pencil, that the Newton diagram of $\bar{f}$ is equal to Δ and that $\bar{f}$ is nondegenerate with respect to Δ.

We sketch the proof that the monodromy transformation of f at x_1

is unipotent. Following [2] to Δ one associates a toric variety $\mathbb{P}(\Delta)$ on which μ_b acts and which is a proper modification of $\mathbb{C}^3$. Let $Y = \mathbb{P}(\Delta)/\mu_b$. Then one has a proper map $\pi: Y \to \mathbb{C}^3/\mu_b \subset X$.
One can show that Y is again a toric variety, that the divisor of $f \circ \pi$ is reduced and that the strict transform of $f^{-1}(0)$ in Y is transverse to the strata of Y. By a slight generalization of A'Campo's formula for the characteristic polynomial of the monodromy (cf.[1]) this implies that the local monodromy of f at x_1 is unipotent.

We omit the detailed proof of the above statements. However one should be aware, that the fact that the divisor of $f \circ \pi$ is reduced is due to a special property of the Newton diagram Δ, which in general is not fulfilled. Namely, for every compact face σ of Δ of codimension 1 there exists a unique positive integer $m(\sigma)$, such that σ is supported by the plane $\alpha u + \beta v + \gamma w = m(\sigma)$ with $\alpha,\beta,\gamma \in \mathbb{N}$ and $\gcd(\alpha,\beta,\gamma) = 1$. The special property of Δ is that $m(\sigma)$ divides b for all σ.

To embed X in a projective space we use a basis of $S(Q)_{abd}$ for some $d \in \mathbb{N}$. If S is a general hyperplane section of X, one has the formula

$$p_g(S) = \dim S(Q)_{abd-a-b-2}$$

so $p_g(S) = 0$ if and only if $d = 1$, $a = 2$ and $b = 3$. Then $b_2(S) = 9$.

Question: give a direct geometric description of S.

We end with an example where three singular points occur.

Let $X = \mathbb{P}(1,7,5,11)$, $x_1 = (0,1,0,0)$, $x_2 = (0,0,1,0)$ and $x_3 = (0,0,0,1)$. Let Δ_i be the Newton diagram, constructed as above for the singular point x_i , $i = 1,2,3$. Then the diagrams Δ_1 and Δ_2 have the special property mentioned above, but Δ_3 has not. The face σ of Δ_3 which is the convex hull of the vertices $(1,3,0)$, $(4,1,0)$ and $(1,0,2)$ has supporting plane $4u+6v+9w = 22$, so $m(\sigma) = 22$ which is not a divisor of 11 .

Still $\rho(\beta_3)$ acts trivially on $\bar{E}^{\perp}$ here, because $\rho(\beta_1)$ and $\rho(\beta_2)$ do. We consider X as embedded in $\mathbb{P}^n$ by a basis of $S(\Omega)_{385d}$, $d \in \mathbb{N}$.

If x_1, x_2, x_3 where on a line, every hyperplane containing x_1 and x_2 would contain x_3 . One obtains a contradiction to this considering the hyperplanes corresponding to the forms z_1^{55d} , z_2^{77d} and z_3^{35d} .

Hence we can apply Theorem (2.2) to $\mathbb{P}(1,7,5,11)$.

REFERENCES

[1] N. A'Campo, La fonction zêta d'une monodromie. Comment. Math. Helv. 50, 233-248 (1975).

[2] V.I. Danilov, Newton polyhedron and vanishing cohomology. Functional analysis and its appl. 13, 32-47 (1979).

[3] P. Deligne, Théorie de Hodge II. Publ. Math. IHES 40, 5-57 (1971)

[4] P. Deligne & N. Katz, Groupes de monodromie en géométrie algébrique. SGA 7 II. Lecture Notes in Math. 340, Springer 1973.

[5] P. Deligne, Proof of Zariski's conjecture, Sem. Bourbaki 1979'80.

[6] C. Delorme, Espaces projectifs anisotropes. Bull. Soc. Math. France 103, 203-223 (1975).

[7] I. Dolgachev, Weighted projective varieties. Mimeographed notes. Moscow State University 1975/1976.

[8] H.A. Hamm & Lê Dung Trang, Un théorème de Zariski du type de Lefschetz. Ann. Sc. ENS 6, 317-366 (1973).

[9] R. Hartshorne, Equivalence relations on algebraic cycles. In: Algebraic Geometry, Arcata 1974. Proc. AMS Symp. Pure Math. Vol. XXIX, 129-164 (1975).

10 E.R. van Kampen, On the fundamental group of an algebraic curve. Amer. J. of Math. 55, 255-260 (1933).

Mathematisch Instituut der,
Rijksuniversiteit Leiden,
Wassenaarseweg 80,
2333 AL Leiden,
The Netherlands.

V A R I E T E S   P O L A I R E S   I I

MULTIPLICITES POLAIRES, SECTIONS PLANES, ET CONDITIONS DE WHITNEY

Bernard TEISSIER

S U M M A R Y

To each reduced equidimensional analytic algebra $\mathcal{O}_{X,x}$, one can associate a sequence of d integers, where $d = \dim \mathcal{O}_{X,x}$:

$$M^*_{X,x} = \{m_x(X), m_x(P_1(X)), \ldots, m_x(P_{d-1}(X))\}$$

where for $0 \leq k \leq d-1$, $P_k(X)$ is a general local polar variety of codimension k of X, as defined by Lê D.T. and myself, and m_x denotes the multiplicity at x.

One can visualize $P_k(X)$ as follows : Pick an embedding $X \subset \mathbb{C}^N$ of a representative of (X,x) and take a general linear projection $p : \mathbb{C}^N \to \mathbb{C}^{d-k+1}$. The closure in X of the critical locus of the restriction $p|X^o$ of p to the non-singular part X^o of X is purely of codimension k or empty, its multiplicity at x is independent of the choice of the general linear projection p and of the embedding. It is denoted by $m_x(P_k(X))$. Note that $P_o(X) = X$. I prove here the

<u>Theorem</u> : <u>Let</u> X <u>be a reduced purely</u> d-<u>dimensional complex-analytic space</u>, <u>and</u> Y <u>a non-singular subspace of</u> X . <u>Given a point</u> $0 \in Y$, <u>the following conditions are equivalent</u> :

i) <u>The pair</u> (X^o, Y) <u>satisfies the Whitney conditions at</u> 0 .

ii) <u>The map from</u> Y <u>to</u> $\mathbb{N}^d$ <u>given by</u> $y \mapsto M^*_{X,y}$ <u>is constant in a neighbourhood of</u> 0 <u>in</u> Y .

Equivalently, (X^o, Y) does not satisfy the Whitney conditions at 0 if and only if one of the general local polar varieties $P_k(X)$ is <u>not</u> equimultiple along Y at 0 .

So the following picture already shows the general phenomenon : here X is the surface in $\mathbb{C}^3$ defined by $y^2 - x^3 - t^2x^2 = 0$, Y is the t-axis, and p is the projection onto the (x,t)-plane : the curve defined by $x + t^2 = 0$, $y = 0$ is a general polar curve for X, it is not equimultiple along Y so (X^{o},Y) does not satisfy the Whitney conditions at 0, which is obvious from the definition (see Chap. III)

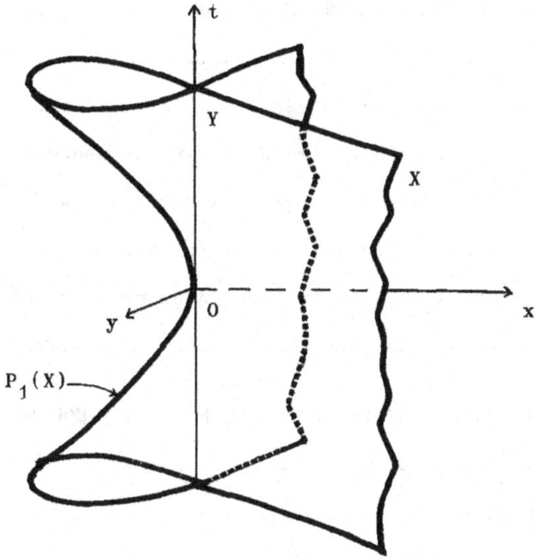

An important feature of the sequence $M^{*}_{X,x}$ is that it is an analytic invariant of the germ (X,x) which can be computed topologically, just like the multiplicity can be computed by counting the number of points of intersection with X of a general (N-d)-plane in $\mathbb{C}^N$ near x (see Chap. IV and VI).

C O N T E N U

***

INTRODUCTION

§ 1.

"Tu n'es qu'un mortel, aussi ton esprit doit-il nourrir deux pensées à la fois."

Bacchylide

Tout ce texte est consacré à la démonstration d'un seul résultat, qui est la réalisation de la partie algébro-géométrique du programme consécutif à [Te 1]. Rappelons ce programme, qui s'inscrit à l'intérieur d'un effort pour comprendre le lien entre la structure algébrique d'un ensemble algébrique ou analytique, et sa structure topologique.

A) Partie algébro-géométrique : Associer à chaque algèbre analytique complexe réduite $\mathcal{O}_{X,x}$ purement de dimension d un élément $M_{X,x}$ d'un ensemble discret (une multiplicité généralisée) de telle façon que, étant donnés un espace analytique complexe réduit purement de dimension d, un sous-espace analytique Y de X et un point non-singulier $0 \in Y$, on ait équivalence entre :

i) L'application $y \mapsto M_{X,y}$ est constante sur Y au voisinage de 0.

ii) Le couple (X^0, Y) satisfait les conditions de Whitney au voisinage 0.

Remarque : Les conditions de Whitney sont des conditions différentielles (pour une définition précise, voir (Chap. III, 2.2.1)), qui d'une part sont impliquées par des conditions algébro-géométriques (cf. Chap. III, 2.3.1) et d'autre part impliquent des résultats de trivialité topologique (Chapitre VI, 4.3.1). Elles ont été introduites par Whitney dans [W] et étudiées par Thom [Th], Mather [Ma], Łojasiewicz et Hironaka [H 1], [H 3].

B) Partie topologique : Donner une interprétation de $M_{X,x}$ en termes de la topologie de l'ensemble analytique X au voisinage de x.

Les résultats de [Te 1] réalisaient ce programme dans le cas particulier où X est une hypersurface de $\mathbb{C}^{d+1}$ dont le lieu singulier Y est non-singulier, le rôle de la multiplicité $M_{X,x}$, pour $x \in Y$, étant tenu par la suite μ^* des nombres de Milnor des sections de X par des plans généraux de $\mathbb{C}^{d+1}$ passant par x et de codimension supérieure ou égale à la dimension de Y.

Pour un énoncé précis du résultat principal de ce travail, voir le Chapitre V, et pour des renseignements sur la partie topologique, voir le Chapitre VI, § 4.

§ 2.

Cette généralisation a été permise par l'introduction dans [Lê-Te] des variétés polaires locales générales associées à un germe d'espace analytique $(X,0) \subset (\mathbb{C}^N,0)$, réduit et purement de dimension d. On peut décrire informellement une variété polaire locale générale de codimension k de X comme ceci : Soit $X \subset \mathbb{C}^N$ un représentant du germe (X,0), et soit $p : \mathbb{C}^N \to \mathbb{C}^{d-k+1}$ une projection linéaire générale. Une variété polaire $P_k(X,0)$ est l'adhérence dans X du lieu critique de la restriction de p à la partie non-singulière X^0 de X. La multiplicité généralisée cherchée est la suite $\{m_0(P_k(X,0) ; 0 \leq k \leq d-1\} \in \mathbb{N}^d$ des multiplicités en 0 des variétés polaires générales de toutes les codimensions k, $0 \leq k \leq d-1$. On montre au Chapitre IV que cette suite d'entiers ne dépend en fait que de l'algèbre analytique $\mathcal{O}_{X,0}$.

Par ailleurs la démonstration est substantiellement plus simple que celle de [Te 1], puisque l'on peut raisonner par récurrence sur $\dim X - \dim Y$ à l'aide de résultats de transversalité. On peut d'ailleurs résumer les moyens techniques employés en disant qu'il s'agit du début d'une théorie systématique de la transversalité des espaces singuliers, si l'on comprend un espace singulier comme ensemble stratifié, muni en outre en chaque point des variétés polaires locales générales des adhérences des strates. Les principaux résultats techniques sont des résultats de transversalité : transversalité d'une variété polaire locale générale au noyau de la projection servant à la définir (cf. Chap. IV) transversalité, sous certaines conditions, d'une hypersurface non-singulière de $\mathbb{C}^N$ générale parmi celles qui contiennent Y à toutes les limites en O d'espaces tangents à X^o, etc. La démonstration du théorème principal (Chap. V) paraîtra d'ailleurs beaucoup plus simple le jour où l'on disposera pour parler de transversalité d'espaces singuliers d'un langage aussi souple et puissant, et sans doute aussi vidé de géométrie, que la topologie algébrique.

§ 3.

"Si le langage était parfait, l'homme cesserait de penser."
Paul Valéry

Provenant d'un cours, ce texte ne va pas droit au but, mais suit un itinéraire que j'aimerais que l'on puisse qualifier de touristique. De plus je me suis efforcé de créer les premiers rudiments d'un langage algébro-géométrique permettant de parler de la transversalité des espaces singuliers , le langage géométrique étant parfois peu commode. Par ailleurs, il me semble possible, et très intéressant, de déformer la rédaction, en respectant sa structure globale, en chacun des deux extrêmes entre lesquels elle se trouve : d'une part une rédaction complètement algébrique, sur un corps de caractéristique zéro algébriquement clos, et d'autre part, mutatis mutandis, une rédaction en géométrie sous-analytique réelle démontrant un analogue du théorème principal.

§ 4.

Les principales sources de ce travail sont d'une part la construction de (Chap. III, 2.3.1) qui d'après une idée de Hironaka permet d'énoncer une condition d'équidimensionnalité des fibres d'un certain morphisme suffisante pour impliquer les conditions de Whitney, et d'autre part la recherche systématique d'invariants de la géométrie d'un espace X au voisinage d'un point x dans la géométrie et la topologie des sections planes générales de X passant par x ou un peu à côté, de toutes les dimensions. La jonction est précisément faite par la théorie des variétés polaires comme le montrent (Chap. IV, 6.1.1 et Chap. VI, 4.2). Cette théorie des variétés polaires elle-même doit beaucoup à la foi communicative de Thom et Zariski en l'utilité des méthodes de contour apparent et de discriminant pour obtenir des informations de nature topologique et algébrique sur les singularités.

Par ailleurs, J.P. Brasselet et M.H. Schwartz ([B-S], Corollaire 10.2) ont prouvé que l'obstruction d'Euler locale $Eu_x(X)$ de MacPherson est constante sur chaque strate d'une stratification de Whitney de X. Puisque Lê D.T. et l'auteur ont prouvé ([Lê-Te]) la formule $Eu_x(X) = \sum_{k=0}^{d-1} (-1)^k m_x(P_k(X,x))$, la somme alternée des multiplicités des variétés polaires de X est constante sur chaque strate, comme l'est la multiplicité de X d'après Hironaka [H 1]. Il faut remarquer cependant que la preuve assez topologique de [B-S] ne permet pas de tirer un résultat de la seule hypothèse que (X^o,Y) satisfait les conditions de Whitney <u>en un point</u> $0 \in Y$. Il faut pour utiliser cette hypothèse a priori bien plus faible une méthode utilisant de l'algèbre, comme le Lemme-clé du Chapitre V, Lemme qui m'a été suggéré par un résultat de Briançon et Henry (cf. [B-H]) concernant les surfaces dans $\mathbb{C}^3$.

§ 5.

Ce texte est la rédaction, fidèle pour les grandes lignes, d'un cours donné à l'Universidad Complutense de Madrid en Septembre 1980, développant une courte note ([Te 5]). Je veux remercier le Professeur Abellanas de m'avoir invité à donner ce cours, l'équipe vivante et sympathique des auditeurs, et particulièrement Maria-Emilia Alonso-Garcia et Ignacio Luengo qui ont rédigé les notes d'une grande partie du cours et ont fait de nombreuses observations pertinentes sur le contenu. Je veux aussi remercier les organisateurs de la Conférence de La Rabida, où j'ai présenté un résumé des résultats, pour l'accueil chaleureux et efficace que tous les participants ont pu apprécier. Pendant la rédaction j'ai bénéficié des conseils éclairés de Lê Dũng Tráng, Michel Merle, J.P.G. Henry et C. Sabbah. En particulier je suis très reconnaissant à Merle et Henry d'avoir tiré la théorie d'une ornière en découvrant qu'il fallait remplacer la modification de Nash par le morphisme conormal dans la démonstration.

Marie-Jo Lécuyer a assuré la frappe avec sa compétence et son amabilité coutumières, et a réalisé les dessins avec talent et imagination.

L'énoncé et les idées principales de la preuve sont nés au cours de séjours à la Fondation des Treilles en 1979-80, et la plus grande partie de la rédaction finale y a été faite. Je remercie Annette Gruner-Schlumberger d'avoir créé un lieu de travail aussi exceptionnel.

***

QUELQUES CONVENTIONS ET NOTATIONS

1) On se permettra souvent de considérer sans le dire explicitement un représentant "assez petit" X d'un germe (X,0) (et de même pour un morphisme). Le sens de "assez petit" sera toujours clair d'après le contexte.

2) Etant donnés un morphisme $f : X \to Y$ et un faisceau d'idéaux I de $\mathcal{O}_Y$ on notera souvent $I\mathcal{O}_X$ ou $I \cdot \mathcal{O}_X$ l'idéal de $\mathcal{O}_X$ image de l'homomorphisme canonique $f^*I \to f^*\mathcal{O}_Y = \mathcal{O}_X$. De même, étant donnés un homomorphisme d'anneaux $\rho : A \to B$ et un idéal I de A, on notera IB ou I . B l'idéal de B engendré par $\rho(I)$.

3) Etant donnés un morphisme $f : X \to S$, un fermé analytique rare $F \subset S$ et un sous-ensemble analytique fermé H de S, on appellera <u>transformé strict</u> de H par f (relativement à F) l'adhérence $\overline{f^{-1}(H) \cap (X - f^{-1}(F))}$ dans X de $f^{-1}(H) \cap (X - f^{-1}(F))$. Nous nous servirons de ceci dans trois cas : celui où f est surjectif et $F \subset S$ est l'ensemble des points $s \in S$ où la dimension de la fibre $f^{-1}(s)$ est strictement supérieure à $\dim X - \dim S$ (cf. Chap. III, § 5), celui où f est un plongement fermé et F rare dans X, le résultat étant alors noté $H \hat{\cap} X$ et appelé intersection stricte, (cf. Chap. IV, 5.4.2) et enfin celui où f apparaît dans un diagramme :

$$\begin{array}{ccc} X_o \underset{S_o}{\times} S & \longrightarrow & X_o \\ \downarrow f & & \downarrow f_o \\ S & \xrightarrow[e]{} & S_o \end{array}$$

où e est un éclatement et $F \subset S$ est l'ensemble exceptionnel de l'éclatement. Le morphisme $f_o' : X \to S$ de l'espace X transformé strict de S par f dans S induit par f est appelé transformé strict de f_o par e .
(Ceci est utilisé dans le Chapitre V).

Dans le cas où f_o est un plongement fermé, on parlera du transformé strict de X par e .

***

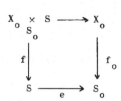

CHAPITRE I

DEPENDANCE INTEGRALE SUR LES IDEAUX

Introduction. Dans ce chapitre, on étudie la situation suivante : soient X un espace analytique complexe réduit et I un faisceau cohérent d'idéaux sur X, définissant un sous-espace analytique fermé $Y \subset X$, que l'on suppose rare dans X. Etant donnée une fonction holomorphe h sur X, on sait que le théorème des zéros de Hilbert permet d'exprimer algébriquement le fait que h s'annule sur $|Y|$, ensemble sous-jacent à Y, par l'inclusion $h \in \sqrt{I}$, où $\sqrt{I}$ est le faisceau racine de I, vérifiant $(\sqrt{I})_x = \sqrt{I_x}$ pour tout $x \in X$, et ce qui nous intéresse ici est la possibilité d'exprimer algébriquement de différentes façons le fait que non seulement h s'annule sur $|Y|$, mais que de plus pour tout choix d'un système de générateurs $g_1, \dots, g_p$ de I au voisinage d'un point $y \in Y$, on a une inégalité $|h(x)| \leq C \operatorname{Sup}_j(|g_j(x)|)$ pour tout x appartenant à un voisinage U assez petit de y, où $C \in \mathbb{R}_+$.

Voici un exemple typique des applications que je ferai de ce type de conditions : soit $f(z_o, \dots, z_d) = 0$ une équation pour une hypersurface complexe dans un ouvert $U \subset \mathbb{C}^{d+1}$. Supposons savoir que $\frac{\partial f}{\partial z_o}(0, z_1, \dots, z_d)$ tend vers 0 au moins aussi vite que les $\frac{\partial f}{\partial z_i}(0, z_1, \dots, z_d)$ $(1 \leq i \leq d)$; ceci signifie que l'hyperplan $z_o = 0$ n'est pas direction limite d'espaces tangents aux hypersurfaces de niveau $f(0, z_1, \dots, z_d) = v$ de la restriction de f à l'hyperplan $z_o = 0$. Et de même, le fait qu'en restriction à l'hypersurface X définie par $f = 0$, on ait au voisinage de 0 l'inégalité $|\frac{\partial f}{\partial z_o}(x)| \leq C \operatorname{Sup}_{i>0} |\frac{\partial f}{\partial z_i}(x)|$ $(x \in X)$ équivaut à ce que $z_o = 0$ ne soit pas direction limite d'hyperplans tangents à X en ses points non-singuliers. La possibilité d'exprimer algébriquement des conditions géométriques de cette espèce est un bon outil permettant en particulier d'énoncer al-

briquement, et de démontrer, des résultats de transversalité fins.

Le dessin suivant aidera peut-être le lecteur :

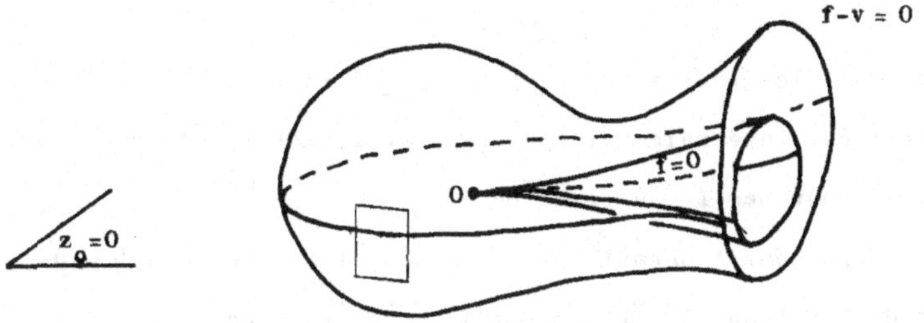

Les ingrédients sont essentiellement l'existence des éclatements et de la normalisation en géométrie analytique complexe et le théorème des singularités inexistantes : sur un espace normal, toute fonction méromorphe et localement bornée est holomorphe.

Bien que ce ne soit pas nécessaire pour la suite, j'ai insisté au § 3 sur l'aspect numérique de la clôture intégrale. Une application qui nous sera utile au chapitre V, la traduction algébrique du concept de fonction méromorphe lipschitzienne, est donnée au § 6.

§ 1. Critères de dépendance intégrale.

1.1 Définition : Soit I un idéal d'un anneau A ; un élément h de A est dit entier sur I si il satisfait une relation de dépendance intégrale de la forme

$$h^k + a_1 h^{k-1} + \dots + a_k = 0 \quad \text{avec } a_i \in I^i .$$

On peut se ramener à la dépendance intégrale sur les anneaux au moyen de l'observation suivante :

l'élément h est entier sur I si et seulement si l'élément $h \cdot T \in A[T]$ est entier sur le sous-anneau $P(I) = \sum_{i \geq 0} I^i T^i$ de $A[T]$.

Le fait (cf. [Bbk 1], § 1.1) que l'ensemble des éléments de A[T] entiers sur P(I) est un sous-anneau de A[T], implique que l'ensemble des éléments de A entiers sur I est un idéal de A ; cet idéal sera noté $\overline{I}$ et appelé clôture intégrale de l'idéal I. Si $\overline{I} = I$, on dira que I est intégralement clos. Il résulte de (Loc. cit.) que $\overline{I}$ est intégralement clos.

<u>1.2</u> <u>Remarques</u> : 1) On a les inclusions $I \subseteq \overline{I} \subseteq \sqrt{I}$, $I \subseteq J$ implique $\overline{I} \subseteq \overline{J}$, et $\overline{I}$ est donc le plus petit idéal intégralement clos de A contenant I, et donc on a l'inclusion $\overline{I} \cdot \overline{J} \subseteq \overline{I \cdot J}$ pour des idéaux I, J de A.

2) Supposons A normal (i.e., réduit et intégralement clos dans son anneau total de fractions),et I inversible, c'est-à-dire principal, engendré par un élément g non-diviseur de 0. Alors $I = \overline{I}$, et réciproquement, si tout idéal inversible d'un anneau réduit A est intégralement clos, A est normal.

<u>Preuve</u> : Puisque I est principal, l'équation $h^k + a_1 h^{k-1} + \dots + a_k = 0$ s'écrit aussi $h^k + \alpha_1 g h^{k-1} + \dots + \alpha_k g^k$ avec $\alpha_i \in A$ et donc, puisque g est non-diviseur de 0 : $(\frac{h}{g})^k + \alpha_1 (\frac{h}{g})^{k-1} + \dots + \alpha_k = 0$, qui est une relation de dépendance intégrale sur A pour l'élément $\frac{h}{g}$ de Tot(A) ; d'où $\frac{h}{g} \in A$ puisque A est normal, et donc $\overline{I} \subseteq I$ La réciproque se vérifie en lisant à l'envers ce qui précède.
En particulier, dans $\mathbb{C}\{t\}$ ou $\mathbb{R}\{t\}$, on a $h \in \overline{(g)}$ si et seulement si l'ordre en t de h est au moins égal à celui de g.

<u>1.3</u> Soient maintenant X un espace analytique complexe réduit et I un faisceau cohérent d'idéaux sur X définissant un sous-espace Y de X rare dans X.

<u>1.3.1</u> <u>Proposition 1</u> : A) <u>Etant donnée une fonction holomorphe</u> h <u>sur</u> X, <u>on a l'inclusion</u> $h_y \subset \overline{I_y}$ <u>en un point</u> $y \in X$ <u>si et seulement si pour tout système de générateurs</u> $(g_1, \dots, g_p)$ <u>de</u> I_y, <u>il existe un voisinage ouvert</u> U <u>de</u> y, <u>dans lequel les</u> g_i <u>convergent</u>, <u>et une constante</u> $C \in \mathbb{R}_+$ <u>tels que l'on ait l'inégalité</u> :

$$|h(x)| \leq C \sup_{1 \leq i \leq p} (|g_i(x)|) \quad \underline{\text{pour tout}}\ x \in U .$$

B) Il existe un unique faisceau cohérent d'idéaux $\overline{I}$ sur X tel que, pour tout $x \in X$ on ait $(\overline{I})_x = \overline{I_x}$ dans $\mathcal{O}_{X,x}$.

Démonstration : Soit $\pi : \overline{X}' \to X$ l'éclatement normalisé de l'idéal I dans X, c'est-à-dire le composé de l'éclatement de I et de la normalisation de l'espace obtenu, qui est réduit puisque X l'est. Le morphisme π est propre et biméromorphe puisque Y est rare, et donc surjectif car c'est un morphisme fermé d'image dense. Le faisceau d'idéaux $I \cdot \mathcal{O}_{\overline{X}'}$ engendré dans $\mathcal{O}_{\overline{X}'}$ par les composés avec π des sections de I est donc un faisceau inversible d'idéaux sur un espace normal.

Soit $h \in \overline{I_y}$; on a par hypothèse une relation de dépendance intégrale $h^k + a_1 h^{k-1} + \dots + a_k = 0$, avec $a_i \in I_y^i$, qui s'étend en une relation de dépendance intégrale pour un représentant, encore noté h, du germe h, c'est-à-dire il existe un voisinage ouvert U de y dans X tel que l'on ait

$$h \in \overline{H^o(U,I)} \subset H^o(U,\mathcal{O}_X) \quad .$$

Ceci implique que $(h \circ \pi)_{x'}$ est entier sur $(I\mathcal{O}_{\overline{X}'})_{x'}$ pour tout $x' \in \pi^{-1}(U)$, et l'on déduit de la remarque 2) que $(h \circ \pi)_{x'} \in (I\mathcal{O}_{\overline{X}'})_{x'}$ pour tout $x' \in \pi^{-1}(U)$. Ceci montre d'une part que $h \in H^o(U, \pi_*(I\mathcal{O}_{\overline{X}'}) \cap \mathcal{O}_X)$ et d'autre part que tout point $x' \in \pi^{-1}(U)$ possède un voisinage ouvert $V_{x'}$ tel qu'il existe $C_{x'} \in \mathbb{R}_+$ tel que, pour tout $w' \in V_{x'}$, on ait $|h \circ \pi(w')| \leq C_{x'} \sup_{1\leq i\leq p} |g_i \circ \pi(w')|$. Le morphisme π étant propre et surjectif, on peut recouvrir $\pi^{-1}(x)$ par un nombre fini d'ouverts $V_{y'_i}$, $y'_i \in \pi^{-1}(y)$, de telle façon que $\cup V_{y'_i} \supset \pi^{-1}(V)$ pour un voisinage V de x dans U. Prenant $C = \sup_i C_{y'_i}$, on en déduit bien l'inégalité $|h(x)| \leq C \sup |g_i(x)|$ pour tout $x \in V$. Réciproquement, supposons une telle inégalité vérifiée, disons pour tout $x' \in V$; la même inégalité est valable pour $h \circ \pi$ et les $g_i \circ \pi$ sur $\pi^{-1}(V)$, mais là, nous savons que $I\mathcal{O}_{\pi^{-1}(V)}$ est inversible et d'après le lemme de Nakayama, si $(g_1, \dots, g_p)$ est un système de générateurs de l'idéal $I_y \subset \mathcal{O}_{X,y}$, chaque point $y' \in \pi^{-1}(y)$ possède un voisinage ouvert $V_{y'}$ tel que $I\mathcal{O}_{\overline{X}',w'} = (g_i \circ \pi)_{w'} \mathcal{O}_{\overline{X}',w'}$ (pour un certain i, $1 \leq i \leq p$). Dire que $(h \circ \pi)_{w'}$

est bornée par les éléments de $I\mathcal{O}_{\overline{X}',w'}$ équivaut donc à dire que la fonction méromorphe $\frac{h \circ \pi}{g_i \circ \pi}$ est bornée au voisinage de y', ce qui signifie puisque X' est normal qu'elle est holomorphe, donc que $(h \circ \pi)\mathcal{O}_{X',w'} \in I\mathcal{O}_{X',w'}$.

On a donc bien équivalence entre

1.3.2
$$\begin{cases} \text{i) } h_y \in \overline{I_y} \\ \text{ii) } |h(x)| \leq C \sup_i |g_i(x)| \quad (g_1,\ldots,g_p)\mathcal{O}_{X,y} = I_y, \text{ et } C \in \mathbb{R}_+ \\ \quad \text{comme ci-dessus.} \\ \text{iii) } h_y \in (\pi_*(I\mathcal{O}_{\overline{X}'}) \cap \mathcal{O}_X)_y . \end{cases}$$

Pour achever la preuve de la proposition, il suffit de remarquer que puisque $I\mathcal{O}_{\overline{X}'} \subset \mathcal{O}_{\overline{X}'}$, on a $\pi_* I\mathcal{O}_{\overline{X}'} \subset \pi_*\mathcal{O}_{\overline{X}'} = \overline{\mathcal{O}_X}$, faisceau des fonctions méromorphes bornées sur X, qui est un faisceau cohérent de $\mathcal{O}_X$-modules d'après un théorème d'Oka (ou le théorème des images directes de Grauert). Le faisceau $\pi_* I\mathcal{O}_{\overline{X}'}$ est cohérent d'après le théorème de Grauert, et donc le faisceau $\overline{I} = \pi_*(I\mathcal{O}_{X'}) \cap \mathcal{O}_X$, intersection de deux sous-faisceaux cohérents d'un faisceau cohérent, est un faisceau cohérent d'idéaux. ∎

<u>1.3.3</u> <u>Corollaire 1</u> : <u>Etant donnés deux faisceaux cohérents d'idéaux</u> I, J, <u>l'ensemble des points</u> $x \in X$ <u>tels que</u> $J_x \subseteq \overline{I}_x$ <u>est le complémentaire d'un fermé analytique de</u> X.

En effet, le faisceau $\overline{I} + J/\overline{I}$ est cohérent puisque $\overline{I}$ l'est, et son support est le fermé cherché.

<u>1.3.4</u> <u>Corollaire 2</u> (Critère valuatif de dépendance intégrale) : <u>Pour que l'on ait</u> $h_y \in \overline{I}$, <u>il faut et il suffit que pour tout morphisme analytique complexe</u> (resp. <u>analytique réel</u>) $\varphi : (\mathbb{D},0) \to (X,y)$ (<u>resp.</u> $\varphi : (\mathbb{I},0) \to (X,y)$) <u>où</u> $\mathbb{D}$ (resp. $\mathbb{I}$) <u>désigne le disque unité de</u> $\mathbb{C}$ (resp. <u>l'intervalle</u> $]-1,+1[$ <u>de</u> $\mathbb{R}$) <u>on ait, en notant</u> $\varphi^* : \mathcal{O}_{X,y} \to \mathcal{O}_{\mathbb{D},0}$ (resp. $\varphi^* : \mathcal{O}_{X,y} \to \mathcal{A}_{\mathbb{I},0}$) <u>le morphisme d'algèbres analytiques complexes</u> (resp. <u>réelles</u>) <u>associé</u>, <u>l'inclusion</u> :

$\varphi^*(h) \subset \varphi^*(I) \cdot \mathcal{O}_{\mathbb{D},0}$ (resp. $\varphi^*(h) \subset \varphi^*(I) \cdot \mathcal{A}_{\mathbb{I},0}$, où $\mathcal{A}_{\mathbb{I},0}$ est l'algèbre des germes en 0 de fonctions analytiques réelles sur $\mathbb{I}$). Ceci signifie que $v(\varphi(h)) \geq v(\varphi(I))$, où v est "l'ordre d'annulation en 0".

La condition est évidemment nécessaire, puisque l'image par un homomorphisme d'anneaux d'une relation de dépendance intégrale est une relation de dépendance intégrale, et que $\mathcal{O}_{\mathbb{D},0}$ et $\mathcal{A}_{\mathbb{I},0}$ sont des anneaux locaux normaux et principaux. Pour montrer que la condition est suffisante, on utilise le critère valuatif de propreté pour le morphisme π, qui affirme qu'il existe $r : (\mathbb{D},0) \to (\mathbb{D},0)$ (resp. $(\mathbb{I},0) \to (\mathbb{I},0)$ et $\varphi' : (\mathbb{D},0) \to (\overline{X}',y')$ (resp. $(\mathbb{I},0) \to (\overline{X}',y')$) tels que $\pi \circ \varphi' = \varphi \circ r$. Il suffit donc de prouver le résultat pour $h \circ \pi$ et $I\mathcal{O}_{\overline{X}'}$ en tout point $y' \in \pi^{-1}(y)$. Puisque $I\mathcal{O}_{\overline{X}'}$ est un faisceau d'idéaux inversible sur un espace normal, on est donc ramené à prouver que si une fonction méromorphe $\frac{h}{g}$ sur un espace normal n'est pas holomorphe en un point, il existe un arc analytique complexe (resp. réel) $\varphi : (\mathbb{D},0) \to (\overline{X}',y')$ (resp. $\varphi : (\mathbb{I},0) \to (\overline{X}',y')$) le long duquel la limite de $\frac{h}{g}$ est infinie. Cela est un résultat bien classique dans le cas où $\overline{X}'$ est non-singulier en y', conséquence facile de la factorialité des anneaux locaux réguliers (cf. [Se], IV-39) et l'on se ramène à ce cas en utilisant l'existence de systèmes de paramètres (morphismes finis) $(\overline{X}',y') \xrightarrow{p} (\mathbb{C}^d,0)$ $(d = \dim_{y'} \overline{X}')$ et le fait que toute fonction méromorphe sur $\overline{X}'$ est algébrique sur le corps des fonctions méromorphes sur $(\mathbb{C}^d,0)$: si une telle fonction $\frac{h}{g}$ n'est pas bornée au voisinage de y', un des coefficients de l'équation unitaire qu'elle satisfait est une fonction méromorphe sur $\mathbb{C}^d$ non bornée au voisinage de 0, donc il existe un arc analytique à valeurs dans $(\mathbb{C}^d,0)$ le long duquel elle tend vers $+\infty$ en module. La propreté de p permet de remonter, après ramification, cet arc en un arc analytique à valeurs dans $(\overline{X}',y')$, le long duquel on vérifie aussitôt que $|\frac{h}{g}|$ tend vers $+\infty$. ∎

1.3.5 Remarque : L'avatar de ce corollaire en algèbre commutative est l'énoncé suivant : soient A un anneau local nœtherien réduit et I un idéal de A. Pour tout anneau de valuation V contenu dans l'anneau total de fractions de A et contenant A, notons I_V l'idéal de V engendré par I ; on a l'égalité

$$\overline{I} = \bigcap_{A \subset V \subset \mathrm{Tot}(A)} I_V$$

(comparer à [Bbk 2], § 1 n° 3).

On a aussi les variantes suivantes : l'élément h appartient à $\overline{I}$ si et seulement si, pour tout homomorphisme $\varphi : h : A \to V$, où V est un anneau de valuation discrète, on a $\varphi(h) \in \varphi(I)$, ou , ce qui est équivalent, $v(h) \geq v(I)$ où v désigne la fonction d'ordre sur A obtenue en composant avec φ la valuation de V. Soient A un anneau nœthérien réduit, I un idéal de A et $\overline{A}$ la fermeture intégrale de A dans son anneau total de fractions. Soit $Z \to \mathrm{Spec}\, A$ un morphisme propre et birationnel tel que Z soit normal et $I\mathcal{O}_Z$ inversible. On a $\overline{I} = H^0(Z, I\mathcal{O}_Z) \cap A$ où $H^0(Z, I\mathcal{O}_Z) \subset H^0(Z, \mathcal{O}_Z) = \overline{A}$.

1.3.6 Corollaire 3 : *Avec I et $h \in H^0(X, \mathcal{O}_X)$ comme ci-dessus, on a l'inclusion $h_y \in \overline{I}_y$ si et seulement si il existe un voisinage ouvert U de y dans X et un morphisme propre et surjectif $\pi : Z \to U$ tel que $h \circ \pi \in I\mathcal{O}_Z$.*

Le fait que la condition soit nécessaire résulte aussitôt de la Proposition, et le fait qu'elle soit suffisante résulte de ce que $h \circ \pi \in I\mathcal{O}_Z$ implique localement sur Z des inégalités comme dans la Proposition, qui se redescendent en inégalités sur U parce que π est propre et surjectif.

En particulier, étant donnés deux faisceaux cohérents d'idéaux I et J sur X, on a l'égalité $\overline{I} = \overline{J}$ si et seulement si il existe un morphisme $\pi : Z \to X$ propre et surjectif tel que $I \cdot \mathcal{O}_Z = J \cdot \mathcal{O}_Z$, et si $\overline{I} = \overline{J}$, ils ont même éclatement normalisé.

1.4 Proposition 2 : *Soit I un faisceau cohérent d'idéaux sur X, définissant un sous-espace rare $Y \subset X$; soit $\pi : Z \to X$ un morphisme propre et surjectif tel que Z soit normal et $I\mathcal{O}_Z$ inversible* (par exemple l'éclatement normalisé de I vu plus haut). *Soit $D = \bigcup_{i \in I} D_i$ la décomposition en composantes irréductibles du sous-espace $D \subset Z$ de codimension 1, diviseur de Cartier dans Z défini par $I\mathcal{O}_Z$.*

A) Chaque D_i contient un ouvert analytique dense (= complémentaire d'un fermé analytique rare) U_i en tout point z duquel Z et D_{red} sont non-singuliers, $(D,z) = (D_i,z)$ et il existe un système de coordonnées locales $w_1,\dots,w_d$ pour Z en z tel que $I\mathcal{O}_{Z,z} = w_1^{v_i}\,\mathcal{O}_{Z,z}$ avec $v_i \in \mathbb{N} - \{0\}$.

B) Etant donné $h \in H^0(X,\mathcal{O}_X)$, il existe sur chaque composante D_i un ouvert analytique dense V_i en tout point z duquel $(h \circ \pi)_z \cdot \mathcal{O}_{Z,z} = w_1^{w_i}\,\mathcal{O}_{Z,z}$ $(w_i \geq 0)$.

C) On a $h_y \in \overline{I}_y$ si et seulement si, pour chaque indice i tel que $D_i \cap \pi^{-1}(y) \neq \emptyset$, on a l'inégalité $w_i \geq v_i$.

Démonstration : Les points A) et B) résultent aussitôt de ce que Z étant normal, le lieu singulier de Z est de codimension au moins 2, et du fait que, D_i étant irréductible, les entiers v_i et w_i, qui sont clairement localement constants hors d'un fermé rare de D_i sont constants parce que le complémentaire de ce fermé est connexe. Le point C) résulte de ce que $I\mathcal{O}_Z$ étant inversible, et Z normal, pour vérifier que $h \circ \pi \in I\mathcal{O}_Z$ dans un voisinage de $\pi^{-1}(y)$, ce qui d'après la Proposition 1 équivaut à l'inclusion $h_y \in \overline{I}_y$, il suffit de vérifier que pour tout $z' \in \pi^{-1}(y)$, $(\frac{h \circ \pi}{g'})_{z'}$, où g' est un générateur local de $I\mathcal{O}_Z$ en z', est holomorphe. Sur un espace normal, le lieu polaire d'une fonction méromorphe est soit de codimension 1, soit vide, comme le montre un argument analogue à celui utilisé dans la preuve du Corollaire 1. Or, le lieu polaire de $\frac{h \circ \pi}{g'}$ qui est clairement contenu dans $D \cap U'$, U' étant un voisinage ouvert de z' où g' engendre $I\mathcal{O}_Z$ et assez petit, contient $D_i \cap U'$ si et seulement si $z' \in D_i$ et $w_i < v_i$, comme on le voit aussitôt. Ainsi, dire que $w_i \geq v_i$ pour tout i, tel que $D_i \cap \pi^{-1}(y) \neq \emptyset$ équivaut à dire que pour tout $z' \in \pi^{-1}(y)$, et tout U' comme ci-dessus, le lieu polaire de $\frac{h \circ \pi}{g'}$ ne contient aucun $D_i \cap U'$, au voisinage de z' donc que ce lieu polaire n'a aucune composante de codimension 1 au voisinage de $\pi^{-1}(y)$ et donc qu'il est vide, c'est-à-dire que $(h \circ \pi)_{z'} \in I\mathcal{O}_{Z,z'}$ pour tout $z' \in \pi^{-1}(y)$, ou encore d'après la Proposition 1, que $h_y \in \overline{I}_y$. ∎

1.4 Remarques : Dans un cas particulier, on peut voir très simplement le rapport entre la Proposition 2 et le critère valuatif de dépendance intégrale, et surtout en tirer un critère valuatif "effectif". (Voir [Te 9] et [Le-T 1], § 5) :

<u>1.4.1</u> Soit I un idéal primaire pour l'idéal maximal, dans une algèbre analytique réduite $\mathcal{O}_{X,x}$. Supposons que I soit entier sur un idéal I_1 engendré par une suite régulière (c'est-à-dire $\mathcal{O}_{X,x}$ Cohen-Macaulay). On peut alors concevoir l'espace obtenu en éclatant I_1 dans X comme une famille de courbes paramétrée par $\mathbb{P}^{d-1}$, où $d = \dim \mathcal{O}$ et munie d'une section σ. En effet si $I_1 = (f_1, \dots, f_d)\mathcal{O}_{X,x}$, l'éclatement de I_1 dans X est le sous-espace X_1 de $X \times \mathbb{P}^{d-1}$ défini par l'idéal engendré par $(f_i T_{i+1} - f_{i+1} T_i \; ; \; 1 \leq i \leq d-1)$ où $(T_i : \dots : T_d)$ est un système de coordonnées homogènes sur $\mathbb{P}^{d-1}$. D'après les résultats généraux sur la normalisation simultanée (cf. [Te 2], I), il existe un ouvert de Zariski dense $U \subset \mathbb{P}^{d-1}$ au-dessus duquel on a normalisation simultanée c'est-à-dire que le morphisme composé $\overline{X}_1 \xrightarrow{n} X_1 \xrightarrow{pr_2} \mathbb{P}^{d-1}$ a la propriété que pour tout $t \in U$, $(pr_2 \circ n)$ est plat au voisinage de tout point de $(pr_2 \circ n)^{-1}(t)$ et que de plus le morphisme induit $\overline{X}_1(t) \to X_1(t)$ sur les fibres est la normalisation. Par conséquent la courbe $\overline{X}_1(t)$ est réunion de germes de courbes non-singuliers, transverses au diviseur exceptionnel ensembliste $D_{red} = (n^{-1}(\{x\} \times \mathbb{P}^{d-1}))_{red}$ qui est lui-même non-singulier en tout point de $(pr_2 \circ n)^{-1}(t)$. On voit que si l'on considère la décomposition en composantes irréductibles $X_1(t) = \bigcup_{j \in J} C_j$, d'une part, et d'autre part la décomposition $D_{red} = \bigcup_{i \in I} D_i$, pour chaque fonction $h \in \mathcal{O}_{X,x}$, il existe un ouvert de Zariski dense $V = V(h)$ de $\mathbb{P}^{d-1}$ tel que, pour $t \in V$, si l'on considère l'application surjective naturelle $\alpha : J \to I$ qui à $j \in J$ associe la composante D_i de D que rencontre la normalisée de la branche C_j de $X_1(t)$, on a, en notant $\tilde{w}_j(h)$ la valuation de l'image de h dans $\mathcal{O}_{C_j}$, l'égalité (notations de la Proposition 2)

$$w_{\alpha(j)} = \tilde{w}_j(h) \quad ,$$

et l'on a aussi, dans $U \cap V(h)$, l'égalité $v_{\alpha(j)} = \tilde{w}_j(I \cdot \mathcal{O}_{X(t)})$. Par conséquent, pour $h \in \mathcal{O}_{X,x}$ donné, il existe un ouvert de Zariski dense $U_h \subset \mathbb{P}^{d-1}$ tel que, si l'on considère pour $t \in U_h$ la courbe C_t <u>contenue dans</u> X définie par l'idéal de $\mathcal{O}_{X,x}$ engendré par $(T_i\ f_{i+1} - T_{i+1}\ f_i\ ;\ 1 \le i \le d-1)$ où $t = (T_1 : \ldots : T_d) \in U_h$ on a, en remarquant que $\pi_o : X_1(t) \to C_t$ est un isomorphisme : $h \in \overline{I}$ si et seulement si pour chaque branche $C_{t,j}$ de C on a, en notant v_j la fonction d'ordre sur $\mathcal{O}_{X,x}$ obtenue en composant les morphismes $\mathcal{O}_{X,x} \to \mathcal{O}_{C_{t,j},0} \to \overline{\mathcal{O}_{C_{t,j},0}} \xrightarrow{v} \mathbb{Z}$ où v est la valuation naturelle, les inégalités

$$v_j(h) \ge v_j(I) \quad .$$

Le schéma suivant aidera peut-être le lecteur.

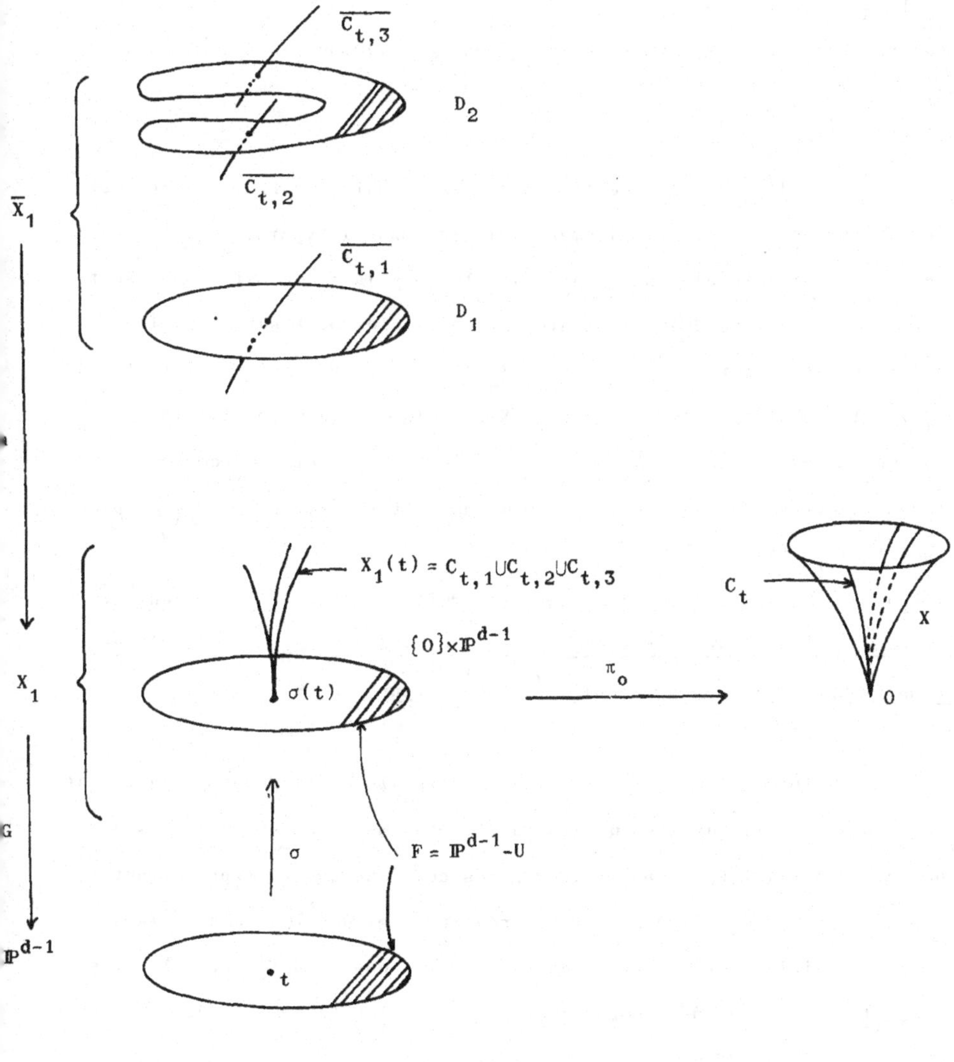

1.4.2 Un avatar algébrique de la Proposition 2 est l'énoncé suivant : Soient A un anneau local nœtherien et $\pi : Z \to \operatorname{Spec} A$ un morphisme propre et surjectif de schémas tel que Z soit normal et $I\mathcal{O}_Z$ inversible. Soit $Z = \cup V_i$ un recouvrement affine de Z, avec $V_i = \operatorname{Spec} B_i$; soient $p_{i,j}$ les idéaux premiers associés à IB_i ; d'après le hauptidealsatz (cf. [Z-S], Ch. VII, § 7, Th. 23), les $p_{i,j}$ sont de hauteur 1 dans l'anneau normal B_i, et donc (cf. [Z-S], ch. VI, § 14, Th. 33) les localisés $(B_i)_{p_{i,j}}$ sont des anneaux de valuation discrète. Soient $v_{i,j}$ les fonctions d'ordre sur A obtenues en composant la valuation de $(B_i)_{p_{i,j}}$ avec l'homomorphisme $A \to (B_i)_{p_{i,j}}$ induit par π. Un élément $h \in A$ est entier sur I si et seulement si $v_{i,j}(h) \geq v_{i,j}(I)$ pour tous les couples (i,j).

1.4.3 Gardant l'hypothèse que l'espace X est réduit, la Proposition 1 et ses corollaires restent essentiellement valables sans l'hypothèse que le sous-espace défini par l'idéal I soit rare dans X, sous la forme suivante : soit X_1 la réunion de celles des composantes irréductibles de X sur lesquelles l'idéal I induit un fermé rare, et soit X_2 la réunion des autres. Notant $\overline{X}'_1 \to X_1$ l'éclatement normalisé de I dans X_1, et $\overline{X}_2 \to X_2$ la normalisation de X_2, on a un morphisme naturel $\pi : \overline{X}'_1 \amalg \overline{X}_2 \to X$ qui est propre et biméromorphe, et tel que l'on ait encore : $h \in \overline{I}_x$ si et seulement si $(h \circ \pi) \in I \cdot \mathcal{O}_{\overline{X}'_1 \amalg \overline{X}_2, z}$ pour tout point $z \in \pi^{-1}(x)$.

Conclusion : Sous l'hypothèse que X est réduit, $h \in \overline{I}_x$ si et seulement si pour tout morphisme propre et surjectif $p : Z \to X$ tel que $I\mathcal{O}_Z$ soit localement principal et Z normal, on a $(h \circ p)_z \in \left(I \cdot \mathcal{O}_Z\right)_z$ pour tout $z \in p^{-1}(x)$.

1.4.4 Le critère valuatif de dépendance intégrale et la Proposition 2 sont les principaux outils que l'on utilisera ici pour montrer des relations de dépendance intégrale. La Proposition 2 a des conséquences du type suivant : soit $\pi : Z \to X$ un morphisme propre et biméromorphe tel que $I\mathcal{O}_Z$ soit un faisceau d'idéaux inversibles, et Z normal. Soit Y le sous-espace défini par X, et soit $h \in H^0(X, \mathcal{O}_X)$. Si chacune des composantes irréductibles de $D = \pi^{-1}(Y)$ s'envoie surjectivement, par π, sur Y, et si $h_x \in \overline{I}_x$ pour tout x hors d'un fermé analytique rare F de Y, alors $h \in \overline{I}_x$ pour tout $x \in Y$, i.e. $h \in H^0(X,\overline{I})$. En effet, $\pi^{-1}(Y - F)$ contiendra, d'après l'hypothèse, un ouvert analytique dense de chaque composante irréductible D_i de D, et l'inclusion $h_x \in \overline{I}_x$ pour $x \in Y - F$ impliquera donc les inégalités $w_i \geq v_i$ de la Proposition 2. (Voir aussi le § 5 ci-dessous).

1.4.5 Exercice (corrigé dans [Te 6]) : Soient $F_1,\ldots,F_c$ et G des éléments de l'idéal maximal de $\mathcal{O}_{N+1} \simeq \mathbb{C}\{z_0,\ldots,z_N\}$. Soient I l'idéal de $\mathcal{O}_{N+1}$ engendré par $F_1,\ldots,F_c$, $\{i_1,\ldots,i_c\} \subset \{0,\ldots,N\}$ un ensemble d'indices et J l'idéal de $\mathcal{O}_{N+1}$ engendré par les éléments $z_j \dfrac{\partial(F_1,\ldots,F_c,G)}{\partial(z_{i_1},\ldots,z_{i_c},z_j)}$, $j \in \{0,\ldots,N\}$. On a l'inclusion :

$$G \cdot \frac{\partial(F_1,\ldots,F_c)}{\partial(z_{i_1},\ldots,z_{i_c})} \cdot \mathcal{O}_{N+1}/I \in \overline{J \cdot \mathcal{O}_{N+1}/I} \quad .$$

1.5 Autres propriétés utiles de la clôture intégrale des idéaux

Lemme de Nakayama intégral : Soient I, I' et J trois idéaux d'un anneau local nœtherien A, tel que $I \subset I'$ et $J \subset \mathfrak{M}$ où $\mathfrak{M}$ est l'idéal maximal de A. Si l'on a $\overline{I + JI'} = \overline{I}'$, alors $\overline{I} = \overline{I}'$.

Pour prouver ce résultat nous aurons besoin de

1.5.1 Proposition : Soient A un anneau nœtherien et I un idéal de A. Pour un élément $h \in A$, les propriétés suivantes sont équivalentes :

i) $h \in \overline{I}$

ii) il existe un A-module de type fini fidèle tel que l'on ait $hM \subset IM$.

i) ⇒ ii). Soit $h^k + a_1 h^{k-1} + \ldots + a_k = 0$ avec $a_i \in I^i$ une équation de dépendance intégrale pour h ; on en déduit que : $h^k \in I \cdot (I + hA)^{k-1}$ d'où $h(I + hA)^{k-1} \subseteq I \cdot (I + hA)^{k-1}$ et le résultat en prenant pour M l'idéal $(I + hA)^{k-1}$.

ii) ⇒ i). Soit $e_1,\ldots,e_p$ un système de générateurs du A-module M. D'après l'hypothèse, on a des égalités $h \cdot e_i = \Sigma \lambda_{ij} e_j$ avec $\lambda_{ij} \in I$, d'où, d'après la règle de Cramer : $\det(h \cdot Id - \Lambda) \cdot M = 0$ où $\Lambda = (\lambda_{ij})$. Puisque M est un A-module fidèle, ceci implique $\det(h \cdot Id - \Lambda) = 0$ ce qui est clairement une relation de dépendance intégrale pour h sur I.

Pour prouver le lemme de Nakayama intégral, remarquons que ce que nous avons à prouver est l'inclusion $\overline{I}' \subset I$. Or l'hypothèse nous donne, d'après la Proposition, l'existence d'un A-module de type fini M tel que l'on ait l'inclusion : $I'.M \subset IM + J.I'M$. Puisque $J \subset \mathfrak{M}$, on en déduit, d'après le Lemme de Nakayama, l'inclusion $I'M \subset IM$, d'où le résultat en appliquant la Proposition.

Remarque : D'après ce qui précède, on peut aussi motiver l'introduction de la clôture intégrale des idéaux par le problème de la "simplification" dans les produits d'idéaux : l'égalité $IJ = I'J$ n'implique pas $I = I'$, mais implique $\overline{I} = \overline{I}'$. (Voir [Sa]).

§ 2. Un exemple : les idéaux engendrés par des monômes.

Considérons le cas particulier d'un idéal I de l'anneau $A = \mathbb{C}\{z_1, \ldots, z_d\}$ engendré par des monômes $z_1^{a_1^{(i)}} \ldots z_d^{a_d^{(i)}}$ $(1 \leq i \leq p)$. D'après le Corollaire 2 du § 1, pour qu'un élément $h \in A$, soit entier sur I il faut et il suffit que, pour tout homomorphisme $\varphi^* : \mathbb{C}\{z_1, \ldots, z_d\} \to \mathbb{C}\{t\}$, posant $v_i = v(\varphi^*(z_i))$ = ordre en t de $\varphi^*(z_i)$, on ait l'inégalité

$$v(\varphi^*(h)) \geq \inf (a_1^{(1)} v_1 + \ldots + a_d^{(1)} v_d, \ldots, a_1^{(p)} v_1 + \ldots + a_d^{(p)} v_d)$$

ce qui implique que toute forme linéaire à coefficients $(v_1, \ldots, v_d)$ prend, sur les exposants des monômes apparaissant dans la série h, une valeur au moins égale au minimum des valeurs qu'elle prend sur les exposants des monômes engendrant I. (On laisse le lecteur vérifier que les annulations possibles par sommes de monômes de h peuvent être évitées par un choix "générique" des coefficients des séries $\varphi^*(z_i) \in \mathbb{C}\{t\}$). Par conséquent, si $h \in \overline{I}$, tous les monômes $z_1^{b_1} \ldots z_d^{b_d}$ apparaissant dans h sont tels que le point $(b_1, \ldots, b_d) \in \mathbb{R}^d$ appartienne à l'enveloppe convexe N(I) dans $\mathbb{R}^d$ de l'ensemble

$$E(I) = \bigcup_{i=1}^{p} (a^{(i)} + \mathbb{R}_+^d) \quad \text{où } a^{(i)} = (a_1^{(i)}, \ldots, a_d^{(i)}).$$

Inversement, si l'exposant $(b_1,\ldots,b_d)$ se trouve dans cette enveloppe connexe, le monôme $z_1^{b_1} \ldots z_d^{b_d}$ est entier sur l'idéal I (exercice : écrire la relation de dépendance intégrale).

Par conséquent :

L'idéal $\overline{I}$ est engendré par les monômes dont les points représentatifs sont situés dans l'enveloppe convexe N(I) de l'ensemble $E(I) = \bigcup_{i=1}^{p} (a^{(i)} + \mathbb{R}_+^d)$ des points représentant les monômes qui appartiennent à I.

Exemple : $d = 2$, $k = 3$. La partie hachurée est $N(I) - E(I)$.

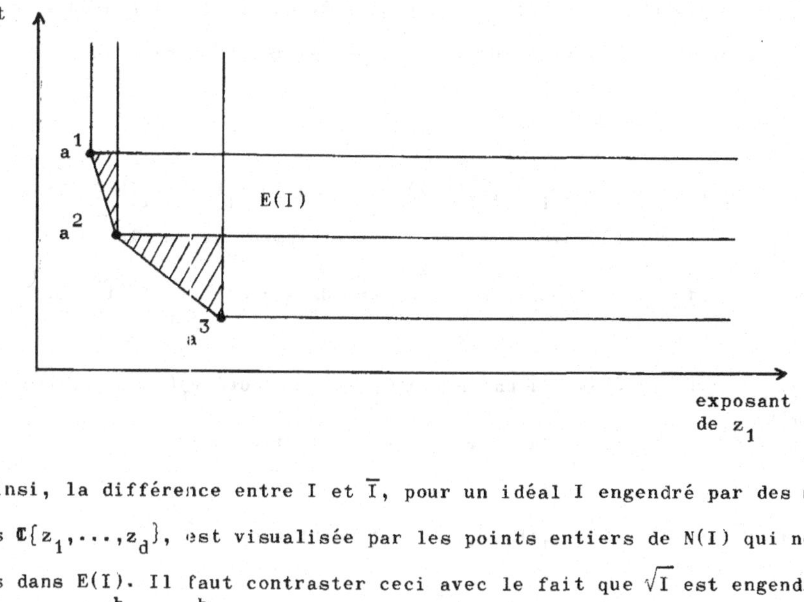

Ainsi, la différence entre I et $\overline{I}$, pour un idéal I engendré par des monômes dans $\mathbb{C}\{z_1,\ldots,z_d\}$, est visualisée par les points entiers de N(I) qui ne sont pas dans E(I). Il faut contraster ceci avec le fait que $\sqrt{I}$ est engendré par les monômes $z_1^{b_1} \ldots z_d^{b_d}$ tels que $(\nu b_1,\ldots,\nu b_d) \in E(I)$ pour un certain $\nu \in \mathbb{N}$, c'est-à-dire appartenant à l'intérieur du cône de sommet 0 sur E(I).

Si l'on observe que pour des idéaux I et J engendrés par des monômes, on a l'égalité $E(I \cdot J) = E(I) + E(J)$, la description classique de l'enveloppe convexe comme ensemble des barycentres implique que, étant donnés d points $b^{(1)},\ldots,b^{(d)}$ dans N(I), leur somme $b^{(1)} + \ldots + b^{(d)}$ appartient à E(I), ce qui se traduit en termes d'idéaux par l'inclusion

$$(\overline{I})^d \subset I \quad .$$

Il se trouve que cette inclusion est vraie pour tous les idéaux dans les annaux de séries convergentes, et en fait dans tout anneau local régulier A, avec $d = \dim A$ (cf. [B-S], [Li-T] et [Li-S]).

§ 3. Dépendance intégrale et inégalité de Łojasiewicz.

Les résultats qui suivent sont des conséquences assez faciles de la Proposition 2 du § 1, et sont prouvés en détail dans ([Le-T 1]).

Définition 1 : Soient I un idéal d'un anneau A. On appelle fonction d'ordre associée à I, et l'on note ν_I l'application $\nu_I : A \to \mathbb{N}$ qui à $h \in A$ associe $\mathrm{Sup}\{\nu \mid h \in I^\nu\}$.

On a : $\nu_I(h+h') \geq \inf(\nu_I(h), \nu_I(h'))$, $\nu_I(h \cdot h') \geq \nu_I(h) + \nu_I(h')$, et en particulier $\nu_I(h^k) \geq k \cdot \nu_I(h)$, avec égalité si et seulement si la classe $\mathrm{in}_I(h)$ de h dans $I^{\nu_I(h)}/I^{\nu_I(h)+1}$ n'est pas nilpotente dans $\mathrm{gr}_I A = \bigoplus_{\nu \geq 0} I^\nu/I^{\nu+1}$.

Définition 2 : On appelle "nu barre" de h, et l'on note $\overline{\nu}_I(h)$ la limite $\lim_{k \to +\infty} \frac{\nu_I(h^k)}{k} \in \mathbb{R}$. (Exercice : montrer que cette limite existe.)

On a : $\overline{\nu}_I(h+h') \geq \inf(\overline{\nu}_I(h), \overline{\nu}_I(h'))$ et maintenant : $\overline{\nu}_I(h^k) = k \cdot \overline{\nu}_I(h)$.

Théorème (Samuel-Nagata-Lejeune-Teissier, cf. [Sa], [Na], [Le-T 1]) : Soit I un idéal dans une algèbre analytique A.

a) Pour tout $h \in A$, $\overline{\nu}_I(h)$ est un nombre rationnel, ou $+\infty$.

b) Si $A = \mathcal{O}_{X,y}$ et $I = I_y$, $h = h_y$ pour un espace analytique réduit X, un faisceau cohérent d'idéaux I sur X, comme dans la Proposition 2 du § 1, et $h \in H^0(X, \mathcal{O}_X)$, $y \in Y$, alors les conditions suivantes sont équivalentes :

i) $\overline{\nu}_I(h) \geq \frac{a}{b}$ $(a, b \in \mathbb{N})$.

ii) Il existe un voisinage U de y dans X et une constante $C \in \mathbb{R}_+$ tels que l'on ait l'inégalité de Łojasiewicz

$$|h(x)|^{b/a} \leq C \operatorname*{Sup}_i |g_i(x)| \qquad (x \in U)$$

(pour un système de générateurs $g_1, \ldots, g_p$ de $H^0(U, I)$ dans $H^0(U, \mathcal{O}_X)$ fixé).

c) Pour tout rationnel a/b, il existe un unique faisceau cohérent d'idéaux intégralement clos $\overline{I^{a/b}}$ tel que pour tout $x \in X$ on ait

$$(\overline{I^{a/b}})_x = \{h \in \mathcal{O}_{X,x} / \overline{\nu}_{I_x}(h) \geq a/b\} \quad .$$

Remarques : 1) L'énoncé a) est vrai plus généralement dès que A est un anneau local nœtherien pseudo-géométrique au sens de Nagata.

2) Un corollaire du Théorème est : $h \in \overline{I}$ si et seulement si $\overline{\nu}_I(h) \geq 1$.

3) Si l'idéal I est engendré par des monômes dans $\mathbb{C}\{z_1, \ldots, z_d\}$, et si h est un monôme, on voit que $(\overline{\nu}_I(h))^{-1}$ est le plus petit nombre réel r tel que $(r\,b_1, \ldots, r\,b_d)$ appartienne à $N(I)$.

4) On a l'égalité $\overline{\nu}_I(h) = \overline{\nu}_{\overline{I}}(h)$.

§ 4. Clôture intégrale d'idéaux et multiplicités.

L'exemple précédent a montré que le passage de I à $\overline{I}$ perd beaucoup moins d'information que le passage de I à $\sqrt{I}$. Dans le cas particulier des idéaux primaires pour l'idéal maximal dans un anneau local A, on peut exprimer numériquement de façon précise l'information qui est conservée dans le passage de I à $\overline{I}$.

Proposition - Définition ([Se], chap. II et chap. V, No 2), [Bbk 3], § 4) : Soient A un anneau, nœtherien et I un idéal de A tel que $\ell g_A\, A/I < +\infty$. Alors, pour tout entier $\nu \geq 1$, $\ell g_A\, A/I^\nu$ est finie, et il existe ν_o et un polynôme $p(\nu)$ de degré $d = \dim A$ à coefficients tel que $\ell g_A\, A/I^\nu = p(\nu)$ pour $\nu \geq \nu_o$. De plus $p(\nu)$ peut être écrit $\frac{e(I)}{d!}\nu^d$ + termes de degré $< \nu$ où $e(I)$ est un entier appelé multiplicité de l'idéal I dans A.

4.1 Théorème (Rees [Re]) : On a l'égalité $e(I) = e(\overline{I})$, (deux idéaux ayant même fermeture intégrale ont donc même multiplicité) et réciproquement, si I et J sont deux idéaux de colongueur finie de A tels que $I \subseteq J$ et $e(I) = e(J)$ on a : $\overline{I} = \overline{J}$, dès que le complété $\hat{A}$ de A est équidimensionnel.

Exemple : Supposons (cf. § 2) l'idéal I engendré par des monômes dans $\mathbb{C}\{z_1,\dots,z_d\}$. La longueur sur A de A/I est égale à $\dim_{\mathbb{C}} A/I$ puisque A est une $\mathbb{C}$-algèbre, et est très clairement égale au nombre de points à coordonnées entières de $\mathbb{R}^d_+ - E(I)$. Nous supposons donc ce nombre fini, ce qui dans le cas $d = 2$ revient à supposer que E(I) rencontre les deux axes de coordonnées, ou que I contient un monôme $z_1^{a_1}$ et un monôme $z_2^{a_2}$.
On a donc $\dim_{\mathbb{C}} \mathbb{C}\{z_1,\dots,z_d\}/I^\nu = \#\{\mathbb{N}^d - E(I^\nu)\}$ et par ailleurs on a, par la description de l'enveloppe convexe comme ensemble de barycentres :

$$\lim_{\nu\to+\infty} \left(\frac{1}{\nu} . E(I^\nu)\right) = N(I)$$

et par conséquent, puisque le calcul classique des volumes donne :

$$\lim_{\nu\to\infty} \frac{\#\{\mathbb{N}^d - E(I^\nu)\}}{\nu^d} = \mathrm{Vol.}(\mathbb{R}^d_+ - N(I))$$

(où le terme de droite est appelé covolume de N(I), et noté Covol.N(I)) on a l'égalité :

$$e(I) = d! \,.\, \mathrm{Covol}.N(I)$$

L'essentiel du théorème de Rees, dans ce cas particulier, revient donc à l'assertion que puisque $I \subseteq J$ (idéaux engendrés par des monômes) implique $N(I) \subseteq N(J)$, l'égalité $\mathrm{Covol}.N(I) = \mathrm{Covol}.N(J)$ implique $N(I) = N(J)$, c'est-à-dire $\overline{I} = \overline{J}$.

en gros l'énoncé revient donc à : si $A \subseteq B$ sont des polytopes convexes et si Vol.(A) = Vol.(B), on a A = B.]

Le Théorème ci-dessous donne un énoncé numérique bien meilleur, en ce qu'il ne suppose pas que l'on a a priori $I \subseteq J$; citons d'abord la :

Proposition : Soient I_1 et I_2 deux idéaux de A, tels que $\ell g_A\, A/I_1$ et $\ell g_1\, A/I_2$ soient finis. Il existe des entiers e_i, $0 \le i \le d = \dim A$ tels que l'on ait, pour ν_1, ν_2 entiers positifs : $e(I_1^{\nu_1} . I_2^{\nu_2}) = \sum_{i=0}^{d} \binom{d}{i} e_i \nu_1^i \nu_2^{d-i}$. Les e_i sont appelés multiplicités mixtes de I_1 et I_2, et interprétés dans ([Te 1], ch. I, § 2). Ils ne dépendent que de $\overline{I}_1$ et $\overline{I}_2$, on note que $e_o = e(I_2)$, $e_d = e(I_1)$.

Dans le cas des idéaux engendrés par des monômes dans $\mathbb{C}\{z_1,\dots,z_d\}$, ceci correspond au résultat suivant, facile à prouver par récurrence sur d :

Etant données deux régions convexes polygonales à sommets entiers $N(I_1)$ et $N(I_2)$ de covolume fini dans $\mathbb{R}_+^d$, il existe des rationnels $w_i \in \mathbb{Q}_+$, $0 \le i \le d$, tels que l'on ait, pour $\nu_1, \nu_2 \in \mathbb{R}_+$:

$$\text{Covol.}(\nu_1 . N(I_1) + \nu_2 . N(I_2)) = \sum_{i=0}^{d} \binom{d}{i} w_i \ \nu_1^i \ \nu_2^{d-i} \quad .$$

Il est facile dans ce cas de vérifier que

$$e_i = d!\ w_i \qquad (0 \le i \le d) \quad .$$

Les w_i sont appelés covolumes mixtes de $N(I_1)$ et $N(I_2)$; on note que $w_o = \text{Covol.}(N(I_2))$, $w_d = \text{Covol.}(N(I_1))$.

Dans ([Te 7], [Te 8], [R-S]) on trouve la preuve de :

4.2 Théorème : Soient A un anneau local nœtherien , I_1 et I_2 deux idéaux de A primaires pour l'idéal maximal. On a les inégalités

$$e_{i-1}^2 \le e_i \cdot e_{i-2} \qquad (2 \le i \le d) \quad .$$

Si l'on suppose de plus A normal et son complété $A^\wedge$ équidimensionnel on a les

<u>égalités</u> $e_o = e_1 = \dots = e_d$ <u>si et seulement si l'on a</u> $\overline{I}_1 = \overline{I}_2$.

<u>Remarques</u> : Ces inégalités impliquent les inégalités analogues $w_{i-1}^2 \leq w_i \cdot w_{i-2}$ entre les covolumes mixtes, qui sont à rapprocher (cf. [Te 7]) des inégalités de Aleksandrov-Fenchel entre les volumes mixtes de convexes dans $\mathbb{R}^d$ que voici : soient K_1 et K_2 deux convexes compacts dans $\mathbb{R}^d$. On peut montrer que l'on a une expression polynomiale pour le volume du convexe $\nu_1 \cdot K_1 + \nu_2 \cdot K_2$ $(\nu_1, \nu_2 \in \mathbb{R}_+)$:

$$\mathrm{Vol.}(\nu_1 K_1 + \nu_2 K_2) = \sum_{i=0}^{d} \binom{d}{i} v_i \, \nu_1^i \, \nu_2^{d-i}$$

$(v_o = \mathrm{Vol.}(K_2),\ v_d = \mathrm{Vol.}(K_1))$,
et l'on a entre les volumes mixtes v_i et K_1 et K_2 les inégalités de Aleksandrov-Fenchel, qui sont une généralisation des inégalités isopérimétriques :

$$v_{i-1}^2 \geq v_i \cdot v_{i-2} \qquad (2 \leq i \leq d)$$

avec l'égalité $v_o = v_1 = \dots = v_d$ si et seulement si $K_1 = K_2$ à translation près (cf. [Al.], [Te 9], [Te 10]).
Ainsi, tandis que l'analogue du théorème de Rees dans la théorie des convexes est : $K_1 \subseteq K_2$ et $\mathrm{Vol.}(K_1) = \mathrm{Vol.}(K_2)$ implique $K_1 = K_2$, l'analogue du Théorème ci-dessus est le résultat d'Aleksandrov-Fenchel. Il faut remarquer que le théorème ci-dessus implique le théorème de Rees puisque l'on vérifie facilement que l'inclusion $I_1 \subseteq I_2$ implique les inégalités : $e(I_1) \geq e_i \geq e(I_2)$.

<u>4.3</u> Rappelons enfin le résultat suivant, d'une nature plus élémentaire que les précédents (cf. [Bbk 3], § 7) :
Soient $\mathcal{O}$ un anneau local nœtherien réduit, $\mathcal{O}'$ un sous-anneau (semi-local) de l'anneau total des fractions de $\mathcal{O}$, contenant $\mathcal{O}$ et entier sur $\mathcal{O}$. Pour tout idéal I de $\mathcal{O}$ primaire pour l'idéal maximal, si l'on note $m'_1, \dots, m'_r$ les idéaux maximaux de $\mathcal{O}'$, et m l'idéal maximal de $\mathcal{O}$, on a l'égalité :

$$e(I) = \sum_{i=1}^{r} e(I\mathcal{O}'_{m'_i}) \cdot \dim_{\mathcal{O}/m} \mathcal{O}'/m'_i \quad .$$

On écrira aussi $e(I\mathcal{O}')$ pour la somme de droite.

Nous retiendrons que si $\pi : X' \to X$ est un morphisme fini biméromorphe d'espaces analytiques, et $I \subset \mathcal{O}_{X,x}$ un idéal primaire pour l'idéal maximal, $I\mathcal{O}_{X'}$ se décompose dans $\mathcal{O}_{X',\pi^{-1}(x)} = \prod_{x'\in\pi^{-1}(x)} \mathcal{O}_{X',x'}$ en idéaux primaires $I'_{x'}$ et l'on a l'égalité : $e(I) = \sum_{x'\in\pi^{-1}(x)} e(I'_{x'}) = e(I\mathcal{O}_{X',\pi^{-1}(x)})$.

§ 5. Le principe de spécialisation de la dépendance intégrale.

Je rappelle ici, sans démonstration, un résultat qui nous servira dans le prochain paragraphe, mais dont nous pourrions nous passer au prix d'une référence à la théorie de la saturation de Zariski. Ce même résultat jouait un rôle crucial dans la démonstration du théorème "μ^* constant implique les conditions de Whitney" (cf. [Te 1], chap. II, § 3) que le résultat principal de ce travail ci généralise. Il est intéressant de noter que dans la preuve que nous verrons au Chapitre V, il est remplacé par un argument de récurrence, reposant essentiellement sur la transversalité et sur un résultat apparenté au théorème de Bertini idéaliste. Le paragraphe se termine pas une caractérisation géométrique de l'équimultiplicité qui sera très utile au Chapitre V.

5.1 Proposition ([Te 2], Appendice I, [Te 1], Chap. II, § 3, Prop. 3.1) : Soit $F : (X,x) \to (S,s)$ un germe de morphisme plat entre espaces analytiques réduits, et soit I un $\mathcal{O}_X$-idéal cohérent de support X, tel que la restriction de F au sous-espace Y de X défini par I soit un morphisme fini $F : Y \to S$, et tel que pour tout représentant assez petit $F : X \to S$ de F, la multiplicité $e(I\mathcal{O}_{X(s')})$ soit indépendante de $s' \in S$ ($e(I\mathcal{O}_{X(s')})$ désigne la somme des multiplicités des idéaux primaires induits par I dans chacun des anneaux locaux $\mathcal{O}_{X(s'),x_i}$, $x_i \in Y(s')$). Soit $\pi : \overline{X}' \to X$ l'éclatement normalisé de l'idéal I, et soit D le diviseur exceptionnel, défini par l'idéal inversible $I\mathcal{O}_{\overline{X}'}$; alors la fibre D(s) au-dessus de $s \in S$ du morphisme composé $D \xrightarrow{\pi|D} X \to S$ est de dimension $\dim X - \dim S - 1$.

Corollaire (Principe de spécialisation de la dépendance intégrale) : La restriction $\pi: D_i \to Y$ de π à chaque composante irréductible D_i de D est surjective, et par conséquent, pour un élément $h \in H^0(X,\mathcal{O}_X)$ les conditions suivantes sont équivalentes

i) Il existe un fermé analytique rare F de S tel que, pour tout $s' \in S - F$, on ait l'inclusion $h\mathcal{O}_{X(s')} \subset \overline{I.\mathcal{O}_{X(s')}}$.

ii) On a l'inclusion : $h \in H^0(X,\overline{I})$, et donc en particulier $h\mathcal{O}_{X(s)} \subset \overline{I.\mathcal{O}_{X(s)}}$.

5.1.1 Remarques : 1) Le point fondamental est que l'équimultiplicité empêche l'existence de "composantes verticales" dans le diviseur exceptionnel de l'éclatement de I dans $\mathcal{O}_X$ (la normalisation n'est là que pour que nous n'ayons pas à surveiller les composantes immergées). Le corollaire s'en déduit facilement au vu du fait que, grâce cette fois à la normalité, le lieu polaire de la "fonction méromorphe" $I^{-1}h$ est soit de codimension 1 (et est alors union de composantes de D) soit vide, et que la surjectivité des morphismes $\pi: D_i \to Y$, (conséquence immédiate du Théorème) jointe à i) implique que ce lieu polaire ne peut contenir de composante de D.

2) Pour une algébrisation et extension de ces résultats, lire [Li].

3) On peut préciser le point 1) sur l'existence de composantes verticales comme ceci : soit $p: X' \to X$ l'éclatement de I, et soit $D = p^{-1}(Y) \simeq \mathrm{Projan}_Y(\bigoplus_{\nu\geq 0} I^{\nu}/I^{\nu+1})$ le diviseur exceptionnel, muni du faisceau inversible $\mathcal{O}_D(1)$ qui est ample par rapport à p et provient du X-plongement local $X' \subset X \times \mathbb{P}^M$, où $M+1$ est le nombre des générateurs de I. On appelle i-ème classe de Segré covariante de Y dans X le cycle $s^i(Y,X) = p_*(c_1(\mathcal{O}_D(1))^{d-1-i})$ où $d = \dim X$, cycle appartenant au groupe $A_i(X)$ des cycles de dimension i de X. On appelle classe de Segre de Y dans X, la somme $s(Y,X) = \sum_{i=0}^{\dim X} s^i(Y,X)$, et en particulier, d'après un résultat prouvé par C.P. Ramanujam ([Ra], voir aussi [Kl 2]) si $|Y|$ est un point $x \in X$, on a $s(Y,X) = e(I.\mathcal{O}_{X,x}).[x]$, et donc $e(I.\mathcal{O}_{X,x}) = \deg_D(\mathcal{O}_D(1)) = \int (c_1(\mathcal{O}_D(1)))^{d-1}$. Le lecteur est prié, en utilisant ce résultat et le théorème de Bezout (plus précisément les propriétés élémentaires

du degré des faisceaux inversibles) de vérifier l'égalité très utile que voici, qui précise ([Te 1], Chap. II, § 3, Prop. 3.1),

Soient $f : (X,0) \to (\mathbb{D},0)$ un morphisme d'espaces réduits, I un idéal de $\mathcal{O}_X$ définissant un sous-espace $Y \subset X$ tel que $f|Y : Y \to \mathbb{D}$ soit fini, $p : X' \to X$ l'éclatement de Y, $D_{vert.}$ la réunion des composantes du diviseur exceptionnel D (non nécessairement réduites) dont l'image ensembliste par p est 0, $\deg D_{vert.} = \deg(\mathcal{O}_{D_{vert.}}(1))$. Pour tout représentant suffisamment petit du germe de f en 0, on a l'égalité

$$\deg D_{vert.} = e(I \cdot \mathcal{O}_{X(0)}) - e(I \cdot \mathcal{O}_{X(s)}) \quad (\text{pour } s \neq 0) \quad .$$

En particulier, on a "$e(I.\mathcal{O}_{X(s)})$ est indépendant de $s \in \mathbb{D}$" si et seulement si $\dim p^{-1}(0) = \dim X - 2$.

1.2 Application : Utilisons ce résultat pour donner l'interprétation géométrique de la multiplicité d'un espace analytique complexe X en un de ses points, c'est-à-dire la multiplicité dans $\mathcal{O}_{X,x}$ de l'idéal maximal $\mathfrak{m}_{X,x}$: d'abord un lemme explicitant ([Li], § 1), et conséquence facile de 1.1 et de Nakayama :

Soient $\mathcal{O}$ une algèbre analytique de dimension d, et soit $\mathfrak{m} = (z_1,\dots,z_N)$ son idéal maximal ; pour une suite $(y_1,\dots,y_d)$ de d éléments de $\mathfrak{m}$ linéairement indépendants modulo $\mathfrak{m}^2$, les conditions suivantes sont équivalentes

i) L'idéal $\mathfrak{m}$ est entier sur l'idéal $(y_1,\dots,y_d)\mathcal{O}$.

ii) Posant $\mathcal{O}_1 = \mathbb{C}\{y_1,\dots,y_d\} \hookrightarrow \mathcal{O}$, et $\mathfrak{m}_1 = (y_1,\dots,y_d)\mathcal{O}_1$, le morphisme naturel d'algèbres graduées

$$\bigoplus_{\nu \geq 0} \mathfrak{m}_1^{\nu}/\mathfrak{m}_1^{\nu+1} = \mathrm{gr}_{\mathfrak{m}_1} \mathcal{O}_1 \longrightarrow \mathrm{gr}_{\mathfrak{m}} \mathcal{O} = \bigoplus_{\nu \geq 0} \mathfrak{m}^{\nu}/\mathfrak{m}^{\nu+1}$$

est fini.

Remarquons tout de suite que la condition ii) est équivalente à la condition géométrique suivante : soit $(X,0) \hookrightarrow (\mathbb{C}^N,0)$ le plongement du germe $(X,0)$ correspondant à l'algèbre $\mathcal{O}$, et soit $\pi : \mathbb{C}^N \to \mathbb{C}^d$ la projection linéaire définie

par $(\mathrm{in}_{\mathfrak{m}}\, y_1,\dots,\mathrm{in}_{\mathfrak{m}}\, y_d)$, où $\mathrm{in}_{\mathfrak{m}}\, y_i \in \mathfrak{m}/\mathfrak{m}^2$ désigne la forme initiale (classe de y_i modulo $\mathfrak{m}^2$), et soit $C_{X,x} \subset \mathbb{C}^N$ le cône tangent $\mathrm{Spec}\, \mathrm{gr}_{\mathfrak{m}}\, \mathcal{O}$ à X en 0. Alors $|\mathrm{Ker}\, \pi \cap C_{X,x}| = \{0\}$, c'est-à-dire que Ker π est transverse à X en 0. On dit aussi que π est une projection transversale. Considérons maintenant le sous-espace Y de X défini par l'idéal I engendré par $d-1$ éléments $(y_1,\dots,y_{d-1})$, et munissons X de la projection $F : X \to \mathbb{C}$ définie par y_d. On suppose que $(y_1,\dots,y_d)$ est un idéal primaire de $\mathcal{O}_{X,x}$.
Soit $p : X' \to X$ l'éclatement de Y dans X ; on a $X' \subset X \times \mathbb{P}^{d-2}$, donc $\dim p^{-1}(0) = d-2$, et d'après le résultat précédent, pour tout représentant assez petit de X on a donc $e(I \cdot \mathcal{O}_{X(0)}) = e(I \cdot \mathcal{O}_{X(\eta_d)})$, pour tout $\eta_d \in \mathbb{C}$ assez petit. En particulier si $(y_1,\dots,y_d)$ définissent une projection transversale $\pi : X \to \mathbb{C}^d$, et si la droite de $\mathbb{C}^d$ définie par $y_1 = \dots = y_{d-1} = 0$ n'est pas contenue dans le discriminant de π, on a $e(I \cdot \mathcal{O}_{X(0)}) = m_0(X)$, multiplicité de X en 0, puisque $\mathfrak{m}$ est entier sur $(y_1,\dots,y_d)\mathcal{O}_X$, (cf. § 4) et pour $\eta_d \neq 0$, le sous-espace de $\mathbb{C}^N$ défini par $y_1 = \dots = y_{d-1} = 0$, $y_d = \eta_d$, coupe X en des points non-singuliers de X, et transversalement ; en chacun de ces points, disons $x_i(\eta_d)$ on a $e(I \cdot \mathcal{O}_{X(\eta_d),x_i}) = 1$, et ces points sont donc au nombre de $m_0(X)$. On a ainsi vérifié en particulier :

5.3 La multiplicité $m_0(X)$ de $X \subset \mathbb{C}^N$ est le nombre des points en lesquels un sous-espace linéaire H_{η_d} de dimension $N-d$, voisin de 0 et parallèle à une direction transverse à X en 0, coupe X, (et qui tendent vers 0, quand $\eta_d \to 0$). C'est aussi le degré local en x d'une projection transversale $\pi : (X,x) \to (\mathbb{C}^d,0)$, c'est-à-dire la multiplicité dans l'anneau local $\mathcal{O}_{X,x}$ de l'idéal engendré par les composées avec π des fonctions coordonnées sur $(\mathbb{C}^d,0)$.

5.4 Corollaire : Etant donné $(X,0) \subset (\mathbb{C}^N,0)$, pour chaque entier k, $0 \leq \dim X$, il existe un ouvert de Zariski dense U de la grassmannienne des plans de codimension k dans $\mathbb{C}^N$ tel que pour tout sous-espace non-singulier de codimension k $(L,0) \subset (\mathbb{C}^N,0)$ tel que $T_{L,0} \in U$, on ait l'égalité $m_0(L \cap X) = m_0(X)$.

C'est pourquoi nous n'aurons aucun mal, dans la preuve ci-dessous, à choisir des rétractions locales $\rho : \mathbb{C}^N \to Y$, $Y \subset X$ sous-espace non-singulier, telles que $m_y(\rho^{-1}(y) \cap X) = m_y(X)$, pour $y \in Y$.

Remarque : L'avatar algébrique, plus précis, du résultat ci-dessus est l'existence d'éléments superficiels (cf. [Bbk 3], § 7).

5.5 Corollaire ("nonséparation", cf. [Te 11], 2.8.5 et Th. 5, [Li 1], 4.3) : Soient X un espace analytique complexe, $Y \subset X$ un sous-espace non-singulier, et $0 \in Y$. Pour tout plongement local $(X,0) \subset (\mathbb{C}^N,0)$, considérons les sous-espaces non-singuliers $(H,0) \subset (\mathbb{C}^N,0)$ contenant Y, de dimension $N - \dim X + \dim Y$ et tels que $\dim(X \cap H) = \dim Y$.

a) Si l'application $Y \to \mathbb{N}$ définie par $y \mapsto m_y(X)$ est localement constante sur Y au voisinage de 0, c'est-à-dire si X est équimultiple le long de Y en 0, pour tout sous-espace non-singulier $H \subset \mathbb{C}^N$ contenant Y et tel que $|T_{H,0} \cap C_{X,0}| = T_{Y,0}$, on a l'égalité

$$|H \cap X| = Y$$

au voisinage de x, et de plus on a $|T_{H,y} \cap C_{X,y}| = T_{Y,y}$ pour tout $y \in Y$ voisin de 0.

b) Inversement, si il existe un ouvert de Zariski dense U dans la grassmannienne des sous-espaces linéaires de $\mathbb{C}^N$ de dimension $N - \dim X + \dim Y$ tel que, pour tout sous-espace non-singulier $(H,0)$ de $(\mathbb{C}^N,0)$ de dimension $N - \dim X + \dim Y$, contenant Y, et tel que $T_{H,0} \in U$, on ait l'égalité $|H \cap X| = Y$, alors X est équimultiple le long de Y en 0.

Prouvons a) : On a $\dim C_{H\cap X,0} = \dim Y$ puisque $C_{H\cap X,0} \subseteq T_{H,0} \cap C_{X,0}$; puisque $(H \cap X,0)$ se spécialise platement sur son cône tangent, on a donc aussi $\dim(H \cap X) = \dim Y$. Nous pouvons donc choisir une rétraction locale $\rho : (\mathbb{C}^N,0) \to (Y,0)$ telle que $\rho^{-1}(0)$ soit transverse à X, c'est-à-dire $m_0(\rho^{-1}(0) \cap X) = m_0(X)$ et que le morphisme $H \cap X \to Y$ induit par ρ soit fini. L'hypothèse implique que $\rho^{-1}(0) \cap H$ est transverse à $\rho^{-1}(0) \cap X$ dans $\rho^{-1}(0)$, car $|T_{\rho^{-1}(0)\cap H,0} \cap C_{\rho^{-1}(0)\cap X,0}| = \{0\}$. Il suffit alors de contempler le diagramme (où on a noté $X(y)$ pour $\rho^{-1}(y)\cap X$)

$$\begin{array}{ccccc} e(I\cdot \mathcal{O}_{X(0),0}) & \geq & e(I\cdot\mathcal{O}_{X(y),y}) & + & \sum_{x_i\in\rho^{-1}(y)\cap H\cap X-Y} e(I\cdot\mathcal{O}_{X(y),x_i}) \\ \| & & \vee\!\!/ & & \\ m_0(X) & = & m_y(X) & , & \end{array}$$

où la première inégalité vient de la semi-continuité de la multiplicité, conséquence directe du résultat précédent. (En fait, dans ce cas-ci, on a même égalité), et l'égalité verticale vient de l'hypothèse de transversalité. Puisque par ailleurs $e(I\cdot\mathcal{O}_{X(y),y}) \geq m_y(X)$, on en conclut $\rho^{-1}(t)\cap H\cap X - Y = \emptyset$ donc $|\rho^{-1}(y)\cap H\cap X| = \{y\}$ pour tout $y\in Y$, c'est-à-dire $|H\cap X| = Y$, et la transversalité.

Prouvons b) : Par semi-continuité de la multiplicité, si X n'était pas équimultiple le long de Y, on pourrait trouver une courbe non-singulière $(C,0)\subset(Y,0)$ telle que, pour une rétraction locale $\rho : (\mathbb{C}^N,0)\to(Y,0)$, vérifiant $m_y(\rho^{-1}(y)\cap X) = m_y(X)$, $y\notin C-\{0\}$, l'espace $X\cap\rho^{-1}(C)$ ne soit pas équimultiple le long de C. On se ramène ainsi à vérifier l'énoncé dans le cas où Y est une courbe. Dans ce cas, $H\cap X$ est défini par dim $X-1$ équations dans X, donc la fibre de son éclatement dans X est de dimension dim $X-2$. Par ailleurs, quitte à rétrécir l'ouvert U, on peut supposer que $\rho^{-1}(y)\cap H$ est transverse à $\rho^{-1}(y)\cap X$ en y pour tout $y\in Y-\{0\}$ (il suffit d'imposer que les transformés stricts H' et X' de H et X par l'éclatement de Y dans $\mathbb{C}^N$ soient disjoints au-dessus de $Y-\{0\}$ ce qui est clairement impliqué par une condition ouverte sur $T_{H,0}$) Puisque pour $T_{H,0}\in U$, on a $|H\cap X| = Y$ par hypothèse, le fait que la fibre au-dessus de 0 de l'éclatement de $H\cap X$ dans X soit de dimension $d-2$ et la transversalité nous donnent pour $y\in Y$, le diagramme suivant

$$\begin{array}{ccc} e(I\cdot\mathcal{O}_{X(0),0}) & = & e(I\cdot\mathcal{O}_{X(y),y}) \\ \vee\!\!/ & & \| \\ m_0(X) & \geq & m_y(X) \end{array}$$

d'où $m_o(X) = m_y(X)$, et le résultat cherché. ∎

Remarque : On pourrait aussi bien prouver ce résultat en considérant une projection générale $p : \mathbb{C}^N \to \mathbb{C}^{d+1}$ où $d = \dim X$; par les techniques exposées au Chapitre IV, on se ramène alors à prouver le résultat pour l'hypersurface $X_1 = p(X)$, et un calcul direct suffit alors.

§ 6. Saturation lipschitzienne.

Je rappelle ici les traits de la théorie, inspirée à F. Pham et moi par la théorie de la saturation de Zariski, qui nous serons utiles plus loin (cf. [P-T], [P], aussi [Li 2], [Bo]), et je signale une erreur d'un travail précédent qui m'a été signalée par Mr. E. Böger.

6.1 Soit X un espace analytique complexe réduit, et notons $n : \overline{X} \to X$ sa normalisation. Considérons l'immersion naturelle $i : \overline{X} \underset{X}{\times} \overline{X} \to \overline{X} \times \overline{X}$ du produit fibré dans le produit. Si l'espace X est séparé, ce que nous pouvons supposer pour simplifier puisque nous ne nous intéresserons qu'à des problèmes de nature locale, l'immersion i est une immersion fermée définie par un faisceau cohérent d'idéaux I. Considérons le sous-faisceau $\mathcal{O}_X^s$ du faisceau $\overline{\mathcal{O}}_X$ [des germes de fonctions méromorphes "faiblement holomorphes" c'est-à-dire localement bornées sur X] associé au préfaisceau défini par $U \to \{h \in \overline{\mathcal{O}}_X(U) / h \otimes 1 - 1 \otimes h \in \bar{I}(U)\}$ pour tout ouvert U de X, où $h \otimes 1 - 1 \otimes h$ désigne la différence des composés de h avec les deux projections $\overline{X} \times \overline{X} \overset{p_1}{\underset{p_2}{\rightrightarrows}} X$, c'est-à-dire $h \circ p_1 - h \circ p_2$, et $\bar{I}$ désigne la fermeture intégrale de l'idéal I.

Lemme : La $\mathcal{O}_X$-algèbre $\mathcal{O}_X^s$ est un $\mathcal{O}_X$-module cohérent.

Notons $\mathcal{O}_{\overline{X}} \hat{\otimes}_{\mathbb{C}} \mathcal{O}_{\overline{X}} = \mathcal{O}_{\overline{X}\times\overline{X}}$ le produit tensoriel analytique ; alors le $\mathcal{O}_X$-module $\mathcal{O}_X^s$ est le noyau du morphisme de $\mathcal{O}_X$-modules

$$\overline{\mathcal{O}}_X \xrightarrow{\varphi} \mathcal{O}_{\overline{X}\times\overline{X}} / \bar{I}$$

défini par $\varphi(f) = $ image de $f \otimes 1 - 1 \otimes f$. Or le $\mathcal{O}_X$-module $\mathcal{O}_{\overline{X}\times\overline{X}}/\overline{I}$ est quotient de $\mathcal{O}_{\overline{X}\times\overline{X}}/I$ par un idéal cohérent de $\mathcal{O}_{\overline{X}\times\overline{X}}/I \simeq \mathcal{O}_{\overline{X}\times_X\overline{X}}$ (puisque l'idéal $\overline{I}$ est cohérent) et la $\mathcal{O}_X$-algèbre $\mathcal{O}_{\overline{X}\times_X\overline{X}}$ est un $\mathcal{O}_X$-module cohérent puisque le produit fibré $\overline{X}\times_X\overline{X}$ est fini au-dessus de X ([Ca], exp. 19, No 5). Ceci prouve que $\mathcal{O}_X^s$ est le conoyau d'un homomorphisme de $\mathcal{O}_X$-modules cohérents, d'où le résultat.

<u>Définition</u> : Le morphisme fini $s : X^s \to X$, où $X^s = \mathrm{Specan}_{\mathcal{O}_X}\, \mathcal{O}_X^s$ est appelé saturation lipschitzienne de l'espace analytique complexe réduit X.

<u>Remarques</u> : X^s est dominé par la normalisation de X, donc s est biméromorphe et par construction du Specan ([Ca], exp. 19) on a un isomorphisme de $\mathcal{O}_X$-modules $s_* \mathcal{O}_{X^s} \simeq \mathcal{O}_X^s$.
La terminologie est justifiée par :

<u>6.1.1 Proposition</u> : <u>L'algèbre $\mathcal{O}_X^s$ est le faiceau des germes de fonctions méromorphes localement lipschitziennes sur</u> X.

<u>Démonstration</u> : Soit x un point de X, et soit $(X,x) \hookrightarrow (\mathbb{C}^N,0)$ un plongement local. Il suffit de prouver que si $h \in (\mathcal{O}_X^s)_x$, on peut trouver un représentant du germe (X,x) sur lequel h s'étend en une fonction méromorphe localement lipschitzienne. Puisque en particulier $h \in \overline{\mathcal{O}_{X,x}}$, on peut étendre h en une fonction méromorphe localement bornée sur X. Nous voulons montrer qu'il existe une constante $C > 0$ telle que pour tout couple $(x_1, x_2) \in X \times X$ on ait $|h(x_1) - h(x_2)| \leqslant C\, \|x_1 - x_2\|$. Notons $z_1, \ldots, z_N$ les fonctions coordonnées sur $\mathbb{C}^N$, et de même leur restriction à X. Au-dessus d'un voisinage de $X \times X$, l'idéal I dans $\mathcal{O}_{\overline{X}\times\overline{X}}$ est engendré par les différences $z_i' - z_i''$, où $z_i' = z_i \circ \overline{p}_1$, $z_i'' = z_i \circ \overline{p}_2$, et $\overline{p}_i : \overline{X} \times \overline{X} \to \overline{X} \to X$. Par définition, et le critère transcendental de dépendance intégrale, dire que $h \in (s_* \mathcal{O}_{X^s})_x$ équivaut à l'existence d'un représentant (X,x) tel que, pour $w \in \overline{X} \times \overline{X}$ on ait :

$$|h\circ\overline{p}_1(w) - h\circ\overline{p}_2(w)| < C \operatorname{Sup}_i |z_i'(w) - z_i''(w)| = C \operatorname{Sup}_i |z_i(\overline{p}_1(w)) - z_i(\overline{p}_2(w)|$$

$$= C\,.\,\|\overline{p}_1(w) - \overline{p}_2(w)\| \;.$$

Puisque le morphisme $\overline{X}\times\overline{X}\to X\times X$ est fini et surjectif, on en déduit aussitôt que h est bien localement lipschitzienne, et qu'inversement, en lisant cette preuve à l'envers, si h est méromorphe et localement lipschitzienne, le germe $(h)_x$ appartient à $(s_* \mathcal{O}_{X^s})_x$.

Exercices : 1) Démontrer de deux façons différentes (l'une fonctorielle et l'autre transcendante) qu'étant donné un morphisme $f : X\to Y$ entre espaces analytiques réduits qui se prolonge en un morphisme $\overline{f} : \overline{X}\to\overline{Y}$ entre les normalisés, il se prolonge en un morphisme $f^s : X^s\to Y^s$ entre les saturés lipschitziens.

2) Montrer que, pour tout $x\in X$, l'anneau $(\mathcal{O}_X^s)_x = \mathcal{O}_{X,x}^s$ est une algèbre analytique, donc en particulier local.

Exemple : Soit $(X,0)\subset(\mathbb{C}^N,0)$ un germe de courbe analytique réduite. Soit $X=\bigcup_i C_i$ sa décomposition en composantes irréductibles, et décrivons $(C_{i,0})$ par une représentation paramétrique, c'est-à-dire comme image réduite d'un morphisme $(\mathbb{C},0)\to(\mathbb{C}^N,0)$ décrit par N fonctions $z_k(t_i)\in\mathbb{C}\{t_i\}$. Alors $\overline{X}=\coprod_i(\mathbb{C},0)$ et l'algèbre de $\overline{X}$ le long de l'ensemble fini $n^{-1}(0)$ est $\mathcal{O}_{\overline{X},n^{-1}(0)}\cong\prod_i\mathbb{C}\{t_i\}$. Par conséquent, l'algèbre $\mathcal{O}_{\overline{X}\times\overline{X},n^{-1}(0)\times n^{-1}(0)}$ est isomorphe à $\prod_{i,j}\mathbb{C}\{t_i,t_j'\}$, l'idéal I définissant le germe de $\overline{X}\times_X\overline{X}$ le long de $n^{-1}(0)\times n^{-1}(0)$ est déterminée par la collection de ses images $I_{i,j}$ dans chacune des algèbres $\mathbb{C}\{t_i,t_j'\}$ et l'on a $I_{i,j}=(\{z_k(t_i)-z_k(t_j')\;;\;1\le k\le N\})\mathbb{C}\{t_i,t_j'\}$ (ici $(\{\ \})$ désigne l'idéal engendré par les éléments de l'ensemble $\{\ \}$). Si l'on considère $h\in\prod\mathbb{C}\{t_i\}\simeq\mathcal{O}_{\overline{X},n^{-1}(0)}$, on a donc $h\in\mathcal{O}_X^s$ si et seulement si, pour tout couple (i,j), on a l'inclusion : $h(t_i)-h(t_j')\in\overline{I_{i,j}}$.

6.2 Soit maintenant $(X,0)\subset(\mathbb{C}^N,0)$ un germe d'espace réduit. Pour étudier la saturation lipschitzienne de X, on est amené à étudier l'éclatement normalisé dans $\overline{X}\times\overline{X}$ de l'idéal I définissant $\overline{X}\times_X\overline{X}$ (cf. I, § 1, Prop. 1). Or, c'est aussi l'éclatement normalisé dans $X\times X$ de l'idéal définissant la diagonale $\Delta_X\subset X\times X$, et cet éclatement-ci peut être décrit comme ceci : choisissons des coordonnées locales $z_1,\dots,z_N$ sur $\mathbb{C}^N$, et considérons le morphisme dir. : $X\times X-\Delta_X\to\mathbb{P}^{N-1}$ qui à $(x,x')\in X\times X-\Delta_X$ associe la direction de la sécante joignant x à x' dans $\mathbb{C}^N$; l'éclatement E_X cherché est isomorphe à l'adhérence dans $X\times X\times\mathbb{P}^{N-1}$ du graphe du morphisme dir., on a donc un diagramme

$$\begin{array}{ccc} E_X & \xrightarrow{\text{dir.}} & \mathbb{P}^{N-1} \\ \downarrow{\scriptstyle b} & & \\ X\times X & & \end{array}$$

avec $\dim E_X=2d$, où $d=\dim X$.

L'ensemble des directions limites en 0 de sécantes peut être identifié au sous-ensemble algébrique $|b^{-1}(0\times 0)|\subset\mathbb{P}^{N-1}$ et l'on a l'inégalité $\dim b^{-1}(0\times 0)\leq 2d-1$ puisque cet ensemble est contenu dans le diviseur exceptionnel. Pour tout ceci, dans le contexte de la saturation lipschitzienne, voir ([P-T]). On a alors :

6.2.1 Proposition : Soit $(X,0)\subset(\mathbb{C}^N,0)$ un germe de courbe réduite et soit $p:\mathbb{C}^N\to\mathbb{C}^2$ une projection linéaire. Notons $X_1=p_*X$ l'image de X dans $\mathbb{C}^2$ (définie par le 0-ième idéal de Fitting $F_0(p_*\mathcal{O}_X)$, (cf. [Te 3]). Les conditions suivantes sont équivalentes :

1) La courbe X_1 est réduite, le morphisme $p|X=\pi: X\to X_1$ est fini biméromorphe, et l'extension naturelle $\pi^s: X^s\to X_1^s$ aux saturés est un isomorphisme.

2) Aucune direction limite en 0 de sécantes à X n'est contenue dans le noyau de la projection p.

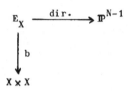

Prouvons 1) ⇒ 2) : Puisque π est fini et biméromorphe, le morphisme π s'étend naturellement en un isomorphisme $\overline{\pi}:\overline{X}\xrightarrow{\sim}\overline{X}_1$ des normalisations. Choisissons des coordonnées locales $(z_1,\dots,z_N)$ sur $\mathbb{C}^N$ telles que la projection p soit

définie par $(z_1,\dots,z_N) \mapsto (z_1,z_2)$. Soit I l'idéal de $\mathcal{O}_{\overline{X}\times\overline{X}}$ engendré par les différences $(z_1 - z'_1,\dots,z_N - z'_N)$, idéal qui définit $\overline{X}\times_X\overline{X}$, et soit I_1 l'idéal engendré par $(z_1 - z'_1, z_2 - z'_2)$, idéal qui définit $\overline{X}\times_{X_1}\overline{X}$. Dire que l'extension π^s est un isomorphisme équivaut donc, au vu de la définition et du fait que l'on peut identifier $\overline{X}$ et $\overline{X}_1$, à dire que l'on a l'égalité de clôtures intégrales $\overline{I} = \overline{I}_1$. Ceci peut se traduire par le critère valuatif (1.3.4), ou comme ceci : il existe une constante $C > 0$ telle que pour toute suite $(p_n,p'_n) \in X\times X - \Delta_X$ de couples de points, tendant vers (0,0), on ait, pour n assez grand, les inégalités

$$|z_k(p_n) - z_k(p'_n)| \le C \cdot \operatorname{Sup}\{|z_1(p_n) - z_1(p'_n)|, |z_2(p_n) - z_2(p'_n)|\}$$

pour $3 \le k \le N$.

Ceci implique que les points de $\mathbb{P}^{N-1}$ représentant les directions des sécantes $\widehat{p_n\,p'_n}$, qui ont pour coordonnées projectives

$$(z_1(p_n) - z_1(p'_n) : z_2(p_n) - z_2(p'_n) : \dots : z_N(p_n) - z_N(p'_n))$$

ne peuvent tendre vers une limite ayant pour coordonnées projectives

$$(0 : 0 : c_3, \dots : c_N)$$

c'est-à-dire ne peuvent tendre vers une limite contenue dans le noyau de p.

Prouvons 2) ⇒ 1) : D'après l'hypothèse 2), tous les couples de points suffisamment voisins de 0 et distincts de 0×0 ont des images distinctes dans $\mathbb{C}^2$ par p, donc π induit un isomorphisme sur un ouvert dense de chaque composante de X, et est en particulier un morphisme fini puisque X est de dimension 1. Par conséquent π est un morphisme fini, son image est réduite puisque c'est une courbe plane qui ne peut avoir de composante immergée parce que X est de Cohen-Macaulay (cf. [Te 3], § 3) et que la définition par idéal de Fitting implique que si une composante irréductible de X_1 n'était pas réduite, au-dessus d'un point général de cette composante il y aurait au moins deux points de X (cf. Loc. cit.). Par conséquent, X_1 est réduite, π est fini et biméromorphe et donc induit un isomorphisme $\overline{\pi} : \overline{X} \xrightarrow{\sim} \overline{X}_1$ des normalisations. Il nous reste à montrer l'égalité $\overline{I}_1 = \overline{I}$ qui impliquera l'isomorphisme des saturations : si cette égalité n'avait pas lieu, d'après le critère valuatif, il existerait un morphisme analytique $(\mathbb{D},0) \xrightarrow{h} \overline{X} \times \overline{X} \xrightarrow{n} X \times X$ tel que $n \circ h(\mathbb{D} - \{0\}) \subset X \times X - \Delta_X$ et que notant $z_i(u)$ et $z_i'(u)$ les éléments de $\mathcal{O}_{\mathbb{D},0} = \mathbb{C}\{u\}$ donnés par $z_i \circ p_1 \circ n \circ h$ et $z_i \circ p_2 \circ n \circ h$ respectivement, on ait, pour au moins un entier k, $3 \leq k \leq N$ l'inégalité de valuations u-adiques

$$v(z_k(u) - z_k'(u)) < \inf\{v(z_1(u) - z_1'(u)), v(z_2(u) - z_2'(u))\}$$

en divisant $(z_1(u) - z_1'(u) : z_2(u) - z_2'(u) : \ldots : z_N(u) - z_N'(u))$ par la plus grande puissance de u qui divise toutes les coordonnées, on voit que la limite dans $\mathbb{P}^{N-1}$, qui correspond à la direction limite des sécantes $\widehat{\underline{z}(u)\underline{z}'(u)}$ est de la forme $(0 : 0 : c_3 : \ldots : c_N)$, donc appartient au noyau de p, d'où la contradiction cherchée. ∎

<u>Définition</u> : Etant donnée une courbe $(X,0) \subset (\mathbb{C}^N,0)$, une projection linéaire $p : \mathbb{C}^N \to \mathbb{C}^2$ est dite générale pour X si elle satisfait les conditions équivalentes de la Proposition. Une projection $p : \mathbb{C}^N \to \mathbb{C}^2$ (i.e., submersion en 0) est dite générale pour une courbe $(X,0) \subset (\mathbb{C}^N,0)$ si elle est conjuguée à une projection linéaire générale pour X par des changements de coordonnées sur $\mathbb{C}^N$ et $\mathbb{C}^2$ tangents à l'identité en 0.

Cette définition est justifiée par le fait que la réunion des limites en 0 de sécantes à X, qui est le cône sur $|b^{-1}(0 \times 0)| \subset \mathbb{P}^{N-1}$ avec les notations introduites plus haut, est de dimension au plus 2, car dim $b^{-1}(0 \times 0) \leq 1$ et donc pour une projection linéaire p générale, le noyau de p qui est de codimension 2 ne rencontre ce cône qu'en 0.

<u>Remarque</u> : On peut paraphraser une partie de ce qui précède en disant que pour une courbe $(X,0) \subset (\mathbb{C}^N,0)$, une projection générale $\pi : X \to X_1 \subset \mathbb{C}^2$ non seulement est un homéomorphisme, mais est un homéomorphisme <u>lipschitzien</u>, c'est-à-dire que le morphisme analytique π admet un inverse lipschitzien. De plus, le

morphisme $s : X^s \to X$ est maximal pour la relation de domination parmi ceux qui factorisent la normalisation et qui sont des projections générales.

6.3 Algèbres analytiques saturées de dimension 1.

Nous allons rappeler ici sans démonstration complète, d'après Zariski ([G-T-S I et II) et Pham-Teissier (cf. [BGG] Appendice), la structure des algèbres analytiques saturées de dimension 1. D'après ce qui précède, toute algèbre analytique saturée de dimension 1 est la saturation de l'algèbre d'un germe de courbe plane réduite.

Soit donc $\mathcal{O} = \mathcal{O}_{X,0}$ l'algèbre d'un germe de courbe plane réduite, d'équation $f(z_1,z_2) = 0$, où $f = f_1 \dots f_r$, chaque f_i étant une série convergente irréductible. Alors la fermeture intégrale de $\mathcal{O}$ dans son anneau total de fractions est isomorphe à un produit direct $\prod_{i=1}^{r} \mathbb{C}\{t_i\}$, et l'on peut choisir les uniformisantes locales t_i de telle façon que l'élément de $\mathcal{O}_{X,0} \hookrightarrow \prod_{i=1}^{r} \mathbb{C}\{t_i\}$ image de z_1 ait pour décomposition $(t_1^{n_1}, t_2^{n_2}, \dots, t_r^{n_r})$ dans $\overline{\mathcal{O}_{X,0}}$, où n_i est la multiplicité en 0 de la i-ème branche (c'est-à-dire composante irréductible) de X en 0, qui est définie par $f_i(z_1,z_2) = 0$. Soit $(z_2(t_1), z_2(t_2), \dots, z_2(t_r))$ la décomposition de l'élément correspondant à l'image de z_2 dans $\mathcal{O}_{X,0}$.

Alors, comme nous l'avons vu plus haut, l'idéal I de $\prod \mathbb{C}\{t_i, t'_j\}$ définissant $\overline{X} \times_X \overline{X}$ dans $\overline{X} \times \overline{X}$ est déterminé par les idéaux

$$I_{i,j} = (t_i^{n_i} - t_j'^{n_j}, z_2(t_i) - z_2(t'_j)) \subset \mathbb{C}\{t_i, t'_j\} .$$

Soit p le p.p.c.m. des multiplicités n_i, et pour chaque i, $1 \leq i \leq r$, soit $(n_i, \beta_1^{(i)}, \dots, \beta_{g_i}^{(i)})$ la suite des exposants caractéristiques de Puiseux du germe de courbe plane irréductible X_i défini par $f_i(z_1,z_2) = 0$. Soit $e_k^{(i)}$ le p.g.c.d. de $(n_i, \beta_1^{(i)}, \dots, \beta_k^{(i)})$ ($e_o^{(i)} = n_i$, $e_{g_i}^{(i)} = 1$) et soit $\widetilde{E}_i \subset \mathbb{N}$ le sous-ensemble

$$(n_i, 2n_i, 3n_i, \dots, \beta_1^{(i)}, \beta_1^{(i)} + e_1^{(i)}, \beta_1^{(i)} + 2e_1^{(i)}, \dots, \beta_2^{(i)}, \beta_2^{(i)} + e_2^{(i)},$$

$$\beta_2^{(i)} + 2e_2^{(i)}, \dots, \beta_3^{(i)}, \dots, \beta_{g_i}^{(i)}, \beta_{g_i}^{(i)} + 1, \dots)$$

où chaque symbole " ... " désigne une suite a priori infinie. En fait notre sous-ensemble contient tous les entiers à partir de $\beta^{(i)}_{g_i}$. Il faut aussi remarquer que c'est un semi-groupe.

D'autre part, pour chaque racine p-ième de l'unité ω et chaque élément $h = (h_1,\dots,h_r) \in \overline{\mathcal{O}_{X,0}}$, on associe à chaque couple i,j l'entier

$m_{i,j,\omega}(h) = v(h_i(\tau^{p/n_i}) - h_j((\omega\tau)^{p/n_j}))$ où v désigne la valuation τ-adique et $h_i(\tau^{p/n_i})$ désigne la série en τ obtenue en substituant τ^{p/n_i} à t_i.

On note $m_{i,j,\omega}$ le nombre $\inf\limits_{h\in\mathcal{O}_{X,0}} (m_{i,j,\omega}(h))$.

<u>6.3.1</u> <u>Théorème</u> (Zariski ([G-T-S II], Th. 3.1), Pham-Teissier (voir [B-G-G], Appendice) : <u>Etant donnée l'algèbre $\mathcal{O}_{X,0}$ d'un germe de courbe plane réduite, une présentation $\overline{\mathcal{O}_{X,0}} = \prod_i \mathbb{C}\{t_i\}$ où l'image de z_1 dans $\overline{\mathcal{O}_{X,0}}$ est $(t_1^{n_1},\dots,t_r^{n_r})$, n_i étant la multiplicité de la i-ème composante irréductible de X, on a</u> :

<u>un élément $h = (h_1,\dots,h_r) \in \overline{\mathcal{O}_{X,0}}$, $h_i = \Sigma h_{i,a} t_i^a$ $(h_{i,a} \in \mathbb{C})$ appartient au saturé $\mathcal{O}^s_{X,0}$ de $\mathcal{O}_{X,0}$ si et seulement si</u> :

α) <u>Pour chaque i, on a $h_{i,a} = 0$ si $a \notin \widetilde{E}_i$</u>.

β) <u>Pour chaque racine p-ième de l'unité ω et chaque couple (i,j), $i \neq j$, on a l'inégalité</u>

$$m_{i,j,\omega}(h) \geq m_{i,j,\omega} .$$

Voici l'esquisse d'une démonstration : puisque chacun des idéaux $I_{i,j} = (t_i^{n_i} - t_j'^{n_j}, z_2(t_i) - z_2(t_j'))$ est engendré par une suite régulière, d'après la Remarque suivant la Proposition 2 du § 1, (1.4)) pour vérifier que l'élément $(h\otimes 1 - 1\otimes h')$ $\mathbb{C}\{t_i,t_j'\}$ est entier sur $I_{i,j}$, il suffit de le vérifier en restriction à une courbe $\lambda(t_i^{n_i} - t_j'^{n_j}) + \mu(z_2(t_i) - z_2(t_j'))$ où $(\lambda:\mu)\in\mathbb{P}^1$ est assez général. Or on vérifie facilement que cette famille de courbes paramétrée par $\mathbb{P}^1$ (qui n'est autre que l'espace total de l'éclaté de $I_{i,j}$ dans $\mathbb{C}^2$) est équisingulière au voisinage de $\lambda = 0$ et que le point $\lambda = 0$ est contenu dans l'ouvert $V(h\otimes 1 - 1\otimes h')$. On est donc ramené à décider si en restriction à la courbe $t_i^{n_i} - t_j'^{n_j} = 0$ de l'élément $(h \otimes 1 - 1 \otimes h')_{i,j}$ est entière sur $I_{i,j}$ (pour tout (i,j)), et le Théorème résulte presque aussitôt de là, en remarquant que $t_i = \tau^{p/n_i}$, $t_j' = (\omega\tau)^{p/n_j}$, $\omega^p = 1$ paramétrise les branches de la courbe $t_i^{n_i} - t_j'^{n_j} = 0$.

3.4 Saturation relative et équisaturation.

Soit $f : X \underset{\sigma}{\leftrightarrows} Y$ un morphisme d'espaces analytiques complexes muni d'une section σ. On suppose X réduit et Y non-singulier. Pour les applications que nous avons en vue, on supposera même que $\dim X = \dim Y + 1$ et que pour tout $y \in Y$, la fibre $(X_y, \sigma(y))$ est de dimension 1 et a une dimension de plongement ≤ 2, (c'est-à-dire que les fibres sont des courbes planes). On suppose pour l'instant f séparé et l'on considère l'idéal I_f de $\mathcal{O}_{\overline{X} \times_Y \overline{X}}$ définissant le plongement fermé $\overline{X} \times_X \overline{X} \hookrightarrow \overline{X} \times_Y \overline{X}$. On considère à nouveau le sous-faisceau de $\overline{\mathcal{O}}_X$ associé au pré-faisceau défini par $U \mapsto \{h \in \overline{\mathcal{O}}_X(U) / h \otimes 1 - 1 \otimes h \in \overline{I}_f(U)\}$ et on le note $\mathcal{O}_X^{s(f)}$: c'est l'algèbre saturée lipschitzienne relative, et l'on vérifie sans mal que c'est le faisceau des germes de fonctions méromorphes sur X qui satisfont localement une inégalité de Lipschitz pour les couples de points situés dans une même fibre de f.

Soit $f : X \underset{\sigma}{\leftrightarrows} Y$ un morphisme plat muni d'une section σ avec Y non-singulier, X réduit, et $\dim X = \dim Y + 1$, et soit $y \in Y$ un point tel que la fibre $X(y) = f^{-1}(y)$ soit une courbe réduite en $\sigma(y)$. Alors d'une part on a pour la normalisation un diagramme

$$\begin{array}{ccc} \overline{X(y)} & \longrightarrow & \overline{X}(y) \\ & \underset{n}{\searrow} & \downarrow \\ & & X(y) \end{array}$$

de morphismes dominants (cf. [Te 2], I ; la preuve donnée dans le cas où $\dim Y = 1$ s'étend aussitôt) correspondant au diagramme d'algèbres suivant, où $\mathcal{O} = \mathcal{O}_{X,\sigma(y)}$, $\mathfrak{m}_y$ est l'idéal définissant $\{y\}$ dans Y au voisinage de $y \in Y$

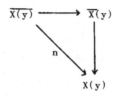

et $\mathfrak{m}_y \cdot \mathcal{O}$ son image par le morphisme $f^* : \mathcal{O}_{Y,y} \to \mathcal{O}_{X,\sigma(y)}$]

$$\begin{array}{ccc} & \overline{\mathcal{O}}/\mathfrak{m}_y \cdot \overline{\mathcal{O}} & \\ \nearrow & & \searrow \\ \mathcal{O}/\mathfrak{m}_y \cdot \mathcal{O} & \longrightarrow & \overline{\mathcal{O}/\mathfrak{m}_y \cdot \mathcal{O}} \end{array}$$

et d'autre part on a un diagramme analogue pour les saturations relatives

$$\begin{array}{ccc} & (s_* \mathcal{O}_{X^{s(f)}(y)})_{\sigma(y)} & \\ \nearrow & & \searrow \\ \mathcal{O}/\mathfrak{m}_y \cdot \mathcal{O} & \longrightarrow & s(y)_*(\mathcal{O}_{(X(y))^s}) \end{array}$$

c'est-à-dire

$$\begin{array}{ccc} X(y)^s & \longrightarrow & X^{s(f)}(y) \\ & {\scriptstyle s(y)} \searrow & \downarrow {\scriptstyle s} \\ & & X(y) \end{array} .$$

La vérification est très facile au vu de (Loc. cit.) et de la définition de la saturation.

<u>6.4.1</u> <u>Définition</u> : On dit que le morphisme $f : X \overset{\sigma}{\leftrightarrows} Y$ est équisaturé au point $\sigma(y) \in X$ si la fibre $X(y)$ est réduite et si l'on a :

a) Le morphisme canonique $X^{s(f)}(y) \to X(y)^s$ est un isomorphisme, et en particulier il y a un seul point $y_1 \in X^{s(f)}$ au-dessus de $\sigma(y)$.

b) Le morphisme composé $X^{s(f)} \xrightarrow{s} X \xrightarrow{f} Y$ est localement analytiquement trivial en y_1.

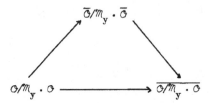

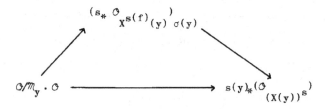

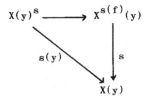

6.4.2 Proposition : Soit $f : X \underset{\sigma}{\rightleftarrows} Y$ un morphisme comme ci-dessus tel que les fibres soient des courbes planes. Il existe un fermé analytique rare $F \subset Y$ tel que le morphisme f soit équisaturé en tout point de $\sigma(Y - F)$.

Démonstration : D'après la semi-continuité du nombre de Milnor (cf. [Te 11], 2.3.1), de la multiplicité des fibres ([Le-T 2]), et enfin par la platitude générique du $\mathcal{O}_Y$-module cohérent $f_*(n_* \mathcal{O}_{\overline{X}} / \mathcal{O}_X)$, il existe un fermé analytique rare F de Y tel qu'au voisinage de tout point $y \in Y - F$, le nombre de Milnor $\mu_{\sigma(y')}(X(y'))$ des fibres (courbes planes réduites) soit constant, (indépendant de y') ainsi que leur "invariant δ", $\delta_{y'} = \dim_{\mathbb{C}}(\overline{\mathcal{O}_{X(y')}} / \mathcal{O}_{X(y')})_{\sigma(y')}$ et la multiplicité $n_{y'}$ au point $\sigma(y')$ de la courbe $X(y')$. D'après l'égalité de Jung-Milnor $2\delta_{y'} = \mu_{\sigma(y')}(X(y')) + r_{y'} - 1$ où $r_{y'}$ est le nombre des composantes irréductibles en $\sigma(y')$ de $X(y')$, ce nombre de composantes est également indépendant de $y' \in Y - F$.

D'après ([Te 2], I), puisque $\delta_{y'}$ est constant sur $\sigma(Y)$ au voisinage de $\sigma(y)$, le morphisme f admet une résolution simultanée très faible donnée par la normalisation $n : \overline{X} \to X$. Puisque le nombre des branches $r_{y'}$ et la multiplicité $n_{y'}$ sont aussi constants, c'est en fait une résolution simultanée forte c'est-à-dire que si nous notons S l'idéal définissant $\sigma(Y)$ dans X, l'idéal $S\mathcal{O}_{\overline{X}}$ est inversible, définit un sous-espace $n^{-1}(\sigma(Y))$ tel que le morphisme induit $|n^{-1}(\sigma(Y))| \to \sigma(Y)$ soit étale, $n^{-1}(\sigma(y))$ contenant $r = r_y$ points, et donc finalement on a un isomorphisme de $\mathcal{O}_{Y,y}$-algèbres

$$\overline{\mathcal{O}_{X,\sigma(y)}} \xrightarrow{\sim} \prod_{i=1}^{r} \mathcal{O}_{Y,y}\{t_i\}$$

tel que, pour un Y-plongement local $X \hookrightarrow Y \times \mathbb{C}^2$ envoyant $\sigma(Y)$ sur $Y \times 0$, et où $\mathbb{C}^2$ est muni de coordonnées (z_1, z_2), l'idéal engendré par les images de (z_1, z_2) soit inversible, engendré disons par z_1. Soit $(\zeta_1, \ldots, \zeta_r) \in \prod_{i=1}^{r} \mathcal{O}_{Y,y}\{t_i\}$ la décomposition dans $\mathcal{O}_{\overline{X},\sigma(y)}$ de l'image de z_1. L'hypothèse d'équimultiplicité implique, comme on le vérifie aussitôt, qu'au prix d'un changement de l'uniformisante locale t_i, l'on peut supposer que $\zeta_i = t_i^{n_i}$.

Considérons maintenant l'idéal I définissant $\overline{X} \times_X \overline{X}$ dans $\overline{X} \times_Y \overline{X}$: clairement l'algèbre de $\mathcal{O}_{\overline{X} \times_Y \overline{X}, n^{-1}(\sigma(y)) \times n^{-1}(\sigma(y))}$ est $\mathcal{O}_{Y,y}$-isomorphe au produit $\prod_{i,j} \mathcal{O}_{Y,y}\{t_i, t'_j\}$ et l'idéal I est déterminé par ses images $I_{i,j} \subset \mathcal{O}_{Y,y}\{t_i, t'_j\}$.

Soit $(H_1, \ldots, H_r) \in \prod_i \mathcal{O}_{Y,y}\{t_i\}$ l'image de z_2 ; $H_i \in \mathcal{O}_{Y,y}\{t_i\}$. L'idéal $I_{i,j}$ est engendré par $(t_i^{n_i} - t_j'^{n_j}, H_i(t_i) - H_j(t'_j))$. Soit maintenant $\mathcal{N}$ l'idéal de $\prod_{i,j} \mathcal{O}_{Y,y}\{t_i, t'_j\}$ défini comme suit : si $i \neq j$, $\mathcal{N}_{i,j} = I_{i,j}$ et si $i = j$, $(t_i - t'_i) \cdot \mathcal{N}_{i,i} = I_{i,i}$. D'après ([Te 2], II, § 5) l'idéal $\mathcal{N}$ est équimultiple le long du sous-espace de $\overline{X} \times_Y \overline{X}$ défini par l'idéal $\prod_{i,j} (t_i, t'_j)$, en ce sens que pour chaque $y' \in Y$ voisin de y, la somme des multiplicités des idéaux $\mathcal{N}_{i,j} \cdot \mathcal{O}_{Y/\mathfrak{m}_{y'}, \mathcal{O}_Y}\{t_i, t'_j\}$ est égale à $2\delta_{y'}$, donc indépendante de $y' \in Y$ voisin de y.

Voyons maintenant comment déterminer si un élément $(h_1, \ldots, h_r) \in \prod \mathcal{O}_{Y,y}\{t_i\}$ appartient au saturé relatif $\mathcal{O}^{s(f)}_{X,\sigma(y)} \subset \prod_i \mathcal{O}_{Y,y}\{t_i\}$. Si nous posons $h_i = \Sigma\, h_{i,a}\, t_i^a$, $h_{i,a} \in \mathcal{O}_{Y,y}$, il faut déterminer si $\Sigma\, h_{i,a}\, t_i^a - \Sigma\, h_{j,a}\, t_j'^a \in \overline{I_{i,j}}$ pour tous les couples (i,j). Remarquons que si $i = j$, il revient au même de déterminer si $\Sigma\, h_{i,a} \dfrac{t_i^a - t_i'^a}{t_i - t'_i}$ appartient à $\overline{\mathcal{N}_{i,i}}$.

Puisque l'idéal $\mathcal{N}$ définit une famille paramétrée par Y d'idéaux, qui est équimultiple le long du sous-espace défini par l'idéal $\prod_{i,j} (t_i, t'_j)$, parce que la multiplicité $e(\mathcal{N}_{y'})$ de l'idéal induit par $\mathcal{N}$ dans la fibre de $\overline{X} \times_Y \overline{X} \to Y$ au-dessus de y' vaut $2\delta_{y'}$, qui est constant (pour tout ceci, voir [Te 2], II), il résulte du principe de spécialisation de la dépendance intégrale (§ 5.1) qu'il suffit de vérifier la relation de dépendance intégrale dans une fibre générale, et de là, d'après le théorème de structure des anneaux saturés vu plus haut, on déduit que, si l'on prend une base $\frac{\partial}{\partial y_1}, \ldots, \frac{\partial}{\partial y_k}$ des dérivations de $\mathcal{O}_{Y,y}$, et si l'on note D_ℓ la dérivation de $\prod \mathcal{O}_{Y,y}\{t_i\}$ obtenue en posant $D_\ell\, t_i = 0$ $(1 \le i \le r)$ et $D_\ell | \mathcal{O}_{Y,y} = \frac{\partial}{\partial t_\ell}$, on a l'inclusion

$$D_\ell(\mathcal{O}^{s(f)}_{X,\sigma(y)}) \subset \mathcal{O}^{s(f)}_{X,\sigma(y)} \qquad (1 \leq \ell \leq k)$$

(en effet $D_\ell(\sum_a h_{i,a}\ t_i^a) = \sum (\frac{\partial}{\partial t_\ell} h_{i,a})\ t_i^a$ ne contient pas d'exposant que $\Sigma\ h_{i,a}\ t_i^a$ ne contienne déjà, et $D_\ell(\omega) = 0$).

Ceci implique, par "intégration des champs de vecteurs correspondant aux dérivations D_ℓ" (cf. [Te 2], I) que $\mathcal{O}^{s(f)}_{X,\sigma(y)}$ est localement analytiquement triviale, en ce sens que l'isomorphisme $\overline{\mathcal{O}_{X,\sigma(y)}} \simeq \prod_i \mathcal{O}_{Y,y}\{t_i\} \simeq \mathcal{O}_{Y,y} \hat{\otimes}_{\mathbb{C}} \overline{\mathcal{O}_{X(y),\sigma(y)}}$ induit un isomorphisme $\mathcal{O}^{s(f)}_{X,\sigma(y)} \simeq \mathcal{O}_{Y,y} \hat{\otimes}_{\mathbb{C}} \mathcal{O}^{s}_{X(y),\sigma(y)}$. Ceci équivaut à l'équisaturation que nous voulions démontrer. ■

6.4.3 <u>Remarques</u> : 1) On peut paraphraser le résultat en disant que, pour $y \in Y - F$, l'algèbre saturée $\mathcal{O}^{s(f)}_{X,\sigma(y)}$ est la sous-algèbre de $\overline{\mathcal{O}_{X,\sigma(y)}} \simeq \prod_{i=1}^{r} \mathcal{O}_{y,y}\{t_i$ formée des éléments $(h_1,\dots,h_r)$, $h_i = \Sigma\ h_{i,a}\ t_i^a$, avec $h_{i,a} \in \mathcal{O}_{Y,y}$ tels que, avec les notations du Théorème ci-dessus

α) l'élément $h_{i,a}$ soit nul si $a \notin \tilde{E}_i$ (où $\tilde{E}_i$ est l'ensemble associé à la i-ème composante de $X(y)$) ;

β) pour tout couple (i,j), $i \neq j$, on a l'inégalité

$$m_{i,j,\omega}(h) \geq m_{i,j,\omega}$$

(même convention).

On voit ainsi pourquoi $D_\ell(\mathcal{O}^{s(f)}_{X,\sigma(y)}) \subset \mathcal{O}^{s(f)}_{X,\sigma(y)}$.

2) On sait (cf. [Te 2], II, § 5) que pour une famille de germes de courbes planes, la constance du nombre de Milnor entraîne non seulement celle de δ_y et de r_y, <u>mais aussi l'équimultiplicité</u> (ce dernier résultat avait été prouvé topologiquement par Lê D.T., [Lê]). Comme me l'a fait observer Mr. E. Böger, la fin de ma démonstration algébrique ([Te 2], II, p. 127) n'est pas complète, puisqu'il n'est pas clair que la dérivation D qui est construite conserve $\mathcal{O}^{s(f)}_X$ au-dessus du point général de Y.

***

CHAPITRE II

IDEAL JACOBIEN, MODIFICATION DE NASH, ET THEOREME DE BERTINI IDEALISTE

Introduction. Dans ce chapitre, on commence à étudier les positions limites des espaces tangents en des points non singuliers tendant vers un point singulier d'un espace analytique X. L'étude géométrique de ces positions limites est celle de la modification de Nash $N(X) \to X$, et l'étude algébrique est celle de l'idéal jacobien de X, c'est-à-dire l'idéal engendré par les mineurs de rang $N+1 - \dim X$ de la matrice jacobienne d'un système de générateurs de l'idéal définissant un plongement local $X \subset \mathbb{C}^{N+1}$. La comparaison de ces deux points de vue permet d'utiliser les théorèmes de finitudes fondamentaux de la géométrie analytique pour prouver une version quantitative du théorème de Bertini, version qui s'exprime par la dépendance intégrale de certains mineurs de la matrice jacobienne sur l'idéal engendré par certains autres, alors que le théorème de Bertini s'exprime par leur appartenance à la racine de cet idéal.

Il nous est très utile de définir l'idéal jacobien,non seulement pour un espace X, mais simultanément pour toutes les fibres d'un morphisme, et de même pour la modification de Nash, c'est-à-dire de définir l'idéal jacobien relatif et la modification de Nash relative associés à un morphisme $X \to S$, puis d'obtenir le cas absolu en prenant S égal à un point.

§ 1. Idéal jacobien relatif.

1.1 Soient $f : X \to S$ un morphisme d'espaces analytiques complexes et Ω^1_f (aussi noté $\Omega^1_{X/S}$) le $\mathcal{O}_X$-module cohérent des 1-formes différentielles relatives (cf. [Ca] ; exposé 14 ; j'utiliserai sans les rappeler les propriétés élémentaires de $\Omega^1_{X/S}$ qui y sont démontrées). Pour chaque entier $d \geq 0$, on peut définir le faisceau cohérent d'idéaux de Fitting $F_d(\Omega_f^1) \subset \mathcal{O}_X$ (cf. [Te 3], § 1, [Pi], [G-R]) qui a la propriété (cf. Loc. cit.) que pour chaque fibre $X(s) = f^{-1}(s)$ on a $F_d(\Omega_f^1) \cdot \mathcal{O}_{X(s)} = F_d(\Omega_{X(s)}^1)$. Ce faisceau d'idéaux peut être décrit localement comme suit : tout point $x \in X$ possède un voisinage ouvert, que nous allons abusivement encore noter X, tel qu'il existe un S-plongement de X dans $S \times \mathbb{C}^N$:

$$\begin{array}{ccc} X & \hookrightarrow & S \times \mathbb{C}^N \\ {\scriptstyle f}\downarrow & \swarrow {\scriptstyle pr_1} & \\ S & & \end{array}$$

Soient $f_1, \dots, f_m \in \mathcal{O}_{S,s}\{z_1, \dots, z_N\} = \mathcal{O}_{S\times\mathbb{C}^N, s\times\{0\}}$ des générateurs de l'idéal I de $\mathcal{O}_{S\times\mathbb{C}^N, s\times\{0\}}$, où $s = f(x)$, définissant le plongement de germes $(X,x) \subset (S \times \mathbb{C}^N, s \times \{0\})$. On a alors la suite exacte

$$I/I^2 \xrightarrow{d} (\Omega^1_{S\times\mathbb{C}^N/S}|X)_x \longrightarrow \Omega^1_{X/S,x} \longrightarrow 0$$

où d désigne la différentielle naturelle $\mathcal{O}_{S\times\mathbb{C}^N, s\times\{0\}} \to (\Omega^1_{S\times\mathbb{C}^N/S})_{s\times\{0\}}$ qui revient à associer à $h \in \mathcal{O}_{S,s}\{z_1, \dots, z_N\}$ l'élément $\sum_1^N \frac{\partial h}{\partial z_i} dz_i$.

Il résulte aussitôt de ceci et de la définition des idéaux de Fitting que $F_d(\Omega_f^1)_x$ est l'idéal de $\mathcal{O}_{X,x}$ engendré par les mineurs de rang $N - d$ de la "matrice jacobienne relative" $(\frac{\partial f_i}{\partial z_i})$ $1 \le i \le m$, $1 \le j \le N$, c'est-à-dire que, <u>pour tout choix</u> d'un S-plongement local comme ci-dessus on a :

$$F_d(\Omega_f^1)_x = \left(\frac{\partial(f_{i_1}, \dots, f_{i_{N-d}})}{\partial(z_{j_1}, \dots, z_{j_{N-d}})} \; ; \{i_1, \dots, i_{N-d}\} \subset \{1, \dots, m\}, \{j_1, \dots, j_{N-d}\} \subset \{1, \dots, N\} \right) \cdot \mathcal{O}_{X,x}$$

(les indices sont supposés tous distincts).

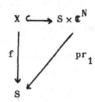

Un cas particulier. Supposons que X est défini dans $S \times \mathbb{C}^N$ par N - d équations $f_1, \dots, f_d$ et que toutes les fibres X(s) sont de dimension d (on dit alors que X est une intersection complète relative). En chaque point non-singulier $x' \in X(s)$, $s = f(x')$ l'espace tangent $T_{X(s),x'}$ à X(s) en x' détermine une direction de sous-espace vectoriel de dimension d, encore notée $T_{X(s),x'}$ dans $\mathbb{C}^N$, c'est-à-dire un point de la grassmannienne $G = G(N-1, d-1)$ des d-1-plans dans $\mathbb{P}^{N-1}$, et par la définition même du plongement de Plücker ([G-H], p. 209) de la grassmannienne G dans l'espace projectif $\mathbb{P}^{\binom{N}{d}-1}$; les déterminants jacobiens $\frac{\partial(f_1,\dots,f_{N-d})}{\partial(z_{j_1},\dots,z_{j_{N-d}})}(x')$, $\{j_1,\dots,j_{N-d}\} \subset \{1,\dots,N\}$ sont les coordonnées homogènes de l'image dans $\mathbb{P}^{\binom{N}{d}-1}$ du point de G déterminé par $T_{X(s),x'}$. Ainsi, notant X^o l'ensemble des points $x' \in X$ qui sont non-singuliers dans leur fibre, nous avons :

L'application analytique ("morphisme jacobien relatif") $X^o \to \mathbb{P}^{\binom{N}{d}-1}$ qui à $x' \in X^o$ associe le point de coordonnées homogènes

$$\frac{\partial(f_1,\dots,f_{N-d})}{\partial(z_{j_1},\dots,z_{j_{N-d}})}(x') \ , \ \{j_1,\dots,j_{N-d}\} \subset \{1,\dots,N\}$$

(les indices j_k sont supposés 2 à 2 distincts)

coïncide avec l'application composée $X^o \xrightarrow{\gamma} G \hookrightarrow \mathbb{P}^{\binom{N}{d}-1}$, où γ est l'application ("morphisme de Gauss relatif") qui à $x' \in X^o$ associe la direction $T_{X(s),x'}$, et $G \hookrightarrow \mathbb{P}^{\binom{N}{d}-1}$ est le plongement de Plücker.

Remarquons qu'en un point $x' \in X^o$, l'un au moins des déterminants jacobiens est non nul, puisque les espaces tangents en x' aux fibres des N-d hypersurfaces $f_i = 0$ doicent être en position générale. Si nous prenons une suite de points $x_i' \in X^o$ tendant vers un point singulier x, la position limite (dans G, ou dans $\mathbb{P}^{\binom{N}{d}-1}$, ce qui revient au même) des espaces tangents $T_{X,x_i'}$ sera en partie déterminé par les vitesses relatives avec lesquelles les divers déterminants jacobiens

$\frac{\partial(f_1,\ldots,f_{N-d})}{\partial(z_{j_1},\ldots,z_{j_{N-d}})}(x_i')$ tendent vers 0. Comme nous le verrons plus bas, c'est pour cette raison que le concept de dépendance intégrale nous sera très utile.

Exemple , montrant que le cas particulier précédent est bien particulier. Soit $f : \mathbb{C}^4 \to \mathbb{C}^4$ le morphisme défini par $u_1 = xz$, $u_2 = xt$, $u_3 = yz$, $u_4 = yt$. Ce morphisme n'est pas surjectif, et $f^{-1}(0)$ est la réunion de deux plans dans $\mathbb{C}^4$ ne se rencontrant qu'en 0. Aussi il n'est pas très étonnant que l'on ait $F_o(\Omega_f^1) = 0$, comme on le vérifie facilement.

1.1.1 Lemme , montrant que si toutes les fibres de f sont purement de la même dimension, le cas particulier précédent permet d'étudier le cas général :

d'après un résultat classique d'algèbre commutative, (cf. [Z-S], II, Th. 22) si nous supposons que toutes les fibres de f sont purement de la même dimension d, sans faire d'hypothèse sur le nombre des générateurs, alors : il existe un ouvert de Zariski dense U de l'espace $\mathbb{C}^{m(N-d)}$ tel que, si $\lambda = (\lambda_{ij}) \in U$, le sous-espace X_1 de $S \times \mathbb{C}^N$ défini au voisinage de $s \times \{0\}$ par les $N-d$ équations

$$g_1 = \sum_{i=1}^{m} \lambda_{1i}\, f_i, \ldots, g_{N-d} = \sum_{i=1}^{m} \lambda_{N-d,i}\, f_i$$

et qui contient évidemment X, a ses fibres de dimension d.

L'on se trouve pour X dans le cas particulier étudié plus haut. De plus si X(s) est réduit, on peut supposer que pour $\lambda \in U$, $X_1(s)$ est réduit, par un argument du type de la preuve du théorème de Bertini classique, et laissé au lecteur.
En particulier, prenant pour S un point, on obtient que tout germe d'espace analytique réduit purement de dimension d peut être défini dans $\mathbb{C}^N$ par un idéal de la forme $(f_1,\ldots,f_{N-d},f_{N-d+1},\ldots,f_m)$ où $(f_1,\ldots,f_{N-d})$ définit une intersection complète réduite X_1 de dimension d : X est réunion de certaines des composantes irréductibles de cette intersection complète, et en particulier

on a l'inclusion $X_1^o \cap X \subseteq X^o$ (où X^o désigne la partie non-singulière de X), $X_1^o \cap X$ est dense dans X, et la restriction à $X_1^o \cap X$ de l'application de Gauss $X^o \to G$ coïncide avec la restriction à $X_1^o \cap X$ de l'application de Gauss $X_1^o \to G$. Sous l'hypothèse que les fibres de $f : X \to S$ sont réduites et purement de dimension d et que $X_1^o \cap X$ est dense dans X, on a un résultat analogue pour le morphisme de Gauss relatif. La manière géométrique d'étudier les rapports des vitesses avec lesquelles les mineurs jacobiens tendent vers 0 est d'étudier l'éclatement de l'idéal jacobien relatif, c'est-à-dire, si nous nous restreignons au cas où toutes les fibres de f sont réduites et de dimension d, l'éclatement $\mathcal{J}_f(X) \to X$ du faisceau cohérent d'idéaux $F_d(\Omega_f^1)$. Localement sur X, c'est-à-dire au-dessus d'un voisinage ouvert assez petit de chaque point $x \in X$, ce morphisme est X-isomorphe au morphisme obtenu comme ceci :

On considère l'adhérence $\mathcal{J}_f(X)$ dans $X \times \mathbb{P}^{\binom{N}{d}-1}$ du graphe $\Gamma \subset X^o \times G \subset X^o \times \mathbb{P}^{\binom{N}{d}-1}$ du morphisme jacobien relatif, et le morphisme $\mathcal{J}_f(X) \to X$ induit par la première projection est X-isomorphe à l'éclatement dans X de l'idéal jacobien relatif.

Cependant cette construction ne décrit avec exactitude les limites d'espaces tangents que dans le cas particulier des intersections complètes relatives, à cause du fait que en dehors de ce cas, le nombre des mineurs considérés est trop grand et contient de regrettables redondances. Il y a cependant une construction, en apparence un peut moins explicite, qui fonctionne toujours : la modification de Nash relative.

1.2 Soit $f : X \to S$ un morphisme entre espaces analytiques (réels ou complexes) tel que le module Ω_f^1 des différentielles relatives soit localement libre de rang $d = \dim X - \dim S$ sur le complémentaire d'un fermé analytique rare F de X. Soit $G_f \xrightarrow{g} X$ la grassmannienne des quotients localement libres de rang d de Ω_f^1, c'est-à-dire (cf. [Ca], exp. XII) que $g^*\Omega_f^1$ a un quotient L localement libre de rang d sur G_f, et que le morphisme g est universel pour cette propriété. D'après cette définition, on a une section bien définie $X - F \xrightarrow{\sigma} G_f$ et d'après un théorème de Remmert, l'adhérence dans G_f de $\sigma(X - F)$ est un sous-espace analytique fermé réduit, noté $N_f(X) \subset G_f$, et g induit un morphisme $\nu_f : N_f(X) \to X$ qui est propre et biméromorphe puisque g est propre, $N_f(X)$ fermé dans G_f et que ν_f y induit un isomorphisme au-dessus de $X - F$. Le morphisme ν_f est donc une modification de X.

Définition : Le morphisme $\nu_f : N_f(X) \to X$ est appelé modification de Nash relative de X/S. Lorsque S est un point, on la note $\nu : N(X) \to X$ et on l'appelle modification de Nash (absolue) de X. La modification de Nash absolue est en particulier définie pour tout espace analytique réduit de dimension pure.

Remarques : 1) On peut aussi caractériser la modification de Nash relative come ceci : sur $X - F$, le $\mathcal{O}_X$-module localement libre $\Omega_f^1 | X - F = \Omega^1_{X-F/S}$ correspond à un fibré vectoriel T_f de rang d qui induit sur chaque fibre $f^{-1}(s) \cap (X - F)$ le fibré tangent à cette fibre. Le fibré vectoriel $\nu_f^* T_f$ sur l'ouvert dense $\nu_f^{-1}(X - F) \subset N_f(X)$ s'étend canoniquement en un fibré vectoriel sur $N_f(X)$ tout entier, correspondant au quotient localement libre $L_f | N_f(X)$ du $\mathcal{O}_{N_f(X)}$-module $\nu_f^* \Omega_f^1$. De plus le morphisme ν_f est minimal pour cette propriété.

2) La construction par adhérence signifie que l'on peut identifier, pour tout $x \in X$, l'ensemble $|\nu_f^{-1}(x)|$ à l'ensemble des directions limites en x d'espaces tangents aux fibres de f en des points de $X - F$. En fait, pour tout S-plongement local $X \subset S \times \mathbb{C}^N$, on peut identifier $N_f(X)$ à l'adhérence dans $X \times G$, où G est la grassmannienne des d-plans de $\mathbb{C}^N$, du graphe du morphisme de Gauss $X - F \xrightarrow{\gamma} G$.

3) La construction précédente est un cas particulier d'une construction valable pour n'importe quel $\mathcal{O}_X$-module M, localement libre de rang constant sur le complémentaire d'un fermé analytique rare de X : il existe une modification $\varphi : X' \to X$ minimale parmi celles qui ont la propriété que φ^*M est localement libre modulo torsion, c'est-à-dire admet un quotient localement libre L ; $0 \to K \to \varphi^*M \to L \to 0$, le noyau K étant de torsion. Ce résultat peut être

vu comme un cas particulier très simple, celui où les fibres sont linéaires, du théorème d'aplatissement de [H-L-T].

Modification de Nash relative et éclatement de l'idéal jacobien relatif.

La comparaison de ces deux morphismes est résumée dans la

1.2.1 Proposition : *Soit* $f : X \to S$ *un morphisme tel que* Ω_f^1 *soit localement libre de rang* $d = \dim X - \dim S$ *en dehors d'un fermé analytique rare* F *de* X ; *soient* $\nu_f : N_f(X) \to X$ *la modification de Nash relative de* X/S *et* $p_f : \mathcal{J}_f(X) \to X$ *l'éclatement dans* X *de l'idéal jacobien relatif de* X/S.

a) *le morphisme* p_f *se factorise canoniquement à travers* ν_f :

$$\begin{array}{ccc} \mathcal{J}_f(X) & \xrightarrow{q_f} & N_f(X) \\ & {\scriptstyle p_f}\searrow \quad \circ \quad \swarrow{\scriptstyle \nu_f} & \\ & X & \end{array}$$

b) *le morphisme* q_f *est un isomorphisme si* X *est localement une intersection complète relative au-dessus de* S, *c'est-à-dire si l'on est dans la situation du cas particulier ci-dessus.*

c) *supposant toutes les fibres de* f *purement de dimension* d *et réduites, il existe localement sur* X *un plongement de* X *dans une intersection complète relative* X_1, *dont toutes les fibres sont purement de dimension* d *et réduites, et la modification de Nash relative de* X/S *est canoniquement* X-*isomorphe à la transformée stricte de* X *par la modification de Nash relative de* X_1/S, *qui d'après le point* b) *coïncide avec l'éclatement dans* X_1 *de l'idéal jacobien relatif de* X_1/S : *finalement, la modification de Nash relative de* X/S *est* X-*isomorphe à l'éclatement de l'idéal de* $\mathcal{O}_X$ *image de l'idéal jacobien relatif de* X_1/S *par la surjection* $\mathcal{O}_{X_1} \to \mathcal{O}_X \to 0$ *correspondant au plongement* $X \hookrightarrow X_1$.

Prouvons a) : D après la définition de $N_f(X)$, il suffit de prouver que

$p_f^*\Omega_f^1$ a un quotient localement libre de rang d. Or, le morphisme p_f est un isomorphisme au-dessus de $X - F$, donc $p_f^*\Omega_f^1$ est localement libre de rang d sur l'ouvert analytique dense $p_f^{-1}(X - F)$. D'autre part, par définition du morphisme p_f, l'idéal $F_d(\Omega_f^1) \cdot \mathcal{O}_{\mathcal{J}_f(X)}$ est inversible. La compatibilité de la formation des idéaux de Fitting au changement de base ([Te 3]), p. 570, [Pi]) implique l'égalité $F_d(p_f^*\Omega_f^1) = F_d(\Omega_f^1) \cdot \mathcal{O}_{\mathcal{J}_f(X)}$ entre idéaux de $\mathcal{O}_{\mathcal{J}_f(X)}$. Le $\mathcal{O}_{\mathcal{J}_f(X)}$-module $p_f^*\Omega_f^1$ est donc localement libre de rang d sur l'ouvert analytique dense $p_f^{-1}(X - F)$ et son d-ième idéal de Fitting est inversible. D'après un lemme de Raynaud ([G-R], 5.4.2 ou [Pi]) ceci implique que $p_f^*\Omega_f^1$ a un quotient localement libre de rang d, d'où a).

Prouvons b) : Si X est localement une intersection complète relative, d'après ce que nous avons vu plus haut, localement sur X, le composé de l'application de Gauss avec le plongement de Plücker, $X - F \to G \to \mathbb{P}^{\binom{N}{d}-1}$ coïncide avec le morphisme $X - F \to \mathbb{P}^{\binom{N}{d}-1}$ défini par les mineurs jacobiens. D'après la remarque 2) ci-dessus, et la description de l'éclatement de l'idéal jacobien comme adhérence dans $X \times \mathbb{P}^{\binom{N}{d}-1}$ du graphe du morphisme $X - F \to \mathbb{P}^{\binom{N}{d}-1}$, cette coïncidence implique aussitôt que la modification de Nash relative coïncide avec l'éclatement de l'idéal jacobien relatif localement, donc aussi globalement puisque les deux vérifient une propriété universelle.

Prouvons c) : D'après le Lemme 1.1.1, il existe localement sur X un plongement de X dans une intersection complète relative X_1, tel que pour chaque $s \in S$, la fibre X(s) soit réunion de certaines des composantes irréductibles de la fibre $(X_1)(s)$ qui de plus est réduite, $(X_1(s))^{\circ} \cap X$ est contenu dans $(X(s))^{\circ}$, est dense dans X(s), et le morphisme de Gauss relatif de X, coïncide avec celui de X_1 sur $(X_1)(s)^{\circ} \cap X(s)$. Le point c) résulte aussitôt de ceci, de la définition de la modification de Nash relative comme adhérence du graphe du morphisme de Gauss, et du point b).

<u>Remarques</u> : 1) En adaptant au cas relatif la preuve donnée dans [Te 3], § 2),

on obtient une preuve de l'énoncé suivant :

Proposition : Si toutes les fibres de f sont réduites, la modification de Nash relative est un isomorphisme si et seulement si le morphisme f est lisse, c'est-à-dire plat et à fibres lisses.

2) Soit N(f) le morphisme composé $N_f(X) \xrightarrow{\nu_f} X \xrightarrow{f} S$; on prendra garde que $N(f)^{-1}(s) = \nu_f^{-1}(X_s)$ ne coïncide pas en général avec la modification de Nash N(X(s)) de X(s), c'est-à-dire que la formation de la modification de Nash relative ne commute pas au changement de base en général.
(Exemple : $x^2 - y^3 + s^2y^2 = 0$ projeté sur l'axe des s, en $s = 0$.) Le lecteur vérifiera par contre sans mal en utilisant la propriété universelle de la modification de Nash, que si les fibres de f sont purement de la même dimension et réduites, pour tout $s \in S$, la transformée stricte par la modification de Nash relative ν_f de la fibre X(s) est canoniquement isomorphe à la modification de Nash de X(s).

3) En route, nous avons montré que la modification de Nash relative comme d'ailleurs toute modification localement projective est localement l'éclatement d'un idéal, à savoir ici l'idéal induit sur X par l'idéal jacobien d'une intersection complète relative X_1 comme ci-dessus. Il paraît très improbable que la modification de Nash, même absolue, soit globalement un éclatement. Voir aussi [No], qui a inspiré une partie de la Proposition.

4) Différence entre l'éclatement jacobien et la modification de Nash. Reprenons l'exemple de la réunion X de deux 2-plans de $\mathbb{C}^4$ se rencontrant seulement en $\{0\}$. L'idéal correspondant peut être écrit $(xz, xt, yz, yt) \subset \mathbb{C}\{x,y,z,t\}$, et l'on vérifie que $F_2(\Omega^1_X)_0$ est le carré $\mathfrak{m}^2$ de l'idéal maximal $\mathfrak{m}$ de $\mathcal{O}_{X,0}$, dont l'éclatement dans X sépare les deux plans et a un diviseur exceptionnel induisant une droite projective $\mathbb{P}^1$ (comptée 2 fois) dans chacun des deux plans séparés, tandis que la modification de Nash de X est le morphisme fini (en fait la normalisation) consistant à séparer les deux plans tout en induisant un isomorphisme de chacun des deux plans séparés sur son image. Le morphisme de factorisation q consiste à éclater dans N(X) le

carré de l'idéal définissant $\nu^{-1}(0)$

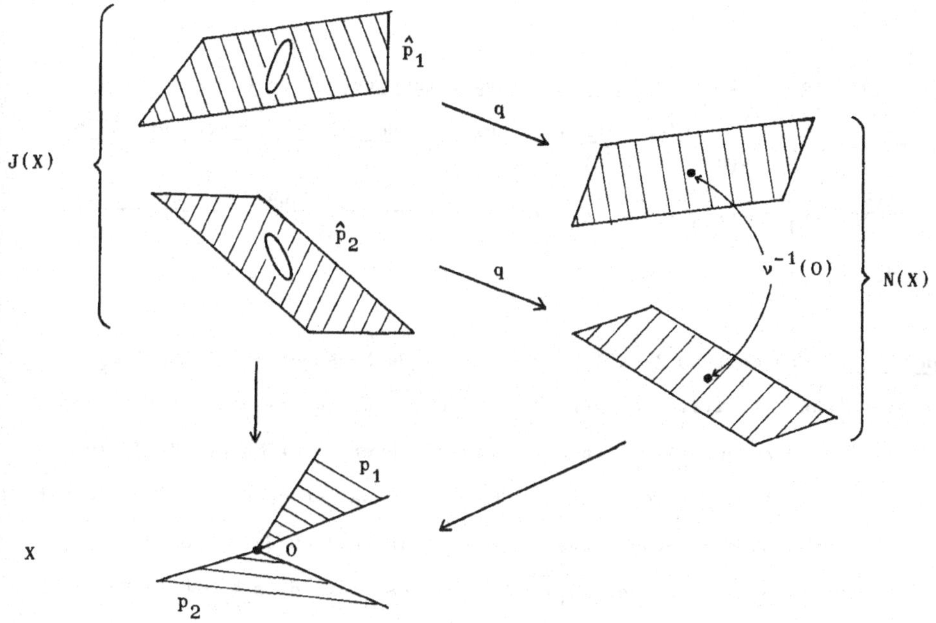

§ 2. Théorème de Bertini idéaliste le long d'une section.

2.1 Soit X un sous-espace analytique de $\mathbb{C}^s \times \mathbb{C}^M$ contenant $\mathbb{C}^s \times \{0\}$. Notant $\Phi : X \to \mathbb{C}^s$ le morphisme induit par la première projection, on suppose $\Omega^1_\Phi = \Omega^1_{X/\mathbb{C}^s}$ localement libre de rang $d = \dim X - s$ dehors d'un fermé analytique rare de X. On suppose enfin X défini dans un ouvert U de $\mathbb{C}^s \times \mathbb{C}^M$ par un idéal engendré par des sections globales $G_1, \dots, G_p \in H^0(U, \mathcal{O}_{\mathbb{C}^s \times \mathbb{C}^M})$.

Pour tout choix de coordonnées $(t_1, \dots, t_s)$ et $(u_1, \dots, u_M)$ sur $\mathbb{C}^s$ et $\mathbb{C}^M$ respectivement, pour tout entier ℓ, $0 \le \ell \le s$ et tout choix d'un sous-ensemble $K = \{k_{\ell+1}, \dots, k_c\} \subset \{1, \dots, M\}$, où $c = \operatorname{codim}_{\mathbb{C}^s \times \mathbb{C}^M} X = M - d$, notant J_K l'idéal de $\mathcal{O}_X$ engendré par les éléments de la forme :

$$u_{k_1}\cdot\ldots\cdot u_{k_\ell}\cdot\frac{\partial(G_{i_1},\ldots,G_{i_c})}{\partial(u_{k_1},\ldots,u_{k_\ell},u_{k_{\ell+1}},\ldots,u_{k_c})}$$

où $\{k_1,\ldots,k_\ell\}\subset\{1,\ldots,M\}$ et $\{i_1,\ldots,i_c\}\subset\{1,\ldots,p\}$ on a :

__2.1.1 Théorème__ : __Il existe un fermé analytique rare__ F __de__ $\mathbb{C}^s\times\{0\}$ __tel que, pour tout point__ $z\in\mathbb{C}^s\times\{0\}-F$, __l'image dans__ $\mathcal{O}_{X,z}$ __de chaque déterminant jacobien de la forme__ $\dfrac{\partial(G_{i_1},\ldots,G_{i_c})}{\partial(t_{j_1},\ldots,t_{j_\ell},u_{k_{\ell+1}},\ldots,u_{k_c})}$ __soit entière dans__ $\mathcal{O}_{X,z}$ __sur l'idéal__ $(J_K)_z=J_K\cdot\mathcal{O}_{X,z}$.

__Démonstration__ : Soit $\pi: Z'\to X$ la modification de X associée à l'idéal J_K comme dans (Chap. I, 1.4.2) et soit Z'' la réunion de celles des composantes irréductibles de Z' où $J_K\cdot\mathcal{O}_{Z'}$ induit un idéal inversible. Soient $D\subset Z''$ le diviseur de Z'' défini par cet idéal et $D=\bigcup_{i\in I} D_i$ sa décomposition (localement finie) en composantes irréductibles. Soit $B\subset X$ la réunion de celles des images $\pi(D_i)$ qui ont la propriété que $\pi(D_i)\cap(\mathbb{C}^s\times\{0\})$ est rare dans $\mathbb{C}^s\times\{0\}$. Par la propriété du morphisme π, B est un sous-ensemble analytique fermé de X, et $B\cap(\mathbb{C}^s\times\{0\})=F$ est un fermé analytique rare de $\mathbb{C}^s\times\{0\}$.

Soit $z\in\mathbb{C}^s\times\{0\}\setminus F$. Puisque l'énoncé du théorème est local sur X, nous pouvons en nous restreignant à un voisinage assez petit de z supposer que $F=\emptyset$, c'est-à-dire que $\pi(D_i)\supset\mathbb{C}^s\times\{0\}$. Le morphisme composé $\Phi\circ\pi|D_i: D_i\to\mathbb{C}^s$ est alors surjectif, et par conséquent il existe un ouvert analytique partout dense $U_i\subset D_i$ en chaque point z' duquel on a les propriétés suivantes :

1) L'espace Z'' et le sous-espace D_{red} sont tous deux non-singuliers en z', et donc D_{red} coïncide avec $D_{i,red}$ au voisinage de z'.

2) Le morphisme $\Phi\circ\pi$ induit en z' une submersion $D_{i,red}\to\mathbb{C}^s$.

3) Le transformé strict par π de chacun des sous-espaces de X définis par $u_j\cdot\mathcal{O}_X$ $(1\le j\le M)$ est vide au voisinage de z'.

En effet, la propriété 1) vient de ce que Z est normal et que D est un diviseur, puisque Z' étant normal est non-singulier en codimension 1. La propriété

2) vient du théorème de Sard et la propriété 3) de la définition du transformé strict.

On peut donc choisir un système de coordonnées locales $(t'_1,\ldots,t'_s,w_1,\ldots,w_d)$ pour Z' en z' tel que

A) On ait $(t_j \circ \pi)_{z'} = t'_j \quad (1 \leq j \leq s)$.

B) Le sous-espace $(D_{i,red})_z$ est défini par $w_1\mathcal{O}_{Z',z'}$.

C) On a $(u_m \circ \pi)_{z'} = A_m\, w_1^{\mu_m}$ où A_m est soit identiquement 0, soit inversible, et $\mu_m \in \mathbb{N}$ $(1 \leq m \leq M)$.

Puisque $G_i \cdot \mathcal{O}_X \equiv 0$ $(1 \leq i \leq p)$ on a $G_i \cdot \mathcal{O}_{Z',z'} \equiv 0$ et en particulier

$$(*) \qquad \frac{\partial}{\partial t'_j}(G_i \circ \pi)_{z'} = \left(\frac{\partial G_i}{\partial t_j} \circ \pi\right)_{z'} + \sum_{m=1}^{M} \left(\frac{\partial G_i}{\partial t'_j} \circ \pi\right)_{z'} \frac{\partial}{\partial t'_j}(u_m \circ \pi)_{z'} \equiv 0$$

et puisque, grâce à la propriété c), nous avons dans $\mathcal{O}_{Z',z'}$:

$\frac{\partial}{\partial t'_j}(u_m \circ \pi) = \frac{\partial A_m}{\partial t'_j} w_1^{\mu_m}$, qui est un multiple de $(u_m \circ \pi)_z$, nous en déduisons que dans $\mathcal{O}_{Z',z'}$ on peut écrire

$$\left(\frac{\partial(G_{i_1},\ldots,G_{i_c})}{\partial(t_{j_1},\ldots,t_{j_\ell},u_{k_{\ell+1}},\ldots,u_{k_c})} \circ \pi\right)_{z'} =$$

$$\sum \varepsilon_{\underline{m}} \frac{\partial(u_{m_1}\circ\pi)_{z'}}{\partial t'_{j_1}} \cdots \frac{\partial(u_{m_\ell}\circ\pi)_{z'}}{\partial t'_{j_\ell}} \left(\frac{\partial(G_{i_1},\ldots,G_{i_c})}{\partial(u_{m_1},\ldots,u_{m_\ell}, u_{k_{\ell+1}},\ldots,u_{k_c})} \circ \pi\right)_{z'} ,$$

où la somme porte sur les $\underline{m} = (m_1,\ldots,m_\ell)$ contenus dans $\{1,\ldots,M\} \setminus (k_{\ell+1},\ldots,k_c)$ et $\varepsilon_{\underline{m}} = \pm 1$,

et donc finalement, puisque $\frac{\partial(u_m \circ \pi)}{\partial t'_j}$ est un multiple de $u_m \circ \pi$, on a montré que chaque composante irréductible D_i de $D \subset Z''$ contient un ouvert analytique dense en chaque point z' duquel on a :

$$\left(\frac{\partial(G_{i_1},\ldots,G_{i_c})}{\partial(t_{j_1},\ldots,t_{j_\ell},u_{k_{\ell+1}},\ldots,u_{k_c})} \circ \pi\right)_{z'} \in J_K \cdot \mathcal{O}_{Z,z'} .$$

Passons maintenant à la réunion $Z^{(''')}$ des composantes de Z' sur lesquelles $J_K \circ \pi$ s'annule identiquement ; par construction, ces composantes sont les normalisées des composantes de X sur lesquelles J_K s'annule identiquement. D'après l'hypothèse faite sur $\Omega^1_{X/\mathbb{C}^s}$, chacune de ces composantes contient un ouvert analytique dense en chaque point z' duquel Z''' est lisse sur $\mathbb{C}^s$ par $\Phi \circ \pi$, et $(u_m \circ \pi)_{z'}$ est soit 0 soit inversible $(1 \leq m \leq M)$. On peut donc choisir un système de coordonnées locales $(t'_1,\ldots,t'_s,w_1,\ldots,w_d)$ sur Z''' en z' tel que

A') $(t_j \circ \pi) = t'_j$

B') $(u_m \circ \pi)_{z'} = 0$ ou bien est inversible ;

et le même calcul que ci-dessus montre que

$$\left(\frac{\partial(G_{i_1},\ldots,G_{i_c})}{\partial(t_{j_1},\ldots,t_{j_\ell},u_{k_{\ell+1}},\ldots,u_{k_c})} \circ \pi\right)_{z'} = 0 ,$$

d'où le résultat, puisque alors cet élément s'annule sur toute la composante irréductible de X considérée.

2.1.2 Remarque (théorème de Bertini idéaliste sans section) : Soit

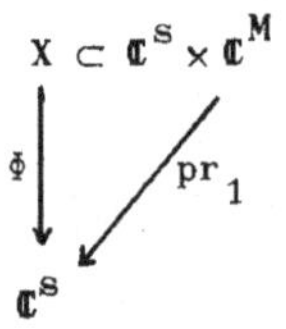

le diagramme décrivant la situation du Théorème précédent, mais ne supposons plus que X contient $\mathbb{C}^s \times \{0\}$ (c'est-à-dire ne supposons plus l'existence d'une section $\sigma : \mathbb{C}^s \to X$ de Φ). La même preuve que ci-dessus montre : (cf. [Te 3], § 2, 2nd part)

2.1.3 Théorème de Bertini idéaliste : Il existe un fermé analytique $B \subset X$, tel que $\Phi(B)$ soit de mesure nulle dans $\mathbb{C}^s$, et tel que en tout point $z \in X - B$

<u>on ait, pour tout entier</u> ℓ, $0 \leq \ell \leq s$, $\frac{\partial(G_{i_1},\ldots,G_{i_c})}{\partial(t_{j_1},\ldots,t_{j_\ell},u_{k_{\ell+1}},\ldots,u_{k_c})}\mathcal{O}_{X,z}$ <u>est entier sur l'idéal de</u> $\mathcal{O}_{X,z}$ <u>engendré par les déterminants jacobiens</u> $\frac{\partial(G_{i_1},\ldots,G_{i_c})}{\partial(u_{k_1},\ldots,u_{k_c})}$.

<u>2.1.4</u> Une autre manière de dire ceci est la suivante : soit $J_X \subset \mathcal{O}_X$ l'idéal jacobien de X (défini de manière intrinsèque comme (d+s)-ième idéal de Fitting du module des différentielles Ω^1_X de X) définissant le <u>sous-espace singulier</u> de X, et soit $J_\Phi = J_{X/\mathbb{C}^s} \subset \mathcal{O}_X$ l'idéal jacobien relatif de $X \xrightarrow{\Phi} \mathbb{C}^s$, (défini de manière intrinsèque comme d-ième idéal de Fitting du module $\Omega^1_{X/\mathbb{C}^s}$ des différentielles relatives) ; l'idéal J_X est, dans la situation du Théorème, engendré par les mineurs jacobiens de la forme $\frac{\partial(G_{i_1},\ldots,G_{i_c})}{\partial(t_{j_1},\ldots,t_{j_\ell},u_{j_{\ell+1}},\ldots,u_{j_c})}$ et les $J_{X/\mathbb{C}^s}$ par les seuls mineurs jacobiens où $\ell = 0$, c'est-à-dire où aucun t_j n'apparaît au "dénominateur". Le théorème de Bertini idéaliste peut s'énoncer en disant que pour tout $z \in X - B$, on a l'égalité de fermetures intégrales d'idéaux de $\mathcal{O}_{X,z}$:

$$\overline{(J_X)_z} = \overline{(J_{X/\mathbb{C}^s})_z} \qquad (z \in X - B) \quad .$$

Le théorème de Bertini-Sard revient à l'égalité des <u>racines</u>

$$\sqrt{(J_X)_z} = \sqrt{(J_{X/\mathbb{C}^s})_z} \qquad (z \in X - B)$$

et dans ([Te 3], § 2) on trouve un exemple montrant qu'il n'existe pas en général de sous-espace $B \subset X$ tel que $\Phi(B)$ soit de mesure nulle dans $\mathbb{C}^s$ et que les idéaux eux-mêmes soient égaux pour $z \in X - B$.

On peut de même donner une formulation indépendante du choix des coordonnées pour le Théorème ci-dessus, mais il faut bien voir que cet énoncé est beaucoup plus faible.

Corollaire : Soit $\Phi : X \underset{\sigma}{\rightleftarrows} \mathbb{C}^s$ un morphisme d'espaces analytiques dont toutes les fibres sont purement de dimension $d = \dim X - s$. Notons S le faisceau d'idéaux de $\mathcal{O}_X$ définissant le sous-espace $\sigma(\mathbb{C}^s)$. Il existe un fermé analytique rare F de $\sigma(\mathbb{C}^s)$ tel que, en tout point $z \in \sigma(\mathbb{C}^s) - F$, on ait dans l'anneau $\mathcal{O}_{X,z}$

$$(J_{X/\mathbb{C}^s})_z \subseteq (J_X)_z \subseteq (J_{X/\mathbb{C}^s})_z + \overline{(S\, .\, J_{X/\mathbb{C}^s})_z}$$

ceci implique en particulier $\overline{(J_X)_z} = \overline{(J_{X/\mathbb{C}^s})_z}$ d'après le Lemme de Nakayama intégral (Chap. I, 1.5).

Considérons maintenant un espace analytique réduit X purement de dimension $d+t$, un sous-espace analytique $Y \subset X$ et un point $y \in Y$ tel que Y soit non-singulier en y. Pour tout plongement local $X \subset \mathbb{C}^M$ d'un voisinage ouvert de y dans X, encore noté X, considérons les rétractions locales $r : \mathbb{C}^M \to Y$ au voisinage de y, c'est-à-dire les rétractions sur Y d'un voisinage ouvert, encore noté $\mathbb{C}^M$, de y dans $\mathbb{C}^M$.
Chacune de ces rétractions induit une rétraction $r|X = \rho : X \to Y$ définie au voisinage de Y, et dont l'inclusion $Y \subset X$ est une section.
Supposons que, par rapport à ρ, on soit dans les conditions du corollaire précédent, c'est-à-dire que l'on ait l'inclusion $(J_X)_y \subset (J_\rho)_y + \overline{(S\, .\, J_\rho)_y}$ où S est l'idéal définissant Y dans X. Alors pour tout autre rétraction locale $r' : \mathbb{C}^M \to Y$, notant $\rho' : X \to Y$ la rétraction induite sur X, on a l'inclusion $(J_X)_y \subset (J_{\rho'})_y + \overline{(S\, .\, J_{\rho'})_y}$.

Preuve : En fait, l'on a l'égalité

$$(J_\rho + \overline{S\, .\, J_\rho})_y = (J_{\rho'} + \overline{S\, .\, J_{\rho'}})_y \quad .$$

En effet, l'hypothèse implique, d'après le Lemme de Nakayama intégral (Chap. I, 1.5) que $(\overline{J_{\rho'}})_y = (\overline{J_X})_y = (\overline{J_\rho})_y$, d'où $(\overline{S\, .\, J_\rho})_y = (\overline{S\, .\, J_{\rho'}})_y$. En faisant un choix de coordonnées sur $\mathbb{C}^M$ adapté à ρ (resp. ρ') un calcul direct permet de vérifier, en utilisant le corollaire précédent que $J_{\rho'} \subseteq J_\rho + \overline{S \cdot J_\rho}$, d'où aussitôt le résultat.

§ 3. Idéal jacobien et transversalité.

Voici maintenant un exemple typique d'utilisation de la dépendance intégrale pour exprimer algébriquement une condition de transversalité "à la limite" pour les espaces tangents.

3.1 Proposition : Soit $f : (X,0) \to (S,0)$ un morphisme comme en (1.2.) avec S non-singulier, gardons les mêmes notations, et fixons un plongement local $(X,0) \subset (\mathbb{C}^M,0)$ au voisinage de 0. Soit $(X,0) \hookrightarrow (S,0) \times (\mathbb{C}^M,0)$ le S-plongement local défini par le graphe de f, et fixons un système de générateurs $f_1,\ldots,f_m$, $f_i \in \mathcal{O}_{S,0}\{z_1,\ldots,z_M\}$ pour l'idéal définissant (X,0) dans $(S,0) \times (\mathbb{C}^M,0)$. Etant donné un sous-espace analytique fermé $(Z,0) \subset (X,0)$, tel que $F \cap Z$ soit rare dans Z, pour un sous-espace vectoriel $D_i \subset \mathbb{C}^M$ de codimension $i \leq d$, les conditions suivantes sont équivalentes :

i) Toute direction T limite en 0 d'espaces tangents à des fibres de f en des points non-singuliers d'icelles contenus dans $Z - F$ est transverse à D_i, en ce sens que $\dim(T \cap D_i) = d - i$.

ii) Choisissant les coordonnées $z_1,\ldots,z_M$ de $\mathbb{C}^M$ de telle façon que D_i soit défini par $z_1 = \cdots = z_i = 0$, on a :

L'idéal $J_{X/S} \cdot \mathcal{O}_{Z,0}$ de $\mathcal{O}_{Z,0}$ engendré par les images des déterminants jacobiens $\dfrac{\partial(f_{j_1},\ldots,f_{j_c})}{\partial(z_{i_1},\ldots,z_{i_c})}$ où $c = M - d$, $\{j_1,\ldots,j_c\} \subset \{1,\ldots,m\}$, $\{i_1,\ldots,i_c\} \subset \{1,\ldots,M\}$ est entier dans $\mathcal{O}_{Z,0}$ sur l'idéal J_0 engendré par les seuls mineurs jacobiens de cette sorte qui sont tels que de plus on ait : $\{i_1,\ldots,i_c\} \subset \{i+1,\ldots,M\}$.

Démontrons i) ⇒ ii) : soit J_p l'idéal de $\mathcal{O}_{Z,0}$ engendré par les mineurs jacobiens au dénominateur desquels apparaissent au plus p variables d'indice non contenu dans $\{i+1,\ldots,M\}$. On a $J_c = J_{X/S} \cdot \mathcal{O}_{Z,0}$ et l'assertion ii) équivaut à l'égalité : $\overline{J_o} = \overline{J_c}$.

Il suffit de prouver que, pour $0 \leq p \leq c-1$, J_{p+1} est entier sur J_p. Pour cela, on utilise le critère valuatif de dépendance intégrale (Chap. I, 1.3.4). Considérons un chemin analytique $h : (\mathbb{D},0) \to (Z,0)$, et soit Δ un des mineurs engendrant J_p et tel que la valuation $v_o(J_p \circ h)$ de J_p selon h soit minime parmi celles des éléments de J_p. Supposons en un point $x \in h(\mathbb{D} - \{0\})$, la fibre de f passant par x est non-singulière et son espace tangent admet pour équations dans $\mathbb{C}^M$:

$$(*) \qquad \Delta \cdot dz_{i_k} = - \sum_{\ell \notin \{i_1,\ldots,i_c\}} \frac{\partial(f_{j_1},\ldots,f_{j_c})}{\partial(z_{i_1},\ldots,z_{i_{k-1}},z_\ell,z_{i_{k+1}},\ldots,z_{i_c})} dz_\ell$$

où tous les mineurs jacobiens apparaissant dans le membre de droite appartiennent à J_{p+1}. Si la valuation d'un élément de J_{p+1} selon h était inférieure à celle de Δ, la limite T des espaces tangents selon h serait contenue dans un hyperplan de la forme $\sum_{\ell \notin \{i,\ldots,N\}} A_\ell \, dz_\ell = 0$, contrairement à l'hypothèse de transversalité de i). Un raisonnement analogue montrerait que si $h^* J_p = 0$, alors $h(\mathbb{D})$ est tout entier formé de points critiques pour f.

Prouvons que ii) ⇒ i) : chaque limite d'espaces tangents étant atteinte le long d'un arc $h : (\mathbb{D},0) \to (Z,0)$ comme ci-dessus puisque le morphisme $\nu_f : N_f(X) \to X$ est une modification, il suffit de lire à l'envers la preuve de i) ⇒ ii).

§ 4. Espace conormal d'un espace analytique plongé.

J'ai été conduit par des suggestions de J.P.G. Henry et M. Merle à considérer aussi la construction suivante, dite "des limites d'hyperplans tangents" ou "construction de l'espace conormal", qui est en gros ce qui subsiste dans le cas local de la construction de la variété duale d'une variété projective (voir [Pi] ou [Kl 2]).

Soit $(X,0) \subset (\mathbb{C}^N,0)$ une immersion fermée d'espaces analytiques réduits pointés. Sur la partie non-singulière X^o de X, on a l'injection naturelle de fibrés vectoriels

$$0 \longrightarrow T_{X^o} \xrightarrow{\ i\ } T_{\mathbb{C}^N} \underset{\mathbb{C}^N}{\times} X^o \quad ,$$

et l'on considère le noyau $T^*_{X^o}\mathbb{C}^N$ de l'homomorphisme dual i^* :

$$0 \longrightarrow T^*_{X^o}\mathbb{C}^N \longrightarrow T^*_{\mathbb{C}^N} \underset{\mathbb{C}^N}{\times} X^o \xrightarrow{\ i^*\ } T^*_{X^o}$$

Le fibré projectif $\mathbb{P}(T^*_{X^o}\mathbb{C}^N)$ associé à $T^*_{X^o}\mathbb{C}^N$ est donc naturellement plongé dans $\mathbb{P}(T^*_{\mathbb{C}^N} \underset{\mathbb{C}^N}{\times} X^o) \simeq X^o \times \check{\mathbb{P}}^{N-1} \subset X \times \check{\mathbb{P}}^{N-1}$, où $\check{\mathbb{P}}^{N-1}$ désigne l'espace des hyperplans de $\mathbb{C}^N$.

<u>.1 Proposition-Définition</u> : <u>On appelle espace conormal à</u> $X \subset \mathbb{C}^N$ <u>l'adhérence</u> $C(X \subset \mathbb{C}^N)$ <u>dans</u> $X \times \check{\mathbb{P}}^{N-1}$ <u>du fibré projectif</u> $\mathbb{P}(T^*_{X^o}\mathbb{C}^N) \subset X \times \check{\mathbb{P}}^{N-1}$. <u>C'est un sous-espace analytique fermé réduit de dimension</u> N-1 <u>de</u> $X \times \check{\mathbb{P}}^{N-1}$.

<u>Preuve</u> : Choisissons un système de coordonnées $z_1,\dots,z_N$ sur $\mathbb{C}^N$. Le problème étant local sur X, nous pouvons supposer X défini dans $\mathbb{C}^N$ par p équations $f_i(z_1,\dots,z_N) = 0$, $1 \le i \le p$. Choisissons les coordonnées naturelles $(a_1 : \dots : a_N)$ sur $\check{\mathbb{P}}^{N-1}$, et considérons le sous-espace analytique fermé W de $X \times \check{\mathbb{P}}^{N-1}$ défini par les équations exprimant les conditions suivantes :

$$\operatorname{rang}\begin{vmatrix} a_1 , & \dots\dots\dots\dots & , a_N \\ \dfrac{\partial f_{i_1}}{\partial z_1} , & \dots\dots , & \dfrac{\partial f_{i_1}}{\partial z_N} \\ & & \\ \dfrac{\partial f_{i_{N-d}}}{\partial z_1} , & \dots , & \dfrac{\partial f_{i_{N-d}}}{\partial z_N} \end{vmatrix} < N-d+1 \quad .$$

D'après la définition, notre espace $C(X)$ n'est autre que l'adhérence dans $X \times \check{\mathbb{P}}^{N-1}$ de $W - (\mathrm{Sing}\, X \times \check{\mathbb{P}}^{N-1})$ d'où aussitôt l'analyticité, et le fait que $C(X \subset \mathbb{C}^N)$ est réduit.

Pour calculer la dimension de $C(X \subset \mathbb{C}^N)$, considérons le morphisme naturel

$$\varkappa : C(X \subset \mathbb{C}^N) \longrightarrow X$$

induit par la première projection $X \times \check{\mathbb{P}}^{N-1} \to X$. Pour tout $x \in X^o$, $\tau^{-1}(x) \subset \check{\mathbb{P}}^{N-1}$ est l'ensemble des hyperplans de $\mathbb{C}^N$ contenant $T_{X,x}$, et l'on a donc $\tau^{-1}(x) = \mathbb{P}^{N-d-1}$ où $d = \dim_x X$. On en déduit bien l'égalité $\dim C(X \subset \mathbb{C}^N) = N-1$.

<u>4.1.1</u> La plupart du temps, nous travaillerons avec un plongement $X \subset \mathbb{C}^N$ fixé et nous noterons simplement

$$\varkappa : C(X) \longrightarrow X$$

l'espace conormal. Noter que le morphisme $\varkappa$ est projectif, et en particulier propre.

<u>4.2</u> <u>Remarques</u> : 1) (On suppose X purement de dimension d.) Pour tout $x \in X$, on a : $N-d-1 \leq \dim \varkappa^{-1}(x) \leq N-2$.

2) On peut penser à un point de $C(X)$ comme à un couple (x,H) où H est une limite $(i \to \infty)$ d'hyperplans H_i tels qu'il existe une suite $x_i \in X^o$, $x_i \to x$, avec $H_i \supset T_{X,x_i}$.

3) Si $N = d+1$, $\varkappa : C(X) \to X$ s'identifie à la modification de Nash de X.

<u>4.3</u> <u>Proposition-Définition</u> : <u>Soit</u> $f : X \to S$ <u>un morphisme satisfaisant les hypothèses de</u> 1.1 <u>et de plus muni d'un</u> $\mathcal{S}$-<u>plongement</u> $X \subset S \times \mathbb{C}^N$. <u>Avec les notations de</u> 1.1, <u>considérons le fibré conormal relatif</u> $T^*_{X-F,f}$ (<u>ou</u> $T^*_{X-F/S}$) <u>défini</u>

comme noyau de l'homomorphisme naturel $T^*_{S\times\mathbb{C}^N/S} \underset{S\times\mathbb{C}^N}{\times} (X-F) \xrightarrow{i^*} T^*_{X-F/S}$. On appelle espace conormal relatif la fermeture dans $X \times \check{\mathbb{P}}^{N-1}$ de $\mathbb{P}(T^*_{X-F/S}) \subset (X-F) \times \check{\mathbb{P}}^{N-1} \subset \check{\mathbb{P}}^{N-1}$. C'est un espace analytique réduit $C_f(X) \xrightarrow{\varkappa_f} X$ propre au-dessus de X, de dimension $N-1+\dim S$.

La preuve est tout-à-fait analogue à celle du cas absolu.

Ici encore, si $N = d+1$, où d est la dimension des fibres de f, supposées réduites et de dimension pure, l'espace conormal relatif peut être naturellement identifié à la modification de Nash relative.

***

CHAPITRE III

STRATIFICATIONS

__Introduction.__ Dans ce chapitre, je commence par montrer comment, étant donnée une condition d'incidence portant sur des couples de sous-espaces non-singuliers (X_α, X_β) d'un espace X, et satisfaisant des hypothèses très simples, il est possible de construire, pour tout espace analytique X, des stratifications $X = \bigcup X_\alpha$ telles que tout couple de strates (X_α, X_β) satisfasse la condition d'incidence donnée.

L'exemple typique pour nous de condition d'incidence, les conditions de Whitney, est introduit et l'on montre comment le théorème de Bertini idéaliste du chapitre précédent implique que cette condition d'incidence est stratifiante.

Au § 3, j'étudie brièvement les stratifications définies par des invariants numériques, puisque le résultat principal de ce travail montre en particulier que la stratification de Whitney "canonique" d'un espace analytique est de cette nature. Ensuite on étudie un peu la transversalité de deux sous-ensembles analytiques stratifiés d'un espace analytique non-singulier, surtout pour montrer des lemmes techniques qui seront utiles dans la suite et énoncer le théorème de généricité par translation de Kleiman. Le chapitre se termine par un résultat facile selon lequel une condition de transversalité implique l'égalité ensembliste du transformé strict et du transformé total d'un sous-espace fermé $Z \subset X$ par un éclatement $p : X' \to X$.

§ 1. Conditions d'incidence.

1.1 Définition : Soit X un espace analytique. Nous dirons qu'un sous-ensemble A de X est localement fermé à la Zariski dans X si il existe deux sous-ensembles analytiques fermés F et G de X tels que $A = F - G$.
Nous dirons que A est constructible dans X si tout point $x \in X$ possède un voisinage ouvert U tel que $A \cap U$ soit combinaison booléenne de sous-ensembles localement fermés à la Zariski dans U.

On remarquera que, d'après [W] , la fermeture dans X d'un sous-ensemble constructible de X est un sous-ensemble analytique fermé de X, ainsi que la fermeture de sa frontière, et que la classe des sous-ensembles constructibles est stable par les opérations booléennes alors que celle des sous-ensembles localement fermés à la Zariski est stable par union finie et intersection finie.

1.2 Définition : Une famille $(A_i)_{i \in I}$ de sous-ensembles d'un espace topologique X est dite localement finie si tout point $x \in X$ possède un voisinage U tel que $\{i \in I / A_i \cap U \neq \emptyset\}$ soit un ensemble fini.

Exercice : Montrer que tout sous-ensemble constructible A d'un espace analytique X est réunion d'une famille localement finie de sous-ensembles localement fermés à la Zariski de X .

1.3 Définition : Soient E et F deux sous-ensembles constructibles (resp. localement fermés à la Zariski) d'un espace analytique X . La frontière de E dans F est le sous-ensemble constructible (resp. localement fermé à la Zariski) dans X et rare dans F :

$$\partial_F(E) = \overline{E-F} \cap F \quad .$$

Dans tout ce qui suit, on peut lire "constructible" à la place de "localement fermé à la Zariski", vocable que l'on abrègera d'ailleurs en "localement fermé" pour alléger.

1.4 Définition : Nous appellerons "condition d'incidence" toute condition portant sur des quadruplets (X,S_1,S_2,x), où X est un espace analytique, S_1 est un sous-ensemble localement fermé de X, S_2 un autre sous-ensemble localement fermé de X contenu dans $\overline{S}_1 - S_1$, enfin x est un point de S_2 ; cette condition est astreinte à vérifier une condition d'hérédité que voici :

(H) Pour tout quadruplet (X,S_1,S_2,x) satisfaisant la condition d'incidence, et tout sous-ensemble $S_2' \subset S_2$ localement fermé dans X et non-singulier en $x \in S_2'$, le quadruplet (X,S_1,S_2',x) satisfait encore la condition d'incidence.

Nous dirons qu'une condition d'incidence est stratifiante si, pour tout triplet (X,S_1,S_2) comme ci-dessus, l'ensemble des points $x \in S_2$ tels que le quadruplet (X,S_1,S_2,x) satisfasse la condition d'incidence contient un sous-ensemble constructible dans X (ou, de façon équivalente, dans $\overline{S}_2$) et dense dans $\overline{S}_2$.

1.5 Proposition : Etant donnée une condition d'incidence stratifiante, pour tout espace analytique X et toute famille localement finie $(X_i)_{i\in I}$ de sous-ensembles localement fermés de X, il existe une partition de X en sous-ensembles localement fermés $(S_\alpha)_{\alpha\in A}$, formant une famille localement finie dans X, telle que chaque S_α soit non-singulier, connexe et non vide, et que de plus la partition $X = \bigcup_{\alpha\in A} S_\alpha$ satisfasse les conditions suivantes :

i) Pour $i \in I$, $S_\alpha \cap \overline{X}_i \neq \emptyset \Rightarrow S_\alpha \subset \overline{X}_i$, et $S_\alpha \cap (\overline{X}_i - X_i) \neq \emptyset \Rightarrow S_\alpha \subset \overline{X}_i - X_i$, c'est-à-dire que pour $i \in I$, $\overline{X}_i$ et X_i sont union de strates.

ii) Pour $\alpha,\beta \in A$, $\alpha \neq \beta$, $\overline{S}_\alpha \cap S_\beta \neq \emptyset \Rightarrow \overline{S}_\alpha - S_\alpha \supseteq S_\beta$ et, pour tout $x \in S_\beta$, le quadruplet (X,S_α,S_β,x) satisfait la condition d'incidence donnée.

Lemme 1 : Soient X un espace analytique, et $(Z_n)_{n\in\mathbb{N}}$ une suite décroissante de sous-ensembles analytiques fermés de X tels que si $Z_n \neq \emptyset$, Z_{n+1} soit rare *dans* Z_n, *pour tout* $n \geq 0$, *alors on a* : $\bigcap_{n\geq 0} Z_n = \emptyset$ *et en fait tout point* $x \in X$ *possède un voisinage ouvert* U *tel que* : *il existe* n_o *tel que* $U \cap Z_n = \emptyset$ *pour* $n \geq n_o$.

Preuve : D'après l'hypothèse, en tout point $x \in X$, on a $\dim_x Z_{n+1} < \dim_x Z_n$. Soit $x \in X$, et posons $d = \dim_x X$. D'après la semi-continuité de la dimension, il existe un voisinage ouvert U de x tel que pour $y \in U$ on ait $\dim_y X \leq d$. D'après la définition topologique de la dimension des espaces analytiques, on a $Z_n \cap U = \emptyset$ pour $n \geq d+1$. ∎

Lemme 2 : *Soient* $Z' \subset Z \subset X$ *deux sous-ensembles analytiques fermés de* X *et* E *un sous-ensemble localement fermé de* X ; *si l'on a* $\partial_{\overline{E}}(Z) \subset Z'$, *alors* $\overline{E} \cap (Z - Z')$ *est ouvert et fermé dans* $Z - Z'$.

Preuve : Soit $x \in \overline{E} \cap (Z - Z')$; d'après la définition de $\partial_{\overline{E}}(Z)$, on a : $x \notin \overline{Z - \overline{E}} \cap \overline{E}$, donc il existe un voisinage ouvert U de x dans X tel que $U \cap E \neq \emptyset$ implique $U \cap (Z - \overline{E}) \neq \emptyset$. Or $x \in U \cap E$, donc $U \cap Z \subset \overline{E}$; on peut supposer que $U \cap Z' = \emptyset$ puisque Z' est fermé et $x \notin Z'$, et on a donc un ouvert U tel que $U \cap Z \subset \overline{E}$ et $U \cap Z' \cap E = \emptyset$, c'est-à-dire que $U \cap Z$ est un voisinage ouvert de x dans $\overline{E} \cap (Z - Z')$. Enfin, $\overline{E} \cap (Z - Z')$ est évidemment fermé dans $Z - Z'$. ∎

Supposons maintenant avoir construit un sous-ensemble localement fermé $Z_n \subset X$, et des sous-ensembles localement fermés $(S_\alpha)_{\alpha \in A}$ de X formant une partition de $X - Z_n$ qui satisfait les conditions de la Proposition. Nous allons construire un sous-ensemble fermé rare $Z_{n+1} \subset Z_n$ et des sous-ensembles $S'_{\alpha'} \subset Z_n - Z_{n+1}$ localement fermés dans X et tels que la collection des S_α, $S'_{\alpha'}$ soit une partition de $X - Z_{n+1}$ satisfaisant aux conditions de la Proposition.

On commencera la construction avec $Z_o = X$, et on construira ainsi par récurrence une collection de strates S_α satisfaisant aux conditions de la Proposition pour $X = X - \bigcap_{n\geq 0} Z_n$ (Lemme 1).

Posons, pour tout $\alpha \in A$, $T_\alpha = \partial_{Z_n}(\overline{S}_\alpha)$ et pour tout $i \in I$, $Y_i = \partial_{Z_n}(\overline{X}_i)$; ce sont deux familles localement finies de sous-ensembles localement fermés de X.

Considérons les triplets $(X,S_\alpha,(\overline{S}_\alpha - S_\alpha)\cap Z_n)$. L'hypothèse que notre condition d'incidence est stratifiante nous assure de l'existence de sous-ensembles denses $V_\alpha \subset (\overline{S}_\alpha - S_\alpha)\cap Z_n$ localement fermés dans X, tels que le quadruplet $(X,S_\alpha,(\overline{S}_\alpha - S_\alpha)\cap Z_n,x)$ satisfasse la condition d'incidence donnée pour tout $x\in V_\alpha$. Soit R_α la frontière totale de V_α dans Z_n, c'est-à-dire

$$R_\alpha = (\overline{V}_\alpha - V_\alpha)\cup(\overline{Z_n - V_\alpha} - (Z_n - V_\alpha)) \quad .$$

La famille des R_α est encore une famille localement finie de sous-ensembles localement fermés.

Posons $Z_{n+1} = \mathrm{Sing}\, Z_n \cup (\bigcup_{i\in I} Y_i)\cup(\bigcup_{\alpha\in A} T_\alpha)\cup(\bigcup_{\alpha\in A} R_\alpha)$; Z_{n+1} est un sous-ensemble analytique fermé de Z_n, donc de X, rare dans Z_n, puisque chacune des familles en vue est une famille localement finie de sous-ensembles analytiques fermés rares de Z_n.

Définissons les $S'_{\alpha'}$ comme étant les composantes connexes de $Z_n - Z_{n+1}$. La fermeture de chaque $S'_{\alpha'}$ est une composante irréductible de Z_n et les $S'_{\alpha'}$ sont donc bien localement fermés dans X. Par ailleurs, la famille des $S'_{\alpha'}$ est localement finie. Pour vérifier que la stratification de $X - Z_{n+1}$ par $(S_\alpha)_{\alpha\in A}$ et $(S'_{\alpha'})_{\alpha'\in A'}$ satisfait la condition d'incidence, on peut, d'après l'hypothèse de récurrence et puisque $\alpha' \neq \beta' \Rightarrow S'_{\alpha'} \cap \overline{S'_{\beta'}} = \emptyset$, se contenter de le vérifier pour une strate S_α et une strate $S'_{\alpha'}$; supposons donc $S'_{\alpha'} \cap \overline{S}_\alpha \neq \emptyset$, et remarquons que $\overline{Z_n - \overline{S}_\alpha} \cap \overline{S}_\alpha = T_\alpha \subset Z_{n+1}$. D'après le Lemme 2, $\overline{S}_\alpha \cap (Z_n - Z_{n+1})$ est ouvert et fermé dans $Z_n - Z_{n+1}$ et donc contient chaque composante connexe de $Z_n - Z_{n+1}$ qu'il rencontre, d'où l'inclusion $\overline{S}_\alpha \supset S'_{\alpha'}$. Le même argument montre que, puisque $Z_{n+1} \supset Y_i$, le sous-ensemble $\overline{X}_i \cap (Z_n - Z_{n+1})$ est ouvert et fermé dans $Z_n - Z_{n+1}$ et donc contient chaque $S'_{\alpha'}$ qu'il rencontre. Vérifions finalement que si $\overline{S}_\alpha \supset S'_{\alpha'}$, la condition d'incidence est satisfaite pour chaque quadruplet $(X,S_\alpha,S'_{\alpha'},x)$ avec $x\in S'_{\alpha'}$. Il suffit pour cela, grâce à la condition (H), de vérifier l'inclusion $S'_{\alpha'} \subset V_\alpha$. Or, puisque Z_{n+1} contient R_α, on a l'égalité $S'_{\alpha'} \cap V_\alpha = S'_{\alpha'} \cap \overline{V}_\alpha$ et par le Lemme 2, $\overline{V}_\alpha \cap (Z_n - Z_{n+1})$ est ouvert et fermé dans $Z_n - Z_{n+1}$, donc contient chaque composante connexe $S'_{\alpha'}$ qu'il rencontre, d'où

$S'_{\alpha'} \subset \overline{V}_\alpha$ et donc $S'_{\alpha'} \subset V_\alpha$.

On a montré toute la proposition sauf le fait que les $\overline{X_i} - X_i$ étaient réunion de strates. Il suffit pour cela d'appliquer la Proposition précédente à la famille localement finie constituée des fermés analytiques $(\overline{X_i}, \overline{X_i} - X_i)_{i \in I}$.

Remarque : Dans le cas où l'on a remplacé "localement fermé" par "constructible", dans la lecture de ce qui précède, il faut remplacer la famille $(\overline{X_i}, \overline{X_i} - X_i)$ par la famille localement finie de sous-ensembles localement fermés à la Zariski que l'on aura construite en résolvant l'exercice proposé plus haut pour chacun des X_i .

§ 2. Conditions de Whitney.

2.1 Considérons l'espace vectoriel $\mathbb{C}^M$ muni de coordonnées $u_1, \ldots, u_M$ et de la métrique hermitienne usuelle ; pour u, v dans $\mathbb{C}^M$, $(u,v) = \Sigma\, u_i\, \overline{v_i}$. Etant donnés deux sous-espaces vectoriels A et B de $\mathbb{C}^M$, on définit la distance de A à B (dans cet ordre) :

$$\operatorname{dist}(A,B) = \sup_{\substack{u \in B^\perp - \{0\} \\ v \in A - \{0\}}} \left(\frac{|u,v|}{\|u\| \cdot \|v\|} \right)$$

où $B^\perp = \{u \in \mathbb{C}^M / (u,b) = 0$ pour tout $b \in B\}$ et $\|u\|^2 = (u,u)$. On notera que $\operatorname{dist}(A,B) = 0$ équivaut à l'inclusion $B^\perp \subset A^\perp$, c'est-à-dire $B \supset A$. Par ailleurs, soit $G = G(M,a)$ la grassmannienne des directions de sous-espaces de dimension a de $\mathbb{C}^M$; posant $a = \dim A$, $b = \dim B$, l'application $G_a \times G_b \to \mathbb{R}$ définie par $(A,B) \mapsto \operatorname{dist}(A,B)$ est une fonction analytique réelle sur le produit $G_a \times G_b$, comme on le vérifie aussitôt. Remarquons aussi que l'inégalité de Schwarz implique $\operatorname{dist}(A,B) \leq 1$.

Soit $p : \mathbb{C}^M \to \mathbb{C}^\ell$ une projection de la forme $(u_1, \ldots, u_M) \to (u_1, \ldots, u_\ell)$ et telle que $\operatorname{Ker} p \cap A = (0)$. On a alors l'inclusion $p(B)^\perp \subset B^\perp$ où $p(B)^\perp$ désigne l'orthogonal de p(B) dans $\mathbb{C}^\ell$ muni de la structure hermitienne définie par les coordonnées $(u_1, \ldots, u_\ell)$: en effet, pour tout $v \in \mathbb{C}^\ell$, on a $(v,b) = (v,p(b))$;

pour la même raison, on a :

$$\sup_{\substack{u'\in p(B)^{\perp}-\{0\}\\ v'\in p(A)-\{0\}}} \left(\frac{|(u',v')|}{\|u'\|\cdot\|v'\|}\right) = \sup_{\substack{u'\in p(B)^{\perp}-\{0\}\\ v\in A-\{0\}}} \left(\frac{|(u',v)|}{\|u'\|\cdot\|v\|}\cdot\frac{\|v\|}{\|p(v)\|}\right).$$

D'après l'hypothèse $\mathrm{Ker}\, p \cap A = (0)$, il existe une constante positive C ne dépendant que de A et de la projection choisie et telle que pour tout $v \in A - \{0\}$ on ait $\frac{\|v\|}{\|p(v)\|} \leq \frac{1}{C} < \infty$; il vient donc

$$\mathrm{dist}\,(p(A),p(B)) \leq \frac{1}{C} \sup_{\substack{u'\in p(B)^{\perp}-\{0\}\\ v\in A-\{0\}}} \left(\frac{|(u',v)|}{\|u'\|\cdot\|v\|}\right) \leq \frac{1}{C}\,\mathrm{dist}\,(A,B)$$

la dernière inégalité provenant de l'inclusion $p(B)^{\perp} \subset B^{\perp}$.

Voici une autre inégalité du même type qui nous servira plus bas : Reprenons la même situation que ci-dessus, mais sans supposer que $\mathrm{Ker}\,(p|A) = 0$. Soit B_1 un sous-espace vectoriel de $\mathbb{C}^{\ell}$. Il existe une constante positive C ne dépendant que de p, A et B_1 et telle que $\mathrm{dist}\,(p(A),B_1) \leq C\,\mathrm{dist}\,(A,p^{-1}(B_1))$. En effet, pour tout $p(a) \in p(A)$, soit $E_a = \{w \in \mathrm{Ker}\, p / p(a) + w \in A\}$; E_a est un sous-espace affine de Ker p de la forme $w_0 + \mathrm{Ker}(p|A)$, et l'application $p(A) \to \mathrm{Ker}\, p / \mathrm{Ker}\,(p|A)$ qui à p(a) associe la classe de w_0 est linéaire. On en déduit que tout élément de p(A) peut s'écrire $p(p(A) + w(a))$ où $\|w(a)\| \leq C_1 \|p(a)\|$, $C_1 \in \mathbb{R}_+$. Or, puisque $(p(p^{-1}(B_1)))^{\perp} = B_1^{\perp}$ est contenu dans $(p^{-1}(B_1))^{\perp}$, on a l'inégalité :

$$\mathrm{dist}\,(A,p^{-1}(B_1)) \geq \sup_{\substack{a\in A-\mathrm{Ker}\, p\\ v\in B_1^{\perp}-\{0\}}} \left\{\frac{|(a,v)|}{\|a\|\cdot\|v\|}\right\}$$

mais on a $(a,v) = (p(a),v)$ puisque $v \in B_1^{\perp} \subset \mathbb{C}^{\ell}$ par définition de $B_1^{\perp}$, et il vient :

$$\sup_{\substack{a\in A-\mathrm{Ker}\,p\\ v\in B_1^{\perp}-\{0\}}} \left\{\frac{|(a,v)|}{\|a\|\cdot\|v\|}\right\} \geq \sup_{\substack{p(a)\in p(A)-\{0\}\\ v\in B_1^{\perp}-\{0\}}} \left\{\frac{|(p(a),v)|}{\|p(a)\|+\|w(a)\|}\right\}$$

$$\geq \left(\frac{1}{1+C_1}\right) \sup_{\substack{p(a)\in p(A)-\{0\}\\ v\in B_1^{\perp}-\{0\}}} \left\{\frac{|(p(a),v)|}{\|p(a)\|\cdot\|v\|}\right\} ,$$

d'où le résultat avec $C = 1+C_1$.

2.2 Soient maintenant X un espace analytique réduit purement de dimension d, Y un sous-espace localement fermé à la Zariski de X et 0 un point non-singulier de Y. Choisissons un plongement local $(X,0)\subset(\mathbb{C}^N,0)$ au voisinage de 0, et une rétraction locale $\rho : (\mathbb{C}^N,0)\to(Y,0)$; à un isomorphisme analytique près, on peut alors identifier Y à (un ouvert de) $\mathbb{C}^k$ et supposer X plongé dans (un ouvert ouvert de) $\mathbb{C}^k\times\mathbb{C}^{n-k}$ de telle façon que la rétraction ρ coïncide avec la première projection.

2.2.1 Définition : On dit que le couple de strates (X^o,Y^o) formé de la partie non singulière X^o de X et de la partie non-singulière Y^o de Y satisfait la condition a) de Whitney en $0\in Y^o$ si il existe un plongement local comme ci-dessus tel que pour toute suite de points $x_i\in X^o$ tendant vers 0, on ait, quitte à extraire une sous-suite telle que $\mathrm{Lim}\, T_{X,x_i}$ existe, l'inclusion

$$\mathrm{Lim}\, T_{X,x_i} \supset T_{Y,0} \qquad \text{(en direction)}$$

c'est-à-dire encore

$$\lim_{x_i\to 0} \mathrm{dist}\,(T_{Y,0},T_{X,x_i}) = 0 .$$

On dit,(après Hironaka [H 1]), que (X^o,Y) satisfait la condition a) de Whitney stricte avec exposant e si e est un nombre réel positif tel qu'il existe un voisinage ouvert U de 0 dans X et un nombre réel positif C tels que pour tout $x\in X^o\cap U$ on ait l'inégalité

$$\text{dit}\,(T_{Y,0}\,,\,T_{X,x}) \leq C\ \text{dist}\,(x,Y)^e$$

où dist(x,Y) désigne la distance de x à Y dans $\mathbb{C}^N$.

On dit que le couple de strates (X^o,Y^o) satisfait la condition b) de Whitney en $0 \in Y^o$ si il existe un plongement local et une rétraction ρ comme ci-dessus tels que pour toute suite de points $x_i \in X^o - Y$, notant $\widehat{x_i\ \rho(x_i)}$ la direction de la droite (sécante) qui joint x_i à $\rho(x_i)$ dans $\mathbb{C}^N$ on ait, quitte à extraire une sous-suite telle que $\lim_{x_i \to 0} T_{X,x_i}$ et $\lim_{x_i \to 0} \widehat{x_i\ \rho(x_i)}$ existent, l'inclusion

$$\lim_{x_i \to 0} T_{X,x_i} \supset \lim_{x_i \to 0} \widehat{x_i\ \rho(x_i)}$$

c'est-à-dire encore

$$\lim_{x_i \to 0} \text{dist}\,(\widehat{x_i\ \rho(x_i)},T_{X,x_i}) = 0 \quad .$$

On dit, après Hironaka (Loc. cit.), que (X^o,Y) satisfait la condition b) de Whitney avec exposant e en 0 si e est un nombre réel positif tel qu'il existe un voisinage ouvert U de 0 dans X et une constante positive C tels que pour tout $x \in X^o \cap U$ on ait l'inégalité

$$\text{dist}\,(\widehat{x\ \rho(x)},T_{X,x}) \leq C\,.\,\text{dist}\,(x,Y)^e \quad .$$

<u>2.2.2 Proposition</u> (Whitney, [W]) : <u>La condition, portant sur des quadruplets</u> (X,S_1,S_2,x) <u>comme plus haut, que voici</u> : x <u>est un point non-singulier de</u> S_2 <u>et le couple de strates</u> (S_1,S_2) <u>satisfait les conditions</u> a) <u>et</u> b) <u>en</u> $x \in S_2$ <u>est une condition d'incidence stratifiante</u>.

<u>Démonstration</u> : Remarquons d'abord que la condition d'hérédité est évidemment satisfaite. Nous allons décrire une condition sur (X^o,Y) qui implique, au voisinage d'un point non-singulier de Y, les conditions de Whitney.

Rappelons tout d'abord qu'à l'immersion $\overline{Y} \hookrightarrow X$ correspond un morphisme surjectif de $\mathcal{O}_Y$-modules

$$\Omega_X^1 | Y \longrightarrow \Omega_Y^1 \longrightarrow 0$$

exprimant l'inclusion de l'espace tangent de Zariski $\mathrm{Specan}_Y\, \mathrm{Sym}_{\mathcal{O}_Y}(\Omega_Y^1)$ de Y dans la restriction à Y de l'espace tangent de Zariski de X.

(On rappelle que l'espace vectoriel relatif associé à un faisceau cohérent tel que Ω_Y^1, qui n'est pas en général un fibré vectoriel, a pour faisceau de sections le faisceau dual du faisceau cohérent donné.

Rappelons aussi que Ω_X^1 est défini comme ceci : soit I le faisceau cohérent définissant la diagonale $X \overset{\delta}{\hookrightarrow} X\times X$ dans l'ouvert de $X\times X$ où elle est fermée ; le $\mathcal{O}_X$-module $\delta^* I/I^2$ est $\mathcal{O}_X$-isomorphe à Ω_X^1. Considérons l'immersion fermée naturelle $id_{\overline{Y}} \times i$ où i désigne l'immersion $\overline{Y} \hookrightarrow X$, et le diagramme

$$\begin{array}{ccc} \overline{Y}\times X & \hookrightarrow & X\times X \\ \uparrow & & \uparrow \\ \overline{Y} & \hookrightarrow & X \end{array}$$

On en déduit un homomorphisme surjectif de $\mathcal{O}_{\overline{Y}}$-modules

$$\Omega_X^1 | \overline{Y} \longrightarrow N_{X,\overline{Y}} \longrightarrow 0$$

où $N_{X,\overline{Y}}$ désigne le faisceau conormal de $\overline{Y}$ dans X, c'est-à-dire le faisceau de $\mathcal{O}_{\overline{Y}}$-modules S/S^2 où S est l'idéal définissant $\overline{Y}$ dans X.

Cet homomorphisme exprime le fait que les limites de sécantes joignant un point de Y à un point de $X - Y$ (pour un plongement local) donnent des directions tangentes à X aux points de Y.

Considérons maintenant le diagramme

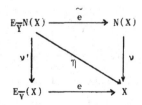

$$\begin{array}{ccc} E_{\overline{Y}}N(X) & \xrightarrow{\ \tilde{e}\ } & N(X) \\ {\scriptstyle \nu'}\downarrow & \searrow^{\eta} & \downarrow{\scriptstyle \nu} \\ E_{\overline{Y}}(X) & \xrightarrow{\ e\ } & X \end{array}$$

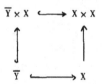

où ν désigne la modification de Nash de X, e l'éclatement de $\overline{Y}$ dans X et $\tilde{e}$ l'éclatement du sous-espace analytique $\nu^{-1}(\overline{Y})$ dans N(X). La propriété universelle de l'éclatement implique alors l'existence d'un morphisme ν' faisant commuter le diagramme et l'on pose $\eta = \nu \circ \tilde{e} = e \circ \nu'$.

Posons $Z = E_{\overline{Y}}N(X)$ et $\mathcal{Y} = \eta^{-1}(\overline{Y})$. Par construction, le $\mathcal{O}_Z$-module $\eta^*\Omega^1_X$ a un quotient localement libre, d'où une suite exacte

$$0 \longrightarrow K \longrightarrow \eta^*\Omega^1_X \longrightarrow L \longrightarrow 0$$

que l'on peut restreindre à $\mathcal{Y}$:

$$0 \longrightarrow K|\mathcal{Y} \longrightarrow \eta^*\Omega^1_X|\mathcal{Y} \longrightarrow L|\mathcal{Y} \longrightarrow 0 \quad .$$

D'autre part, on a sur $\mathcal{Y}$ la suite exacte

$$\eta^*\Omega^1_X|\mathcal{Y} \longrightarrow \eta^*\Omega^1_{\overline{Y}} \longrightarrow 0 \quad .$$

De même, si l'on note $\mathcal{J}$ l'idéal de Z définissant $\mathcal{Y}$, on a une surjection naturelle de $\mathcal{O}_{\mathcal{Y}}$-modules :

$$\eta^* N_{X,\overline{Y}} \longrightarrow \mathcal{J}/\mathcal{J}^2 \longrightarrow 0$$

et une surjection

$$\eta^*\Omega^1_X|\mathcal{Y} \longrightarrow \eta^* N_{X,\overline{Y}} \longrightarrow 0 \quad .$$

Soit maintenant $y \in Y$. Si en tout point $y' \in \eta^{-1}(y)$ l'homomorphisme

$$(\eta^*\Omega^1_X|\mathcal{Y})_{y'} \longrightarrow (\eta^*\Omega^1_{\overline{Y}})_{y'} \longrightarrow 0$$

se factorise par $(L|\mathcal{Y})_{y'}$, grâce à la propreté de η, il existe un voisinage ouvert U de y dans $\overline{Y}$ tel que l'on ait une surjection

$$L|\eta^{-1}(U) \longrightarrow \eta^*\Omega^1_{\overline{Y}}|\eta^{-1}(U) \longrightarrow 0$$

ce qui signifie précisément, d'après la définition de L, que la condition a) de Whitney est vérifiée pour tout plongement local de X dans un espace affine au voisinage d'un point $y_1 \in U$. En effet, pour un plongement local $X \subset \mathbb{C}^N$, on peut comme nous l'avons vu plus haut, identifier le fibré vectoriel sur N(X) associé à L à la restriction à N(X) du fibré $X \times \mathcal{C}$ sur $X \times G$, où $\mathcal{C}$ est le fibré tautologique sur la grassmannienne G des d-plans de $\mathbb{C}^N$, et la surjection ci-dessus se traduit par une injection $\eta^*(T_{Y^o}) \hookrightarrow X \times \mathcal{C} | \eta^{-1}(Y)$.

Soit $B_1 \subset \mathcal{Y}$ le support dans $\mathcal{Y}$ du $\mathcal{O}_{\mathcal{Y}}$-module cohérent image de $K|\mathcal{Y}$ par l'homomorphisme :

$$\eta^*\Omega^1_X|\mathcal{Y} \longrightarrow \eta^*\Omega^1_{\overline{Y}} \longrightarrow 0 \quad .$$

Il est clair que la condition ci-dessus est satisfaite pour tout point y de l'ouvert (éventuellement vide) $Y^o - \eta(B_1)$.

Soit de même B_2 le support dans $\mathcal{Y}$ du $\mathcal{O}_{\mathcal{Y}}$-module cohérent image de $K|\mathcal{Y}$ par l'homomorphisme composé

$$\eta^*\Omega^1_X|\mathcal{Y} \longrightarrow \eta^* N_{X,\overline{Y}} \longrightarrow \mathcal{I}/\mathcal{I}^2 \quad .$$

Il est clair que tout point y' de l'ouvert (éventuellement vide) $Y^o - \eta(B_2)$ possède un voisinage ouvert U dans $\overline{Y}$ tel que l'on ait une surjection

$$L|\eta^{-1}(U) \longrightarrow \mathcal{I}/\mathcal{I}^2|\eta^{-1}(U) \longrightarrow 0$$

ce qui, en utilisant un plongement local et le fait que la fibre du fibré en droites sur $\mathcal{Y}$ correspondant au faisceau inversible de $\mathcal{O}_{\mathcal{Y}}$-modules $\mathcal{I}/\mathcal{I}^2$ en un point $y' \in \mathcal{Y}$ n'est autre que la direction limite en $y = \eta(y')$ de directions de sécantes qui correspond au point $\nu'(y') \in E_{\overline{Y}}(X)$, implique que la condition b) de Whitney est satisfaite pour tout plongement local $X \subset \mathbb{C}^N$ et en tout point de U, puisque la surjection ci-dessus implique une injection, pour toute suite de points $x_i \in X^o$ tendant vers $y = \eta(y')$, que voici :

$$\{(y', \text{limite } (x_i \to y) \text{ en } y \text{ de direction de sécantes } \widehat{x_i \rho(x_i)})\}$$
$$\subset \{(y', \text{limite en } y \text{ de direction d'espaces tangents } T_{X,x_i})\} \quad .$$

Par conséquent la condition d'incidence est vérifiée en tout point de l'ensemble $\overline{Y} - (\text{Sing } \overline{Y}) \cup \eta(B_1) \cup \eta(B_2)$ qui est localement fermé à la Zariski dans $\overline{Y}$ puisque B_1 et B_2 sont des sous-ensembles analytiques fermés de $\mathcal{Y}$ et que η est propre.

Inversement, étant donné un point $y \in Y^o$ tel que le couple (X^o, Y^o) satisfasse les conditions a) et b) de Whitney en tout point d'un voisinage U de y dans Y, on peut remonter l'argument précédent pour prouver que y appartient à $Y^o - \eta(B_1) \cup \eta(B_2)$.

Il ne nous reste plus qu'à montrer que le sous-ensemble $\overline{Y} - (\text{Sing } Y) \cup \eta(B_1) \cup \eta(B_2)$ est dense dans $\overline{Y}$. Pour cela, il suffit de prouver que l'ouvert formé des points de Y^o au voisinage desquels les conditions a) et b) de Whitney sont satisfaites est dense dans Y^o.

Soit donc 0 un point de Y^o. Choisissons un plongement local $(X,0) \subset (\mathbb{C}^{N-k} \times \mathbb{C}^k, 0)$ envoyant Y sur $0 \times \mathbb{C}^k$, et la rétraction locale $\rho : X \to Y$ induite par la projection $\mathbb{C}^{N-k} \times \mathbb{C}^k \to \mathbb{C}^k$. Munissons $\mathbb{C}^{N-k} \times \mathbb{C}^k$ des coordonnées $z_1, \dots, z_{N-k}, y_1, \dots, y_k$ et essayons, pour un point $x \in X^o$, d'estimer la distance de $T_{Y,0}$ (resp. de la droite $\widehat{x\,\rho(x)}$) à $T_{X,x}$. Supposons X défini dans $\mathbb{C}^N$ au voisinage de 0 par l'idéal engendré par $(f_1, \dots, f_m)$, f_i holomorphe sur $\mathbb{C}^N$ au voisinage de 0. L'espace tangent $T_{X,x}$ peut être défini dans $T_{\mathbb{C}^N,x}$, identifié à $\mathbb{C}^N$ muni des coordonnées $dz_1, \dots, dz_{N-k}, dy_1, \dots, dy_k$, par les N-d équations :

$$(E_i) \qquad \frac{\partial(f_1,\dots,f_{N-d})}{\partial(z_1,\dots,z_{N-d})}(x)dz_i = \sum_{\ell=1}^{k} \varepsilon_\ell \frac{\partial(f_1,\dots,f_{N-d})}{\partial(y_\ell, z_1,\dots,\hat{z}_i,\dots,z_{N-d})}(x)\, dy_\ell$$

$$+ \sum_{j=N-d}^{N-k} \varepsilon_j \frac{\partial(f_1,\dots,f_{N-d})}{\partial(z_j, z_1,\dots,\hat{z}_i,\dots,z_{N-d})}(x)\, dz_j$$

$(1 \leq i \leq N-d)$ où les ε valent ± 1, pourvu que le mineur jacobien $\frac{\partial(f_1,\dots,f_{N-d})}{\partial(z_1,\dots,z_{N-d})}(x)$ soit non nul. Comme nous l'avons vu au chapitre précédent, on peut toujours choisir un système de générateurs ayant cette propriété et même, étant donné un arc analytique $h : (\mathbb{D},0) \to (X,0)$ tel que $h(\mathbb{D} - \{0\}) \subset X^o$, tel que ce mineur jacobien ne s'annule en aucun point de $h(\mathbb{D} - \{0\})$.

Ainsi, l'espace vectoriel $T^{\perp}_{X,x}$ perpendiculaire à $T_{X,x}$ dans $T_{\mathbb{C}^N,0}$ est engendré par les N-d vecteurs w_i dont les coordonnées sont les complexes conjugués des déterminants jacobiens apparaissant dans l'équation (E_i). Par définition on a donc

$$\text{dist }(T_{Y,0},T_{X,x}) = \sup_{\substack{dy\in\mathbb{C}^k-\{0\}\\ \underline{\lambda}\in\mathbb{C}^{N-d}-\{0\}}} \left\{ \frac{|\sum_i \overline{\lambda}_i \ (\sum_{\ell=1}^{n} \varepsilon_\ell \frac{\partial(f_1,\dots,f_{N-d})}{\partial(y_\ell,z_1,\dots,\hat{z}_i,\dots,z_{N-d})} dy_\ell)|}{\|dy\| \cdot \|\sum \lambda_i \ w_i\|} \right\}$$

Or, l'interprétation transcendantale de la dépendance intégrale (Chap. I, 1.3.1) et l'énoncé du théorème de Bertini idéaliste avec section impliquent l'existence, dans un voisinage ouvert U de 0 dans Y, d'un fermé analytique rare F tel que tout point $y\in Y-F$ possède un voisinage ouvert V dans X tel qu'il existe une constante positive C telle que pour tout point $x\in X^o\cap V$ on ait les inégalités $(1\leq \ell\leq h)$:

$$(*) \quad \left|\frac{\partial(f_1,\dots,f_{N-d})}{\partial(y_\ell,z_1,\dots,\hat{z}_i,\dots,z_{N-d})}(x)\right| \leq$$

$$C \sup_{1\leq j\leq N-k} |z_j(x)| \ . \sup_{\{i_1,\dots,i_{N-d}\}\subset\{1,\dots,N-k\}} \left\{\left|\frac{\partial(f_1,\dots,f_{N-d})}{\partial(z_{i_1},\dots,z_{i_{N-d}})}(x)\right|\right\}$$

On peut supposer, sans perte de généralité, que le supremum des $\left|\frac{\partial(f_1,\dots,f_{N-d})}{\partial(z_{i_1},\dots,z_{i_{N-d}})}(x)\right|$ est atteint par $\left|\frac{\partial(f_1,\dots,f_{N-d})}{\partial(z_1,\dots,z_{N-d})}(x)\right|$. Un calcul sans mystère montre alors l'inégalité

$$\|\sum \lambda_i \ w_i\| \geq (\sum_{i=1}^{N-d} |\lambda_i|^2)^{1/2} . \left|\frac{\partial(f_1,\dots,f_{N-d})}{\partial(z_1,\dots,z_{N-d})}(x)\right|$$

$$\geq \sup |\lambda_i| \ . \left|\frac{\partial(f_1,\dots,f_{N-d})}{\partial(z_1,\dots,z_{n-d})}(x)\right|$$

et l'inégalité triangulaire, jointe à l'inégalité (*) nous donne aussitôt que pour $x\in X^o\cap V$ on a l'inégalité

$$\text{dist }(T_{Y,0},T_{X,x}) \leq C' \text{ dist }(x,Y)$$

où dist (x,Y) désigne la distance (par exemple $\sup |z_j(x)|$) de x à Y dans $\mathbb{C}^N$.
Nous avons donc montré <u>la condition a) de Whitney stricte avec exposant 1</u>, et

en particulier la condition a) de Whitney, en $y \in U - F$.

Avant de traiter la condition b) remarquons que, au voisinage de $0 \in Y$, l'espace $Z = E_{\overline{Y}}N(X)$ peut être construit comme adhérence dans $X \times \mathbb{P}^{N-k-1} \times G$, où G désigne la grassmannienne des d-plans dans $\mathbb{C}^N$, du graphe du morphisme $X^o - Y \to \mathbb{P}^{N-k-1} \times G$ défini par $x \mapsto (\overline{x\,\rho(x)}, T_{X,x})$. Nous avons donc un diagramme commutatif :

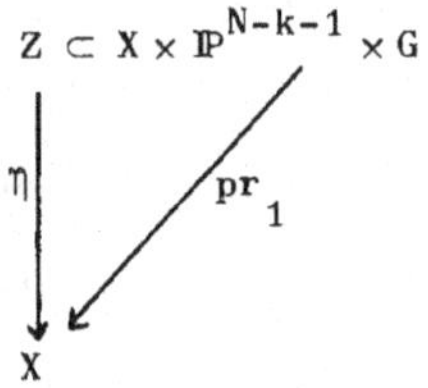

et si nous notons $\delta_b : \mathbb{P}^{N-k-1} \times G \to \mathbb{R}$ la fonction $(\ell, T) \mapsto \operatorname{dist}(\ell, T)$, vérifier la condition b) en $y \in Y$ revient à vérifier que $|\eta^{-1}(y)| \subset \{y\} \times \mathbb{P}^{N-k-1} \times G$ est contenu dans $\{y\} \times \delta_b^{-1}(0)$, et la condition b) de Whitney stricte avec exposant revient à vérifier que au voisinage de $\eta^{-1}(y)$, la fonction $\delta_b | Z$ est majorée par $C \cdot \operatorname{dist}(\eta(y'), Y)^e$. On peut (voir [H 3], § 9) assurer la condition b) de Whitney stricte avec un exposant e non précisé, essentiellement en utilisant la première inégalité de Łojasiewicz, pourvu que l'on sache que la fonction δ s'annule en tout point de $\eta^{-1}(Y)$, au-dessus d'un voisinage de 0, c'est-à-dire que la condition b) est vérifiée en tout point d'un voisinage de 0 dans Y.

Nous allons prouver ici que le couple de strates (X^o, Y^o) satisfait la condition b) en tout point $y \in U - F$, renvoyant à (Loc. cit.) pour la condition b) stricte. Puisque la restriction à Z de la fonction distance δ est analytique réelle, si δ_b ne s'annule pas identiquement sur $|\eta^{-1}(y)|$, on peut trouver $z \in \eta^{-1}(y)$ et un arc analytique réel $h : (\Pi, 0) \to (Z, z)$ (où $(\Pi =]-1, 1[)$ tel que la limite de δ_b le long de $h(\Pi)$ (i.e. $\lim_{t \to 0} (\delta \circ h)(t)$) soit différente de 0. Ceci résulte aussitôt du lemme des petits chemins (cf. [B-C], [H 3]). Montrons que, au contraire, cette limite est nulle, ce qui prouvera que la condition b) est satisfaite. Il suffit de montrer que pour le chemin $\eta \circ h : (\Pi, 0) \to (X, y)$ la limite en question est nulle. Pour un point $x(t) = \eta \circ h(t) \in X^o - Y$ de coordonnées $z_1(t),\dots,z_{N-k}(t),y_1(t),\dots,y_k(t)$, estimons la distance dist $(\widehat{x(t)\ \rho(x(t))},T_{X,x(t)})$: on a $(\delta_b \circ h)(t) = \mathrm{dist}(x(t),T_{X,x(t)})$ qui est le supremum pour $\underline{\lambda} \in \mathbb{C}^{N-d} - \{0\}$ des quotients

$$\left\{\frac{\left|\sum_i \overline{\lambda}_i \left(\sum_{j=N-d}^{N-k} \varepsilon_j \frac{\partial(f_1,\dots,f_{N-d})}{\partial(z_j,z_1,\dots,\hat{z}_i,\dots,z_{N-d})}(x(t))z_j(t) - \frac{\partial(f_1,\dots,f_{N-d})}{\partial(z_1,\dots,z_{N-d})}(x(t)) \cdot z_i(t)\right)\right|}{\|z(t)\| \cdot \|\Sigma\, \lambda_i\ w_i\|}\right\}.$$

Notant v la valuation t-adique, il nous suffit de montrer que la valuation du numérateur est supérieure à celle du dénominateur : nous pouvons, sans perte de généralité, supposer que l'infimum des valuations des mineurs jacobiens $\frac{\partial(f_1,\dots,f_{N-d})}{\partial(z_j,z_1,\dots,\hat{z}_i,\dots,z_{N-d})}(x(t))$ est atteint par le mineur $\frac{\partial(f_1,\dots,f_{N-d})}{\partial(z_1,\dots,z_{N-d})}(x(t))$.

Soit b cette valuation, et écrivons $\frac{\partial(f_1,\dots,f_{N-d})}{\partial(z_1,\dots,z_{N-d})}(x(t)) = c_o\ t^b + \dots$, et

$\frac{\partial(f_1,\dots,f_{N-d})}{\partial(z_j,z_1,\dots,\hat{z}_i,\dots,z_{N-d})}(x(t)) = c_j\ t^b + \dots$ $(N-d \le j \le N-k)$. Ecrivons aussi

$z_i = d_o\ t^a + \dots$, et $z_j = d_j\ t^a + \dots$, avec $a > 0$ et au moins un des d_j non nul.

Or, nous pouvons réecrire chacune des équations (E_i) en remplaçant dz_i (resp. dy_j) par $\frac{dz_i}{dt}$ (resp. $\frac{dy_j}{dt}$) et donc chacun des coefficients de $\overline{\lambda}_i$ au numérateur de l'expression précédente est égal à :

$$-\sum_{\ell=1}^{k} \varepsilon_\ell \frac{\partial(f_1,\dots,f_{N-d})}{\partial(y_\ell,z_1,\dots,\hat{z}_i,\dots,z_{N-d})}(x(t)) \frac{dy_\ell}{dt}$$

par conséquent l'inclusion du théorème de Bertini idéaliste (chap. II, § 2) jointe au critère valuatif de dépendance intégrale nous donne

$$v\left(\left(\sum_{j=N-d}^{N-k} \varepsilon_j\ c_j\ d_j\ a - c_o\ d_o\ a\right)\ t^{a+b-1} + \dots\right) \ge a+b$$

d'où

$$\sum_{j=N-d}^{N-k} \varepsilon_j\ c_j\ d_j\ a - c_o\ d_o\ a = 0 \quad \text{donc} \quad \Sigma\, \varepsilon_j\ c_j\ d_j - c_o\ d_o = 0\ .$$

La valuation du numérateur de l'expression donnant la distance est donc au moins égale à $a+b+1$ puisque le coefficient de t^{a+b} y est nul, tandis que la valuation du dénominateur est égale à $a+b$ comme on le vérifie aussitôt en utilisant l'inégalité $\|\Sigma \lambda_i w_i\| \geq \operatorname{Sup}_i |\lambda_i| \cdot \left|\frac{\partial(f_1,\ldots,f_{N-d})}{\partial(z_1,\ldots,f_{N-d})}(x(t))\right|$ comme plus haut. Donc $\lim_{t\to 0} (\delta \circ h)(t) = 0$.

Ceci achève la démonstration de la Proposition, et donc la preuve de l'existence de stratifications de Whitney, si l'on remarque qu'il suffit d'appliquer le résultat précédent avec $X = \overline{S_1}$, $Y = S_2$ pour obtenir le fait que la condition d'incidence est stratifiante.

Exercice : Vérifier que si les conditions de Whitney sont réalisées pour un plongement local $X \subset \mathbb{C}^N$, elles le sont pour tous.

2.3 Gardons les notations de 2.2 et considérons un plongement local $(X,0) \subset (\mathbb{C}^N,0)$ au voisinage d'un point $0 \in Y^o$, et le diagramme commutatif

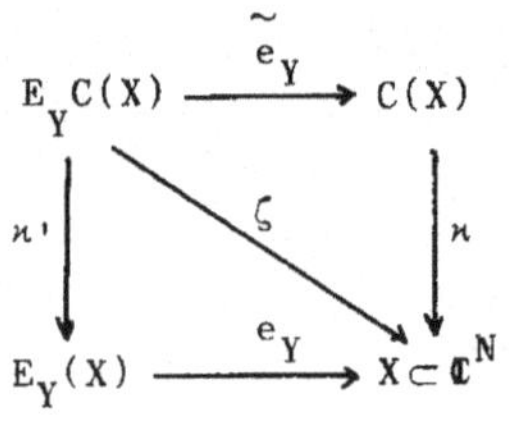

où $\kappa : C(X) \to X$ est l'espace conormal (Chap. II, § 4), e_Y est l'éclatement de Y dans X, $\tilde{e}_Y$ celui de $\tau^{-1}(Y)$ dans C(X) et κ' le morphisme donné par la propriété universelle de l'éclatement. Posons $\zeta = \kappa \circ \tilde{e}_Y$.

2.3.1 Proposition (essentiellement due à Hironaka, [H 1] et [H 2]) : Si l'on a l'égalité $\dim \zeta^{-1}(0) = N-2-\dim Y$, le couple de strates (X^o,Y) satisfait la condition a) de Whitney stricte avec exposant 1, et la condition b) stricte avec un exposant non précisé, au voisinage de 0.

<u>Preuve</u> : Remarquons d'abord qu'il résulte des définitions (2.1)

$$\mathrm{dist}(T_{Y,0}, T_{X,x}) = \sup_{H\supset T_{X,x}} \mathrm{dist}(T_{Y,0}, H)$$

$$\mathrm{dist}(\widehat{x\rho(x)}, T_{X,x}) = \sup_{H\supset T_{X,X}} \mathrm{dist}(x\rho(x), H)$$

H parcourant l'ensemble des hyperplans de $\mathbb{C}^N$ contenant (en direction) $T_{X,x}$. Reprenons maintenant les notations de 2.2.2, et remarquons que $E_Y C(X)$ est naturellement plongé dans $X \times \check{\mathbb{P}}^{N-1} \times \mathbb{P}^{N-t-1}$, où $t = \dim Y$, dès que nous avons choisi une rétraction locale $\mathbb{C}^N \to Y$, et un plongement $X \subset Y \times \mathbb{C}^{N-t}$ compatible avec cette rétraction. Notons $y_1, \ldots, y_t, z_1, \ldots, z_{N-t}$ un système de coordonnées sur $Y \times \mathbb{C}^{N-t}$, et $(b_1 : ,\ldots, : b_t ; a_1 : \ldots : a_{N-t})$ (resp. $(Z_1 : \ldots : Z_{N-t})$) les coordonnées correspondantes sur $\check{\mathbb{P}}^{N-1}$ (resp. $\mathbb{P}^{N-t-1}$). On a alors, pour $H \in \check{\mathbb{P}}^{N-1}$,

$$\mathrm{dist}(T_{Y,0}, H) = \sup_{dy\in\mathbb{C}^t - \{0\}} \left\{ \frac{\left|\sum_1^t b_i \, dy_i\right|}{\|dy\| \sqrt{\Sigma\,|b_j|^2 + \Sigma\,|a_i|^2}} \right\},$$

et pour $\ell \in \mathbb{P}^{N-t-1}$,

$$\mathrm{dist}(\ell, H) = \sup_{Z\in\mathbb{C}^{N-t} - \{0\}} \left\{ \frac{\left|\Sigma\, a_i Z_i\right|}{\|Z\| \sqrt{\Sigma\,|b_j|^2 + \Sigma\,|a_i|^2}} \right\}.$$

L'hypothèse implique que <u>pour tout fermé analytique rare</u> $F \subset Y$, <u>l'image réciproque</u> $\zeta^{-1}(F)$ <u>est rare dans</u> $\zeta^{-1}(Y)$; le théorème de Bertini idéaliste nous donne, après une petite traduction (cf. 2.2) un fermé analytique rare $F \subset Y$ tel que en tout point $z \in \zeta^{-1}(Y - F)$, on ait pour $1 \le j \le t$, $b_j \, \mathcal{O}_{E_Y C(X), z} \in (\mathfrak{J}(1))_z$, où $\mathcal{J}$ désigne l'idéal, inversible par construction, de $E_Y\, C(X)$ engendré par $(z_1 \circ \zeta, \ldots, z_{N-t} \circ \zeta)$ et $\mathfrak{J}(1)$ est son produit avec l'image réciproque de $\mathcal{O}_{\check{\mathbb{P}}^{N-1}}(1)$ par la projection naturelle sur $\check{\mathbb{P}}^{N-1}$. On en déduit, par une variante de l'argument de (Chap. I, 1.4), que l'on a $b_j \cdot \mathcal{O}_{E_Y C(X)} \in \mathfrak{J}(1)$, et ceci étudié au voisinage de $\kappa^{-1}(0)$ implique la condition a) de Whitney stricte avec exposant 1 ; on traite la condition b) stricte comme en 2.2.

(Remarque : On peut remplacer l'espace conormal par l'éclatement de l'idéal jacobien ou la modification de Nash.)

Voici un schéma illustrant la situation que l'hypothèse fait éviter :

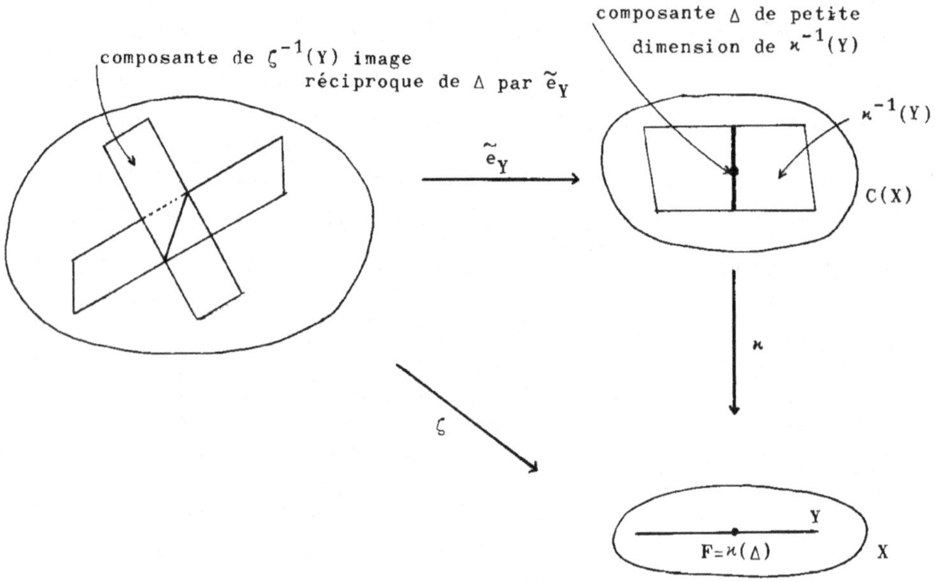

L'hypothèse faite implique qu'il ne peut y avoir de composante de $\zeta^{-1}(Y)$ s'envoyant tout entière dans F, fermé analytique strict de Y.

Remarque : Le lecteur pourrait s'étonner que nous fassions porter les conditions d'incidence que des quadruplets (X,S_1,S_2,x) alors que pour tout ce qui précède, il suffirait de considérer des triplets (S_1,S_2,x) puisque l'assertion "(S_1,S_2) satisfait les conditions de Whitney en $x \in S_2 \subset \overline{S_1}$" a un sens. La raison en est que nous voulons pouvoir énoncer des résultats ou X joue effectivement un rôle, comme le suivant, qui est une conséquence facile de la résolution des singularités à la Hironaka :

Proposition : La condition, portant sur des quadruplets (X,S_1,S_2,x), que voici : Il existe une résolution des singularités $\pi : X' \to X$ de X telle que la transformée stricte $(\overline{S_1})'$ de $\overline{S_1}$ par π soit non-singulière et que le morphisme $\pi|(\overline{S_1})')^{-1}(S_2) \to (S_2)$ induit par π soit localement (sur $(\pi|(\overline{S_1})')^{-1}(S_2))$ analytiquement trivial en tout point de l'image inverse de x, est une condition d'incidence stratifiante.

§ 3. Stratifications définies par des conditions numériques.

Soit $\mathcal{O}$ l'ensemble des classes d'algèbres analytiques réduites et équidimensionnelles. Soient E un ensemble et $M : \mathcal{O} \to E$ une application ("multiplicité généralisée") ; on notera $M_{X,x} \in E$ l'image dans E de la classe de l'algèbre locale $\mathcal{O}_{X,x}$ associée à un germe (X,x). On fait l'hypothèse de constructibilité suivante : étant donné un espace analytique réduit X purement de dimension d, et un sous-ensemble analytique fermé $Y \subset X$, il existe un fermé analytique rare $F \subset Y$ tel que l'application $Y - F \to E$ définie par $y \mapsto M_{X,y}$ soit localement constante.

Proposition : Pour toute application $M : \mathcal{O} \to E$ satisfaisant l'hypothèse ci-dessus, la condition sur (X,S_1,S_2,x) que voici : "$\overline{S_1}$ est équidimensionnel en x et l'application $S_2 \to E$ qui à $y \in S_2$ associe $M_{\overline{S_1},y}$ est localement constante sur S_2 au voisinage de x", est une condition d'incidence stratifiante.

Preuve : La condition d'hérédité est évidemment satisfaite, et la seconde condition résulte aussitôt du fait qu'un ensemble analytique est équidimensionnel hors d'un fermé analytique rare, et de l'hypothèse faite sur M.

Corollaire : Etant donnée une application M comme ci-dessus, tout espace analytique X peut être stratifié en $X = \bigcup S_\alpha$, où chaque S_α est localement fermé à la Zariski dans X, et pour chaque couple (S_α, S_β) tel que $S_\beta \subset \overline{S_\alpha}$, $\overline{S_\alpha}$ est équidimensionnel en chaque point de S_β et l'application $S_\beta \to E$ définie par $y \mapsto M_{\overline{S_\alpha},y}$ est localement constante.

<u>Remarques</u> : 1) La situation ici a un avantage marqué sur celle du § 1 : parmi les stratifications d'un espace X dont l'existence est assurée par le Corollaire, il en existe une qui est moins fine que toutes les autres, c'est-à-dire telle que pour chaque autre stratification (T_β) ayant les mêmes propriétés, chaque strate S_α soit réunion de strates T_{β_i}. Voici comment l'on peut la construire en spécialisant la construction du § 1 (cf. aussi [Lê-T], 6.1.5). Définissons par récurrence une suite de sous-espaces fermés emboîtés de X comme ceci : $F_o = X$ et $F_{i+1} = \{x \in F_i$ / Il existe j, $0 \le j \le i$ tel que F_j n'est pas équidimensionnel en x, ou l'application $x \mapsto M_{F_j,x}$ n'est pas localement constante sur F_i au voisinage de x$\}$. Alors F_{i+1} est un fermé analytique rare de F_i et les $S_i = F_i - F_{i+1}$ sont les strates de la stratification minimale cherchée.

2) Un des principaux résultats de ce travail est la définition d'une multiplicité généralisée $M : \mathcal{O} \to \mathbb{N}^{\mathbb{N}}$ telle que la stratification de X qui lui est associée soit une stratification de Whitney, et qu'inversement si (S_1, S_2) satisfait les conditions de Whitney en $x \in S_2$, l'application $x \to M_{\overline{S_1},x}$ soit localement constante sur S_2 au voisinage de x, c'est-à-dire une description numérique des conditions de Whitney (cf. Chap. V).

§ 4. Stratifications et transversalité.

<u>4.1.1 Lemme</u> : <u>Soient T_1 et T_2 deux sous-espaces vectoriels de $\mathbb{C}^N$. La suite</u>

$$0 \longrightarrow \mathbb{C}^N/T_1 \cap T_2 \xrightarrow{a} \mathbb{C}^N/T_1 \oplus \mathbb{C}^N/T_2 \xrightarrow{b} \mathbb{C}^N/T_1 + T_2 \longrightarrow 0$$

<u>est exacte, où</u> $a(u \bmod T_1 \cap T_1) = (u \bmod T_1,\ u \bmod T_2)$ <u>et</u>
$b(u \bmod T_1, v \bmod T_2) = u - v \bmod(T_1 + T_2)$.

<u>Preuve</u> : C'est clair.

<u>Définition</u> : Les deux sous-espaces vectoriels de $\mathbb{C}^N$ sont transverses en 0 si l'une des deux conditions suivantes est réalisée :

i) On a l'égalité $T_1 + T_2 = \mathbb{C}^N$

ii) On a $\dim(T_1 \cap T_2) = \dim T_1 + \dim T_2 - N$.

Il résulte aussitôt du Lemme précédent que ces deux conditions sont équivalentes.

Remarques : On dit parfois que T_1 et T_2 sont "en position générale" si l'on a l'égalité : $\dim(T_1 + T_2) = \dim T_1 + \dim T_2$. Par exemple, deux droites distinctes de $\mathbb{C}^3$ sont en position générale sans être transverses. (La condition équivaut à : $T_1 \cap T_2 = (0)$, seule condition de transversalité raisonnable lorsque $\dim T_1 + \dim T_2 \leq N$.)

4.1.2 Lemme : *Posons* $t_1 = \dim T_1$, $t_2 = \dim T_2$, *et soit* G_1 (resp. G_2) *la grassmannienne des sous-espaces vectoriels de dimension* t_1 (resp. t_2) *de* $\mathbb{C}^N$. *L'ensemble des couples* $(T_1,T_2) \in G_1 \times G_2$ *tels que* T_1 *soit transverse à* T_2 *est un ouvert de Zariski de* $G_1 \times G_2$, *qui est dense si* $t_1 + t_2 \geq N$ *et vide sinon.*

Preuve : Soit $E_1 \subset G_1 \times \mathbb{C}^N$ (resp. $E_2 \subset G_2 \times \mathbb{C}^N$) l'espace total du fibré tautologique sur G_1 (resp. G_2). L'addition des fibres donne un morphisme linéaire de fibrés au-dessus de $G_1 \times G_2$

$$\begin{array}{ccc} E_1 \times E_2 & \longrightarrow & G_1 \times G_2 \times \mathbb{C}^N \\ & \searrow \quad \swarrow & \\ & G_1 \times G_2 & \end{array}$$

et l'ouvert cherché est celui formé des points $x \in G_1 \times G_2$ tels que le morphisme $E_1(x) \times E_2(x) \to \mathbb{C}^N$ soit surjectif, c'est-à-dire le complémentaire du support du conoyau du morphisme des fibrés ci-dessus. Il s'agit donc bien d'un ouvert de Zariski, et clairement il n'est vide que si $t_1 + t_2 < N$.

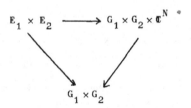

Corollaire : *Si* T_1 *et* T_2 *sont transverses, étant donnés* $t'_1 \geq t_1$ *et* $t'_2 \geq t_2$, *il existe un nombre réel* $\varepsilon > 0$ *tel que pour tout couple de sous-espaces vectoriels* T'_1 *et* T'_2 *de* $\mathbb{C}^N$ *tel que* $\dim T'_i = t'_i$, $i = 1,2$, *et* $\operatorname{dist}(T_1,T'_1) < \varepsilon$, dist $(T_2,T'_2) < \varepsilon$, T'_1 et T'_2 soient transverses.

En effet, si $T'_1 \supset T_1$ et $T'_2 \supset T_2$, c'est évident et on applique ensuite le lemme précédent.

Définition : Soient X et Y deux sous-ensembles constructibles non singuliers et de dimension pure dans un espace analytique non singulier connexe Z. On dit que X et Y sont transverses dans Z en un point $z \in Z$ si

ou bien $\dim_z X + \dim_z Y < \dim_z Z$ et alors $(X \cap Y)_z = \emptyset$

ou bien $\dim_z X + \dim_z Y \geq \dim_z Z$ et alors $T_{X,z}$ et $T_{Y,z}$ sont transverses dans $T_{Z,z}$.

On dit que X et Y sont transverses dans Z si ils sont transverses en tout point $z \in Z$.

4.2.1 Théorème (des fonctions implicites) : Si X et Y sont transverses dans Z comme ci-dessus, $X \cap Y$ est un sous-ensemble constructible non-singulier de Z, de dimension $\dim X + \dim Y - \dim Z$, ou vide.

Inversement, si X et Y sont des sous-espaces analytiques non singuliers localement fermés et si le sous-espace $X \cap Y$ défini par la somme des idéaux est non-singulier de dimension $\dim X + \dim Y - \dim Z$, X et Y sont transverses.

L'assertion est locale sur Z, et résulte aussitôt du théorème des fonctions implicites en prenant des coordonnées locales et des équations locales pour X et Y.

4.2.2 Lemme : Soit A un espace analytique non singulier et soient $X = \bigcup_{\alpha \in A} X_\alpha$ et $Y = \bigcup_{\beta \in B} Y_\beta$ deux sous-ensembles fermés (analytiques, ou sous-ensemble fermés (analytiques, ou sous-analytiques dans le cas réel) munis de stratifications de Whitney. Supposons que pour tout $\alpha \in A$ et $\beta \in B$, les strates X_α et X_β soient transverses dans Z. Alors la décomposition $\bigcup_{\substack{\alpha \in A \\ \beta \in B}} (X_\alpha \cap Y_\beta)$ de $X \cap Y$ est une stratification de Whitney de $X \cap Y$.

Démonstration (due à D. Cheniot [Ch]) : Montrons que l'on a l'égalité $\overline{X_\alpha \cap Y_\beta} = \overline{X_\alpha} \cap \overline{Y_\beta}$; l'inclusion $\overline{X_\alpha \cap Y_\beta} \subseteq \overline{X_\alpha} \cap \overline{Y_\beta}$ étant évidente, il suffit de prouver l'inclusion inverse. Soit donc $z \in \overline{X_\alpha} \cap \overline{X_\beta}$ et supposons $z \notin \overline{X_\alpha \cap X_\beta}$. Il existe un voisinage ouvert U de z dans Z tel que $U \cap (X_\alpha \cap X_\beta) = \emptyset$ et par conséquent $(\overline{X_\alpha} \cap \overline{Y_\beta}) \cap U = (\overline{X_\alpha} \cap \overline{Y_\beta} - (X_\alpha \cap Y_\beta)) \cap U = ((\overline{X_\alpha} - X_\alpha) \cap \overline{Y_\beta} \cap U) \cup (\overline{X_\alpha} \cap (\overline{Y_\beta} - Y_\beta) \cap U)$. Or puis que $\overline{X_\alpha} \cap \overline{Y_\beta} \neq \emptyset$, on a $\dim (\overline{X_\alpha} \cap \overline{Y_\beta}) \geq \dim X_\alpha + \dim Y_\beta - \dim Z$. Par ailleurs nous avons $\dim (\overline{X_\alpha} - X_\alpha) < \dim X_\alpha$ et $\dim (\overline{Y_\beta} - Y_\beta) < \dim Y_\beta$, et l'hypothèse de transversalité implique encore : $\dim (\overline{X_\alpha} - X_\alpha) \cap \overline{Y_\beta} = \dim (\overline{X_\alpha} - X_\alpha) + \dim Y_\beta - \dim Z$ et de même $\dim (\overline{X_\alpha} \cap (\overline{Y_\beta} - Y_\beta)) = \dim (\overline{Y_\beta} - Y_\beta) + \dim X_\alpha - \dim Z$. Chacun de ces deux termes étant strictement inférieur à $\dim X_\alpha + \dim Y_\beta - \dim Z$ nous obtenons la contradiction cherchée.

Cette égalité implique que la partition de $X \cap Y$ par les $X_\alpha \cap Y_\beta$ vérifie la condition de frontière (la frontière d'une strate est une union de strates) : en effet, si l'on a $(X_{\alpha'} \cap Y_{\beta'}) \cap (\overline{X_\alpha \cap Y_\beta}) \neq \emptyset$, l'égalité précédente montre que ceci implique $X_{\alpha'} \cap \overline{X_\alpha} \neq \emptyset$ et $Y_{\beta'} \cap \overline{Y_\beta} \neq \emptyset$ d'où $X_{\alpha'} \subset \overline{X_\alpha}$, $Y_{\beta'} \subset \overline{Y_\beta}$ et donc $X_{\alpha'} \cap Y_{\beta'} \subset \overline{X_\alpha} \cap \overline{Y_\beta} = \overline{X_\alpha \cap Y_\beta}$ et l'inclusion cherchée.

D'après le théorème des fonctions implicites, les $X_\alpha \cap Y_\beta$ sont des sous-ensembles constructibles non singuliers de Z. Vérifions maintenant les conditions de Whitney : soit $z \in X_{\alpha'} \cap Y_{\beta'} \subset \overline{X_\alpha \cap Y_\beta}$ et soit $(x_i)_{i \in \mathbb{N}}$ une suite de points de $X_\alpha \cap Y_\beta$ tendant vers z. D'après les hypothèses, si $T_\alpha = \lim_{x_i \to z} T_{X_\alpha, x_i}$ et $T_\beta = \lim_{x_i \to z} T_{X_\beta, x_i}$, on a : $T_\alpha \supset T_{X_{\alpha'}, z}$ et $T_\beta \supset T_{X_{\beta'}, z}$ donc T_α et T_β sont transverses dans $T_{Z,z}$, ce qui implique par raison de dimension, au vu de la transversalité, les égalités

$$T_\alpha \cap T_\beta = \lim_{x_i \to z} T_{X_\alpha, x_i} \cap T_{Y_\beta, x_i} = \lim_{x_i \to z} T_{X_\alpha \cap Y_\beta, x_i} ,$$

et donc $\lim_{x_i \to z} T_{X_\alpha \cap Y_\beta, x_i} \supset T_{X_{\alpha'}, z} \cap T_{Y_{\beta'}, z} = T_{X_{\alpha'} \cap Y_{\beta'}, z}$ et la condition a).

Considérons maintenant une carte locale de Z autour de z permettant d'identifier un voisinage de z dans Z à $\mathbb{C}^N$, et une rétraction locale $\rho : Z \to X_{\alpha'} \cap Y_{\beta'}$.

Si $\ell = \operatorname{Lim} \widehat{x_i\ \rho(x_i)}$, on a $\ell \subset T_{X_\alpha',z} \cap T_{Y_\beta',z} = \operatorname{Lim} T_{(X_\alpha \cap Y_\beta),z}$ d'où la condition b) de Whitney, ce qui achève la démonstration. (Notons que nous nous sommes permis ici, et nous permettrons encore, d'extraire sans prévenir des sous-suites de (x_i) pour faire converger les directions d'espaces tangents et de sécantes.) ∎

4.2.3 Remarque : L'égalité $\overline{X_\alpha} \cap \overline{Y_\beta} = \overline{X_\alpha \cap Y_\beta}$ implique que si X^o (resp. Y^o) est une strate dense dans X (resp. Y), l'intersection $X^o \cap Y^o$ est une strate dense dans $X \cap Y$.

4.2.4 Lemme : Soient $X = \bigcup_\alpha X_\alpha$, $Y = \bigcup_\beta Y_\beta$ deux espaces analytiques réduits munis de stratifications de Whitney. La partition $X \times Y = \bigcup_{\alpha,\beta} X_\alpha \times Y_\beta$ est une stratification de Whitney du produit $X \times Y$.

La démonstration est un exercice.

Remarques : 1) On dit que deux morphismes $f : X \to Z$ et $g : Y \to Z$ d'espaces non-singuliers sont transverses si pour $t \in X \times_Z Y$, les images dans $T_{Z,z}$ des applications tangentes à f en $p_1(t)$ et $p_2(t)$, où $z = f(p_1(t)) = g(p_2(t))$, sont transverses. On peut utiliser les lemmes précédents pour prouver que si $f : X \to Z$ et $g : Y \to Z$ sont des morphismes d'espaces stratifiés $X = \bigcup X_\alpha$ et $Y = \bigcup Y_\beta$ dans un espace non-singulier Z tel que pour tout couple (α,β), $f|X_\alpha : X_\alpha \to Z$ et $g|Y_\beta : Y_\beta \to Z$ soient transverses, les produits fibrés $X_\alpha \times_Z Y_\beta$ forment une stratification de Whitney du produit fibré $X \times_Z Y$. On peut aussi généraliser à des produits fibrés d'un nombre fini de morphismes, etc. Nous ne nous servirons pas de ces résultats dont la preuve est au pire fastidieuse.

4.2.5 Lemme : Soient K un espace analytique compact non-singulier, U un ouvert de $\mathbb{C}^M$ contenant 0, X et Y deux sous-espaces analytiques fermés réduits du produit $K \times U$. Soit $X = \bigcup X_\alpha$ une stratification de Whitney de X telle que $X_o = X \cap (K \times \{0\})$ soit réunion de strates, et supposons que Y soit muni d'une *stratification* $Y = \bigcup Y_\beta$ *à chaque strate de laquelle* $K \times \{0\}$ *est transverse dans* $K \times U$. *Alors*

1) *La décomposition* $Y = (\bigcup_\beta (Y_\beta - Y_{o,\beta})) \cup (\bigcup_\beta Y_{o,\beta})$, *où* $Y_{o,\beta} = Y_\beta \cap (K \times \{0\})$ *est une stratification de Whitney de* Y, *telle que* $Y_o = Y \cap (K \times \{0\}) = \bigcup Y_{o,\beta}$ *soit réunion de strates.*

2) *Si, pour tout couple* (α,β) *tel que* $X_\alpha \subset X_o$, *les strates* X_α *et* $Y_{o,\beta}$ *sont transverses dans* K, *il existe un voisinage* U' *de* 0 *dans* U *tel que deux strates* $X_\alpha \cap (K \times U')$ *et* $Y_\beta \cap (K \times U')$ *quelconques de* $X \cap (K \times U')$ *et* $Y \cap (K \times U')$ *soient transverses dans* $K \times U'$.

Démonstration : L'assertion 1) résulte du Lemme 4.2.2. Prouvons donc 2). Tout d'abord, puisque K est compact on peut, quitte à rétrécir U en un voisinage ouverts U_1' de 0 dans U, supposer que toutes les strates de X ont une adhérence qui rencontre X_o. Etant données deux strates X_α et X_β, montrons que si $\dim X_\alpha + \dim Y_\beta < \dim K + N$, on a $X_\alpha \cap Y_\beta = \emptyset$. Soit en effet $X_{\alpha'}$ une strate contenue dans $\overline{X}_\alpha \cap X_o$. On a $\dim X_{\alpha'} \leq \dim X_\alpha$ et par ailleurs, d'après la structure des Y_β, $\dim(\overline{Y}_\beta \cap (K \times \{0\}) = \dim Y_\beta - N$ donc pour chaque strate $Y_{o,\beta'} \subset \overline{Y}_\beta \cap (K \times \{0\})$, on a $\dim X_{\alpha'} + \dim Y_{\beta'} \leq \dim X_\alpha + \dim Y_\beta - N < \dim K$, donc d'après l'hypothèse de transversalité, $X_{\alpha'} \cap Y_{\beta'} = \emptyset$. Or, l'inclusion évidente $\overline{X_\alpha \cap Y_\beta} \subseteq \overline{X_\alpha} \cap \overline{Y_\beta}$ nous fournit l'inclusion
$\overline{X_\alpha \cap Y_\beta} \cap (K \times \{0\}) \subseteq \overline{X_\alpha} \cap (K \times \{0\}) \cap (\overline{Y_\beta} \cap K \times \{0\})$ qui est vide d'après ce qui précède, donc $\overline{X_\alpha} \cap \overline{Y_\beta} \cap (K \times \{0\}) = \emptyset$ et par conséquent il existe un voisinage U_2' de 0 dans U_1' tel que $\overline{X_\alpha \cap Y_\beta} \cap (K \times U_2') = \emptyset$ pour tout couple (α,β) tel que $\dim X_\alpha + \dim Y_\beta < \dim K + N$, puisque l'ensemble des couples de strates concernés est fini.

Supposons maintenant $\dim X_\alpha + \dim Y_\beta \geq \dim K + N$. Si $X_\alpha \cap Y_\beta = \emptyset$, il n'y a rien à démontrer. Supposons donc $X_\alpha \cap Y_\beta \neq \emptyset$ et considérons le sous-ensemble analytique $B \subset X_\alpha \cap Y_\beta$ formé des points z en lesquels X_α et Y_β ne sont pas transverses dans $K \times U_2'$. Démontrons par l'absurde que $\overline{B} \cap (K \times \{0\}) = \emptyset$. Soit en effet $z_o \in \overline{B} \cap (K \times \{0\})$ et considérons la strate $X_{\alpha'} \subset K \times \{0\}$ qui contient z_o et la strate $Y_{o,\beta'}$ qui contient z_o. D'après la condition a) de Whitney, étant donnée une carte locale $K \times U_2' \simeq$ (ouvert de) $\mathbb{C}^\ell \times \mathbb{C}^N$ et un nombre $\varepsilon > 0$, il existe un voisinage ouvert V de z_o dans $K \times U_2'$ tel que $dist(T_{X_{\alpha'},z_o}, T_{X_\alpha,z}) < \varepsilon$ et $dist(T_{Y_{o,\beta'},z_o}, T_{Y_\beta,z}) < \varepsilon$ pour tout $z \in X_\alpha \cap Y_\beta \cap V$, puisque z_o appartenant à $Y_{\overline{\beta}}$, on a $Y_{0,\beta'} \subset \overline{Y_\beta}$. Il résulte maintenant de l'hypothèse de transversalité des strates dans $K \times \{0\}$ et du Lemme 4.1.2 que pour tout point z du voisinage V de z_o, les espaces $T_{Y_\beta,z}$ et $T_{X_\alpha,z}$ sont transverses dans $T_{K\times U,z}$, d'où la contradiction cherchée. Ainsi $\overline{B} \cap K \times \{0\} = \emptyset$ et quitte à remplacer U_2' par un voisinage encore plus petit U' de O dans $\mathbb{C}^N$, on obtient $B \cap K \times U' = \emptyset$ pour tout B associé à un couple de strates X_α, X_β, et le résultat cherché. ∎

Voici maintenant un résultat, dû à S. Kleiman, qui permet de créer beaucoup de situations où deux sous-ensembles algébriques d'une variété algébrique X ont des stratifications transverses au sens du Lemme. Nous allons énoncer le résultat seulement dans un cas particulier, et indiquer les grandes lignes de la démonstration pour la commodité du lecteur.

4.3 Théorème (Kleiman [Kl 1], 2. Theorem) : Soient Γ un groupe algébrique défini sur $\mathbb{C}$, X une variété algébrique intègre (= réduite et irréductible) munie d'une action transitive du groupe Γ ; $\Gamma \times X \to X$ notée $(\gamma,x) \to \gamma \cdot x$. Soient $f: E \to X$ et $g: Z \to X$ deux morphismes algébriques entre variétés algébriques réduites et équidimensionnelles. Pour chaque point $\gamma \in \Gamma$, on note γE la variété E munie du morphisme $E \to X$ défini par $e \mapsto \gamma \cdot f(e)$. Alors, il existe un ouvert de Zariski dense U de Γ tel que pour tout $\gamma \in U$, le produit fibré $\gamma E \times_X Z$ soit vide, ou équidimensionnel de dimension égale à $\dim E + \dim Z - \dim X$.
De plus si E et Z sont non-singuliers, on peut choisir U de telle façon que pour $\gamma \in U$, le produit fibré $\gamma E \times_X Z$ soit non-singulier.

<u>Esquisse de démonstration</u> : Considérons le morphisme $q : \Gamma \times E \to X$ défini par $(\gamma, e) \to \gamma \cdot f(e)$, et le diagramme :

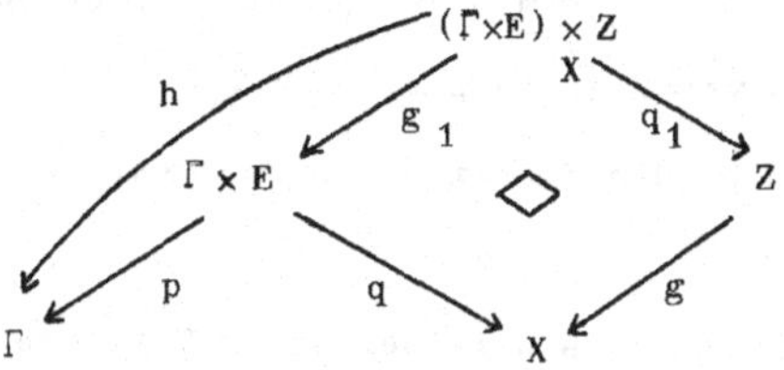

où p désigne la première projection.

Tout d'abord, puisque Γ et E sont réduits et équidimensionnels, il en est de même de $\Gamma \times E$. Montrons que le morphisme q est plat : d'une part d'après le théorème de platitude générique il existe un ouvert de Zariski V dense dans X tel que le morphisme induit $q : q^{-1}(V) \to V$ soit plat, et d'autre part le morphisme q est équivariant pour l'action naturelle de Γ sur $\Gamma \times E$ et l'action donnée de Γ sur X, donc pour tout $\gamma \in \Gamma$, le morphisme $q^{-1}(\gamma V) \to \gamma V$ est isomorphe au morphisme plat $q^{-1}(V) \to V$. Puisque Γ agit transitivement, les γV recouvrent X, d'où le résultat. Si de plus E est non-singulier, le produit $\Gamma \times E$ est non-singulier donc d'après le théorème de lissité générique, (ou le théorème de Bertini-Sard) les fibres $q^{-1}(x)$ sont non-singulières pour $x \in X - F$, où F est un sous-ensemble strict de X, et la transitivité implique aussitôt que toutes les fibres $q^{-1}(x)$ sont non-singulières.

On en déduit, puisque la platitude est conservée par changement de base, que le morphisme q_1 est plat et donc par les propriétés de la dimension $(\Gamma\times E) \times_X Z$ est équidimensionnel de dimension $\dim q_1^{-1}(z) + \dim Z$ (pour tout $z \in Z$), c'est-à-dire

$$\dim q^{-1}(g(z)) + \dim Z = \dim \Gamma \times E - \dim X + \dim Z \quad .$$

Si de plus E est non-singulier, le morphisme q_1 est plat et à fibres non-singulières puisque c'est le cas pour q, en particulier si Z est aussi non-singulier, l'espace $(\Gamma\times E)\times_X Z$ est non-singulier. Enfin par platitude générique (resp. et lissité générique) il existe un ouvert de Zariski dense U tel que pour tout $\gamma \in U$, la fibre $h^{-1}(\gamma)$ soit ou bien vide, ou bien équidimensionnelle de dimension égale à $\dim (\Gamma\times E)\times_X Z - \dim \Gamma$ (resp. et de plus non-singulière). Si l'on remarque que $h^{-1}(\gamma) = \gamma E\times_X Z$ d'une part et que $\dim (\Gamma\times E)\times_X Z - \dim \Gamma = \dim E + \dim Z - \dim X$, on voit que le théorème est démontré.

Voici sous quelle forme nous utiliserons le Théorème de Kleiman :

<u>4.3.1 Corollaire</u> : <u>Soient</u> $E = \bigcup_{\alpha\in A} E_\alpha$ <u>et</u> $Z = \bigcup_{\beta\in B} Z_\beta$ <u>deux sous-ensembles algébriques fermés de la grassmannienne</u> G <u>des</u> d-<u>plans de</u> $\mathbb{C}^N$ <u>munis de décompositions finies en sous-ensembles localement fermés à la Zariski non singuliers</u> (<u>par exemple des stratifications de Whitney</u>). <u>Il existe un ouvert de Zariski dense</u> U <u>du groupe algébrique</u> $\Gamma = GL(N,\mathbb{C})$ <u>tel que pour tout</u> $\gamma\in U$ <u>et tout</u> $\alpha\in A$ <u>le translaté</u> γE_α <u>de</u> E_α <u>par l'action naturelle</u> (<u>transitive</u>) <u>de</u> Γ <u>sur</u> G <u>soit transverse dans</u> G <u>à tous les sous-espaces non-singuliers</u> Z_β.

<u>Démonstration</u> : On applique le théorème précédent aux injections $E_\alpha \to G$, $Z_\beta \to G$ et l'on trouve pour chaque couple (α,β) un ouvert de Zariski dense $U_{\alpha\beta}$ tel que $\gamma E_\alpha \times_G Z_\beta = \gamma E_\alpha \cap Z_\beta$ soit vide ou non-singulier de dimension $\dim E_\alpha + \dim Z_\beta - \dim G$, ce qui signifie que l'intersection est transversale. Les α et β étant en nombre fini, l'ouvert $\cap U_{\alpha\beta} = U$ convient. ∎

<u>Remarque</u> : En particulier, si E et Z sont équidimensionnels, $\gamma E\cap Z$ est équidimensionnel si $\gamma\in U$, et si une strate E_{α_o} (resp. Z_{β_o}) est dense dans E (resp. Z), alors pour $\gamma\in U$, $\gamma E_{\alpha_o}\cap Z_{\beta_o}$ est de dimension $\dim E + \dim Z - \dim G$ et dense dans $E\cap Z$. Enfin si les UE_α et UZ_β sont des stratifications de Whitney, pour $\gamma\in U$, les $(\gamma E_\alpha)\cap Z_\beta$ forment une stratification de Whitney de $(\gamma E)\cap Z$.

§ 5. Stratifications, transversalité, et éclatement.

Nous utiliserons le théorème de Kleiman aussi pour mettre certains sous-espaces d'un espace non singulier Z en position générale par rapport à un morphisme $f : Y \to Z$, au sens défini ci-dessous.

5.1 Etant donné un morphisme propre $p : X' \to X$ entre espaces, par exemple un éclatement, l'existence de stratifications de Samuel relatives (Le-T 2) implique que pour chaque entier i, le sous-ensemble $F_i = \{x' \in X' \mid \dim_{x'} p^{-1}(p(x')) \geq i\}$ est un sous-ensemble analytique fermé de X. Puisque p est propre, l'image B $B_i = p(F_i)$ de chaque F_i est un sous-ensemble analytique fermé de X, et l'on a l'inégalité $\dim B_i \leq \dim X' - i$. Par exemple si p est un éclatement de centre rare, on a $B_o = X$ et B_1 est contenu dans le centre d'éclatement, mais ne lui est pas nécessairement égal.

5.2 Proposition (d'après [Lê-T], 5.1.3.2) : Soient Z un espace analytique non singulier, X un sous-espace analytique fermé équidimensionnel de Z et $p : X' \to X$ un morphisme propre tel que X' soit équidimensionnel et que la dimension des fibres $p^{-1}(x)$ soit constante pour $x \in X - B_k$, où B_k est rare dans X et $p^{-1}(B_k)$ rare dans X'. Soit $Z = \cup Z_\alpha$ une stratification de Whitney de Z tel que X et chacun des sous-ensembles B_i de X associés à p comme ci-dessus soient unions de strates, ainsi que chacune des différences $B_i - B_{i+1}$. Soit H un sous-espace non singulier de Z transverse à chacune des strates Z_α. L'image inverse $p^{-1}(H \cap X)$ par p de l'intersection $H \cap X$ coïncide ensemblistement avec l'adhérence dans X' de $p^{-1}(H \cap (X - B_k))$.

Démonstration : Il suffit de prouver qu'en tout point $x' \in p^{-1}(H \cap X)$, $\dim_{x'} p^{-1}(H \cap B_k)$ est strictement inférieur à la plus petite des dimensions des composantes irréductibles non immergées locales en x' de $p^{-1}(H \cap X)$: il en résultera en effet que $p^{-1}(H \cap B_k)$ est rare dans $p^{-1}(H \cap X)$, donc que $p^{-1}(H \cap (X - B_k)) = p^{-1}(H \cap X - H \cap B_k)$ est dense dans $p^{-1}(H \cap X)$. Or on a :

$p^{-1}(H\cap B_k) = \bigcup_{j\geq k} p^{-1}(H\cap (B_j - B_{j+1}))$ et l'on a les inégalités :
$\dim B_j \leq \dim p^{-1}(B_k) - j$ puisque $p^{-1}(B_j)\subset p^{-1}(B_k)$, pour $j\geq k$. Puisque H est transverse aux strates, on a $\dim(H\cap B_j) = \dim B_j - h \leq \dim p^{-1}(B_k) - j - h$, où h est la codimension de H dans Z, d'où pour tout $x'\in p^{-1}(B_k)$,
$\dim_{x'} p^{-1}(H\cap B_j) \leq \dim_{x'} p^{-1}(B_k) - h$ pour tout $j\geq k$. Or H est défini localement par h équations dans Z et donc, d'après ([Bbk 3], § 3, No 1, Prop. 2) on a pour chaque composante irréductible locale non immergée $p^{-1}(H\cap X)_i$ de $p^{-1}(H\cap X)$ en x' l'inégalité $\dim_{x'} p^{-1}(H\cap X)_i \geq \dim_{x'} X' - h$. Par ailleurs, puisque par hypothèse $p^{-1}(B_k)$ est rare dans X', on a l'inégalité $\dim_{x'} p^{-1}(B_k) < \dim_{x'} X'$ ce qui donne finalement : $\dim_{x'} p^{-1}(H\cap B_k) \leq \dim_{x'} p^{-1}(B_k) - h < \dim_{x'} p^{-1}(H\cap X)_i$, ce qu'il fallait démontrer. ■

<u>5.2</u> <u>Application</u> : Reprenons la situation de (Chap. I, 5.1) où l'on éclate dans X un sous-espace Y défini par un idéal I, tel que X soit équimultiple" le long de Y relativement à un morphisme $F:(X,x)\to(S,s)$. Le lecteur vérifiera que la conclusion de (Loc. cit.) et la proposition 5.1 ci-dessus impliquent que

<u>Le transformé strict par l'éclatement</u> $\pi_0 : X'\to X$ <u>de</u> Y <u>de la fibre spéciale</u> $X(s) = F^{-1}(s)$ <u>coïncide ensemblistement avec</u> $\pi_0^{-1}(X(s))$, <u>et de même pour l'éclatement normalisé</u> π.

Ainsi, par l'intermédiaire de Chap. I, 5.1 et Chap. III, 5.1, une condition numérique implique l'égalité d'un transformé total et d'un transformé strict.

<u>5.3</u> <u>Résolution simultanée forte et conditions de Whitney</u>.

<u>5.3.1</u> <u>Définition</u> (cf. [Te 2], II) : Soient X un espace analytique réduit et équidimensionnel, $Y\subset X$ un sous-espace et $0\in Y$ un point non-singulier de Y. On dit que X admet une résolution simultanée forte le long de Y en 0 si il existe un morphisme $\pi : X'\to X$ qui soit résolution des singularités de X, c'est-à-dire

que X' est non-singulier, que π est propre et induit un isomorphisme $\pi : X' - \pi^{-1}(\mathrm{Sing}\ X) \xrightarrow{\sim} X - \mathrm{Sing}\ X$, morphisme tel que l'on ait :

Le morphisme induit $\pi^{-1}(Y) \to Y$ est localement trivial en tout point $x' \in \pi^{-1}(Y)$, c'est-à-dire que l'on a un Y-isomorphisme local $(\pi^{-1}(Y),x') \simeq (\pi^{-1}(\pi(x')),x') \times (Y,\pi(x'))$.

5.3.2 Remarques : 1) C'est la condition que l'on a affirmée être stratifiante en (2.3.1, Remarque).

5.3.3 Proposition (cf. [Te 2], II) : Soient X un espace analytique réduit équidimensionnel, Y un sous-espace analytique fermé de X et $0 \in Y$ un point non-singulier de Y. Si X admet une résolution simultanée forte le long de Y en 0, le couple (X^0,Y) satisfait la condition a) de Whitney stricte avec exposant 1, et la condition b) de Whitney stricte en 0.

Démonstration : Soit $\pi : X' \to X$ la résolution simultanée, et soit $y_1,\dots,y_t$ un système de coordonnées locales pour Y en 0. D'après l'hypothèse, on peut trouver en tout point $x' \in \pi^{-1}(0)$ un système de coordonnées locales $y'_1,\dots,y'_t$, $w_1,\dots,w_{d-t}$ pour X' en x' tel que $(y_i \circ \pi)_{x'} = y'_i$ pour $1 \le i \le t$. Choisissons un plongement local $(X,0) \subset (\mathbb{C}^N,0)$ et des coordonnées $(y_1,\dots,y_t,z_1,\dots,z_{N-t})$ sur $(\mathbb{C}^N,0)$. Soient $(f_1,f_2,\dots,f_m)$ des générateurs pour l'idéal de $\mathbb{C}\{y_1,\dots,y_t,z_1,\dots,z_{N-t}\}$ définissant $(X,0)$ dans $(\mathbb{C}^N,0)$.
L'hypothèse de résolution simultanée forte implique (critère différentiel de lissité) que la dérivation $\frac{\partial}{\partial y_i}$ de $\mathcal{O}_{Y,0}$ dans lui-même s'étend, par le morphisme naturel $\mathcal{O}_{Y,0} \xrightarrow{\pi|Y} \mathcal{O}_{\pi^{-1}(Y),x'}$ en une dérivation de $\mathcal{O}_{\pi^{-1}(Y),x'}$, c'est-à-dire encore que dans le morphisme naturel $\mathcal{O}_{Y,0} \hookrightarrow \mathcal{O}_{X',x'} = \mathbb{C}\{y'_1,\dots,y'_t,w_1,\dots,w_{d-t}\}$, envoyant y_i sur y'_i, on peut étendre la dérivation $\frac{\partial}{\partial y_i}$ en une dérivation D_i de $\mathcal{O}_{X',x'}$ qui respecte l'idéal $S' = ((z_1 \circ \pi)_{x'},\dots,(z_{N-t} \circ \pi)_{x'})\mathcal{O}_{X',x'}$ définissant $\pi^{-1}(Y)$ dans X' en x'.
On a donc $D_j((z_i \circ \pi)_{x'} \in S'$, pour $1 \le j \le t$, $1 \le i \le N-t$, et $D_j\ y'_r = \delta_{jr}$ (symbole

de Kronecker).

Puisque f_k est nul sur X, nous avons $(f_k \circ \pi)_{x'} = 0$ dans $\mathcal{O}_{X',x'}$, d'où d'après la règle de Leibniz :

$$0 = D_j(f_k \circ \pi) = \sum_{i=1}^{N-t} \left(\frac{\partial f_k}{\partial z_i} \circ \pi\right)_{x'} \cdot D_j(z_i \circ \pi)_{x'} + \left(\frac{\partial f_k}{\partial y_j} \circ \pi\right)_{x'} .$$

Par conséquent, chaque mineur $\dfrac{\partial(f_{j_1},\ldots,f_{j_c})}{\partial(y_j,z_{i_2},\ldots z_{i_c})}$ composé avec π et localisé en x', vérifie :

$$\left(\frac{\partial(f_{j_1},\ldots,f_{j_c})}{\partial(y_j,z_{i_2},\ldots,z_{i_c})} \circ \pi\right)_{x'} = -\sum_{i=1}^{N-t} \left(\frac{\partial(f_{j_1},\ldots,f_{j_c})}{\partial(z_i,z_{i_2},\ldots,z_{i_c})} \circ \pi\right)_{x'} D_j(z_i \circ \pi)_{x'} .$$

Puisque d'après l'hypothèse, $D_j(z_i \circ \pi)_{x'} \in S'$, on en déduit que si l'on note S l'idéal de $\mathcal{O}_{X,0}$ engendré par $z_1,\ldots,z_{N-t}$, l'on a en tout point $x' \in \pi^{-1}(0)$:

$$\left(\frac{\partial(f_{j_1},\ldots,f_{j_c})}{\partial(y_j,z_{i_2},\ldots,z_{i_c})} \circ \pi\right)_{x'} \in S \cdot J_{X/Y} \cdot \mathcal{O}_{X',x'} ,$$

où $J_{X/Y}$ désigne l'idéal engendré par les seuls mineurs jacobiens au dénominateur desquels aucun y n'apparaît et donc, d'après (Chap. I, 1.3.6)

$$\frac{\partial(f_{j_1},\ldots,f_{j_c})}{\partial(y_j,z_{i_2},\ldots,z_{i_c})} \mathcal{O}_{X,0} \in \overline{S \cdot J_{X/Y}} \quad \text{dans } \mathcal{O}_{X,0} .$$

Le fait que (X^o,Y) satisfasse les conditions de Whitney strictes annoncées résulte alors de la preuve de 2.2.2. ■

<u>5.3.4</u> <u>Remarque</u> : Le contenu nouveau de l'article [Ve] de Verdier consiste en les deux intéressants résultats suivants :

i) La condition a) de Whitney stricte avec exposant 1 implique la condition b). Dans le cadre où nous sommes, on peut s'en convaincre assez facilement en examinant la preuve de 2.2.2 ci-dessus.

ii) Si (X^o,Y) satisfait la condition a) de Whitney stricte avec exposant 1, tout champ de vecteurs analytique réel sur Y peut s'étendre localement en un champ de vecteurs analytique réel sur X^o et satisfaisant au voisinage de tout point de Y une condition de "rugosité" qui implique que ce champ de vecteurs est localement intégrable. Son intégration donne une trivialisation locale "rugueuse", et en particulier topologique, de X le long de Y, dès que Y est une strate d'uns stratification de X telle que chaque couple de strates satis fasse la condition a) de Whitney stricte avec exposant 1.

***

C H A P I T R E I V

VARIETES POLAIRES

Introduction. Dans ce chapitre, j'introduis le principal concept nouveau qui a permis de donner une caractérisation numérique des conditions de Whitney : celui de variété polaire locale, dont voici une description intuitive : étant donné un représentant $(X,0) \subset (\mathbb{C}^N,0)$ d'un germe d'espace analytique réduit équidimensionnel, considérons pour chaque entier k, $0 \leq k \leq d-1$, où $d = \dim X$, une projection linéaire $p : \mathbb{C}^N \to \mathbb{C}^{d-k+1}$, et le lieu critique de la restriction $p|X^o$ de p à la partie non-singulière X^o de X. Si la projection linéaire p est "assez générale", l'adhérence dans X de ce lieu critique sera un sous-espace analytique réduit $P_k <p>$ purement de codimension k ou vide, appelé "variété polaire locale de X associée à p" . Comme on le voit, même en géométrie algébrique, parler de "la variété polaire générale de codimension k de $(X,0) \subset (\mathbb{C}^N,0)$ " est abusif en ceci que cette variété polaire n'est pas définie sur le corps de base (disons $\mathbb{C}$), même pour un plongement $X \subset \mathbb{C}^N$ fixé, mais au mieux sur le corps des fonctions de l'espace des projections linéaires de $\mathbb{C}^N$ sur $\mathbb{C}^{d-k+1}$. Pis encore, la "variété polaire générale de codimension k de X" , c'est-à-dire rendue indépendante du plongement, n'est définie que sur une extension du corps de base de degré de transcendance infini. Néanmoins, pour un plongement local fixé, il existe un ouvert de Zariski dense U de l'espace des projections linéaires tel que pour $p \in U$, la multiplicité $m_o(P_k <p>)$ de $P_k <p>$ en 0 soit indépendante de $p \in U$, et de plus ce nombre ne dépend en fait que de l'algèbre locale $\mathcal{O}_{X,0}$, c'est-à-dire est en fait un invariant analytique du germe (X,0), invariant que l'on note $m_o(P_k(X,0))$. Un des principaux résultats de ce chapitre est donc d'associer à chaque algèbre analytique réduite puremen de dimension d une "multiplicité généralisée" qui est la suite d'entiers $M^*_{(X,0)} = (m_o(P_o(X,0)), m_o(P_1(X,0)), \ldots, m_o(P_{d-1}(X,0)))$ dont le premier terme est d'ailleurs la multiplicité de X en 0 puisqu'il résulte aussitôt des définitions que $P_o(X,0) = (X,0)$. Le paradigme de ce genre de construction est la construction qui associe à un germe d'hypersurface à singularité isolée la suite $(\mu^{(i+1)} + \mu^{(i)})_{0\leq i\leq d-1}$, où $\mu^{(i)}$ est le nombre de Milnor de l'intersection de l'hypersurface avec un plan général de dimension i passant par 0. C'est d'ailleurs le souci de faire le lien entre les variétés polaires d'un germe introduites par Lê et moi dans [Lê-T 1] et les variétés polaires introduites dans [Te 12] qui m'a conduit à introduire (cf. [Te 5]) les variétés polaires relatives associées à un morphisme $f : X \to S$. Ces variétés polaires relatives sont en général bien plus que la collection des variétés polaires des fibres de f, et ont un comportement bien plus turbulent. C'est dans ce contexte qu'il faut prouver les résultats de "transversalité dynamique" essentiels pour la théorie, affirmant essentiellement que lorsque $\dim S \leq 1$, les variétés polaires relatives générales sont transverses au noyau des projections servant à les définir.

La définition la plus opératoire des variétés polaires n'est pas toujours celle qui est décrite ci-dessus mais la définition donnée ci-dessous, au § 1, (voir le Corollaire 1.3.1).

§ 1. Définitions des variétés polaires.

1.1 Rappels du Chapitre II.

Soit $f : X \to S$ un morphisme d'espaces analytiques réduits tel que le module Ω^1_f des différentielles relatives soit localement libre de rang $d = \dim X - \dim S$ sur le complémentaire d'un fermé analytique rare F de X. On peut alors (Chap. I, 1.2) définir la modification de Nash relative $\nu_f : N_f(X) \to X$, et pour tout point $0 \in X$ et toute installation locàle d'un représentant assez petit du germe de f en 0,

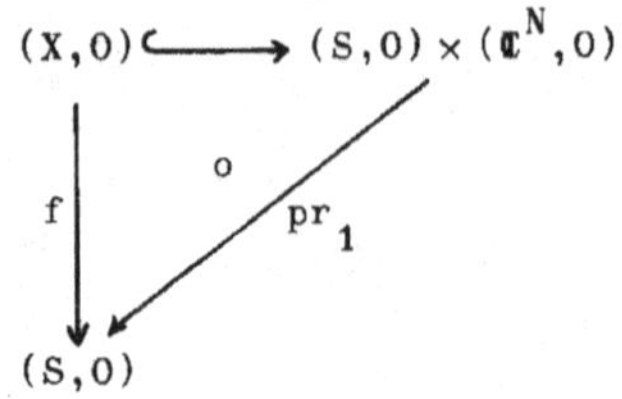

on peut décrire la modification de Nash relative comme ceci : soit G la grassmannienne des d-plans de $\mathbb{C}^N$, et soit, pour un représentant X assez petit, $N_f(X) \subset X \times G$ l'adhérence du graphe du morphisme $X - F \to G$ qui à $x \in X-F$ associe la direction de l'espace tangent en x à la fibre $X(f(x)) = f^{-1}f(x)$. On peut donc définir un morphisme de Gauss relatif γ_f dans le diagramme suivant :

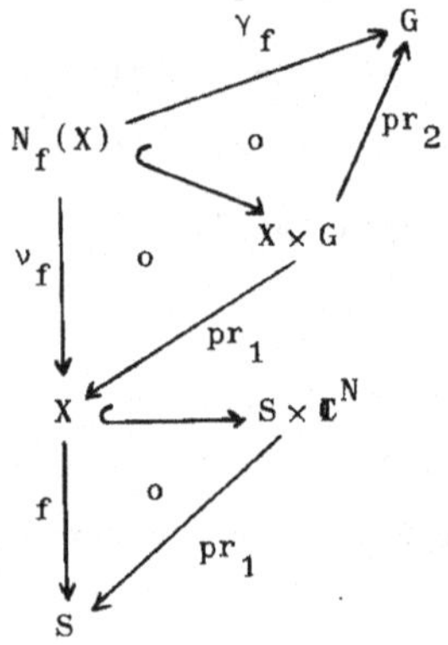

1.2 Proposition-Définition 1 (Schubert) : Soit N un entier, et soit

$$(\mathcal{D}) \; : \; (0) \subset D_{N-1} \subset D_{N-2} \subset \ldots \subset D_1 \subset D_o = \mathbb{C}^N$$

un drapeau de sous-espaces vectoriels de $\mathbb{C}^N$, avec codim $D_i = i$. Pour chaque entier k, $0 \le k \le d$, on appelle k-ième variété de Schubert projective associée à $\mathcal{D}$, et l'on note $c_k(\mathcal{D})$ le sous-ensemble de la grassmannienne G des d-plans de $\mathbb{C}^N$ défini par

$$c_k(\mathcal{D}) = \{T \in G / \dim(T \cap D_{d-k+1}) \ge k\} \quad .$$

Pour tout $\mathcal{D}$, l'ensemble $c_k(\mathcal{D})$ est muni naturellement d'une structure de sous-variété algébrique réduite de G, purement de codimension k dans G (cf. [G.H], Chap. I, § 5).

Remarque 1 : La sous-variété $c_k(\mathcal{D})$ de G ne dépend en fait que de $D_{d-k+1} \subset \mathbb{C}^N$, et pour cette raison on l'écrira aussi volontiers $c_k(D_{d-k+1})$.

Remarque 2 : Pour un drapeau $\mathcal{D}$ et une suite d'entiers $a = (a_1, \ldots, a_d)$ considérons le sous-ensemble

$\sigma_a(\mathcal{D}) = \{T \in G/\dim(T \cap D_{d+a_i-i}) \geq i\}$. C'est une sous-variété algébrique de G, de codimension $\Sigma\, a_i$, appelée variété de Schubert associée à a et $\mathcal{D}$, (voir [G-H], Chap. I, § 5). Les $c_k(\mathcal{D})$ sont un cas très particulier de cette construction ($a_i = 0$ pour $i > k$, $a_i = 1$ pour $i \leq k$). Dans cette rédaction, nous n'utiliserons que les $c_k(\mathcal{D})$ et omettrons l'épithète "projective".

1.3 Proposition 2 : Etant donnés un (représentant d'un germe de) morphisme $f : (X,0) \to (S,0)$ comme ci-dessus, avec S non-singulier, et un S-plongement local $(X,0) \subset (S,0) \times (\mathbb{C}^N,0)$, pour tout entier k, $0 \leq k \leq d = \dim X - \dim S$, il existe un ouvert de Zariski dense W_k de la grassmannienne G_k des sous-espaces de codimension d-k+1 de $\mathbb{C}^N$ tel que pour tout $D_{d-k+1} \in W_k$ on ait :

i) $\gamma_f^{-1}(c_k(D_{d-k+1})) \cap \nu_f^{-1}(X - F)$ est dense dans $\gamma_f^{-1}(c_k(D_{d-k+1}))_{red}$, et ce dernier espace est vide ou de codimension pure k dans $N_f(X)$.

ii) L'égalité $\dim(\nu_f^{-1}(0) \cap \gamma^{-1}(c_k(D_{d-k+1})) = \dim \nu_f^{-1}(0) - k$ a lieu si l'intersection n'est pas vide.

Démonstration : Nous allons en fait prouver un résultat plus précis : Soient $c_k(D_{d-k+1}) = \cup\, \sigma_\alpha$ une stratification de Whitney fixée de la variété de Schubert qui nous intéresse ; pour chaque élément $\mu \in \Gamma = GL(N,\mathbb{C})$, l'image $\mu \cdot c_k(D_{d-k+1})$ de notre variété de Schubert par l'action naturelle de μ sur G est munie de la stratification de Whitney $\cup\, \mu.\sigma_\alpha$ (cf. Chap. III, 2.2, Exercice) et d'autre part est égale à $c_k(\mu^{-1} \cdot D_{d-k+1})$ en utilisant cette fois-ci l'action naturelle de

μ^{-1} sur G_k.

Fixons une stratification de Whitney $N_f(X) = \cup Z_\beta$ de $N_f(X) \subset S \times \mathbb{C}^N \times G$ telle que $\nu_f^{-1}(0)$ et $\nu_f^{-1}(F)$ soient réunions de strates (cf. Chap. III, 1.5 et 2.2.2). D'après le Corollaire du Théorème de Kleiman (cf. Chap. III, 4.3), il existe un ouvert de Zariski dense $U \subset \Gamma$ tel que pour tout $\mu \in U$, chacune des strates $0 \times 0 \times \mu.\sigma_\alpha$ soit transverse à chacune des strates Z_β contenues dans $\nu_f^{-1}(0)$. D'après le Lemme de (Chap. III, 4.2.5), la stratification de Whitney $(S \times \mathbb{C}^N \times \mu.\sigma_\alpha)$ de $S \times \mathbb{C}^N \times \mu.c_k(D_{d-k+1})$ est donc transverse, dans $S \times \mathbb{C}^N \times G$ à la stratification fixée Z_β. Puisque $\nu_f^{-1}(X - F)$ est un ouvert analytique dense dans $N_f(X)$ et réunion de strates, l'intersection $\mu.c_k(D_{d-k+1})^o \cap \nu_f^{-1}(X - F)$, qui est contenue dans $\gamma_f^{-1}(\mu.c_k(D_{d-k+1})) \cap \nu_f^{-1}(X - F) = \gamma_f^{-1}(c_k(\mu^{-1}.D_{d-k+1})) \cap \nu_f^{-1}(X - F)$ est donc dense dans $\gamma_f^{-1}(c_k(\mu^{-1}.D_{d-k+1})) = S \times \mathbb{C}^N \times c_k(\mu^{-1}.D_{d-k+1}) \cap N_f(X)$. La proposition résulte alors aussitôt du fait que l'action de Γ sur G_k est transitive, le point ii) provenant de la transversalité dans G des strates $0 \times 0 \times \mu.\sigma_\alpha$ avec les strates Z_β contenues dans $\nu_f^{-1}(0)$. ∎

1.3.1 Remarque : La même assertion, avec la même preuve, est valable pour toutes les variétés de Schubert, et permet de définir, pour chaque suite d'entiers $(a_1,\dots,a_d)$ une variété polaire locale associée à $(a_1,\dots,a_d)$ et à un drapeau $\mathcal{D}$ assez général, définie par : $\nu_*(\gamma^{-1}(\sigma_a(\mathcal{D})))$. C'est un sous-espace analytique réduit, de codimension Σa_i ou vide, de X, dont le transformé strict par le morphisme ν est égal ensemblistement à $\gamma^{-1}(\sigma_a(\mathcal{D}))$. Ces variétés polaires sont sûrement destinées à jouer un rôle important dans l'étude locale des singularités, mais nous n'en avons pas besoin ici.

1.3.2 Corollaire : Supposons que f soit un morphisme lisse, c'est-à-dire plat et à fibres lisses, en tout point $x \in X\text{-}F$. Soit $p : \mathbb{C}^N \to \mathbb{C}^{d-k+1}$ une projection linéaire telle que Ker $p = D_{d-k+1}$ appartienne à l'ouvert W_k. Pour $x \in X\text{-}F$, la fibre $X(f(x))$ est non-singulière en x, contenue dans $\{f(x)\} \times \mathbb{C}^N$. et l'on notera $\pi_x : X(f(x)) \to \mathbb{C}^{d-k+1}$ la restriction à $X(f(x))$ de la projection p. Soit $P_k\langle f;p\rangle^o$ l'ensemble des points $x \in X - F$ tels que x soit critique pour π_x.

Alors

i) $P_k\langle f;p\rangle^o = \nu_f(\gamma_f^{-1}(c_k(D_{d-k+1}) \cap \nu_f^{-1}(X-F))$.

ii) L'adhérence $P_k\langle f;p\rangle$ de $P_k\langle f;p\rangle^o$ dans X est un sous-espace analytique fermé de X, purement de codimension k dans X ou vide, égal à l'image réduite $\nu_f(\gamma_f^{-1}(c_k(D_{d-k+1}))$.

iii) Le transformé strict de $P_k\langle f;p\rangle$ par le morphisme ν_f est égal à $\gamma_f^{-1}(c_k(D_{d-k+1}))$, ensemblistement.

Preuve : Il suffit de remarquer que x est critique pour π_x si et seulement si l'on a $\dim(T_{X(f(x)),x} \cap D_{d-k+1}) \geq k$, ce qui donne i). Le reste est conséquence immédiate de la Proposition, et du fait que le morphisme ν_f est propre. ∎

1.4 Définition : Etant donnés un morphisme $f:(X,0) \to (S,0)$ comme ci-dessus, muni d'une S-installation $(X,0) \subset (S,0) \times (\mathbb{C}^N,0)$, et un sous-espace linéaire $D_{d-k+1} \subset \mathbb{C}^N$ de codimension d-k+1 contenu dans W_k, on appelle variété polaire locale relative de codimension k associée à f et à D_{d-k+1} le sous-espace analytique fermé $P_k\langle f;p\rangle$ de X, qui est purement de codimension k dans X, ou vide. On le note aussi $P_k\langle f;D_{d-k+1}\rangle$, ou $P_k((X,0),D_{d-k+1})$ si S est un point. On note de même son germe en 0. Lorsqu'aucun risque de confusion n'est présent, on notera aussi $P_k\langle p\rangle$ ou $P_k\langle D_{d-k+1}\rangle$. Lorsque S est un point, on parlera de variété polaire absolue, mais dans tous les cas, on omettra souvent les adjectifs "relatif" ou "absolu".

1.4.1 Premier avatar : Supposons le S-plongement fermé $(X,0) \subset (S,0) \times (\mathbb{C}^N,0)$ défini par l'idéal $I = (f_1,\dots,f_m)$ de $\mathcal{O}_{S,0}\{z_1,\dots,z_N\}$. Soit $D_{d-k+1} \in W_k$ et choisissons les coordonnées de telle façon que D_{d-k+1} soit défini par $z_1 = \dots = z_{d-k+1} = 0$. Posons $d = \dim X - \dim S$ et $c = N-d$. Notons J l'idéal de $\mathcal{O}_{X,0}$ engendré par les déterminants jacobiens $\frac{\partial(f_{i_1},\dots,f_{i_c})}{\partial(z_{j_1},\dots,z_{j_c})}$ avec $\{i_1,\dots,i_c\} \subset \{1,\dots,m\}$ et $\{j_1,\dots,j_c\} \subset \{1,\dots,N\}$, et notons $J\langle D_{d-k+1}\rangle \subset J$ l'idéal de $\mathcal{O}_{X,0}$ engendré par les seuls déterminants jacobiens qui sont tels que $\{j_1,\dots,j_c\}\subset\{d-k+2,\dots,N\}$. La variété polaire $P_k(f;D_{d-k+1})$ est définie dans X par l'idéal

$$\pi<D_{d-k+1}> = (J : \sqrt{J<D_{d-k+1}>}) = \{h\in \mathcal{O}_{X,0} / h . J\subset\sqrt{J<D_{d-k+1}>}\} .$$

Remarque : Comme le calcul des idéaux résiduels (I : J) est assez impraticable en général, cet avatar ne fait guère que décrire la variété polaire $P_k(f;D_{d-k+1})$ dans le langage de l'algèbre.

"Le secret de la pensée solide est dans la défiance des langages. Les spéculations bien séparées des notations sont les plus puissantes."

Paul Valéry. Cahiers

1.4.2 Second avatar : Soient k un corps de caractéristique zéro, et $\varphi : R\to A$ un homomorphisme de k-algèbres nœthériennes locales réduites complètes équidimensionnelles. On suppose que les extensions résiduelles sont triviales, que R est régulier et que $\Omega^1_{A/R}\otimes_A \mathrm{Tot}(A)$ est libre de rang $d=\dim A-\dim R$. On fixe un système $z_1,\dots,z_N$ d'éléments de l'idéal maximal m_A de A qui engendrent m_A modulo $m_R . A$, et un entier k, $0\le k\le d-1$.

Soient λ_{ij}, $1\le j\le N$ des indéterminées et posons $z_i^* = \sum_{j=1}^{N} \lambda_{ij} z_j$. Notons K^* le corps $k(\{\lambda_{ij}\})$ et posons $R^* = R\otimes_k K^*$, $A^* = A\otimes_k K^*$; soit enfin R_k^* la R^*-algèbre complète $R^*[[z_1^*,\dots,z_{d-k+1}^*]]$. On a donc des homomorphismes : $R^*\hookrightarrow R^*_{d-k+1}\xrightarrow{\varphi_{d-k+1}} A^*$. Considérons les idéaux premiers p^* de A^* qui sont minimaux parmi ceux qui satisfont les deux conditions suivantes :

i) L'anneau $A^*_{p^*}$ est géométriquement régulier.

ii) L'homomorphisme naturel $(R^*_{d-k+1})_{\varphi^{-1}_{d-k+1}(p^*)}\to A^*_{p^*}$ n'est pas formellement lisse.

Alors ces idéaux premiers sont de hauteur k, en nombre fini, et leur intersection est l'idéal définissant dans Spec A^* la variété polaire locale (générique) associée à φ et au choix de $z_1,\dots,z_N$.

§ 2. Exemples.

2.1 Soit $f:\mathbb{C}^{d+1}\to\mathbb{C}$ un (germe de) morphisme analytique, que l'on installe par le plongement $\mathbb{C}^{d+1}\hookrightarrow\mathbb{C}\times\mathbb{C}^{d+1}$ défini par l'idéal engendré par $v-f(z_o,\dots,z_d)$ Si 0 est un point critique isolé de f, la k-ième variété polaire $p_k\langle f;D_{d-k+1}\rangle$ de f associée au sous-espace D_{d-k+1}, que l'on peut supposer défini par $z_o=\dots=z_{d-k}=0$ est le sous-espace de $\mathbb{C}^{n+1}$ défini par l'idéal de $\mathbb{C}\{z_o,\dots,z_n\}$ engendré par $\left(\frac{\partial f}{\partial z_{d-k+1}},\dots,\frac{\partial f}{\partial z_d}\right)$. Dans [Te 12], il est prouvé que pour D_{d-k+1} assez général, la multiplicité en 0 de $P_k\langle f;D_{d-k+1}\rangle$ est égale au nombre de Milnor $\mu^{(k)}$ de l'intersection de l'hypersurface $X=f^{-1}(0)$ avec un plan de dimension k général de $\mathbb{C}^{d+1}$ passant par 0.

2.2 Examinons maintenant, dans la même situation, les variétés polaires (absolues) du germe d'hypersurface $X=f^{-1}(0)$ associées a des D_{d-k+1} : elles sont définies par les idéaux suivants de $\mathcal{O}_{X,0}$:

$$\pi_k\langle D_{d-k+1}\rangle = \left(\left(\frac{\partial f}{\partial z_o},\dots,\frac{\partial f}{\partial z_1}\right)\cdot\mathcal{O}_X : \sqrt{\left(\frac{\partial f}{\partial z_{d-k+1}},\dots,\frac{\partial f}{\partial z_d}\right)\cdot\mathcal{O}_X}\right)$$

et si X est à singularité isolée en 0, on a

$$\pi_k\langle D_{d-k+1}\rangle = \left(\frac{\partial f}{\partial z_{d-k+1}},\dots,\frac{\partial f}{\partial z_d}\right)\mathbb{C}\{z_o,\dots,z_d\}/(f)$$

on en déduit que l'on a l'égalité

$$P_k\langle f;D_{d-k+1}\rangle\cap f^{-1}(0) = P_k(f^{-1}(0);D_{d-k+1}\rangle\ .$$

Dans (loc. cit.), on a montré que la multiplicité en 0 de $P_k\langle X;D_{d-k+1}\rangle$ était, pour D_{d-k+1} assez général, égale à $\mu^{(k)}+\mu^{(k+1)}$ pour $0\le k\le d-1$.

Le dessin suivant aidera peut-être le lecteur ; ici X est une surface réduite de $\mathbb{C}^3$, pas nécessairement à singularité isolée.

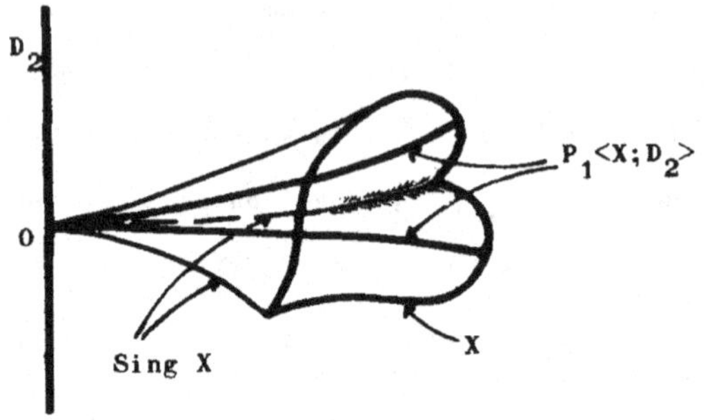

2.3 Cas particulier servant de note historique : Supposons maintenant que f soit un polynôme homogène de degré m, définissant un cône réduit X, cône sur une variété projective réduite $V \subset \mathbb{P}^d$. Choisir une droite $D_d \subset \mathbb{C}^{d+1}$ revient à choisir un point $(X_0 : \dots : X_d) \in \mathbb{P}^d$ et la variété polaire relative $P_1\langle f;D_d\rangle$ est contenue dans l'hypersurface $\sum_0^d X_i \frac{\partial f}{\partial z_i} = 0$, et lui est égale si V est non-singulière. Par ailleurs, dans ce cas $P_k\langle X;D_d\rangle = P_k\langle f;D_{d-k}\rangle \cap X$. La sous-variété projective de $\mathbb{P}^d$ correspondant à $P_1\langle f;D_d\rangle$ sera appelée variété polaire à la Poncelet associée à f. On notera qu'elle est définie dans le cas où V est non singulière par l'équation obtenue en polarisant le polynôme f par rapport à un point suffisamment général $(X_0 : \dots : X_d) \in \mathbb{P}^d$. Son intersection avec V sera appelée variété polaire à la Todd de V associée à $D_d \in \mathbb{P}^d$. On peut bien sûr faire de même pour toutes les variétés projectives, et l'on voit que la théorie des variétés polaires locales absolues contient la théorie des variétés polaires à la Todd des variétés projectives, telle qu'elle a été développée par R. Piene [Pi] : c'est tout simplement le cas où notre germe est un cône. On peut aussi consulter cet excellent travail de Piene, ou [Kl 2], pour des références historiques.

Exercice : Soit $X \subset \mathbb{C}^3$ le cône sur une courbe projective plane réduite $V \subset \mathbb{P}^2$. On sait que la multiplicité en 0 de X est égale au degré de la courbe V. Montrer que la multiplicité en 0 de la courbe polaire $P_1\langle X; D_2\rangle$ de X pour D_2 assez général, est égale à la classe de la courbe projective V, c'est-à-dire au degré de la courbe duale $V^\vee \subset \check{\mathbb{P}}^2$.

§ 3. Multiplicité des variétés polaires.

3.1 Théorème : Soit $f : (X,0) \to (S,0)$ un morphisme d'espaces analytiques comme en 1.1.

i) Pour tout S-plongement local $(X,0) \subset (S,0) \times (\mathbb{C}^N,0)$, tout choix de coordonnées sur $\mathbb{C}^N$, et pour tout entier k, $0 \leq k \leq d = \dim X - \dim S$, il existe un ouvert de Zariski dense V_k contenu dans l'ouvert W_k de la Proposition 2 (1.3) tel que la multiplicité $m_0(P_k\langle f; D_{d-k+1}\rangle)$ en 0 de la variété polaire $P_k\langle f; D_{d-k+1}\rangle$ soit indépendante de $D_{d-k+1} \in V_k$.

ii) Pour chaque k, $0 \leq k \leq d$, cette multiplicité ne dépend en fait que de la classe d'isomorphisme de l'homomorphisme d'algèbres $\mathcal{O}_{S,0} \to \mathcal{O}_{X,0}$ associé à f.

Démonstration : Décrivons d'abord une construction : Donnons-nous une famille à un paramètre de projections $\mathbb{C}^N \to \mathbb{C}^{d-k+1}$, que nous supposons décrite par

$$(*) \qquad z_i^* = z_i + \sum_{j=d-k+1}^{N} \gamma_{ij}(t) z_j + \sum_{|A| \geq 2} c_{i,A}(t)\, z^A ,$$

pour $1 \leq i \leq d-k+1$, avec $\gamma_{ij}(t)$ et $c_{i,A}(t)$ dans $t \cdot \mathbb{C}[t]$, dans des coordonnées $z_1, \dots, z_N$ fixées sur $\mathbb{C}^N$. Notons $\mathbb{A}$ la droite affine, et considérons le morphisme (où $t \in \mathbb{A}$)

$$p^* : \mathbb{C}^N \times \mathbb{A} \longrightarrow \mathbb{C}^{d-k+1} \times \mathbb{A}$$

décrit par (*) et l'identité de $\mathbb{A}$.

Considérons le diagramme suivant :

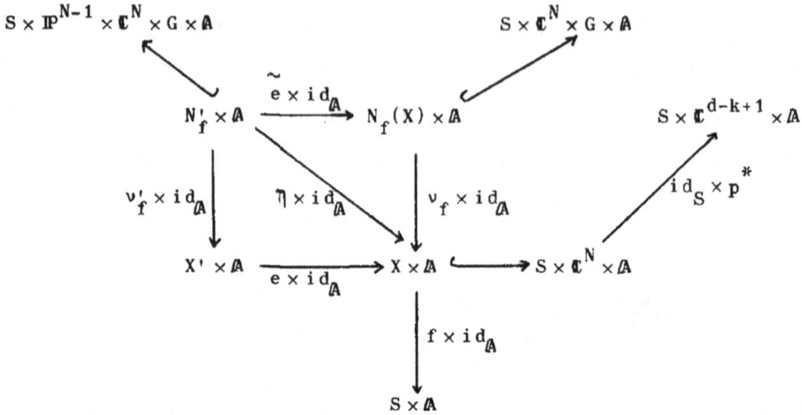

où $\nu_f \times \mathrm{id}_{\mathbb{A}}$ n'est autre que la modification de Nash relative associée à $f \times \mathrm{id}_{\mathbb{A}}$, $e \times \mathrm{id}_{\mathbb{A}}$ est l'éclatement dans $X \times \mathbb{A}$ du sous-espace $0 \times \mathbb{A}$, $\tilde{e} \times \mathrm{id}_{\mathbb{A}}$ est l'éclatement dans $N_f(X) \times \mathbb{A}$ du sous-espace $\nu_f^{-1}(0) \times \mathbb{A}$, enfin $\nu' \times \mathrm{id}_{\mathbb{A}}$ est le morphisme dû à la propriété universelle de l'éclatement.

Notons $dp_t^*(z)$ l'application tangente à la projection $p_t^* = p^* | \mathbb{C}^N \times \{t\}$ au point z, et considérons le sous-espace du produit $\mathbb{C}^N \times G \times \mathbb{A}$ défini comme ceci

$$P = \{(z,T,t) \,/\, \dim(T \cap \mathrm{Ker}\, dp_t^*(z)) \geq k\} \quad .$$

Il est facile de voir que P est un sous-espace analytique fermé de $\mathbb{C}^N \times G \times \mathbb{A}$, défini localement par les conditions :

$$\mathrm{rang} \begin{vmatrix} a_{1,1} & ,\dots\dots\dots\dots, & a_{1,N} \\ \vdots & & \vdots \\ a_{c,1} & ,\dots\dots\dots\dots, & a_{c,N} \\ \dfrac{\partial z_1^*}{\partial z_1} & ,\dots\dots\dots\dots, & \dfrac{\partial z_1^*}{\partial z_N} \\ & & \\ \dfrac{\partial z_{d-k+1}^*}{\partial z_1} \;,\dots, & \dfrac{\partial z_{d-k+1}^*}{\partial z_{d-k+1}} & ,\dots,\; \dfrac{\partial z_{d-k+1}^*}{\partial z_N} \end{vmatrix} < c+d-k+1 \quad , \quad \text{où } c = N-d$$

où les équations $\Sigma a_{i,j} z_j = 0$ $(1 \leq i \leq c)$ définissent $T \subset \mathbb{C}^N$.

Le point est maintenant que, d'après les résultats généraux (cf. [Ke] (§ 3)) sur les variétés déterminantielles, l'espace analytique P admet une stratification $P = \cup P_\alpha$ le long de chaque strate de laquelle on a résolution simultanée forte pour les strates adjacentes, et qui est donc en particulier une stratification de Whitney (cf. Chap. III, 5.2.2). L'existence d'une telle stratification provient de la résolution des singularités explicite donnée dans (Loc. cit.).

Détaillons un peu : d'une facon générale, soit U l'espace affine ayant pour coordonnées U_{ij}, $1 \leq i \leq n$, $1 \leq j \leq p$. Considérons le sous-espace fermé de $U \times (\mathbb{C}^n - \{0\})$ défini par les équations

$$(E_j) : \Sigma U_{ij} X_i = 0 .$$

Cet espace est non-singulier, et son image Z dans $U \times \mathbb{P}^{n-1}$ est un sous-espace fermé non-singulier de $U \times \mathbb{P}^{n-1}$, dont on calcule facilement que la dimension vaut np-p+n-1 .

<u>Proposition</u> (Kempf, [Ke], § 3) : <u>Le morphisme</u> $Z \to U$ <u>induit par la première projection est une résolution rationnelle des singularités de son image, qui est le sous-espace de</u> U <u>défini par les équations exprimant que</u> : $\mathrm{rang}(U_{ij}) < p$.

Dans le cas qui nous intéresse, nous utilisons le fait que puisque la matrice $\left(\frac{\partial z_i^*}{\partial z_j}\right)$ $(1 \leq i,\ j \leq d-k+1)$ est inversible, nous pouvons utiliser les d-k+1 dernières lignes du système d'équations linéaires

$$\begin{aligned}
a_{1,1} Z_1 + \cdots\cdots + a_{1,N} Z_N &= 0 \\
\vdots \qquad\qquad & \\
a_{c,1} Z_1 + \cdots\cdots + a_{c,N} Z_N &= 0 \\
\frac{\partial z_1^*}{\partial z_1} Z_1 + \cdots\cdots + \frac{\partial z_1^*}{\partial z_N} Z_N &= 0 \\
\vdots \qquad\qquad & \\
\frac{\partial z_{d-k+1}^*}{\partial z_1} Z_1 + \cdots + \frac{\partial z_{d-k+1}^*}{\partial z_N} Z_N &= 0
\end{aligned}$$

pour exprimer linéairement $Z_1,\dots,Z_{d-k+1}$ en fonction de $Z_{d-k+2},\dots,Z_N$, et nous ramener à écrire qu'un système de c équations en $N-(d-k+1)$ inconnues a une solution non triviale. Le sous-espace $\widetilde{P}$ de $\mathbb{C}^N \times G \times \mathbb{A} \times \mathbb{P}^{N-(d-k+1)-1}$ défini localement par les c équations obtenues en substituant dans les c premières lignes du système ci-dessus les valeurs de $Z_1,\dots,Z_{d-k+1}$ obtenues à partir des d-k+1 dernières, muni du morphisme $\widetilde{P} \to \mathbb{C}^N \times G \times \mathbb{A}$ induit par la projection sur les trois premiers facteurs, a pour image P et en est une résolution des singularités. On prend pour P_α les images des strates d'équidimensionalité des fibres de cette résolution $\widetilde{P} \to P$, et on vérifie la résolution simultanée. D'après la forme des équations définissant $\widetilde{P}$, chaque strate P_α est envoyée submersivement sur la droite A par la projection $\mathbb{C}^N \times G \times \mathbb{A} \to \mathbb{A}$. La fibre de P au-dessus de $0 \in \mathbb{A}$ (par cette projection) n'est autre que $\mathbb{C}^N \times c_k(\mathrm{Ker}\ p_o)$ comme on le voit. Enfin, l'image par $\nu_f \times id_{\mathbb{A}}$ de l'intersection $(S \times P) \cap (N_f(X) \times \mathbb{A})$ (dans $S \times \mathbb{C}^N \times G \times \mathbb{A}$) est un sous-espace analytique fermé de $X \times \mathbb{A}$, que nous noterons $P_k(\mathrm{Ker}\ p^*)$, qui a la propriété que pour chaque $t \in \mathbb{A}$, $P_k(\mathrm{Ker}\ p^*) \cap S \times \mathbb{C}^N \times \{t\}$ est la variété polaire relative de X, muni des coordonnées $z_1^*(t),\dots,z_{d-k+1}^*(t)$, $z_{d-k+2},\dots,z_N$ correspondant à $z_1^*(t) = \dots = z_{d-k+1}^*(t) = 0$.

<u>Lemme</u> : <u>Il existe un ouvert de Zariski dense</u> U_k <u>de la grassmannienne des</u> $D_{d-k+1} \subset \mathbb{C}^N$ <u>tel que si</u> $\mathrm{Ker}\ p_o \in U_k$, <u>l'espace</u> $P_k(\mathrm{Ker}\ p^*)$ <u>soit équimultiple le long de</u> $0 \times \mathbb{A}$.

<u>Preuve</u> : Choisissons une stratification de Whitney $N_f(X) = \cup Z_\beta$ de $N_f(X)$ telle que non seulement $\nu_f^{-1}(0)$ soit réunion de strates, mais encore pour chaque i, l'image par $\widetilde{e}$ de l'ensemble $F_i = \{x' \in N_f'(X) / \dim \widetilde{e}^{-1}(\widetilde{e}(x')) \geq i\}$ soit réunion de strates. Cela fait, choisissons p_o de telle façon que $c_k(\mathrm{Ker}\ p_o)$ (muni de sa stratification naturelle) soit transverse (en tant qu'ensemble stratifié, c-à-d., strate par strate) à $\nu_f^{-1}(0)$ dans G. D'après le lemme 4.2.5 du Chapitre III, l'ensemble stratifié $S \times P$ est transverse à $N_f(X) \times \mathbb{A}$ dans $S \times \mathbb{C}^N \times G \times \mathbb{A}$, et d'autre part, d'après (Chap. III, 5.1), puisque $c_k(\mathrm{Ker}\ p_o)$ est transverse à $\nu_f^{-1}(0)$, l'ensemble $((\widetilde{e} \times id_{\mathbb{A}})^{-1}((S \times P) \cap N_f(X) \times \mathbb{A})_{red}$ est le

transformé strict de $(S\times P)\cap(N_f(X)\times \mathbb{A})$ par l'éclatement $\tilde{e}\times id_{\mathbb{A}}$.

D'après le Corollaire 1.3.2, cet ensemble est le transformé strict par le morphisme $\eta\times id_{\mathbb{A}}=(\nu_f\circ\tilde{e})\times id_{\mathbb{A}}$ du sous-espace analytique $P_k(\mathrm{Ker}\ p^*)$ de $X\times\mathbb{A}$, et nous le noterons $P_k(\mathrm{Ker}\ p^*)'^{\wedge}$. D'après l'hypothèse de transversalité, et la preuve de (Chap. III, 5.1), l'intersection de $P_k(\mathrm{Ker}\ p^*)'^{\wedge}$ avec $(\eta\times id_{\mathbb{A}})^{-1}(0\times\mathbb{A})$ a pour dimension $d+\dim S-k$, et sa fibre au-dessus de $0\in\mathbb{A}$ a pour dimension $d+\dim S-k-1$. C'est donc a fortiori le cas pour le diviseur exceptionnel du transformé strict $P_k(\mathrm{Ker}\ p^*)'$ de $P_k(\mathrm{Ker}\ p^*)$ par l'éclatement $e\times id_{\mathbb{A}}$ de $0\times\mathbb{A}$ dans $X\times\mathbb{A}$, c'est-à-dire par l'éclatement de $0\times\mathbb{A}$ dans $P_k(\mathrm{Ker}\ p^*)$. L'équimultiplicité résulte alors de (Chap. I, Remarque 5.1.1, 3)). ∎

Pour achever la preuve du théorème, on commence par comparer les variétés polaires correspondant à des projections linéaires dans deux S-plongements $X\subset S\times\mathbb{C}^N$, $X\subset S\times\mathbb{C}^{N'}$, $\mathbb{C}^N$ et $\mathbb{C}^{N'}$ étant munis de coordonnées $z_1,\dots,z_N$ et $z'_1,\dots,z'_{N'}$, respectivement. En considérant $\mathbb{C}^N\times\mathbb{C}^{N'}$, on se ramène aussitôt à prouver que les multiplicités des variétés polaires de X correspondant à des projections linéaires générales de $\mathbb{C}^N$ et de $\mathbb{C}^{N+N'}$ sur $\mathbb{C}^{d-k+1}$ sont égales. Supposant que $z_1,\dots,z_{d-k+1}$ définit une projection linéaire générale pour $\mathbb{C}^N$ on peut, au prix d'un changement linéaire de coordonnées sur $\mathbb{C}^{N'}$ supposer que $z_i+tz'_i$, $1\le i\le d-k+1$, définit pour $t\neq 0$, assez petit, une projection linéaire générale pour $\mathbb{C}^{N+N'}$, car les sous-espaces de codimension d-k+1 de $\mathbb{C}^N\times\mathbb{C}^{N'}$ qui rencontrent $\mathbb{C}^N\times\{0\}$ et $\{0\}\times\mathbb{C}^{N'}$ en codimension d-k+1 forment un ouvert dense de la grassmannienne. Il suffit alors d'appliquer le Lemme à la famille $z^*_i=z_i+tz'_i$.

Il nous reste à étudier le cas de deux systèmes de coordonnées différents dans $\mathbb{C}^N$, disons $z_1,\dots,z_N$ et $z'_1,\dots,z'_N$. Au prix d'un changement linéaire de coordonnées, on peut se ramener au cas où l'on a une expression

$$z_i' = z_i + \sum_{|A|\geq 2} c_{i,A} z^A \qquad (1 \leq i \leq d-k+1), \quad c_{i,A} \in \mathbb{C}$$

et il suffit d'appliquer le Lemme à la famille

$$z_i^* = z_i + \sum_{|A|\geq 2} c_{i,A} t^{|A|-1} z^A$$

qui pour tout $t \neq 0$ donne des coordonnées homothétiques aux z_i' , donc fournissant les mêmes variétés polaires. ∎

3.2 Remarques : 1) En fait (comme l'a remarqué aussi V. Navarro (cf. [Na 2] pour un très bon exposé), la modification de Nash n'est qu'un cas particulier de la construction suivante : Etant donné un faisceau F, localement libre, disons de rang f, sur le complémentaire d'un fermé analytique rare B de X, il existe une modification $\nu(F):\tilde{X}\to X$ telle que $\nu(F)^*(F)$ ait un quotient localement libre de rang f sur $\tilde{X}$, et minimale pour cette propriété. (La modification de Nash est le cas où $F = \Omega_X^1$.) On peut au moyen d'une présentation locale $\mathcal{O}_{X|U}^M \to F_{|U} \to 0$ plonger $\nu(F)^{-1}(U)$ dans $U \times G$, où G est la grassmannienne des f-plans de $\mathbb{C}^M$. On peut ainsi définir les variétés polaires locales de F en tout point $x \in X$, en utilisant le même argument de mise en position générale qu'en 1.3 et ce pour tout symbole de Schubert $(a_1,\ldots,a_n)$. Le rôle du plongement local $X \subset \mathbb{C}^N$ n'est que de donner une présentation locale $\mathcal{O}_X^N \to \Omega_X^1 \to 0$. On peut même, utilisant l'argument de [Lê-Te], montrer que les multiplicités en chaque point de ces variétés polaires locales ne dépendent que de F. Je me suis abstenu de démontrer le Théorème par cette voie pour deux raisons ; un peu pour éviter une généralisation pour le moment inutile, mais surtout parce que la démonstration géométrique donnée ci-dessus, convenablement raffinée permet de prouver le résultat suivant :

Théorème : La classe d'équisingularité (au sens des conditions de Whitney) des variétés polaires projectives relatives générales associées à un S-plongement $X \subset S \times \mathbb{C}^N$ ne dépend que du type d'isomorphisme de l'homomorphisme d'algèbres $\mathcal{O}_{S,0} \to \mathcal{O}_{X,0}$ correspondant à f.

et ce résultat cadre très bien avec le point de vue selon lequel on doit s'efforcer de remplacer des résultats numériques (ou de classe d'équivalence rationnelle) en théorie des intersections par des résultats concernant la classe d'équisingularité. D'ailleurs il est probable que la démonstration ci-dessus peut être un peu modifiée pour prouver un résultat analogue pour toutes les variétés polaires locales d'un faisceau cohérent F.

2) En fait, comme on le verra au Chapitre VI, les multiplicités des variétés polaires locales projectives sont des invariants de nature essentiellement topologique (au sens de la topologie des sous-ensembles stratifiés de $\mathbb{C}^N$), ce qui généralise la présentation des $\mu^{(k)}$ de 2.1 donnée dans [Te 1].

<u>3.2 Corollaire</u> : <u>A toute algèbre analytique locale</u> $\mathcal{O}_{X,0}$ <u>réduite et purement de dimension</u> d, <u>on peut associer une suite d'entiers</u>

$$M^*_{X,x} = (m_0(P_0(X,0)), m_0(P_1(X,0)), \ldots, m_0(P_{d-1}(X,0)))$$

<u>où</u> $m_0(P_k(X,0))$ <u>est la multiplicité en</u> 0 <u>de la variété polaire projective</u> $P_k\langle(X,0), D_{d-k+1}\rangle$ <u>calculée au moyen d'un plongement</u> $(X,0) \subset (\mathbb{C}^N,0)$ <u>et d'un sous-espace</u> D_{d-k+1} <u>assez général</u>.

<u>3.3 Remarque</u> : Nous nous permettrons dorénavant de parler de "la" variété polaire (sous-entendu "projective générale") de codimension k de X et même de la noter $P_k(X,0)$. Noter que d'après 1.3, ii), $P_d(X,0)$ est vide.

<u>3.4 Exercice</u> : Soit X la surface dans $\mathbb{C}^3$ d'équation $x^2 - y^2 z = 0$; montrer que $M^*_{X,0} = (2,1)$.

§ 4. <u>Variétés polaires et espace conormal</u>.

Voici une autre construction possible des variétés polaires projectives, où la modification de Nash est remplacée par le morphisme conormal. C'est la forme locale de la construction des cycles polaires des variétés projectives qui permet d'étudier le comportement de ces cycles par dualité (cf. [Pi]). Je remercie J.P.G Henry et M. Merle de m'avoir signalé son intérêt.

4.1 Reprenons la situation de 1.1. Soit $f : X \to S$ un morphisme, comme en 1.1, et supposons fixé un S-plongement $X \subset S \times \mathbb{C}^N$. Considérons l'espace conormal relatif

$$\begin{array}{ccc} & & \check{\mathbb{P}}^{N-1} \\ & \overset{\lambda_f}{\nearrow} & \\ C_f(X) \subset X \times \check{\mathbb{P}}^{N-1} & & \\ \big\downarrow{\scriptstyle \varkappa_f} & & \\ X & & \end{array}$$

et notons $\lambda_f : C_f(X) \to \check{\mathbb{P}}^{N-1}$ le morphisme induit par la seconde projection $X \times \check{\mathbb{P}}^{N-1} \to \check{\mathbb{P}}^{N-1}$. ($\lambda_f$ joue le rôle du morphisme de Gauss.)

4.1.1 Proposition (à comparer à la Prop. 2 du § 1, et son Corollaire 1.3.1) : Pour tout entier k, $0 \leq k \leq d = \dim X - \dim S$, il existe un ouvert de Zariski dense U_k de la grassmannienne des sous-espaces projectifs de dimension $d-k$ de $\check{\mathbb{P}}^{N-1}$ tel que pour tout $L^{d-k} \in U_k$ on ait :

1) $\lambda_f^{-1}(L^{d-k}) \cap \varkappa_f^{-1}(X - F)$ est dense dans $\lambda_f^{-1}(L^{d-k})$ et ce dernier espace est vide ou de codimension pure $N-1-d+k$ dans $C_f(X)$.

2) L'égalité $\dim(\varkappa_f^{-1}(0) \cap \lambda_f^{-1}(L^{d-k})) = \dim \varkappa_f^{-1}(0) - N+1+d-k$ a lieu si l'intersection n'est pas vide.

3) L'intersection D_{d-k+1} dans $\mathbb{C}^N$ de tous les hyperplans de $\mathbb{C}^N$ représentés par des points de L^{d-k} est un sous-espace de codimension $d-k+1$ de $\mathbb{C}^N$ appartenant à l'ouvert W_k de la Prop. 2 du § 1 et l'on a :

$$(*) \qquad (\varkappa_f(\lambda_f^{-1}(L^{d-k})))_{red} = P_k\langle f ; D_{d-k+1}\rangle \quad .$$

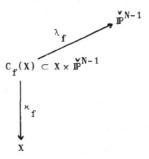

Preuve : Les points 1) et 2) peuvent être démontrés de façon tout à fait analogue aus points 1) et 2) de la Prop. 2 du § 1. Dans le point 3), l'assertion de dimension vient de ce que D_{d-k+1} est l'intersection des hyperplans correspondant à d-k+1 points en position générale dans $\mathbb{P}^{d-k}$, la condition que D_{d-k+1} soit dans W_k est une condition ouverte sur L^{d-k} comme on le voit facilement. Enfin, d'après les points 1) et 2) il suffit de vérifier l'égalité (*) en tout point x de $X - F$, c'est-à-dire que, notant $T \subset \mathbb{C}^N$ l'espace tangent en x à la fibre $X(f(x))$ de f passant par x, on doit prouver l'équivalence $T \subset H \in L^{d-k} \Leftrightarrow \dim(T \cap D_{d-k+1}) \geq k$. Si $T \subset H \in L^{d-k}$, on a $\dim(T \cap D_{d-k+1}) \geq d+N-d+k-1-N+1 = k$ puisque $D_{d-k+1} \subset H$, et inversement, si $\dim(T \cap D_{d-k+1}) \geq k$, on a $\dim(T + D_{d-k+1}) \leq N-1$, d'où l'implication inverse puisque L^{d-k} est l'ensemble des hyperplans contenant D_{d-k+1}. ∎

§ 5. Transversalité des variétés polaires.

5.1 Théorème : Soit $f : (X,0) \to (S,0)$ un morphisme d'espaces analytiques comme en 1.1, où de plus S est non-singulier de dimension au plus 1. Pour tout représentant assez petit de f, tout plongement local $(X,0) \subset (\mathbb{C}^M,0)$, tout entier k, $0 \leq k \leq d = \dim X - \dim S$, il existe un ouvert de Zariski dense T_k de la grassmannienne G_k des sous-espaces vectoriels de codimension d-k+1 dans $\mathbb{C}^M$, contenu dans l'ouvert W_k de 1.4 et tel que, pour tout $D_{d-k+1} \in T_k$, on ait :

A) Le sous-espace D_{d-k+1} est transverse en 0, dans $\mathbb{C}^M$, à la variété polaire relative $P_k\langle f;D_{d-k+1}\rangle$, en ce sens que l'on a

$$|D_{d-k+1} \cap C_o(P_k\langle f;D_{d-k+1}\rangle)| = \{0\}$$

où $C_o(\ .\ )$ désigne le cône tangent en 0, et l'intersection est prise dans $T_{\mathbb{C}^M,0}$.

B) Les directions limites en 0 d'espaces tangents aux fibres de f en des points non-singuliers d'icèlles contenus dans $X \cap D_{d-k+1}$ sont transverses à D_{d-k+1} dans $\mathbb{C}^M$, en ce sens que pour une telle limite T, on a $\dim(T \cap D_{d-k+1}) = k-1$.

Démonstration : Ces deux résultats sont d'une certaine manière duaux l'un de l'autre : considérons le diagramme

$$\begin{array}{ccc} N'_f(X) & \xrightarrow{\tilde{e}} & N_f(X) \\ \downarrow{\scriptstyle \nu'_f} & & \downarrow{\scriptstyle \nu_f} \\ X' & \xrightarrow{e} & X \end{array}$$

déjà utilisé au paragraphe précédent, et posons $X_{d-k+1} = (X \cap D_{d-k+1})_{red}$.

5.1.1 L'assertion A) équivaut à la suivante :

A') Les transformés stricts $P_k\langle f;D_{d-k+1}\rangle'$ de $P_k\langle f;D_{d-k+1}\rangle$ et X'_{d-k+1} de X_{d-k+1} par l'éclatement e de 0 dans X sont disjoints :

$$P_k\langle f;D_{d-k+1}\rangle' \cap X'_{d-k+1} = \emptyset \ .$$

L'assertion B) équivaut à la suivante :

B') Les transformés stricts $\widehat{P_k\langle f;D_{d-k+1}\rangle}$ de $P_k\langle f;D_{d-k+1}\rangle$ et $\hat{X}_{d-k+1}$ de X_{d-k+1} par la modification de Nash relative ν_f sont disjoints :

$$\widehat{P_k\langle f;D_{d-k+1}\rangle} \cap \hat{X}_{d-k+1} = \emptyset \ .$$

La vérification de ces équivalences est immédiate au vu des définitions.

Traduisons maintenant les deux assertions en termes de coordonnées. Supposons les coordonnées de $\mathbb{C}^M$ choisies de telle façon que D_{d-k+1} soit défini par $z_1 = \cdots = z_{d-k+1} = 0$; posons $P_k = P_k\langle f;D_{d-k+1}\rangle$. D'après (Chap. I, 5.2), l'assertion A) équivaut à la suivante :

A") Pour tout i tel que $d-k+1 < i \leq M$, l'élément $z_i \cdot \mathcal{O}_{P_k,0}$ image de z_i dans $\mathcal{O}_{P_k,0}$ est entier, dans $\mathcal{O}_{P_k,0}$, sur l'idéal $(z_1,\ldots,z_{d-k+1})\cdot\mathcal{O}_{P_k,0}$ de $\mathcal{O}_{P_k,0}$ engendré par les images de $z_1,\ldots,z_{d-k+1}$.

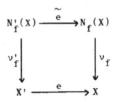

D'après (Chap. II, 3.1), l'assertion B) équivaut à la suivante, pour tou choix d'un système de générateurs $(f_1,\dots,f_p)$ pour l'idéal définissant $(X,0)$ dans $(S,0)\times(\mathbb{C}^M,0)$:

B") L'image $J_{X/S}\cdot\mathcal{O}_{X_{d-k+1},0}$ dans $\mathcal{O}_{X_{d-k+1},0}$ de l'idéal jacobien relatif $J_{X/S}$, engendré par les images des déterminants jacobiens de rang $c = M-d$:

$$\frac{\partial(f_{j_1},\dots,f_{j_c})}{\partial(z_{i_1},\dots,z_{i_c})} \quad \{j_1,\dots,j_c\}\subset\{1,\dots,p\},\ \{i_1,\dots,i_c\}\subset\{1,\dots,M\}$$

est entière dans $\mathcal{O}_{X_{d-k+1},0}$ sur l'idéal J_o engendré dans $\mathcal{O}_{X_{d-k+1},0}$ par ceux de ces déterminants jacobiens qui sont tels que l'on ait :

$$\{i_1,\dots,i_c\}\subset\{d-k+1,\dots,M\} \quad .$$

Démontrons maintenant le Théorème. Choisissons un système $f_1,\dots,f_{p-1}$ de générateurs pour l'idéal définissant $(X,0)\subset(\mathbb{C}^M,0)$. Plaçons-nous dans le cas où $\dim S = 1$; on peut supposer $(X,0) = (\mathbb{C},0)$ et que $f : X\to S$ est la restriction à X de $f_p : (\mathbb{C}^M,0)\to(\mathbb{C},0)$. Le S-plongement $(X,0)\subset(S,0)\times(\mathbb{C}^M,0)$ peut alors être défini par l'idéal $(f_1,\dots,f_{p-1},v-f_p)$ de $\mathbb{C}\{v,z_1,\dots,z_M\}$ où v est une coordonnée locale sur $(S,0)$.

Considérons l'ouvert affine $A=\mathbb{C}^\alpha$ de G_k, où $\alpha=(d-k+1)(M-d+k-1)$, de la grassmannienne G_k muni des coordonnées $a_j^{(i)}$, $1\le i\le d-k+1$, $d-k+2\le j\le M$, et le sous-espace analytique fermé de $\mathbb{C}\times\mathbb{C}^M\times A$ défini (au voisinage de 0) par l'idéal de $\mathbb{C}\{v,z_1,\dots,z_M,(a_j^{(i)})\}$ engendré par $((f_1)_a,\dots,(f_{p-1})_a,v-(f_p)_a)$ où pour un élément $h\in\mathbb{C}\{z_1,\dots,z_M\}$ la notation $(h)_a$ désigne l'élément de $\mathbb{C}\{z_1,\dots,z_M,(a_j^{(i)})\}$ obtenu en substituant $z_i+\sum_{j=d-k+2}^{M} a_j^{(i)} z_j$ à z_i dans h, pour $1\le i\le d-k+1$. Il suffit de prouver que A") et B") sont réalisées en tout point d'un ouvert analytique dense de A. Appliquons le Théorème de Bertini idéaliste (Chap. II, 2.1) au morphisme $Z\overset{\sigma}{\leftrightarrows}A$ induit par la projection $\mathbb{C}\times\mathbb{C}^M\times A\to A$, muni de la section $\sigma : A\xrightarrow{\sim}0\times0\times A\subset Z$. Il existe un ouvert analytique dense $U=A-B$ dans A tel que pour $\underline{a}\in U$, nous ayons les relations de dépendance intégrale suivantes : pour tout entier ℓ, $1\le\ell\le M-d+k-2$ et pour un choix fixe de

$\Delta=\{i_{\ell+1},\dots,i_c\}\subset\{1,\dots,M\}$, notons J_Δ l'idéal de $\mathcal{O}_{Z,\sigma(a)}$ engendré par les éléments de la forme :

$$z_{m_1}\cdots z_{m_\ell}\frac{\partial((f_{j_1})_a,\dots,(f_{j_c})_a)}{\partial(z_{m_1},\dots,z_{m_\ell},z_{i_{\ell+1}},\dots,z_{i_c})}\qquad \{j_1,\dots,j_c\}\subset\{1,\dots,p\}$$

et les éléments de la forme :

$$v\cdot z_{m_2}\cdots z_{m_\ell}\frac{\partial(v-(f_p)_a,(f_{j_2})_a,\dots,(f_{j_c})_a)}{\partial(v,z_{m_2},\dots,z_{m_\ell},z_{i_{\ell+1}},\dots,z_{i_c})}\qquad \{j_2,\dots,j_c\}\subset\{1,\dots,p-1\}$$

Alors on a, pour $\{j_1,\dots,j_\ell\}\subset\{d-k+2,\dots,M\}$, $\{k_1,\dots,k_\ell\}\subset\{1,\dots,d-k+1\}$

$$(*)\qquad \frac{\partial((f_{j_1})_a,\dots,(f_{j_c})_a)}{\partial(a_{j_1}^{(k_1)},\dots,a_{j_\ell}^{(k_\ell)},z_{i_{\ell+1}},\dots,z_{i_c})}\cdot\mathcal{O}_{Z,\sigma(a)}\in\overline{J}_\Delta\ .$$

Remarquons que, puisque sur Z on a $v=(f_p)_a$, il résulte de (Chap. I, 1.4.5) que les éléments engendrant J_Δ et où v apparaît au dénominateur sont entiers dans $\mathcal{O}_{Z,\sigma(a)}$, sur l'idéal engendré par ceux du groupe précédent. A dépendance intégrale près, nous pouvons donc supposer que J_Δ est engendré par les seuls éléments du premier groupe. (N.B. C'est ici que nous utilisons l'hypothèse $\dim S\le 1$).

Supposons avoir l'inclusion : $\{i_{\ell+1},\dots,i_c\}\subset\{1,\dots,d-k+1\}$. On a alors les identités

$$(**)\qquad \frac{\partial((f_{j_1})_a,\dots,(f_{j_c})_a)}{\partial(a_{j_1}^{(k_1)},\dots,a_{j_\ell}^{(k_\ell)},z_{i_{\ell+1}},\dots,z_{i_c})}=$$

$$z_{j_1}\cdots z_{j_\ell}\left(\frac{\partial(f_{j_1},\dots,f_{j_c})}{\partial(z_{k_1},\dots,z_{k_\ell},z_{i_{\ell+1}},\dots,z_{i_c})}\right)_a.$$

Remarquons qu'au prix d'un changement linéaire des coordonnées, nous pouvons supposer que tous les $a_j^{(i)}$ sont nuls au point $\sigma(a)\in Z$ que nous considérons. Notons $\mathfrak{a}$ l'idéal engendré par les $(a_j^{(i)})$ dans $\mathcal{O}_{Z,z}$.

Sans plus faire l'hypothèse que $\{i_{\ell+1},\dots,i_c\}\subset\{1,\dots,d-k+1\}$, on a la congruence modulo $\mathfrak{a}$ correspondant à (**), que nous noterons (**) mod. $\mathfrak{a}$, et de même on a les congruences

$$\frac{\partial((f_{j_1})_a,\dots,(f_{j_c})_a)}{\partial(z_{m_1},\dots z_{m_\ell},z_{i_{\ell+1}},\dots,z_{i_c})}\equiv\left(\frac{\partial(f_{j_1},\dots,f_{j_c})}{\partial(z_{m_i},\dots,z_{m_\ell},z_{i_{\ell+1}},\dots,z_{i_c})}\right)_a \mod \mathfrak{a} .$$

Démontrons l'assertion A") : Sur la variété polaire $P_k=P_k<f;D_{d-k+1}>$, où D_{d-k+1} est défini par $z_1=\dots z_{d-k+1}=0$, on a d'après 1.4.1 :

$$\frac{\partial(f_{j_1},\dots,f_{j_c})}{\partial(z_{i_1},\dots,z_{i_c})}\mathcal{O}_{P_k,0}=0 \quad \text{si } \{i_1,\dots,i_c\}\subset\{d-k+1,\dots,M\} .$$

Par ailleurs, les relations de dépendance intégrale (*) sur Z donnent des relations de dépendance intégrale par restriction en sous-espace de $(Z,\sigma(a))$ défini par l'idéal $\mathfrak{a}$, sous-espace qui n'est autre que X.

Prenant le cas particulier où $\ell=1$, et utilisant (**) mod. $\mathfrak{a}$, il vient donc, au vu de la remarque suivant (*), que pour tout $j\le d-k+2$, et tout k, $1\le k\le d-k+1$, pour tout $\{i_1,\dots,i_c\}\subset\{1,\dots,M\}$, on a :

L'élément $z_j\cdot\frac{\partial(f_{j_1},\dots,f_{j_c})}{\partial(z_k,z_{i_2},\dots,z_{i_c})}\cdot\mathcal{O}_{P_k,0}$ est entier, dans $\mathcal{O}_{P_k,0}$, sur l'idéal de $\mathcal{O}_{P_k,0}$, engendré par les éléments de la forme :

$z_m\frac{\partial(f_{j_1},\dots,f_{j_c})}{\partial(z_m,z_{i_2},\dots,z_{i_c})}$ où $\{m,i_2,\dots,i_c\}\not\subset\{d-k+2,\dots,M\}$. Nous pouvons choisir $\{i_2,\dots,i_c\}\subset\{d-k+2,\dots,M\}$, ce qui force notre idéal à être engendré par les éléments tels que l'on ait $1\le m\le d-k+1$, et implique, comme on le voit aussitôt, par exemple en utilisant le critère valuatif, que chacun des $z_j\,\mathcal{O}_{P_k,0}$, $d-k+1<j\le M$ est entier sur l'idéal engendré par les $z_m\cdot\mathcal{O}_{P_k,0}$, $1\le m\le d-k+1$, c'est-à-dire que A") est réalisée pour $a\in U$.

Pour prouver B"), prenons $\{i_1,\dots,i_c\}\subset\{1,\dots,M\}$ et posons $\{i_1,\dots,i_c\}\cap\{1,\dots,d-k+1\}=\{n_1,\dots,n_\ell\}$. On peut donc écrire $\{i_1,\dots,i_c\}=\{n_1,\dots,n_\ell,i_{\ell+1},\dots,i_c\}$, au prix d'une permutation. Nous allons utiliser le critère valuatif de dépendance intégrale : soit

$h : (\mathbb{D},0) \to (X_{d-k+1},0)$ un arc, et supposons que l'élément de valuation selon h minimale parmi tous les éléments de la forme

$$z_{m_1} \cdot \ldots \cdot z_{m_\ell} \frac{\partial(f_{j_1},\ldots,f_{j_c})}{\partial(z_{m_1},\ldots,z_{m_\ell},z_{i_{\ell+1}},\ldots,z_{i_c})}$$

soit l'élément

$$z_{r_1} \cdot \ldots \cdot z_{r_\ell} \frac{\partial(f_{j_1},\ldots,f_{j_c})}{\partial(z_{r_1},\ldots,z_{r_\ell},z_{i_{\ell+1}},\ldots,z_{i_c})} .$$

Puisque $z_1 = \ldots = z_{d-k+1} = 0$ sur X_{d-k+1}, on a $\{r_1,\ldots,r_\ell\} \subset \{d-k+2,\ldots,M\}$. En appliquant le Théorème de Bertini idéaliste via (*) et en utilisant (**) mod. $\mathfrak{a}$ sur $X_{d-k+1} \subset X$, on obtient que

$$\frac{\partial(f_{j_1},\ldots,f_{j_c})}{\partial(a_{r_1}^{(n_1)},\ldots,a_{r_\ell}^{(n_\ell)},z_{i_{\ell+1}},\ldots,z_{i_c})} = z_{r_1} \cdots z_{r_\ell} \frac{\partial(f_{j_1},\ldots,f_{j_c})}{\partial(z_{n_1},\ldots,z_{n_\ell},z_{i_{\ell+1}},\ldots,z_{i_c})}$$

a une valuation au moins égale à celle de l'élément isolé ci-dessus, et donc $\dfrac{\partial(f_{j_1},\ldots,f_{j_c})}{\partial(z_{n_1},\ldots,z_{n_\ell},z_{i_{\ell+1}},\ldots,z_{i_c})}$ a une valuation au moins égale à celle de $\dfrac{\partial(f_{j_1},\ldots,f_{j_c})}{\partial(z_{r_1},\ldots,z_{r_\ell},z_{i_{\ell+1}},\ldots,z_{i_c})}$ ce qui suffit, d'après le critère valuatif, à prouver que l'assertion B") est vérifiée pour $a \in U$. ∎

La preuve dans le cas "absolu" où S est un point est identique à ceci près que l'on n'a pas besoin d'utiliser (Chap. I, 1.4.5).

<u>Remarque</u> : L'assertion A) généralise le Théorème 1, p. 269 de [Te 1] concernant les hypersurfaces à singularité isolée. Ce résultat a déjà été généralisé par M. Giusti et J.P.G. Henry ([G-H]) aux courbes polaires d'intersections complètes à singularité isolée. Enfin, le Théorème 5.1 est une conséquence d'un

théorème de J.P.G. Henry et M. Merle dans [H-M], qui généralise ([Te 1],I, 2.7 à 2.9).

5.2 Dans le cas absolu, on a un résultat de transversalité plus précis.

5.2.1 Proposition (Variante du Lemme 4.1.8 de [Lê-Te]) : *Soient* $X \subset \mathbb{C}^N$ *un espace analytique complexe réduit purement de dimension* d, Y *un sous-espace non-singulier de dimension* 1 *de* X *et* $0 \in Y$. *Il existe un ouvert de Zariski dense* W *de l'espace des drapeaux*

$$\mathcal{D} : (0) \subset D_{N-1} \subset \ldots\ldots \subset D_2 \subset D_1 \subset D_o = \mathbb{C}^N$$

où D_i *est de codimension* i, *tel que pour tout* $\mathcal{D} \in W$, *on ait*

$$|D_{d-k} \cap C_y(P_k\langle X,y\rangle;D_{d-k+1}\rangle| = \{0\} \quad \text{(intersection dans } T_{\mathbb{C}^N,y}) \quad ,$$

en tout point $y \in Y - \{0\}$ *assez proche de* 0, *et* $0 \leq k \leq d-1$.

Preuve : Nous allons d'abord montrer que pour toute rétraction locale $\rho : (\mathbb{C}^N,0) \to (Y,0)$, l'ensemble des drapeaux tels que $D_1' = T_{\rho^{(1)}(0),0}$ contient un ouvert de Zariski dense de drapeaux satisfaisant la condition de la Proposition. Considérons l'éclatement $e_Y : E_Y(X) \to X$ de Y dans X, et la réunion F_o des composantes irréductibles du diviseur exceptionnel $e_Y^{-1}(Y)$ qui sont envoyés surjectivement sur Y par e_Y (c'est le sous-espace de $e_Y^{-1}(Y)$ défini par l'idéal engendré par les éléments qui sont annulés par une puissance de l'idéal maximal $m_{Y,0}$ de $\mathcal{O}_{Y,0}$). Considérant le plongement $E_Y(X) \subset X \times \mathbb{P}^{N-2}$ défini par un choix de coordonnées sur la fibre $\rho^{-1}(0)$, on voit que $F_o \cap (\{0\} \times \mathbb{P}^{N-2})$ est de dimension d-2. On peut donc choisir un sous-espace D_d de codimension d-1 dans $T_{\rho^{-1}(0),0} = D_1$ et tel que $\mathrm{Proj}\, D_d \cap (F_o \cap (\{0\} \times \mathbb{P}^{N-2})) = \emptyset$. Supposons maintenan avoir choisi $D_d \subset D_{d-1} \subset \ldots \subset D_{i+1}$ dans D_1' de telle façon que, notant $F_{d-j} \subset X \times \mathbb{P}^{N-2}$ la réunion des composantes irréductibles du diviseur exceptionnel de l'éclatement $E_Y\, P_{d-j}\langle X;D_{j+1}\rangle \to P_{d-j}\langle X;D_{j+1}\rangle$ de Y dans la variété polaire

P_{d-j} associée à D_{j+1}, on ait Proj $D_j \cap (F_{d-j} \cap \{0\} \times \mathbb{P}^{N-2}) = \emptyset$ pour $i+1 \leq j \leq d$. Remarquons que puisque $P_{d-i}<X;D_{i+1}>$ est contenu dans $P_{d-i-1}<X;D_{i+2}>$, on a l'inclusion $F_{d-i} \subset F_{d-i-1}$, et donc en posant $F_j(0) = F_j \cap \{0\} \times \mathbb{P}^{N-2}$, on a $F_{d-i}(0) \subset F_{d-i-1}(0)$, d'où Proj $D_{i+1} \cap F_{d-i}(0) = \emptyset$. Comme par construction, on a dim $F_{d-i}(0) = i-2$, on voit que l'ensemble des sous-espaces vectoriels D_i de codimension i-1 de D_1', contenant D_{i+1} et tels que Proj $D_i \cap F_{d-i}(0) \neq \emptyset$ est de dimension au plus i-2 (image de $F_{d-i}(0)$ dans Proj D_1'/D_{i+1}) alors que l'ensemble des sous-espaces vectoriels D_i de codimension i-1 de D_1 contenant D_{i+1} est de dimension i-1. Nous pouvons donc choisir D_i contenant D_{i+1} et tel que Proj $D_i \cap F_{d-i}(0) = \emptyset$. On construit ainsi un drapeau $\mathcal{D}$ ayant la propriété que Proj $D_i \cap F_{d-i}(0) = \emptyset$ pour $2 \leq i \leq d$, et finalement on peut par le même argument choisir D_1 contenant D_2 et tel que $|D_1 \cap C_y(P_{d-1}(D_2))| = \{0\}$.
Utilisons maintenant le fait qu'en tout point $y \in Y - \{0\}$ assez proche de 0, chaque $P_k<X;D_{d-k+1}>$ est <u>normalement plat</u> le long de Y, ce qui implique (cf. [Le-Te 2]) que, en notant $N_k \to Y$ le cône normal de P_k le long de Y, on a en tout $y \in Y - \{0\}$ une <u>suite exacte de cônes</u> (cf. Loc. cit.)

$$0 \longrightarrow T_{Y,y} \longrightarrow C_y(P_k<X;D_{d-k+1}>) \longrightarrow N_k(y) \longrightarrow 0$$

ce qui signifie que les translations (dans $T_{\mathbb{C}^N,y}$) laissent invariant le cône tangent, et que le quotient par cette action s'identifie à la fibre $N_k(y)$ du cône normal. Par notre choix de $\mathcal{D}$, nous avons Proj $D_{d-k} \cap F_k(0) = \emptyset$, donc Proj $D_{d-k} \cap F_k(y) = \emptyset$ pour y assez voisin de 0, mais par la définition de l'éclatement, nous avons pour $y \in Y - \{0\}$, $F_k(y) = \text{Proj } N_k(y)$, et cela donne $|D_{d-k} \cap N_k(y)| = \{0\}$, et en fait $|T_{Y,y} \times D_{d-k}) \cap C_y(P_k)| - |Y|$ puisque d'après la suite exacte, on a $C_y(P_k) = T_{Y,y} \times N_k(y)$. Finalement on a bien $|D_{d-k} \cap C_y(P_k<X;D_{d-k+1}>)| = \{0\}$. ∎

<u>5.2.2</u> <u>Corollaire</u> (Lê-Teissier, [Lê-Te]) : <u>Pour</u> $(X,0) \subset (\mathbb{C}^N,0)$ <u>et un drapeau</u> $\mathcal{D}$ <u>assez général</u>, <u>on a</u> :

$$|D_{d-k} \cap P_k<(X,0);D_{d-k+1}>| = \{0\} \quad (0 \le k \le d-1) \quad .$$

<u>Preuve</u> : Appliquer la Proposition à $X \times \mathbb{C}$ le long de $0 \times \mathbb{C}$.

<u>5.3</u> Revenons au cas relatif, pour interpréter la transversalité B) du Théorème 4.1 comme finitude de certains morphismes.

Soit $f : (X,0) \to (S,0)$ comme en 1.1, et supposons donné un plongement local $(X,0) \subset (\mathbb{C}^M,0)$; soit D_i un sous-espace vectoriel de codimension i de $\mathbb{C}^M$ tel que $F \cap D_i$ soit rare dans $X \cap D_i$ et que $X_i = (X \cap D_i)_{red}$ soit de dimension (pure) d-i . Considérons la modification de Nash relative et l'espace conormal relatif de $X \subset S \times \mathbb{C}^M$ (plongé par le graphe de f) et notons $\hat{X}_i$ le transformé strict de $X_i \subset X$ par la modification de Nash relative, et $\tilde{X}_i$ le transformé strict par le morphisme conormal relatif κ_f, c'est-à-dire l'adhérence dans $C_f(X \subset S \times \mathbb{C}^M)$ de $\kappa_f^{-1}(X_i - F)$. On obtient le diagramme :

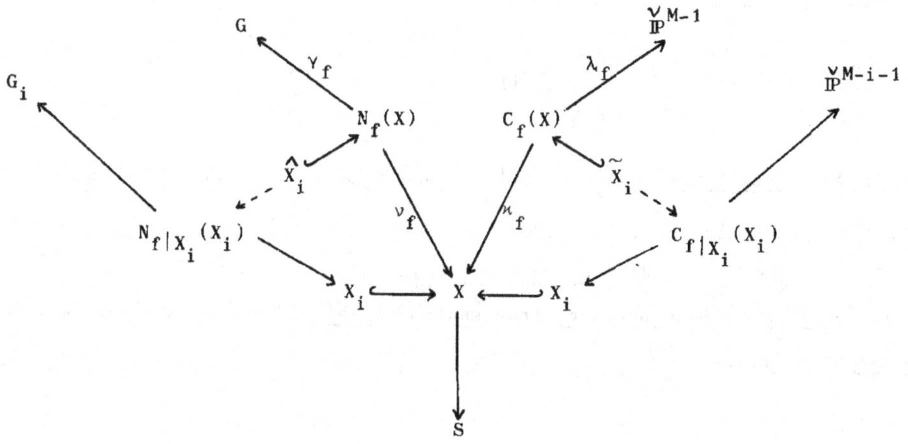

5.3.1 Proposition : Les conditions suivantes sont équivalentes, si l'on fait précéder chacune de "pour tout représentant suffisamment petit de X".

i) On a $\gamma_f^{-1}(c_{d-i+1}(D_i)) \cap \hat{X}_i = \emptyset$.

ii) Le morphisme de X_i^o dans $X_i \times G_i$, où G_i désigne la grassmannienne des (d-i)-plans de D_i , défini par $x \to (x, T_{f^{-1}(f(x))} \cap T_{D_i,x})$, s'étend en un morphisme de $\hat{X}_i$ dans $X_i \times G_i$, induit par le morphisme naturel de $G - c_{d-i-1}(D_i)$ dans G_i qui à T associe $T \cap D_i$, et dont l'image dans $X_i \times G$ est le modifié de Nash relatif $N_{f|X_i}(X_i)$ de X_i .

iii) Le morphisme précédent $\hat{X}_i \to N_{f|X_i}(X_i)$ non seulement est défini, mais encore est un morphisme fini.

iv) Pour tout choix d'un système de coordonnées locales $(z_1,\dots,z_M)$ sur $\mathbb{C}^M$ tel que D_i soit défini par $z_1 = \dots = z_i = 0$ et d'un système de générateurs $F_1,\dots,F_m \in \mathcal{O}_{S,s}\{z_1,\dots,z_M\}$ de l'idéal définissant $X \subset S \times \mathbb{C}^M$ en 0, l'idéal engendré par les $\dfrac{\partial(F_{j_1},\dots,F_{j_{M-d}})}{\partial(z_{i_1},\dots,z_{i_{M-d}})}$ est entier dans $\mathcal{O}_{X_i,0}$ sur l'idéal engendré par les seuls mineurs de cette forme tels que $\{i_1,\dots,i_{M-d}\} \subset \{i+1,\dots,M\}$.

v) Pour toute limite T d'espace tangents à X^o en des points de $X_i - F$, on a $\dim(T \cap D_i) = d-i$.

i') Notant $L^{i-1} \subset \mathbb{P}^{M-1}$ l'espace projectif formé des hyperplans qui contiennent D_i , on a :

$$\lambda_f^{-1}(L^{i-1}) \cap \tilde{X}_i = \emptyset \quad .$$

ii') Le morphisme de $\tilde{X}_i - \lambda_f^{-1}(L^{i-1})$ dans $X_i \times \mathbb{P}^{M-1-i}$, où $\mathbb{P}^{M-1-i}$ est l'espace des hyperplans de D_i , qui à (x,H) associe $(x, H \cap D_i)$, s'étend en un morphisme $\tilde{X}_i \to X_i \times \mathbb{P}^{M-1-i}$ dont l'image est $C_{f|X_i}(X_i)$.

iii') Le morphisme précédent non seulement est défini, mais encore est un morphisme fini.

Démonstration : i) et v) sont clairement équivalent d'après la définition de la modification de Nash. D'autre part l'application $T \mapsto T \cap D_i$ est précisément

définie de $G - C_{d-i+1}(D_i)$ dans G_i , ce qui montre aussitôt que i) $\Leftrightarrow$ ii). Nous avons vu l'équivalence de v) et iv) en (Chap. II, 3.1), et il nous suffit de montrer que ii) $\Rightarrow$ iii) puisque l'implication inverse est évidente. Pour ce faire, nous pouvons plonger localement X dans une intersection complète relative Z (Chap. II, 1.1.1) définie disons par $F_1,\ldots,F_{M-d}$, et calculer $N_f(X)$ (resp. $\hat{X}_i$) comme transformés stricts de X (resp. X_i) par l'éclatement de l'idéal jacobien relatif $J_{Z/S}$ de X_1 . Mais l'hypothèse de ii) implique précisément que $J_{Z/S}\cdot\mathcal{O}_{X_i,0}$ est entier sur l'idéal $J_{Z\cap D_i/S}\cdot\mathcal{O}_{X_i,0}$ image dans $\mathcal{O}_{X_i,0}$ de l'idéal jacobien relatif $J_{Z\cap D_i/S}$ de l'intersection complète relative $Z\cap D_i$.
Comme $N_f(X_i)$ est l'éclatement dans X_i de $J_{Z\cap D_i/S}\cdot\mathcal{O}_{X_i,0}$, que $\hat{X}_i \to X_i$ est l'éclatement dans X_i de $J_{Z/S}\cdot\mathcal{O}_{X_i,0}$, et que ces deux idéaux ont la même fermeture intégrale (puisque $J_{Z\cap D_i/S}\cdot\mathcal{O}_{X_i,0}\subset J_{Z/S}\cdot\mathcal{O}_{X_i,0}$ et que $J_{Z/S}\cdot\mathcal{O}_{X_i,0}\subset\overline{J_{Z\cap D_i/S}\cdot\mathcal{O}_{X_i,0}}$), le résultat cherché est donné par (Chap. I, 1.3.6).
L'équivalence de v) et i)' résulte aussitôt de ce que nous avons vu en 4.1.1.
Pour prouver l'équivalence de i)' et ii)', remarquons que l'application $H \to H\cap D_i$ est définie de $\check{\mathbb{P}}^{M-1} - L^{i-1}$ dans $\check{\mathbb{P}}^{M-1-i}$, et il nous reste à prouver que ii)' $\Rightarrow$ iii)' : les fermetures des fibres de la projection $\pi_i : \check{\mathbb{P}}^{M-1} - L^{i-1} \to \check{\mathbb{P}}^{M-1-i}$ sont des espaces projectifs $\mathbb{P}^i$ dont L^{i-1} est un hyperplan, et si le morphisme $\tilde{X}_i \to C_f(X_i)$ n'était pas fini , cela signifierait que la restriction de π_i à $\kappa_f^{-1}(0)$ n'est pas finie, donc qu'il existe une fibre $\mathbb{P}^i$ telle que $\mathbb{P}^i\cap\kappa_f^{-1}(0)$ soit un sous-ensemble algébrique de dimension ≥ 1 (puisque $\kappa_f^{-1}(0)$ est fermé dans $\mathbb{P}^{M-1}$, donc algébrique) et l'adhérence dans $\mathbb{P}^i$ de ce sous-ensemble rencontrerait L^{i-1} d'après le théorème de Bezout, d'où une contradiction avec i)'. ∎

5.3.2 Il résulte du Théorème 5.1, B) que les conditions de la Proposition 5.3.1 sont satisfaites, lorsque dim $S \leq 1$, pour tout D_i appartenant à un ouvert de Zariski dense de la grassmannienne des plans de codimension i de $\mathbb{C}^M$.

5.4 Variétés polaires et sections planes.

5.4.1 Reprenons la situation de 1.1, avec les mêmes notations

$$\begin{array}{ccc} (X,0) & \hookrightarrow & (S\times \mathbb{C}^N,0) \\ \downarrow & & \\ (S,0) & & \end{array}$$

Soit G_o un sous-espace vectoriel de $\mathbb{C}^N$, posons $G = S\times G_o \subset S\times \mathbb{C}^N$ et $[X\cap G] = (X\cap G)_{red}$. Supposons que G_o soit transverse aux limites d'espace tangents aux fibres de f en des points de $X\cap G$, c'est-à-dire que d'une part $F\cap G$ est rare dans $[X\cap G]$, et d'autre part , pour une telle limite T, on a $\dim T\cap G_o = d+g-N$, où $g = \dim G_o$. On peut aussi exprimer cette transversalité de la manière suivante : soient $\varkappa_f : C_f(X)\to X$ l'espace conormal relatif de X et $\widetilde{[X\cap G]}$ le transformé strict de $[X\cap G]$ par $\varkappa_f$, c'est-à-dire l'adhérence dans $C_f(X)$ de $\varkappa_f^{-1}([X\cap G]-F\cap G)$. Soit $\mathbb{P}^{N-g-1}\subset \check{\mathbb{P}}^{N-1}$ l'ensemble des hyperplans de $\mathbb{C}^N$ contenant G_o. La transversalité s'exprime par : $\varkappa_f^{-1}(0)\cap \widetilde{[X\cap G]}\cap \mathbb{P}^{N-g-1} = \emptyset$, c'est-à-dire $\widetilde{[X\cap G]}\cap \Pi_f^{-1}(\mathbb{P}^{N-g-1}) = \emptyset$ pour un représentant assez petit. Remarquons que dans le cas où S est un point, puisque $\dim \varkappa^{-1}(0) < N-1$, il existe toujours des hyperplans G (avec $g = N-1$) transverses à toutes les limites d'espaces tangents à X^o (et pas seulement aux limites d'espaces tangents à X^o en des points de $X\cap G$).

Quoi qu'il en soit, étant donné un tel sous-espace vectoriel G_o, soit $D_{d-k+1}\subset G_o$ un sous-espace de G_o, de codimension $d-k+1$ dans $\mathbb{C}^N$.

5.4.2 Proposition : *Sous les hypothèses précédentes, si de plus le sous-espace vectoriel D_{d-k+1} appartient à l'ouvert W_k associé en 1.3 (Prop. 2) à $f|[X\cap G]$, et appartient à l'ouvert correspondant associé à f, on a l'égalité :*

$$P_k\langle f;D_{d-k+1}\rangle \,\hat{\cap}\, G = P_k\langle f|[X\cap G]\,;\,D_{d-k+1}\rangle \;,$$

où le terme de gauche désigne l'adhérence dans X de $(X-F)\cap P_k\langle f;D_{d-k+1}\rangle\cap G$.

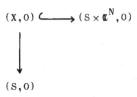

<u>Démonstration</u> : D'après l'hypothèse de transversalité, en tout point $x \in (X - F) \cap G$, voisin de 0, on a : $T_{X(f(x)),x}$
déduit d'une part que $\Omega^1_{f|[X \cap G]}$ est localement libre de rang d+g-N en tout point de l'ouvert analytique dense $(X - F) \cap G$ de $[X \cap G]$, et d'autre part (cf. 5.3, iii)') que la projection $\check{\mathbb{P}}^{N-1} - \mathbb{P}^{N-1-g} \to \check{\mathbb{P}}^{g-1}$ donnée par $H \mapsto H \cap G_o$ induit un morphisme <u>fini</u> $\widetilde{[X \cap G]} \to C_{f|[X \cap G]}([X \cap G])$.

On a donc le diagramme suivant :

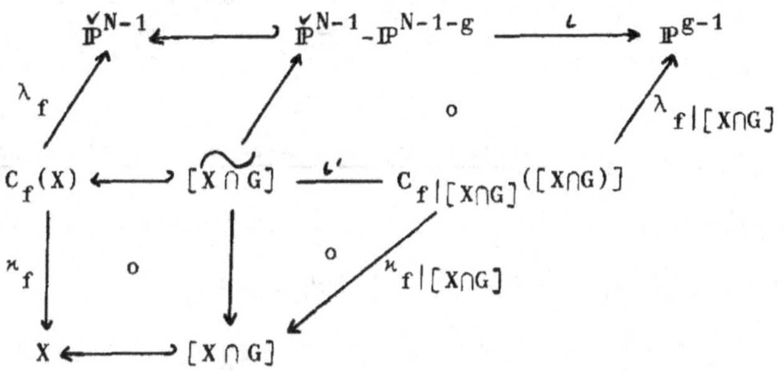

et puisque D_{d-k+1} est contenu dans G_o, le sous-espace $L^{d-k} \subset \mathbb{P}^{N-1}$ représentant l'ensemble des hyperplans de $\mathbb{C}^N$ contenant D_{d-k+1} est la fermeture dans $\check{\mathbb{P}}^{N-1}$ de l'image réciproque par la projection ι de centre $\mathbb{P}^{N-1-g}$ donnée par $H \mapsto H \cap G$ du sous-espace $L^{d-k-N-g} \subset \check{\mathbb{P}}^{g-1}$ représentant les hyperplans de G_o qui contiennent D_{d-k+1}. On a donc l'égalité

$$(*) \qquad \widetilde{[X \cap G]} \cap \lambda_f^{-1}(L^{d-k}) = \iota'^{-1}[C_{f|[X \cap G]}([X \cap G]) \cap \lambda^{-1}_{f|[X \cap G]}(L^{d-k-N+g})] \quad .$$

D'après les hypothèses, le morphisme ι' est fini (cf. 5.3.1), il est biméromorphe puisqu'il induit un isomorphisme au-dessus de $[X \cap G] - F \cap G$ dont l'image réciproque est partout dense dans $\widetilde{[X \cap G]}$ (resp. $C_{f|[X \cap G]}([X \cap G])$), donc il est surjectif. De plus les hypothèses impliquent (cf. 4.1.1) que l'image réduite du terme de gauche de (*) est $P_k < f ; D_{d-k+1} > \hat{\cap} G$, tandis que l'image réduite du terme entre crochets à droite est $P_k < f|[X \cap G] ; D_{d-k+1} >$, d'où le résultat. ∎

5.4.3 Corollaire : *Plaçons-nous dans la situation de* 5.1. *Supposons* S *non-singulier de dimension* ≤ 1 *et soit* $(Y,0) \subset (X,0)$ *un sous-espace non-singulier tel que* $f|Y : (Y,0) \to (S,0)$ *soit une submersion. Posons* $t = \dim Y - \dim S$, *et supposons que l'espace projectif* $\check{\mathbb{P}}^{M-1-t}$ *des hyperplans de* $\mathbb{C}^M$ *qui contiennent* $T_{Y(0),0}$ *contienne un ouvert de Zariski dense* V *formé d'hyperplans transverses en* 0 *aux limites en* 0 *d'espaces tangents aux fibres de* f. *Soit* k *un entier* $\leq d-t$. *Il existe un ouvert de Zariski dense de l'espace*

$$\{(D_{d-k+1}, H_o) \mid D_{d-k+1} \subset H_o\} \subset G_k \times \check{\mathbb{P}}^{M-1-t}$$

tel que, pour (D_{d-k+1}, H_o) *appartenant à cet ouvert de Zariski dense on ait, en posant* $H = S \times H_o$,

a) $(P_k\langle f;D_{d-k+1}\rangle \hat{\cap} H)_{red} = P_k\langle f|[X \cap H] ; D_{d-k+1}\rangle$

b) $m_o(P_k\langle f;D_{d-k+1}\rangle \hat{\cap} H) = m_o(P_k\langle f;D_{d-k+1}\rangle) = m_o\ P_k\langle f|[X \cap H];D_{d-k+1}\rangle)$

où m_o *désigne la multiplicité à l'origine.*

Si $\dim S = 1$, *ce nombre est égal au nombre d'intersection* $(P_k\langle f;D_{d-k+1}\rangle, D_{d-k+1})$ *en* 0.

Démonstration : Nous reprenons les notations du § 1. L'espace I_Y est muni de deux projections naturelles :

$$\begin{array}{ccccc} & & I_Y & & \\ & {}^{p_1}\swarrow & & \searrow^{p_2} & \\ G_k & & & & \check{\mathbb{P}}^{M-1-t} \end{array}$$

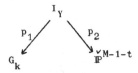

La projection p_2 en fait un fibré en grassmanniennes sur $\check{\mathbb{P}}^{M-1-t}$, donc I_Y est irréductible. Puisque $k \leq d-t$, la projection p_1 est surjective, car $\dim(D_{d-k+1} + T_{Y(0)}, 0) \leq M-1$. Soit $W \subset G_k$ le constructible de Zariski dense formé des D_{d-k+1} tels que $D_{d-k+1} \in W_k$, que la multiplicité en 0 de $P_k\langle f;D_{d-k+1}\rangle$ soit la multiplicité d'une variété polaire générique, et que les conditions de 1.3 et 5.1 soient vérifiées. Le sous-ensemble $p_1^{-1}(W) \subset I_Y$ est constructible dense, et rencontre $p_2^{-1}(V)$ selon un sous-ensemble constructible dense de I_Y. Par ailleurs, pour chaque $H \in V$, l'ensemble des $D_{d-k+1} \subset H_o$ qui apprtiennent aux ouverts W_k (cf. 1.3) et T_k (cf. 5.1) associés à $f|[X \cap H]$ est, d'après 1.3.1 un ouvert de Zariski dense de $p_2^{-1}(H_o)$, et d'après la manière dont W_k et T_k sont construits, on vérifie rapidement que la réunion pour $H_o \in V$ de ces ouverts de Zariski denses des $p_2^{-1}(H)$ contient un ouvert de Zariski dense W' de I_Y. D'après 5.4.2 et 5.1, A) l'intérieur de $p_1^{-1}(W) \cap W'$ a toutes les propriétés demandées ; en effet l'assertion a) résulte de 5.4.2, la première égalité de b) est immédiate et la seconde résulte de la première et de la définition de W_k. En effet d'après 5.1 et (Chap. I, 5.2) puisque D_{d-k+1} est transverse à $C_o(P_k\langle f;D_{d-k+1}\rangle$ et à $C_o(P_k\langle f|[X \cap H];D_{d-k+1}\rangle)$ l'idéal $\mathfrak{m}_1$ définissant D_{d-k+1} dans $\mathbb{C}^N$ induit dans les anneaux locaux de $P_k\langle f;D_{d-k+1}\rangle$ et $P_k\langle f|[X \cap H];D_{d-k+1}\rangle$ des idéaux ayant même fermeture intégrale que l'idéal maximal, (cf. 5.1, A") donc même multiplicité (Chap. I, 4.1). Enfin dans le cas où $\dim S = 1$, $P_k\langle f;D_{d-k+1}\rangle$ et D_{d-k+1} sont de dimension complémentaire dans $\mathbb{C}^M$, ne se coupent qu'en $\{0\}$ et leur nombre d'intersection est cette multiplicité d'après [Se].

5.4.4 Remarques : 1) Si $F \cap P_k\langle f;D_{d-k+1}\rangle \cap G$ est rare dans $P_k\langle f;D_{d-k+1}\rangle \cap G$, on a l'égalité $P_k\langle f;D_{d-k+1}\rangle \hat{\cap} G = (P_k\langle f;D_{d-k+1}\rangle \cap G)_{red}$.

2) Dans la situation de 5.4.3, X étant plongé dans $\mathbb{C} \times \mathbb{C}^M$ par le graphe de f, $[X\cap H]$ est le graphe de $f|[X\cap H_o]$, donc $[X\cap H] \simeq [X\cap H_o]$ et de même $P_k\langle f;D_{d-k+1}\rangle \hat{\cap} H \simeq P_k\langle f;D_{d-k+1}\rangle \hat{\cap} H_o$.

5.5 Variétés polaires et projections.

5.5.1 Proposition (généralisant [Lê-Te], 4.2.1, i) et 4 2.3) : Soit $f:(X,0) \to (S,0)$ un morphisme comme en 1.1 avec S non-singulier de dimension ≤ 1 muni d'un S-plongement $(X,0) \subset (S \times \mathbb{C}^N,0)$. On suppose que toutes les fibres de f sont réduites et purement de dimension d. Il existe un ouvert de Zariski dense U de l'espace des projections linéaires $p:\mathbb{C}^N \to \mathbb{C}^{d+1}$ tel que, pour toute projection $p \in U$, on ait, en notant encore p la S-projection, $id_S \times p$.

i) Le germe image $(X_1,0) = (p(X),0)$ est une hypersurface de $S \times \mathbb{C}^{d+1}$, dont toutes les fibres sont réduites de dimension d, et le morphisme $p: X \to X_1$ est fini et biméromorphe. Notons f_1 la restriction à X_1 de la projection

$S \times \mathbb{C}^{d+1} \to S$.

ii) Il existe un ouvert de Zariski dense $W_{1,k}$ de la grassmannienne des sous-espaces de codimension d-k+1 de $\mathbb{C}^{d+1}$ tel que, pour $(D_1)_{d-k+1} \in W_{1,k}$ on ait :

$$(p(P_k\langle f;p^{-1}((D_1)_{d-k+1}\rangle))_{red} = P_k\langle f_1;(D_1)_{d-k+1}\rangle$$

et ces deux variétés polaires ont la même multiplicité en 0, et ont la multiplicité d'une variété polaire générale (cf. 3.1) de X et X_1 respectivement.

Preuve : Donnons-la dans le cas où S est un point. Le cas où S est une courbe et analogue. On suit la preuve de (Loc. cit.).

Prouvons i) : Il existe un ouvert U_o de projections linéaires $q : \mathbb{C}^N \to \mathbb{C}^d$ tel que si $q \in U_o$, la restriction de q à (X,0) soit un morphisme fini (mise en position d'un germe). D'après le théorème de Bertini-Sard, pour des représentants X et V de (X,0) et $(\mathbb{C}^d,0)$, le morphisme $q : X \to V$ est fini et il existe un fermé analytique rare $F_1 \subset V$ tel que q induise un morphisme lisse : $q^{-1}(V - F_1) \to V - F_1$. Soit $g \in V - F_1$. Il existe un ouvert dense de l'espace des projections linéaires z de $\mathbb{C}^N$ sur $\mathbb{C}$ tel que la restriction de z à $q^{-1}(y) \cap X$ soit injective. On vérifie sans peine que l'ensemble F_2 des points $y \in V$ tels que la restriction de z à $q^{-1}(y) \cap X$ ne soit pas injective est un fermé analytique rare de V. Pour un tel z, la projection $p = (q,z) : \mathbb{C}^N \to \mathbb{C}^{d+1}$ induit un isomorphisme de l'ouvert analytique dense $X - q^{-1}(F_1 \cup F_2)$ sur son image X_1 dans $\mathbb{C}^{d+1}$, ce qui prouve que cette image, qui est une hypersurface par un argument de profondeur déjà utilisé, est réduite [cf. [Te 3], § 3 : il faut plonger localement dans une intersection complète Z, appliquer l'argument et remarquer que p(X) est réunion de composantes irréductibles de p(Z), qui est une hypersurface]. Le morphisme $\pi : X \to X_1$ induit par p est bien fini et biméromorphe, d'où i).

Prouvons ii) : Notons G (resp. G_1) la grassmannienne des sous-espaces de codimension d-k+1 de $\mathbb{C}^N$ (resp. $\mathbb{C}^{d+1}$). Soit Z le sous-espace de $G \times U$ formé des couples (D_{d-k+1},p) tels que $\operatorname{Ker} p \subset D_{d-k+1}$ et soit W_1 le sous-ensemble de $G_1 \times U$ formé des couples $((D_1)_{d-k+1},p)$ tels que $(D_1)_{d-k+1}$ appartienne à l'ouvert W_k de

1.3 et que $P_k\langle(p(X),0);(D_1)_{d-k+1}\rangle$ ait la multiplicité d'une variété polaire générale de $(p(X),0)$. On prouve que ce sous-ensemble est constructible et dense dans $G\times U$ par un argument semblable à celui utilisé en 5.4.3. On a le diagramme

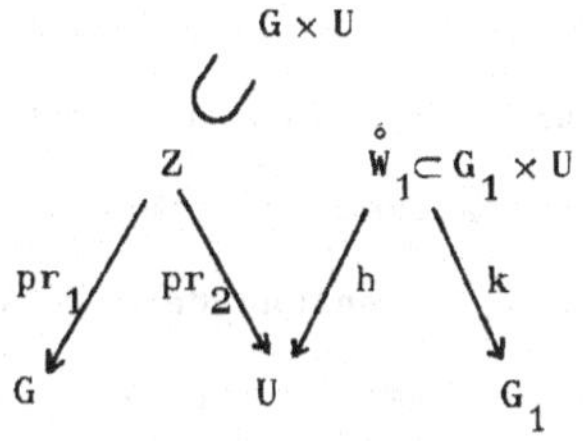

où h est induit par la seconde projection $G\times U\to U$, et est d'image dense. On remarque que pr_2 est une fibration algébrique, donc un morphisme ouvert pour la topologie de Zariski. D'après 5.2.2, il existe un ouvert de Zariski dense $T_o\subset G\times U$ tel que pour $(D_{d-k+1},p)\in T_o$, le noyau Ker p de p soit transverse à $C_o(P_k\langle(X,0);D_{d-k+1}\rangle)$. Soit $V\subset G$ un ouvert de Zariski dense tel que pour $D_{d-k+1}\in V$, la variété polaire $P_k\langle(X,0);D_{d-k+1}\rangle$ ait en 0 la multiplicité générale. Considérons enfin la construction suivante. Soit $\nu: N(X)\to X$ la modification de Nash de X ; et considérons le morphisme de Gauss $\gamma: N(X)\to G_d$, où G_d est la grassmannienne des d-plans dans $\mathbb{C}^N$. Pour toute projection linéaire $p:\mathbb{C}^N\to\mathbb{C}^{d+1}$, la variété de Schubert $\Sigma(p)=\{T\in G_d/\dim(\operatorname{Ker}p\cap T)\geq 1\}$ est de codimension 2 dans G_d. Soit R l'ouvert de Zariski (dense grâce au théorème de Kleiman, Chap. III, 4.3) de $G\times U$ formé des couples (D_{d-k+1},p) tel que $\gamma^{-1}(\Sigma(p))\cap\gamma^{-1}(c_k(D_{d-k+1}))$ soit de codimension 2 dans $\gamma^{-1}(c_k(D_{d-k+1}))$. Montrons que l'intérieur U_o du constructible dense $h(\mathring{W}_1)\cap(pr_2(V)\cap T_o\cap R))$ convient. Pour cela, il suffit de montrer que si $p\in U_o$, on a $p((P_k\langle(X,0);D_{d-k+1}\rangle)_{red})=\mathring{P}_k\langle(X_1,0);p_1(D_{d-k+1})\rangle$. (Rappelons-nous que $D_{d-k+1}\supset\operatorname{Ker}p$, donc $p(D_{d-k+1})=(D_1)_{d-k+1}$ est de codimension d-k+1 dans $\mathbb{C}^{d+1}$.) D'après la définition des variétés polaires locales, et le fait que $p\in h(W_1)$ $P_k\langle(X,0);p(D_{d-k+1})\rangle\cap X_1^o$ est dense dans $P_k\langle(X_1,0);p(D_{d-k+1})\rangle$. Si cette intersection n'est pas vide, il résulte aussitôt de 1.3.1 et du fait que $p\in pr_2(pr_1^{-1}(V))$ que $\pi^{-1}(P_k\langle(X_1,0);p(D_{d-k+1})\rangle)$ est réunion de composantes irréductibles de $P_k\langle(X,0);D_{d-k+1}\rangle$. Montrons qu'en fait il y a égalité. Si P est une composante irréductible de $P_k\langle(X,0);D_{d-k+1}\rangle$ qui n'est pas dans $\pi^{-1}(P_k\langle(X,0);p(D_{d-k+1}))$, P est de codimension k puisque $p\in pr_2(pr_1^{-1}(V))$ et puisque $p\in pr_2(R)$, il existe un ouvert analytique dense P^o de P en chaque point x duquel X est non-singulier et Ker p transverse à $T_{X,x}$. L'image par p de $T_{X,x}$ est donc une limite d'espaces tangents en p(x) à X_1. Si (P) rencontre X_1^o, $\pi(P)$ est une composante de $P_k\langle(X_1,0);p(D_{d-k+1})\rangle$ et donc $P\subset\pi^{-1}(P_k\langle(X_1,0);p(D_{d-k+1})\rangle)$ contrairement à l'hypothèse. Donc $\pi(P)$ est un sous-espace de codimension k de X_1, contenu dans le lieu singulier de X_1, et en tout point duquel il existe une limite T_1 d'espaces tangents à X_1^o telle que $\dim(T_1\cap p(D_{d-k+1}))\geq k$. Ceci montre que $\pi(P)$ est une composante de dimension d-k de $\nu(\gamma_1^{-1}(c_k(p(D_{d-k+1}))$, où $\gamma_1 : N(X_1)\to\mathbb{P}^d$ est l'application de Gauss, et ceci contredit, par 1.3, le fait que $p\in h(\mathring{W}_1)$. Ainsi, une telle composante ne peut exister, et l'on a bien $\pi^{-1}(P_k\langle(X_1,0);p(D_{d-k+1})\rangle=P_k\langle(X,0);D_{d-k+1}\rangle$. Finalement l'assertion sur les multiplicités résulte de ce que puique $p\in pr_2(T_o)$, Ker p est transverse à $C_o(P_k\langle(X,0);D_{d-k+1}\rangle)$ et du fait que la multiplicité est conservée par projection transversale (cf. Chap. I, 5.2 et 4.1). ∎

Remarques : 1) L'assertion iii) du Théorème 4.2.1 de [Lê-Te] est inexacte, comme l'ont montré J.P.G. Henry et M. Merle. Il en est de même du Corollaire 4.2.4 du même travail. Ces résultats n'étaient utilisés nulle part dans [Lê-Te].

2) L'étude des variétés polaires relatives de $f : X\to S$ est compliquée, dans le cas dim $S\geq 2$, par le fait que $f|P_k\langle f;D_{d-k+1}\rangle$ peut avoir de l'éclatement.

§ 6. Mini-formulaire pour les variétés polaires.

Ici l'on s'autorise à écrire $P_k(X,0)$, ou $P_k(f,0)$ pour "la" variété polaire générale de codimension k, dont seule la classe d'équisingularité, voire la multiplicité, est définie sur $\mathbb{C}$. Le lecteur que cela gêne est prié de penser à la variété polaire générique, définie sur une extension de $\mathbb{C}$ (cf. § 0 et § 1, 1.4)

6.1 Quelques inégalités utiles.

6.1.1 Proposition : Avec les notations introduites ci-dessus au § 1, et au § 4.

a) On a l'équivalence

$$P_k(f,0) = \emptyset \Longleftrightarrow \dim \varkappa_f^{-1}(0) < N-1-d+k \quad .$$

b) Tout point $x \in X$ possède un voisinage ouvert U tel que pour tout $x' \in U$ on ait, pour $0 \leq k \leq d-1$:

$$m_{x'}(P_k(f,x')) \leq m_x(P_k(f,x)) \quad .$$

Démonstration : a) est une conséquence immédiate du théorème de Bezout et de 4.1.1. Démontrons b) : d'après 1.3, pour x fixé, l'ensemble des $D_{d-k+1} \in G_k$ (grassmannienne des plans de codimension d-k+1 de $\mathbb{C}^N$) tels que $\dim(P_k<f;D_{d-k+1}>) = d-k$, si $P_k(f,0) \neq \emptyset$ est un ouvert de Zariski dense de G_k ; on construit facilement une famille $\pi : \mathcal{P}_k \rightleftarrows W_k$ telle que $\pi^{-1}(D_{d-k+1})$ ne diffère de $P_k<f;D_{d-k+1}>$ que par des composantes immergées. Puisque la dimension des fibres de cette famille est constante, on a par semi-continuité de la multiplicité $m_o(P_k<f;D_{d-k+1}>) \geq m_o(P_k(f,0))$ pour tout $D_{d-k+1} \in W_k$.
Si $P_k(f,x) = \emptyset$, d'après a) et la semi-continuité de la dimension des fibres d'un morphisme, on a $P_k(f,x') = \emptyset$ pour tout x' assez voisin de x. Supposons $P_k(f,x) \neq \emptyset$; en un point $x' \in X - P_k(f,x)$ on a $\varkappa_f^{-1}(x') \cap \lambda_f^{-1}(L^{d-k}) = \emptyset$ pour un certain L^{d-k}, donc par Bezout, $\dim \varkappa_f^{-1}(x') < N-1-d+k$ et $P_k(f,x') = \emptyset$.
Si $x' \in P_k(f,x)$ et est assez voisin de x, on a par semi-continuité de la multiplicité de $P_k(f,x)$, l'inégalité

$$m_x(P_k(f,x)) \geq m_{x'}(P_k(f,x))$$

mais puisque $P_k(f,x)$ est de dimension d-k, le germe de $P_k(f,x)$ en x' est de dimension d-k et est un $P_k<f;D_{d-k+1}>$ en x', d'où l'inégalité

$$m_{x'}(P_k(f,x)) \geq m_{x'}(P_k(f,x'))$$

d'après ce que nous avons vu plus haut, et le résultat. ■

6.1.2 Proposition : Etant donné un sous-ensemble analytique fermé Y de X, il existe un fermé analytique rare $F \subset Y$ tel que l'application $y \to M^*_{X,y}$ de Y dans $\mathbb{N}^d$ définie en 3.2 soit localement constante sur $Y - F$.

Démonstration : Considérons le diagramme (cf. § 3)

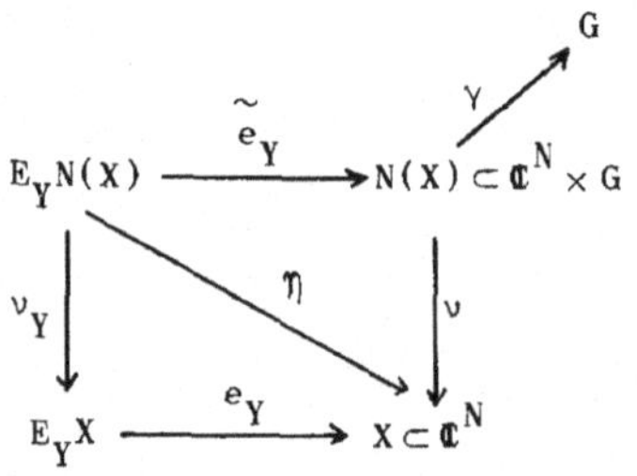

où ν est la modification de Nash, e_Y l'éclatement de Y dans X, $\tilde{e}_Y$ l'éclatement de $\nu^{-1}(Y)$ dans $N(X)$ et ν'_Y le transformé strict de ν par e_Y. Soit $Y = \bigcup Y_\ell$ la décomposition de Y en composantes irréductibles et soit $B_\ell = \{z \in \eta^{-1}(Y_\ell) / \dim_z \eta^{-1}(\eta(z)) > d-1-\dim Y_\ell\}$. Puisque $\dim \eta^{-1}(Y) = d-1$ et que η est propre, le sous-ensemble analytique $F_o = \eta(\bigcup B_\ell) \subset Y$ est rare. Soit F_1 le lieu singulier de Y. En posant $F = F_o \cup F_1$, on voit que l'on est ramené à prouver que dans le cas où Y est non-singulier, et $y \in Y$ tel que $\dim \eta^{-1}(y) = d-1-\dim Y$, l'application $y \mapsto M^*_{X,y}$ est constante au voisinage de y. Soit $N(X) = \bigcup Z_\alpha$ une stratification de Whitney de $N(X)$ telle que $\nu^{-1}(y)$, $\nu^{-1}(Y)$, et les images par $\tilde{e}_Y$ des strates d'équidimensionnalité relatives de $\tilde{e}_Y$ soient réunion de strates (cf. Chap. III, § 5). Alors pour $0 \leq k \leq d-1$ il existe un ouvert de Zariski dense U_k de la grassmannienne des plans de codimension $d-k+1$ de $\mathbb{C}^N$ tel que pour $D_{d-k+1} \in U_k$, la stratification naturelle de $c_k(D_{d-k+1})$ soit transverse dans $0 \times G$ aux strates Z_α contenues dans $\nu^{-1}(0)$. D'après (Chap. III,

4.2.5 et 5.1) on en déduit que $\tilde{e}^{-1}(\gamma^{-1}(c_k(D_{d-k+1}))$ coïncide ensemblistement avec le transformé strict de $\gamma^{-1}(c_k(D_{d-k+1}))$ par $\tilde{e}_Y$, et que la dimension de sa fibre au-dessus de 0 vaut d-k-dim Y. C'est donc a fortiori le cas pour la fibre au-dessus de 0 de l'éclatement de Y dans $P_k\langle(X,0);D_{d-k+1}\rangle$, donc $P_k\langle(X,0);D_{d-k+1}\rangle$ est équimultiple le long de Y (cf. [Li]) au voisinage de y pour $D_{d-k+1} \in U_k$. Comme par ailleurs on a, grâce à la transversalité, l'égalité $(P_k\langle(X,y');D_{d-k+1}\rangle,y') = (P_k\langle(X,y);D_{d-k+1}\rangle,y')$ pour y' assez voisin de y, on obtient le résultat cherché. ∎

6.2 Quelques formules, dans le cas absolu.

6.2.1 Pour un plan H_i de codimension i dans $\mathbb{C}^N$ et assez général, on a, en posant $X_i = (X \cap H_i)_{red}$:

$$(P_k(X,0) \widehat{\cap} H_i)_{red} = P_k(X_i,0) \quad \text{si } 0 \leq k \leq d-i-1 \quad ,$$

et pour les mêmes valeurs de k, on a :

$$m_o(P_k(X_i,0)) = m_o(P_k(X,0)) \quad .$$

Ceci résulte aussitôt de 5.4.

En particulier, la multiplicité de $P_k(X,0)$ est celle de la courbe polaire $P_k(X_{d-1-k},0)$, ce qui permet de ramener les calculs de multiplicités de variétés polaires à des calculs de multiplicités de courbes.

6.2.2 Soit $X_1 \subset \mathbb{C}^{d+1}$ une projection hyperplane générale de X, on a l'égalité

$$m_o(P_k(X,0)) = m_o(P_k(X_1,0)) \quad \text{pour } 0 \leq k \leq d-1 \quad .$$

Ceci résulte de 5.5.1, ii).

***

CHAPITRE V

LE THEOREME PRINCIPAL

Introduction. Soient X un espace analytique complexe réduit purement de dimension d, soit Y un sous-espace analytique complexe de X, non-singulier et de dimension t, et soit 0 un point de Y. Ce chapitre contient la démonstration du résultat principal de ce travail, affirmant que le fait que le couple de strates (X^o, Y) satisfasse en 0 en les conditions de Whitney (Chap. III, § 2) équivaut à ce que l'application $y \to M^*_{X,y}$ de Y dans $\mathbb{N}^d$ définie par la multiplicité généralisée (Chap. IV, 3.2) soit localement constante sur Y au voisinage de 0.

La méthode de démonstration est la récurrence sur d-t : on va couper X par une hypersurface non-singulière H contenant Y et "assez générale". Pour l'essentiel, la démonstration va consister à vérifier que chacune des conditions dont on veut vérifier l'équivalence implique que pour H assez générale, l'espace tangent $T_{H,0}$ à H en 0 est transverse à toutes les limites en 0 d'espaces tangents à X^o, et que $(X \cap H)_{red}$ satisfait encore la même condition le long de Y.

§ 1. La démonstration.

1.1 Un outil technique important est le diagramme commutatif suivant, associé à un plongement $X \subset \mathbb{C}^N$, auquel il sera fait référence dans tout le chapitre :

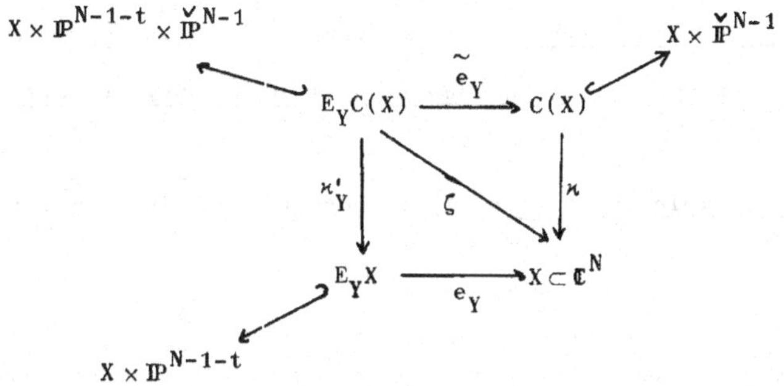

où $\varkappa$ est l'espace conormal de X dans $\mathbb{C}^N$, e_Y l'éclatement de Y dans X, $\tilde{e}_Y$ l'éclatement de $\varkappa^{-1}(Y)$ dans $C(X)$ et $\varkappa'_Y$ vient de la propriété universelle de l'éclatement. On pose $\zeta = \varkappa \circ \tilde{e}_Y$. Je remercie J.P.G. Henry et M. Merle de m'avoir montré que la modification de Nash de X utilisée dans [Te 5] devait être remplacée par l'espace conormal.

1.1.1 Remarque : Dans le diagramme ci-dessus, le morphisme $\varkappa'_Y$ est le transformé strict par l'éclatement e_Y du morphisme $\varkappa$, c'est-à-dire que l'on peut identifier $E_Y C(X)$ au sous-espace de $E_Y X \times_X C(X)$ défini par l'idéal cohérent du produit fibré dont le germe en un point de ce produit est engendré par les fonctions holomorphes sur ce produit annulées par une puissance de l'équation du diviseur exceptionnel de $E_Y X$ via l'homomorphisme $\mathcal{O}_{E_Y X} \xrightarrow{pr_1^*} \mathcal{O}_{E_Y X \times_X C(X)}$ correspondant à la première projection (cf. [H.L.T.]).

1.2 Théorème : *Soient X un espace analytique complexe réduit purement de dimension d, Y un sous-espace analytique de X, purement de dimension t, et 0 un point non singulier de Y. Les conditions suivantes sont équivalentes* :

i) *L'application de Y dans $\mathbb{N}^d$ définie par*

$y \mapsto M^*_{X,y} = (m_y(X), m_y(P_1((X,y))), \ldots, m_y(P_{d-1}(X,y)))$ *est constante au voisinage de* 0.

ii) *Pour tout plongement local* $(X,0) \subset (\mathbb{C}^N, 0)$, *on a l'égalité*

$$\dim \zeta^{-1}(0) = N-2-t \quad \text{si } \zeta^{-1}(0) \text{ n'est pas vide}.$$

iii) Le couple de strates (X^o, Y) vérifie en 0 la condition a) de Whitney stricte avec exposant 1 et la condition b) de Whitney stricte avec un exposant non précisé.

iv) Le couple de strates (X^o, Y) vérifie en 0 les conditions a) et b) de Whitney.

Démonstration : Tout d'abord, on peut supposer, et l'on supposera, Y plongé dans $\mathbb{C}^N$ comme sous-espace linéaire.

i) ⇒ ii). Supposons d'abord $d-t>0$ et considérons l'espace conormal $\varkappa : C(X) \to X$. Puisque $d-t>0$, Y est rare dans X et $\varkappa^{-1}(Y)$ est rare dans $C(X)$, donc $\dim \varkappa^{-1}(Y) \leq N-2$ et il existe (cf. Chap. III, § 5) un fermé analytique strict $F \subset Y$ tel que pour tout $y \in Y - F$ on ait l'inégalité $\dim \varkappa^{-1}(y) \leq N-2-t$. D'après (Chap. IV, 6.1, a)) ceci équivaut à : $P_k(X,y) = \emptyset$ pour $k \geq d-t$, c'est-à-dire $m_y(P_k(X,y)) = 0$ pour $k \geq d-t$. Par l'hypothèse d'équimultiplicité on a donc $m_o(P_k(X,0)) = 0$, c'est-à-dire $P_k(X,0) = \emptyset$ pour $k \geq d-t$ et d'après (Loc. cit.) $\dim \varkappa^{-1}(0) \leq N-2-t$.

Démontrons maintenant l'assertion par récurrence sur d-t : Supposons d'abord $d-t=0$; dans ce cas, Y est une composante irréductible de X, et dire que X est équimultiple le long de Y équivaut à dire que $Y = X$, donc $\varkappa^{-1}(Y) = C(X)$ et $E_Y C(X) = \emptyset$, d'où $\zeta^{-1}(0) = \emptyset$.

Supposons maintenant $d-t \geq 1$. Remarquons d'abord que $\zeta^{-1}(Y)$ est un diviseur dans $E_Y C(X)$ qui est de dimension N-1 puisque $\dim C(X) = N-1$, donc $\dim \zeta^{-1}(Y) = N-2$. La fibre de $\zeta | \zeta^{-1}(Y) : \zeta^{-1}(Y) \to Y$ au-dessus d'un point général $y \in Y$ est de dimension N-2-t et par semi-continuité de la dimension des fibres d'un morphisme, on a $\dim \zeta^{-1}(0) \geq N-2-t$.

Examinons le cas où $d-t=1$: D'après l'hypothèse i), comme nous l'avons vu, nous avons $\dim \varkappa^{-1}(0) \leq N-2-t$. Par ailleurs, l'équimultiplicité de X le long de Y implique (Chap. I, 5.1) que e_Y est un morphisme fini. On en déduit aussitôt, puisque $\zeta^{-1}(0)$ est contenu dans $e_Y^{-1}(0) \times \varkappa^{-1}(0)$, que $\dim \zeta^{-1}(0) \leq N-2-t$, d'où l'égalité, puisque l'inégalité inverse est un fait général comme nous venons de voir.

Supposons maintenant le résultat démontré pour $d-t \leq c-1$ et démontrons-le pour $d-t = c$.

On peut choisir une rétraction locale $\rho : (\mathbb{C}^N,0) \to (Y,0)$ et considérer $(X,0)$ comme plongé dans $Y \times \mathbb{C}^{N-t}$. Ainsi on peut considérer $E_Y(X)$ comme plongé dans $Y \times \mathbb{C}^{N-t} \times \mathbb{P}^{N-t-1}$. Considérons une stratification $E_Y(X) = \cup X'_\alpha$ de $E_Y(X)$ telle que $e_Y^{-1}(0)$ soit réunion de strates ainsi que chacune des images par $\kappa'_Y : E_Y C(X) \to E_Y(X)$ des strates d'équidimensionnalité relative de κ'_Y (cf. Chap. III, § 5). Remarquons que puisque l'hypothèse i) implique $\dim \kappa^{-1}(0) \leq N-2-t$, il existe un ouvert de Zariski dense $U = \check{\mathbb{P}}^{N-1-t} - \kappa^{-1}(0)$ de l'espace $\check{\mathbb{P}}^{N-1-t} \subset \check{\mathbb{P}}^{N-1}$ des hyperplans de $\mathbb{C}^N$ contenant $T_{Y,0}$ tel que pour $H \in U$, H soit transverse à toutes les limites en 0 d'espaces tangents à X^0. Soit $H_0 \subset \mathbb{C}^{N-t}$ un hyperplan, et soit H l'hyperplan $Y \times H_0 \subset Y \times \mathbb{C}^{N-t}$. La transformée stricte par e_Y de l'hypersurface H est $Y \times H'_0$ où $H'_0 \subset \mathbb{C}^{N-t} \times \mathbb{P}^{N-t-1}$ est l'espace éclaté de H_0 en 0, dont la fibre au-dessus de $0 \in \mathbb{C}^{N-t}$ est l'hyperplan $\mathrm{Proj}\, T_{H_0,0}$ de $\mathbb{P}^{N-t-1}$ correspondant à $T_{H_0,0}$.
Choisissons H_0 de tèlle facon que d'une part $\mathrm{Proj}\, T_{H_0,0}$ soit transverse dans $\mathbb{P}^{N-t-1}$ à toutes les strates de la stratification de $E_Y(X)$ décrite ci-dessus qui sont contenues dans $e_Y^{-1}(0) \subset \mathbb{P}^{N-t-1}$, que d'autre part l'espace tangent $T_{H,0}$ en 0 à $H = Y \times H_0$ appartienne à l'ouvert U déterminé ci-dessus et pour toût k, $0 \leq k < d-t$, à l'image par la projection p_2 de l'ouvert de Zariski de I_Y dont l'existence a été prouvée en (Chap. IV, 5.4.3).

D'après (Chap. III, 4.2.5), pour tout représentant assez petit du germe $(X,0)$ le sous-espace $Y \times H'_0$ de $Y \times \mathbb{C}^{N-t} \times \mathbb{P}^{N-t-1}$ est transverse à toutes les strates X'_α de $E_Y(X)$ et par conséquent, d'une part le transformé strict $E_Y[X \cap H]$ de $[X \cap H] = (X \cap H)_{red}$ par e_Y coïncide avec $E_Y X \cap (Y \times H'_0)$ (intersection dans $Y \times \mathbb{C}^{N-t} \times \mathbb{P}^{N-t-1}$), d'autre part, d'après (Chap. III, 5.1) le transformé strict de $E_Y[X \cap H]$ par le morphisme κ'_Y coïncide ensemblistement avec $\kappa'^{-1}_Y([X \cap H])$.
Soit $\mathcal{Y}_0$ une composante irréductible de dimension maximale de $\zeta^{-1}(0)$, et soit $\mathcal{V}_0 \subset \mathbb{P}^{N-t-1}$ l'image par κ'_Y de $\mathcal{Y}_0$. Si $\dim \mathcal{V}_0 = 0$, puisque, par construction, $\mathcal{Y}_0 \subseteq \mathcal{V}_0 \times \kappa^{-1}(0)$, on a $\dim \mathcal{Y}_0 \leq N-2-t$ et par conséquent l'égalité $\dim \mathcal{Y}_0 = N-2-t$, puisque l'égalité inverse a toujours lieu. Si $\dim \mathcal{V}_0 > 0$, d'après le théorème

de Bezout, Proj $T_{H_o,O} \cap \mathcal{V}_o \neq \emptyset$ et donc $\mathcal{V}_o \cap E_Y[X \cap H] \neq \emptyset$. Puisque l'image réciproque par κ'_Y de $E_Y[X \cap H]$ coïncide ensemblistement avec son transformé strict, on en déduit que, notant $[X \cap H]'^{\sim}$ le transformé strict de $[X \cap H]$ par le morphisme ζ, qui n'est autre que le transformé strict par κ'_Y du transformé strict $E_Y[X \cap H]$ de $[X \cap H]$ par e_Y, on a :

$$\mathcal{Y}_o \cap [X \cap H]'^{\sim} \neq \emptyset \ .$$

Puisque $T_{H,O} \in U$, d'après (Chap. IV, 5.3.1), le morphisme naturel

$$\rho : [X \cap H]^{\sim} \longrightarrow C([X \cap H])$$

du tranformé strict de $[X \cap H]$ par κ dans l'espace conormal de $[X \cap H]$ est <u>fini</u>. Considérons maintenant le diagramme

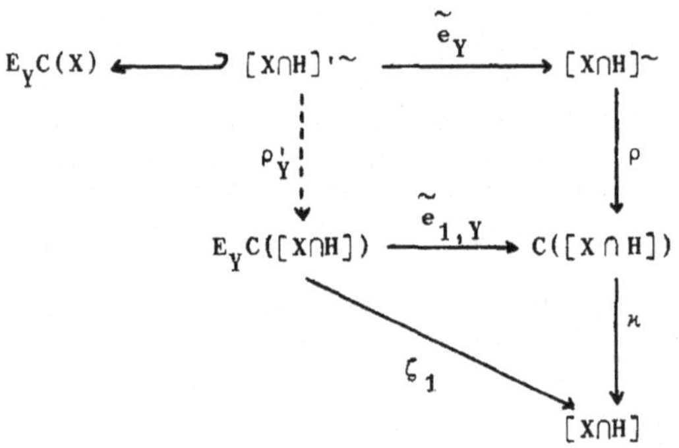

où κ est le morphisme conormal de $[X \cap H] \subset H$, $\tilde{e}_{1,Y}$ est l'éclatement de $\kappa^{-1}(Y)$ dans $C([X \cap H])$ et $\tilde{e}_Y$ est le morphisme induit par le morphisme d'éclatement $E_Y C(X) \to C(X)$. En utilisant la propriété universelle de l'éclatement, on obtient une factorisation ρ'_Y, qui est en fait le transformé strict du morphisme ρ par l'éclatement $\tilde{e}_{1,Y}$, donc est un morphisme <u>fini</u> puisque ρ est fini. De plus, posant $\mathcal{Y}(0) = \zeta^{-1}(0)$ et $\mathcal{Y}_1(0) = \zeta_1^{-1}(0)$, on a l'égalité :

(*) $$\rho'^{-1}_Y(\mathcal{Y}_1(0)) = \mathcal{Y}(0) \cap [X \cap H]'^{\sim} \ .$$

Montrons maintenant que $[X \cap H]$ satisfait aussi l'hypothèse de i), c'est-à-dire l'équimultiplicité des variétés polaires. Par notre choix de H_o, pour un D_{d-k+1} assez général parmi ceux qui sont contenus dans H, on peut prendre $P_k(X,0) = P_k\langle(X,0);D_{d-k+1}\rangle$ et on a $m_o(P_k(X,0)) = m_o(P_k(X,0) \hat{\cap} H)$ (cf. Chap. IV, 5.4.3). Montrons que l'hypothèse d'équimultiplicité implique l'égalité $m_y(P_k(X,y)) = m_y(P_k((X,0);D_{d-k+1}))$ pour $y \in Y$: considérons le diagramme d'inégalités :

$$\begin{array}{ccc} m_o(P_k(X,0)) & = & m_y(P_k(X,y)) \\ \| & & \wedge\!\!\backslash \\ m_o(P_k\langle(X,0);D_{d-k+1}\rangle) & \geq & m_y(P_k\langle(X,0);D_{d-k+1}\rangle) \end{array}$$

où l'égalité horizontale vient de l'hypothèse, l'égalité verticale du choix de D_{d-k+1}, l'inégalité horizontale de la semi-continuité de la multiplicité, et l'inégalité verticale de l'argument de (Chap. IV, 6.1). Le résultat cherché en découle. De même le diagramme d'inégalités :

$$\begin{array}{ccc} m_o(P_k\langle(X,0);D_{d-k+1}\rangle) & = & m_y(P_k\langle(X,0);D_{d-k+1}\rangle) \\ \| & & \wedge\!\!\backslash \\ m_o(P_k\langle(X,0);D_{d-k+1}\rangle \hat{\cap} H) & \geq & m_y(P_k\langle(X,0);D_{d-k+1}\rangle \hat{\cap} H) \end{array}$$

dont la justification est analogue nous montre que d'une part $P_k\langle(X,0);D_{d-k+1}\rangle \hat{\cap} H$ est équimultiple le long de Y, et d'autre part H est transverse, au sens des multiplicités, à $P_k(X,y)$ en tout point $y \in Y$ assez voisin de 0. Comme H reste transverse aux limites en y d'espaces tangents à X^o pour y assez proche de 0 et contient D_{d-k+1}, on a

$$P_k\langle(X,y);D_{d-k+1}\rangle \hat{\cap} H = P_k\langle([X \cap H],y);D_{d-k+1}\rangle$$

d'après (Chap. IV, 5.4.2).

Finalement, on a les égalités

$$m_y(P_k([X\cap H],y)) = m_y(P_k(X,y)\,\hat{\cap}\,H) = m_y(P_k<(X,0);D_{d-k+1}>\,\hat{\cap}\,H) =$$

$$m_y(P_k<(X,0);D_{d-k+1}>) = m_o(P_k<(X,0);D_{d-k+1}>) = m_o(P_k(X,0)) = m_o(P_k([X\cap H],0))$$

d'où le résultat cherché.

Par l'hypothèse de récurrence, et puisque $\dim[X\cap H] = d-1$ et que $[X\cap H]\subset H$, on en déduit que l'on a l'égalité :

$$\dim \zeta_1^{-1}(0) = N-3-t \quad .$$

D'après l'égalité (*) ci-dessus, et la finitude du morphisme ρ'_Y, on en déduit : $\dim \mathcal{Y}(0)\cap[X\cap H]'^{\sim} = N-3-t$, et donc en particulier, en revenant à notre composante irréductible $\mathcal{Y}_o$ de $\mathcal{Y}(0)$, puisque nous savons que $\mathcal{Y}_o$ rencontre $[X\cap H]'^{\sim}$, nous avons

$$\dim(\mathcal{Y}_o\cap[X\cap H]'^{\sim}) \leq N-3-t \quad .$$

Par ailleurs, $[X\cap H]'^{\sim}$ est une union de composantes d'un sous-espace de $E_YC(X)$ défini par une seule équation : celle de H composée avec ζ, par conséquent on a ([Bbk 3], § 3, No 1, Prop. 2) l'inégalité

$$\dim(\mathcal{Y}_o\cap[X\cap H]'^{\sim}) \geq \dim \mathcal{Y}_o - 1$$

d'où finalement

$$\dim \mathcal{Y}_o \leq N-2-t \quad ,$$

et le résultat cherché. ■

1.1.2 Remarque : Soient X et Y comme en 1.1. Nous venons de voir que si $M^*_{X,y}$ est constant le long de Y au voisinage de 0, il existe un ouvert de Zariski dense $A_k \subset G_k$ tel que pour $D_{d-k+1} \in A_k$, $P_k\langle(X,0);D_{d-k+1}\rangle$ soit équimultiple le long de Y. La réciproque de cet énoncé est vraie aussi. Soit en effet (cf. Chap. IV, 6.1.2) $F_k \subset Y$ un fermé analytique rare tel que $m_y(P_k(X,y))$ soit constante pour $y \in Y - F_k$. Si F_k est non vide, on coupe toute la situation par un sous-espace lisse de codimension dim Y - 1 de $\mathbb{C}^N$ rencontrant F_k en $\{0\}$, et l'on se ramène au cas où dim Y = 1. On montre par un argument analogue à celui de (Chap. IV, 5.2.1) qu'il existe un ouvert de Zariski dense $D_k \subset G_k$ tel que si $D_{d-k+1} \in D_k$, on ait $m_y(P_k\langle(X,y);D_{d-k+1}\rangle) = m_y(P_k(X,y))$ pour $y \in Y - \{0\}$ assez proche de 0. Comme il existe un ouvert de Zariski dense $C_k \subset G_k$ tel que la même égalité ait bien lieu pour y = 0, en prenant $A_k = D_k \cap C_k$, on obtient un ouvert de Zariski dense tel que

$$m_y(P_k\langle(X,0);D_{d-k+1}\rangle) = m_y(P_k(X,y)) \quad \text{pour } y \in Y$$

d'où le résultat. ∎

L'implication ii) ⇒ iii) n'est autre que le résultat de (Chap. III, 2.3.1).

L'implication iii) ⇒ iv) est évidente.

Prouvons iv) ⇒ i). Pour cela, nous allons tout d'abord montrer :

1.2.1 Proposition : Si le couple de strates (X^o,Y) satisfait les conditions de Whitney en 0, on a l'inégalité : $\dim \varkappa^{-1}(0) \leq N-2-t$.

Démonstration : D'après (Chap. IV, 6.1), il suffit de prouver $P_{d-t}(X,0) = \emptyset$, et pour un plan H de $\mathbb{C}^N$ de codimension t-1 passant par 0 et au demeurant assez général, d'une part l'intersection $Y \cap H$ sera non-singulière de dimension 1, d'autre part d'après (Chap. III, 4.2.2), le couple $([X \cap H]^o, Y \cap H))$ satisfera en 0 les conditions de Whitney, et enfin (cf. Chap. IV, 6.2) on aura l'égalité de multiplicités : $m_o(P_{d-t}(X)) = m_o(P_{d-t}(X \cap H)_{red})$. Il suffit donc de prouver que $P_{d-t}((X \cap H)_{red},0)$ est vide, et puisque $\dim(X \cap H)_{red} = d-t$, nous voyons que nous avons ramené la preuve de la Proposition à sa preuve dans le cas particulier où $t = 1$. Tout va reposer sur le lemme suivant :

1.2.2 Lemme-clé : Soient $(X,0) \subset (\mathbb{C}^N,0)$ un espace analytique complexe réduit purement de dimension d, et $(Y,0) \subset (X,0)$ une courbe non-singulière en 0. Dans l'espace $\mathcal{P}$ des projections linéaires $p : \mathbb{C}^N \to \mathbb{C}^2$, il existe un ouvert constructible dense V tel que pour toute projection $p \in V$ on ait, en notant $P_{d-1}\langle p\rangle$ la fermeture dans X de l'ensemble des points critiques de $p|X^o$:

i) L'ensemble $P_{d-1}\langle p\rangle$ est, au voisinage de 0, un ensemble analytique fermé de dimension 1, qui est une courbe polaire générale de $(X,0)$, et $|P_{d-1}\langle p\rangle \cap Y| = \{0\}$.

ii) La projection p est générale pour la courbe $P_{d-1}\langle p\rangle \cup Y$ en 0, au sens de (Chap. I, § 6, Prop. 6.2.1).

Démonstration : On suppose toujours $\mathbb{C}^N$ muni des coordonnées $z_1,\dots,z_N$ et l'on munit $\mathbb{C}^2$ de coordonnées x et y. On peut pour prouver le Lemme se contenter de considérer les projections qui sont, à un automorphisme linéaire de $\mathbb{C}^2$ près, de la forme : $x = z_1 + \sum_3^N a_i z_i$, $y = z_2 + \sum_3^N b_i z_i$. On considère donc la famille de projections ci-dessus comme un morphisme $\mathbb{C}^N \times \mathbb{C}^{2(N-2)} \to \mathbb{C}^2 \times \mathbb{C}^{2(N-2)}$. D'après (Chap. IV, 1.3.1), il existe un fermé de Zariski $B \subset \mathbb{C}^{2(N-2)}$ tel que pour $(\underline{a},\underline{b}) \in \mathbb{C}^{2(N-2)} - B$, le lieu critique de la restriction à X^o de la projection correspondante soit une courbe, dont l'adhérence dans X est une courbe polaire générale.

Par ailleurs, remarquons que si Y est contenu dans le lieu singulier de X, on a $P_{d-1}\langle p\rangle \cap Y = \{0\}$ puisque d'après la définition des variétés polaires (Chap. IV, 1.3) et le fait que $P_{d-1}\langle p\rangle$ est une courbe, on a $P_{d-1}\langle p\rangle \cap \operatorname{Sing} X = \{0\}$. Si Y n'est pas contenu dans Sing X, soit T la limite en 0 des espaces tangents à X^o aux points de $Y - \{0\}$. Les projections p telles que $\operatorname{Ker} p \cap T$ ne soit pas de codimension 2 dans T forment un fermé analytique rare B' de $\mathbb{C}^{2(N-2)}$, et pour tout projection $p \in \mathbb{C}^{2(N-2)} - (B \cup B')$, on aura

$P_{d-1}\langle p\rangle \cap Y = \{0\}$. Ceci achève d'établir le point i).

Grâce au lemme de transversalité des variétés polaires (Chap. IV, 5.1) il existe dans $\mathbb{C}^{2(N-2)} - (B \cup B')$ un ouvert V_1, constructible et dense dans $\mathbb{C}^{2(N-2)}$ tel que, pour tout $p \in V_1$, le noyau Ker p de p soit transverse au cône tangent en 0 à la courbe $P_{d-1}\langle p\rangle \cup Y$, c'est-à-dire ne contienne aucune des droites qui constitutent le cône tangent. Comme nous l'avons déjà vu, nous pouvons choisir un système $f_1, \ldots, f_m$, $f_i \in \mathcal{O}_{\mathbb{C}^N,0}$ de générateurs de l'idéal définissant $(X,0) \subset (\mathbb{C}^N,0)$ tel que dans un voisinage de $0 \times V_1$ dans $\mathbb{C}^N \times V_1$, la réunion de Sing $X \times V_1$ avec $\bigcup_{p \in V_1} P_{d-1}\langle p\rangle$ soit définie par l'idéal engendré par les $(f_i)_{\underline{a},\underline{b}}$ et par les déterminants jacobiens $\dfrac{\partial((f_1)_{\underline{a},\underline{b}}, \ldots, (f_{N-d})_{\underline{a},\underline{b}})}{\partial(z_{i_1}, \ldots, z_{i_{N-d}})}$, tels que $(i_1, \ldots, i_{N-d}) \subset \{3, \ldots, N\}$, où pour $h \in \mathcal{O}_{\mathbb{C}^N,0}$ on a noté $h_{a,b}$ la fonction obtenue en substituant $x - \sum_3^N a_i z_i$ à z_1 et $y - \sum_3^N b_i z_i$ à z_2, et l'on a pris pour coordonnées sur $\mathbb{C}^N \times V_1$ les fonctions $x, y, z_3, \ldots, z_N, a_3, \ldots, a_N, b_2, \ldots, b_N$. Nous avons ainsi une famille analytique $\mathfrak{X}$ de courbes de $\mathbb{C}^N$ paramétrée par V_1, dont l'image par le morphisme $\mathbb{C}^N \times V_1 \to \mathbb{C}^2 \times V_1$ qui à $(x,y,z_3,\ldots,z_N,\underline{a},\underline{b})$ associe $(x,y,\underline{a},\underline{b})$ est une famille analytique $\mathfrak{X}_1$ de courbes planes paramétrée par V_1 (l'image est toujours définie (cf. [Te 3], § 3) ici par l'idéal de Fitting $F_0(\mathcal{O}_{\mathfrak{X}})$ du $\mathcal{O}_{\mathbb{C}^2 \times V_1}$-module $\mathcal{O}_{\mathfrak{X}}$).

D'après les résultats généraux (cf. [Te 2], I) sur la résolution simultanée, il existe un fermé analytique rare A de V_1 tel que au voisinage de tout point de $V_1 - A$, la famille $\mathfrak{X} \overset{\sigma}{\rightleftarrows} V_1$ (où σ est la section qui pique 0 dans chaque fibre) admette une normalisation simultanée qui de plus a la propriété que $|n^{-1}(\sigma(V_1))| \to \sigma(V_1)$ est étale, où $n: \bar{\mathfrak{X}} \to \mathfrak{X}$ est la normalisation.

Par ailleurs, d'après (Chap. I, 6.4.2), il existe un fermé analytique rare B de V_1 tel que la famille de courbes planes $\rho_1 : \mathfrak{X}_1 \overset{\sigma_1}{\rightleftarrows} V_1$ (où σ_1 pique 0 dans chaque fibre) soit équisaturée en tout point de $V_1 - B$.

Examinons ce qui se passe au voisinage d'un point p_0 de $V_1 \setminus (A \cup B)$. La famille des courbes polaires $P_{d-1}\langle p\rangle$ admet une paramétrisation simultanée, c'est-à-dire

qu'elle peut être représentée comme l'image réduite d'un morphisme $(\coprod_{\rho=1}^{r}(\mathbb{C},0))\times V_1 \to \mathbb{C}^N \times V_1$ où r est le nombre des composantes irréductibles de $P_{d-1}\langle p_o\rangle \cup Y$ (ou de $(\mathfrak{X},\sigma(p_o))$, à cause du fait que $|n^{-1}(\sigma(V_1))| \to \sigma(V_1)$ est étale). Donc si nous faisons un changement de coordonnées de sorte que les coordonnées $\underline{a}$, $\underline{b}$ du point $p_o \in V_1$ soient nulles, et notons dorénavant 0 pour p_o, chaque composante irréductible $\mathfrak{X}_\ell$ de $\mathfrak{X}$ au voisinage de 0 peut être représentée paramétriquement comme suit : (cf. Chap. I, § 6) : $x = t_\ell^{n_\ell}$, $y = \nu(t_\ell,\underline{a},\underline{b})$, $z_i = \zeta_i(t_\ell,\underline{a},\underline{b})$ $(3 \le i \le N)$ et $1 \le \ell \le r$. Noter que nous utilisons ici le fait que la projection de $P_{d-1}\langle p\rangle$ sur son image dans $\mathbb{C}^2$ est transversale pour $p \in V_1$, et que $\mathfrak{X}_1$ est équimultiple le long de $\sigma_1(V_1 - A\cup B)$. Nous nous proposons de prouver que <u>dans l'anneau</u> $\mathcal{O}_{\mathfrak{X},0}$ on a les identités : $z_k = \frac{\partial \nu}{\partial b_k}$ pour $3 \le k \le N$. Il suffit de prouver que ces identités sont vérifiées dans le normalisé $\overline{\mathcal{O}_{\mathfrak{X},0}} = \prod_{\ell=1}^{r}\{t_\ell,\underline{a},\underline{b}\}$ de $\mathcal{O}_{\mathfrak{X},0}$.
Remarquons tout d'abord que sur la composante $Y \times V_1$ de $\mathfrak{X}$, les identités sont vérifiées puisque la paramétrisation correspondante est donnée par des fonctions ζ_i, ξ, ν indépendantes de $\underline{a}$ et $\underline{b}$ et que par conséquent on est ramené à dériver par rapport à b_j l'identité $\nu(t) = z_2(t) + \sum_3^N b_j\, z_j(t)$ où $z_1(t), z_2(t), \ldots, z_N(t)$ est une paramétrisation de Y.
Ecrivons maintenant les équations satisfaites par les autres composantes de $\mathfrak{X}$, en utilisant la paramétrisation :

$$f_j(t_\ell^{n_\ell} - \sum_3^N a_i\, \zeta_i(t_\ell,\underline{a},\underline{b}), \nu(t_\ell,\underline{a},\underline{b}) -$$
$$\sum_3^N b_i\, \zeta_i(t_\ell,\underline{a},\underline{b}),\ \zeta_3(t_\ell,\underline{a},\underline{b}), \ldots, \zeta_N(t_\ell,\underline{a},\underline{b})) = 0$$

pour $1 \le j \le m$ et chaque ℓ,

et, par définition des courbes polaires, sur ces composantes de $\mathfrak{X}$ s'annulent aussi des déterminants jacobiens :

$$\frac{\partial((f_1)_{\underline{a},\underline{b}}, \ldots, (f_{N-d})_{\underline{a},\underline{b}})}{\partial(z_{i_1}, \ldots, z_{i_{N-d}})} \quad \text{avec} \quad \{i_1, \ldots, i_{N-d}\} \subset \{3, \ldots, N\}$$

dérivons chacune des N-d premières équations du premier groupe par rapport à b_k : nous obtenons un système d'équations linéaires que nous pouvons écrire :

$$\sum_{i=3}^{N} \left(\left(\frac{\partial (f_j)_{\underline{a},\underline{b}}}{\partial z_i}\right)\circ \lambda\right) \frac{\partial \zeta_i}{\partial b_k} + \left(\left(\frac{\partial (f_j)_{\underline{a},\underline{b}}}{\partial y}\right)\circ \lambda\right)\left(\zeta_k - \frac{\partial \upsilon}{\partial b_k}\right) = 0 \qquad (1 \leq j \leq N-d)$$

où λ désigne la paramétrisation de la ℓ-ième composante de $\mathcal{Z}$.
Utilisant l'annulation des déterminants jacobiens, la règle de Cramèr nous donne alors, pour chaque $\{i_2,\dots,i_{N-d}\} \subset \{3,\dots,N\}$, l'équation (dans $\mathcal{O}_{\overline{\mathcal{Z}},0}$)

$$\left(\frac{\partial((f_1)_{\underline{a},\underline{b}},\dots,(f_{N-d})_{\underline{a},\underline{b}})}{\partial(y,z_{i_2},\dots,z_{i_{N-d}})}\circ\lambda\right).\left(\zeta_k - \frac{\partial \upsilon}{\partial b_k}\right) = 0 \quad .$$

Montrons qu'il est impossible que tous les mineurs jacobiens apparaissant dans ces équations s'annulent sur la composante de $\mathcal{Z}$ que nous considérons. Si en effet tel était le cas, les règles de Cramèr nous montrent, au vu de la nullité des déterminants jacobiens qui s'annulent sur $\mathcal{Z}$, que l'espace tangent à X au point général de la ℓ-ième composante de la courbe polaire $\mathcal{I}(p_o)$ correspondant à notre point 0 serait contenu dans l'hyperplan $dx = 0$; en particulier, l'espace tangent à la courbe polaire elle-même serait contenu dans $dx = 0$ en ce point, ce que contredit aussitôt le Lemme de transversalité des variétés polaires (Chap. IV, 5.1). Ainsi nous avons bien, sur chaque composante de $\mathcal{Z}$, et donc dans l'anneau $\mathcal{O}_{\mathcal{Z},0}$, les égalités

$$z_k = \frac{\partial \upsilon}{\partial b_k} \qquad (3 \leq k \leq N) \quad .$$

Achevons maintenant la démonstration du Lemme-clé :
Au voisinage de tout point de $\sigma_1(V_1 - A \cup B)$, la famille de courbes planes $\mathcal{Z}_1 \to V_1$ est représentée paramétriquement par $x = t_\ell^{n_\ell}$, $y = \upsilon(t_\ell,\underline{a},\underline{b})$ $(1 \leq \ell \leq r)$ et est équisaturée. D'après ce que nous vu en (Chap. I, 6.4) les dérivations $\frac{\partial}{\partial b_k}$ de $\mathbb{C}\{\underline{a},\underline{b}\}$ s'étendent en des dérivations D_k de $\overline{\mathcal{O}_{\mathcal{Z}_1,p_o}} \simeq \prod_{\ell=1}^{r} \mathbb{C}\{t_\ell,\underline{a},\underline{b}\}$ conservant l'anneau $\mathcal{O}_{\mathcal{Z}_1,0}^{s(\rho_1)}$ saturé relatif, et vérifiant $D_k\, a_i = 0$, $D_k\, t_\ell = 0$. Puisque bien

sûr l'image $\prod_{1}^{r} \nu(t_\ell,\underline{a},\underline{b})$ de y dans $\overline{\mathcal{O}_{\mathfrak{X}_1,0}}$ appartient au saturé relatif, nous avons ainsi prouvé les inclusions :

$$z_k \cdot \overline{\mathcal{O}_{\mathfrak{X}_1,0}} \in \mathcal{O}^{s(\rho_1)}_{\mathfrak{X}_1,0} \qquad (3 \leq k \leq N)$$

ce qui implique que le sous-anneau de $\overline{\mathcal{O}_{\mathfrak{X}_1,0}}$ engendré par $\mathcal{O}_{\mathfrak{X}_1,0}$ et les $z_k \cdot \overline{\mathcal{O}_{\mathfrak{X}_1,0}}$, sous-anneau qui n'est autre que $\mathcal{O}_{\mathfrak{X},0}$, a le même saturé relatif que $\mathcal{O}_{\mathfrak{X}_1,0}$, comme on le vérifie aussitôt. Nous avons donc démontré que le morphisme $\mathfrak{X} \xrightarrow{p} \mathfrak{X}_1$ qui est fini, est biméromorphe au voisinage de $\sigma(0)$, et induit un isomorphisme de saturés relatifs (à ρ_1 et $\rho = \rho_1 \circ p$). Par passage aux fibres au-dessus de 0, on obtient donc (cf. Chap. I, 6.4.1) que le morphisme de la courbe $P_{d-1}\langle p_0\rangle \cup Y$ sur son image par p_0 dans $\mathbb{C}^2$ s'étend en un isomorphisme des saturés, ce qui achève la démonstration du Lemme-clé.

<u>Démontrons maintenant la Proposition</u>, par l'absurde. Comme nous l'avons remarqué, il suffit de prouver que la variété polaire $P_{d-1}(X,0)$ est vide, avec l'hypothèse : dim $Y = 1$ et (X^0,Y) satisfait en 0 les conditions de Whitney. Choisissons une projection linéaire $p : \mathbb{C}^N \to \mathbb{C}^2$ assez générale pour satisfaire aux conclusions du Lemme-clé ci-dessus, et choisissons une rétraction locale $\rho : (\mathbb{C}^N,0) \to (Y,0)$ telle que, notant T_ρ l'application tangente à ρ en 0, on ait l'inclusion Ker $p \subset$ Ker T_ρ. D'après le Lemme-clé, on a $P_{d-1}(X,0) = P_{d-1}\langle p\rangle$ et si cette courbe n'est pas vide, soit T une direction limite en 0 d'espace tangents à X^0 le long d'une des composantes irréductibles de cette courbe. Puisque pour chaque point $x \in P_{d-1}(X,0) \cap X^0$, on a l'inégalité $\dim(T_{X,x} \cap \text{Ker } p) \geq d-1$, on a également l'inégalité $\dim(T \cap \text{Ker } p) \geq d(1$. Soit ℓ une direction limite de sécantes $\widehat{x_i\ \rho(x_i)}$ où x_i tend vers 0 le long de la même composante irréductible de $P_{d-1}(X,0)$.

D'après le lemme-clé, ni ℓ ni $T_{Y,0}$ n'est contenu dans Ker p, <u>d'après la condition</u> a) <u>de Whitney</u>, $T_{Y,0} \subset T$, et <u>d'après la condition</u> b) <u>de Whitney</u>, $\ell \subset T$. Il en résulte d'une part que $\dim(T \cap \text{Ker } p) = d-1$, car sinon $T \cap \text{Ker } p = T$, et on aurait $\ell \subset$ Ker p ce qui est exclu, et d'autre part les égalités :

$T = (T \cap \mathrm{Ker}\ p) + \mathbb{C} \cdot T_{Y,0}$ et $T = (T \cap \mathrm{Ker}\ p) + \mathbb{C} \cdot \ell$. Mais par construction de ℓ, nous avons $\ell \subset \mathrm{Ker}\ T_\rho$, donc $\mathrm{Ker}\ T_\rho = \mathrm{Ker}\ p + \mathbb{C} \cdot \ell$, d'où l'on déduit aussitôt les égalités $\mathrm{Ker}\ p + \mathbb{C} \cdot \ell + \mathbb{C} \cdot T_{Y,0} = \mathrm{Ker}\ T_\rho + \mathrm{Im}\ T_\rho = T_{\mathbb{C}^N,0}$, qui excluent que l'on ait l'inclusion $\ell \subset \mathrm{Ker}\ p + \mathbb{C} \cdot T_{Y,0}$, et <u>a fortiori</u> l'inclusion $\ell \subset (T \cap \mathrm{Ker}\ p) + \mathbb{C} \cdot T_{Y,0}$. Nous aboutissons donc à une contradiction, ce qui implique que $P_{d-1}(X,0)$ doit être vide, et achève la démonstration de la Proposition. ∎

<u>Remarque</u> : On peut aussi comme dans ([Te 5]) commencer par démontrer, en utilisant la première inégalité métrique du Chapitre III, que si (X^o,Y) satisfait en 0 les conditions de Whitney, pour une projection $p : \mathbb{C}^N \to \mathbb{C}^{d+1}$ assez générale, le couple de strates $(p(X)^o, p(Y))$ satisfait en 0 les conditions de Whitney. On se ramène au cas des hypersurfaces, qui est un peu plus simple.

<u>Terminons maintenant la preuve de l'implication</u> iv) ⇒ i) <u>du Théorème</u>. Montrons d'abord que nous pouvons nous ramener à prouver l'implication dans le cas particulier où $t = 1$.

D'une part, d'après la semi-continuité de la multiplicité des variétés polaires (Chap. IV, 6.1), pour chaque k, il existe un fermé analytique strict F_k d'un voisinage de 0 dans Y tel que, pour $y \in F_k$, $m_y(P_k(X,y))$ soit strictement supérieur à la valeur $m_{y'}(P_k(X,y'))$ pour $y' \in Y - F_k$, cette dernière ("multiplicité générique") étant indépendante de $y' \in Y - F_k$. D'autre part, d'après (Chap. IV, 5.4.2) et le lemme de transversalité des variétés polaires (Chap. IV, 5.2.1 et 5.4.3) pour tout couple j, k d'entiers tels que $j+k \leq d-1$ et tout plan D_j de codimension j dans $\mathbb{C}^N$ assez général, passant par 0, on a égalité $P_k(X,0) \hat{\cap} D_j = P_k([X \cap D_j], 0)$ et $m_0(P_k([X \cap D_j], 0)) = m_0(P_k(X,0))$. Or puisque d'après la Proposition et (Chap. IV, 6.1), les variétés polaires $P_k(X,0)$ sont vides pour $k \geq d-t$, nous ne perdons aucune information en coupant X par un plan D_{t-1} de codimension t-1 général, qui aura en particulier la propriété que pour tout k, $|D_{t-1} \cap F_k|$ est soit vide (si F_k l'est) soit réduit à $\{0\}$. D'après (Chap. III, 4.2.2), le couple de strates $((X \cap D_{t-1})^o, Y \cap D_{t-1})$ satisfera encore les conditions de Whitney en 0, et les variétés polaires de X seront équimultiples le long de Y si et seulement si les variétés polaires de $[X \cap D_{t-1}]$ sont équimultiples le long de $Y \cap D_{t-1}$, qui est non-singulier de dimension 1, puisque les multiplicité des variétés polaires $P_k(X,0)$ de X sont nulles pour $k \geq d-t$, et égales à celles de $(X \cap D_{t-1})_{red}$ pour $k < d-t$. ∎

Plaçons-nous donc dans le cas où $t = 1$. Nous allons terminer la démonstration par récurrence sur d ; si $d = 1$, Y est une des composantes irréductibles de la courbe X, et nous allons voir que si $X \neq Y$, les conditions de Whitney provoquent une contradiction. Soit en effet Γ une composante de X distincte de Y, et soit T la limite de long de Γ des tangentes à Γ. Soit $\rho : (\mathbb{C}^N, 0) \to (Y,0)$ une rétraction locale, et soit ℓ la limite le long de Γ des sécantes $\widehat{x\,\rho(x)}$: la condition a) implique que $T = T_{Y,0}$ alors que la condition b) implique que $T = \ell$, ce qui est impossible puisque $\ell \subset \operatorname{Ker} T_\rho$ qui est un supplémentaire de $T_{Y,0}$. Examinons maintenant (par curiosité) le cas où $d = 2$. Dans ce cas, la Proposition nous fournit le fait que $P_1(X) = \emptyset$, c'est-à-dire l'équimultiplicité de $P_1(X,y)$ le long de Y, ou encore le fait que la dimension de la fibre $\varkappa^{-1}(0)$ du morphisme conormal $\varkappa : C(X) \to X$ est $\leq N-3$. Il nous reste à prouver que X est équimultiple le long de Y : D'après le Lemme 5.1.1, il existe un ouvert de Zariski dense V de l'espace $\check{\mathbb{P}}^{N-2}$ des hyperplans de $\mathbb{C}^N$ contenant $T_{Y,0}$ tel que, pour $H \in V$, H ne contienne aucune limite en 0 d'espaces tangents à X. Nous allons montrer que pour tout $H \in V$, on a $|H \cap X| = Y$, ce qui prouvera l'équimultiplicité, d'après (Chap. I, 5.5). Supposons le contraire, et soit Γ une composante irréductible, distincte de Y, de la courbe $|H \cap X|$; soit $\rho : (\mathbb{C}^N, 0) \to (Y,0)$ une rétraction locale, posons $T = \lim_{\substack{x \in X^0 \cap \Gamma \\ x \to 0}} T_{X,x}$ et $\ell = \lim_{\substack{x \in X^0 \cap \Gamma \\ x \to 0}} \widehat{x\,\rho(x)}$. D'après la condition a) de Whitney, T contient $T_{Y,0}$, et comme $T \cap H$ est de dimension 1, il faut que $T \cap H = T_{Y,0}$, mais d'après la condition b) de Whitney, ℓ appartient aussi à $T \cap H$, d'où une contradiction puisque $\ell \subset \operatorname{Ker} T_\rho$, et que ce dernier est supplémentaire de $T_{Y,0}$. Il faut donc que l'on aie l'égalité $|H \cap X| = Y$.

Supposons maintenant le résultat iv) ⇒ i) démontré pour les espaces de dimension au plus d-1, où $d = \dim X$.

Pour notre espace X, la Proposition 1.2.1 nous fournit l'inégalité $\dim \varkappa^{-1}(0) \le N-3$ et par conséquent $U = \check{\mathbb{P}}^{N-2} - \varkappa^{-1}(0)$ est un ouvert de Zariski dense de l'espace projectif $\check{\mathbb{P}}^{N-2}$ des hyperplans de $\mathbb{C}^N$ contenant $T_{Y,0}$ tel que tout $H \in U$ soit transverse à toutes les limites en 0 d'espaces tangents à X^o. Il nous faut montrer que les multiplicités $m_y(P_k(X,y))$ sont constantes sur Y au voisinage de 0 pour $0 \le k \le d-2$, puisque nous savons déjà que $m_0(P_{d-1}(X,0)) = 0$. D'après (Chap. IV, 5.4.3), il existe un ouvert de Zariski dense de couples (D_{d-k+1}, H) tels que $D_{d-k+1} \subset H = H_o \times T_{Y,0}$, que $P_k\langle(X,0);D_{d-k+1}\rangle$ ait la multiplicité de la variété polaire générique, et que l'on ait

$P_k\langle(X,0);D_{d-k+1}\rangle \hat{\cap} H = P_k\langle([X \cap H],0);D_{d-k+1}\rangle$ <u>pourvu que</u> $k \le d-2$. On peut supposer D_{d-k+1} général dans H, et donc $P_k\langle[X \cap H],0);D_{d-k+1}\rangle = P_k([X\cap H],0)$ et par ailleurs puisque H est transverse aux limites en 0 d'espaces tangents à X, le couple $([X \cap H]^o, Y)$ satisfait les conditions de Whitney. Par l'hypothèse de récurrence, $P_k\langle[X \cap H];D_{d-k+1}\rangle$ est équimultiple le long de Y au voisinage de 0 (cf. Remarque 1.1.2 plus haut). D'après (Chap. IV, 5.2.1 et 5.4.3) on a $m_y(P_k\langle(X,0);D_{d-k+1}\rangle) = m_y(P_k\langle(X,0);D_{d-k+1}\rangle \hat{\cap} H)$ pour $y \in Y$ et donc pour $k < d-2$, $P_k\langle(X,0);D_{d-k+1}\rangle$ est équimultiple au voisinage de 0. Il nous reste à prouver que $P_{d-2}\langle(X,0);D_3\rangle$ est équimultiple le long de Y au voisinage de 0 pour presque tout D_3. Pour cela, il suffit (Chap. I, 5.5) de prouver que l'on a $|P_{d-2}\langle(X,0);D_3\rangle \cap H| = Y$ pour presque tout hyperplan H contenant Y, et en fait (cf. la preuve de loc. cit.) il suffit que cette égalité ait lieu pour un hyperplan H tel qu'il existe une projection linéaire $\rho : \mathbb{C}^N \to Y$ telle que pour tout $y \in Y - \{0\}$ voisin de 0, $\rho^{-1}(y) \cap H$ soit transverse à $\rho^{-1}(y) \cap P_{d-2}\langle(X,0);D_{d-k+1}\rangle$ en y. Or, d'après (Chap. IV, 5.2.1) il existe un ouvert de Zariski dense W_{d-2} dans la grassmannienne des plans de codimension 2 dans $\mathbb{C}^N$ tel que pour $D_2 \in W_{d-2}$, on ait, pour $y \in Y - \{0\}$, $|D_2 \cap C_y(P_{d-2}\langle(X,y);D_{d-k+1})| = \{0\}$, et nous pouvons supposer $D_3 \subset D_2$. Posons $H = D_2 \times Y$, et choisissons D_2 de telle façon que H ne contienne aucune limite en 0 d'espaces tangents à X^o. Soit $\rho : \mathbb{C}^N \to Y$ une projection telle que $\operatorname{Ker} \rho \supset D_2$.

Puisque $D_3 \subset D_2$, on a $P_{d-2}\langle(X,0);D_3\rangle \cap H \cap X^o = P_{d-2}\langle([X\cap H],0);D_3\rangle \cap X^o$

mais puisque $([X \cap H]^o, Y)$ satisfait les conditions de Whitney, la courbe polaire de $[X \cap H]$ en 0 est vide, donc le terme de droite est vide. Nous avons donc un ouvert de Zariski W'_{d-2} tel que pour $D_2 \in W'_{d-2}$, on ait

$$P_{d-2}\langle(X,0);D_3\rangle \cap (D_2 \times Y) \cap X^o = \emptyset \ .$$

Puisque l'ensemble $(\mathrm{Sing}\ X) \cap P_{d-2}\langle(X,0);D_3\rangle$ est de dimension ≤ 1 (Chap. IV, 1.3) et que D_3 ne rencontre son cône tangent qu'en $\{0\}$ (Chap. IV, 5.1), il est possible de choisir $D_2 \supset D_3$ de telle façon que

$P_{d-2}\langle(X,0);D_3\rangle \cap (D_2 \times Y) \cap \mathrm{Sing}\ X \subset Y$, et par conséquent

$P_{d-2}\langle(X,0);D_3\rangle \cap (D_2 \times Y) \cap X^o = \emptyset$ implique que $|P_{d-2}\langle(X,0);D_3\rangle \cap (D_2 \times Y)| = Y$,

ce qu'il fallait démontrer.

1.3 Remarque : En chemin, nous avons vu que si (X^o, Y) satisfait les conditions de Whitney en $0 \in Y$, pour tout plongement local $(X,0) \subset (\mathbb{C}^N, 0)$ et toute hypersurface non-singulière $H \subset \mathbb{C}^N$ contenant Y et assez générale, le couple $([X \cap H]^o, Y)$ satisfait encore les conditions de Whitney. Ceci répond affirmativement à une question de [Té 11], à laquelle V. Navarro avait déjà apporté une réponse affirmative dans le cas particulier où $\dim Y = 1$ (cf. [Na 1]) .

§ 2. Version relative.

2.1 Théorème : Soit $f : (X,0) \overset{\sigma}{\rightleftarrows} (\mathbb{C},0)$ un morphisme muni d'une section σ, dont toutes les fibres sont réduites et purement de dimension d. Posons $Y = \sigma(\mathbb{C}) \subset X$ et supposons donné un $\mathbb{C}$-plongement local :

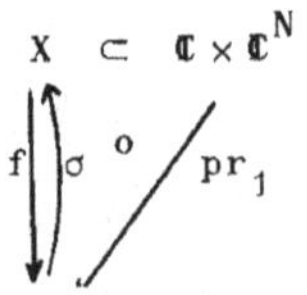

tel que $Y = \mathbb{C} \times \{0\}$.

Les conditions suivantes sont équivalentes :

i) L'hyperplan $K = 0 \times \mathbb{C}^N$ est transverse à toutes les limites en 0 d'espaces tangents à X^o et l'application $Y \to \mathbb{N}^{d+1}$ définie par $y \to (m_y(X), m_y(P_1(f,y)), \dots, m_y(P_d(f,y)))$ est constante sur Y au voisinage de 0.

ii) Le couple de strates (X^o, Y) satisfait les conditions de Whitney en 0 (resp. <u>les conditions de Whitney strictes au voisinage de</u> 0, <u>comme en</u> 1.2, ii)).

Prouvons i) ⇒ ii). Soit $\kappa_f : C_f(X) \to X$ le morphisme conormal associé à f est au plongement local $X \subset \mathbb{C} \times \mathbb{C}^N$. Puisque $\dim \kappa_f^{-1}(Y) < \dim C_f(X)$, on a $\dim \kappa_f^{-1}(Y) \leq N-1$ et donc pour $Y \in Y - \{0\}$ assez proche de 0, $\dim \kappa_f^{-1}(y) \leq N-2$, donc (cf. Chap. IV, 6.1) $P_d(f,y) = \emptyset$. Par l'hypothèse d'équimultiplicité, on a donc $P_d(f,0) = \emptyset$ d'où $\dim \kappa_f^{-1}(0) \leq N-2$. Ainsi il existe un ouvert de Zariski dense $U = \check{\mathbb{P}}^{N-1} - \kappa_f^{-1}(0)$ de l'espace $\check{\mathbb{P}}^{N-1}$ des hyperplans de $\mathbb{C}^N$ tel que tout $H_o \in U$ soit transverse à toutes les limites en 0 d'espaces tangents aux fibres de f. Montrons d'abord que les hypothèses impliquent $P_d(X,0) = \emptyset$. Considérons le morphisme conormal absolu

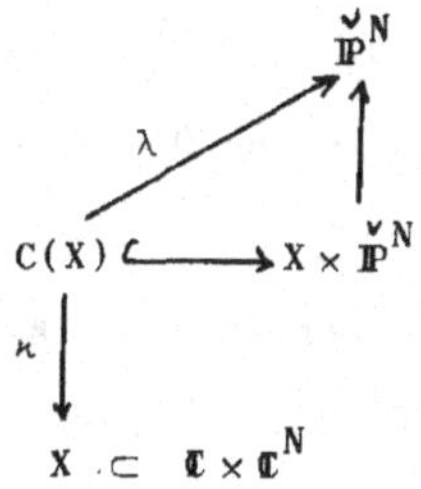

D'après (Chap. IV, § 4) il nous faut montrer que $\lambda^{-1}(L^1) = \emptyset$, où L^1 est une droite projective de $\check{\mathbb{P}}^N$ assez générale pour être transverse à toutes les strates d'une certaine stratification de $\kappa^{-1}(0)$. Or l'hypothèse implique que le point $K \in \check{\mathbb{P}}^N$ correspondant à l'hyperplan $0 \times \mathbb{C}^N$ n'est pas contenu dans $\kappa^{-1}(0)$. Par conséquent une droite L^1 de $\check{\mathbb{P}}^N$ contenant K et au demeurant assez générale sera transverse aux strates d'une stratification donnée de $\kappa^{-1}(0)$. Cela signifie que nous pouvons calculer $P_d(X,0)$ comme $P_d\langle(X,0);D_2\rangle$ où D_2 est un plan de codimension 2 de $\mathbb{C} \times \mathbb{C}^N$ <u>contenu dans</u> $0 \times \mathbb{C}^N$. Or si $P_2\langle(X,0);D_2\rangle$ n'était pas vide, il existerait un chemin analytique $h : (\mathbb{D}, 0) \to (X,0)$, tel que $h(\mathbb{D} - \{0\}) \subset X^o$ et que en tout point $x(t) \in h(\mathbb{D} - \{0\})$ on ait

$$\dim(T_{X,x(t)} \cap D_2) \geq d-1 \quad .$$

Mais puisque $K = 0 \times \mathbb{C}^N$ n'est pas limite d'espaces tangents à X, et que $D_2 \subset 0 \times \mathbb{C}^N$, ceci équivaut à

$$\dim(T_{X,x(t)} \cap K \cap D_2) \geq d-1 \quad , \qquad \text{c'est-à-dire}$$

$$\dim(T_{f^{-1}(f(x(t))} \cap D_2) \geq d-1 \qquad \text{(et } x(t) \text{ non critique pour } f\text{)}$$

ce qui implique, puisque D_2 est général, $P_d(f,0) \neq \emptyset$ contrairement à l'hypothèse. D'après (Chap. IV, 6.1) nous avons donc

$$\dim \varkappa^{-1}(0) \leq N-2$$

ce qui implique que l'ensemble $V = \check{\mathbb{P}}^{N-1} - \varkappa^{-1}(0)$ des hyperplans de $\mathbb{C} \times \mathbb{C}^N$ contenant $\mathbb{C} \times \{0\}$ et transverses aux limites en 0 d'espaces tangents à X^o est un ouvert de Zariski dense.

Montrons maintenant que pour un hyperplan assez général $D_2 = H_o \subset \mathbb{C}^N$ (de codimension 2 dans $\mathbb{C} \times \mathbb{C}^N$), posant $H = \mathbb{C} \times H_o$, l'espace $[X \cap H] = (X \cap H)_{red}$, satisfait encore les hypothèses de i) le long de $\mathbb{C} \times \{0\}$. Du fait que H et K sont transverses aux limites en 0 d'espaces tangents à X^o, et que H_o est transverse aux limites d'espaces tangents aux fibres de f, on déduit que $0 \times H_o$ est transverse aux limites en 0 d'espaces tangents à $[X \cap H]^o$. En effet une telle limite peut s'écrire $T \cap H$ où T est une limite en 0 d'espaces tangents à X^o. Puisque K est transverse à T, $T \cap K$ est une limite d'espaces tangents aux fibres de f, donc on a $\dim(T \cap K \cap H_o) = \dim(T \cap H_o) \leq d-1$. Par contre si H_o contenait $T \cap H$ on aurait $\dim(T \cap H_o) = d$, d'où la contradiction voulue. Les variétés polaires $P_k(f|[X \cap H],y)$ sont équimultiples le long de Y pour $0 \leq k \leq d-1$ d'après (Chap. IV, 5.4.3), et d'après le même résultat, l'on a $P_k([X \cap H],0) = P_k(X,0) \cap H$ pour $0 \leq k \leq d-1$, et les multiplicités sont préservées. Il nous suffit maintenant de prouver par récurrence sur d que l'on a l'égalité

$$m_o(P_k(f,0)) = m_o(P_k(X,0)) \qquad \text{pour } 0 \leq k \leq d \quad .$$

Si $d = 1$, d'après ce que nous venons de voir, nous avons $P_o(f,0) = P_o(X,0)$ et $P_1(f,0) = P_1(X,0) = \emptyset$ d'où le résultat. Supposons l'avoir prouvé pour les morphismes dont les fibres sont de dimension $\leq d-1$. D'après ce que nous venons de voir, nous avons

$$m_o(P_k(f,0)) = m_o(P_k(f|[X\cap H],0) = m_o(P_k([X\cap H],0) = m_o(P_k(X,0))$$

pour $0 \leq k \leq d-1$, la seconde égalité venant de l'hypothèse de récurrence, et $P_d(X,0) = P_d(f,0) = \emptyset$ d'après ce que nous avons vu au début.
L'espace X satisfait donc l'hypothèse i) du Théorème 5.2, d'où ii) d'après le Théorème.

Prouvons ii) $\Rightarrow$ i). La condition a) de Whitney implique aussitôt que toute limite en 0 d'espaces tangents à X^o est transverse à K. Par le même argument que plus haut on peut calculer $P_k(X,0)$ comme $P_k\langle(X,0);D_{d-k+1}\rangle$ avec $D_{d-k+1} \subset K$, et pour $D_{d-k+1} \subset K$ assez général, on a l'égalité $P_k\langle f;D_{d-k+1}\rangle = P_k\langle X;D_{d-k+1}\rangle$ comme on le vérifie aussitôt, d'où le résultat, puisque d'après le Théorème 5.2, les $P_k\langle X,D_{d-k+1}\rangle$ sont équimultiples le long de Y. ∎

***

CHAPITRE VI

CONSEQUENCES

Introduction. Dans ce chapitre, j'indique brièvement quelques-unes des conséquences des résultats du chapitre précédent. La plus importante me semble être la "réciproque du théorème de Thom-Mather".

§ 1. Exemples.

1.1 Le cas des familles d'hypersurfaces à singularité isolée.

Soit $F(y, z_1, \dots, z_{d+1}) = 0$ une équation pour une hypersurface $X \subset \mathbb{C} \times \mathbb{C}^{d+1}$ telle que chacune des fibres $X(y) = (\{y\} \times \mathbb{C}^{d+1}) \cap X$ soit une hypersurface à singularité isolée en 0. Soit $D_{d-k+1} \subset \mathbb{C}^{d+1}$ un sous-espace vectoriel de codimension $d-k+1$. Choisissons les coordonnées de façon qu'il soit défini par $z_1 = \dots = z_{d-k+1} = 0$. Notons $f : X \to \mathbb{C}$ la projection induite par $pr_1 : \mathbb{C} \times \mathbb{C}^{d+1} \to \mathbb{C}$ et σ la section qui pique 0 dans chaque fibre. L'hypothèse implique que $(\frac{\partial F}{\partial z_1}, \dots, \frac{\partial F}{\partial z_{d+1}}) \supset (z_1, \dots, z_{d+1})^N$ pour N assez grand, et que $(\frac{\partial F}{\partial z_1}, \dots, \frac{\partial F}{\partial z_{d+1}})$ définit un sous-espace de codimension $d+1$ de $\mathbb{C}^{d+2}$, donc est une suite régulière. Pour un choix assez général de D_{d-k+1}, on vérifie d'après (Chap. IV, 1.3) que la variété polaire relative $P_k\langle f; D_{d-k+1}\rangle$ pour $0 \leq k \leq d-1$ est le sous-espace de X défini dans $\mathbb{C}^{d+1}$ par l'idéal engendré par $(F, \frac{\partial F}{\partial z_{d-k+2}}, \dots, \frac{\partial F}{\partial z_{d+1}})$. Pour la même raison que plus haut, si $P_k\langle f; D_{d-k+1}\rangle \neq \emptyset$, ce système de générateurs est une suite régulière. Pour $k = d$, la variété polaire relative est la réunion des composantes irréductibles de la courbe Γ définie par l'idéal $(F, \frac{\partial F}{\partial z_2}, \dots, \frac{\partial F}{\partial z_{d+1}})$ qui ne sont pas contenues dans $\mathbb{C} \times \{0\}$. Aucune de ces composantes irréductibles ne peut être contenue dans l'hyperplan $0 \times \mathbb{C}^{d+1}$.

Si tel était le cas, on aurait en effet un arc analytique $h : (\mathbb{D},0) \to (X(0),0)$, non constant et tel qu'en un point général x de son image $T_{X(0),x} \subset D_1$, où D_1 est un hyperplan général de $\mathbb{C}^{d+1}$; on vérifie cela en utilisant le fait que $F(0,z_1,\ldots,z_{d+1}) = 0$ est à singularité isolée. Or un hyperplan général de $\mathbb{C}^{d+1}$ ne peut être limite d'espaces tangents à $X(0)^o$, puisque la fibre de la modification de Nash de X(0) est de dimension $\leq d-1$. Par conséquent la suite $(F, \frac{\partial Z}{\partial z_2}, \ldots, \frac{\partial F}{\partial z_{d+1}}, y)$ est encore une suite régulière, et le morphisme $\varphi : \Gamma \to \mathbb{C}$ induit par y est un morphisme fini et plat, et la somme des longueurs d'anneaux artiniens

$$\sum_{x \in \varphi^{-1}(y)} \dim_{\mathbb{C}}(\mathcal{O}_{\varphi^{-1}(y),x})$$

est indépendante de $y \in \mathbb{C}$.

D'après [Te 1, Chap. II], pour $x \in \varphi^{-1}(y) \cap (\mathbb{C} \times \{0\})$, on a

$$\dim_{\mathbb{C}}(\mathcal{O}_{\varphi^{-1}(y),x}) = \mu^{(d+1)}(X(y)) + \mu^{(d)}(X(y))$$

où $\mu^{(i)}(X(y))$ est le nombre de Milnor en 0 de l'intersection de X(y) avec un plan de dimension i assez général de $\mathbb{C}^{d+1} \simeq \{y\} \times \mathbb{C}^{d+1}$ passant par 0. Par différence il vient

$$m_o(P_d(f,0)) = \mu^{(d+1)}(X(0)) + \mu^{(d)}(X(0)) - (\mu^{(d+1)}(X(y)) + \mu^{(d)}(X(y)))$$

pour $y \neq 0$ assez petit.

Un argument analogue permet de vérifier que pour $y \in \mathbb{D}$ on a, en notant $\underline{y}$ le point $\{y\} \times 0$ de $\mathbb{C} \times \{0\}$

$$m_{\underline{y}}(P_k(f;0)) = \mu^{(k+1)}(X(y)) + \mu^{(k)}(X(y)) \quad .$$

On voit que dans ce cas particulier le Théorème 2.1 du Chapitre V signifie que l'<u>on a équivalence entre</u>

1) La suite $\mu^*_y = (\mu^{(d+1)}(X(y)),\dots,\mu^{(0)}(X(y)))$ est indépendante de $y \in \mathbb{C}$ au voisinage de 0 .

2) Le couple $(X^o, \mathbb{C}\times\{0\})$ satisfait les conditions de Whitney en 0 (resp. les conditions de Whitney strictes au voisinage de 0).

1.2 Le cas d'une famille de courbes où la dimension de plongement saute.

Considérons la surface $X \subset \mathbb{C}\times\mathbb{C}^3$ définie par les deux équations

$$F_1 = z_2^2 - z_1^3 + yz_3 = 0$$

$$F_2 = z_3^2 - z_1^5 z_2 = 0 \quad .$$

Soit $f : X \overset{\sigma}{\rightleftarrows} \mathbb{C}$ le morphisme induit par la projection $pr_1 : \mathbb{C}\times\mathbb{C}^3 \to \mathbb{C}$ et soit σ la section définie par $\sigma(y) = (y,0,0,0)$.

La fibre X(0) est la courbe monomiale donnée paramétriquement par $z_1 = t^4$, $z_2 = t^6$, $z_3 = t^{13}$, alors que pour tout $y \neq 0$, la fibre X(y) est isomorphe à la courbe plane d'équation $(z_2^2 - z_1^3)^2 - z_1^5 z_2 = 0$ qui a deux exposants caractéristiques de Puiseux (3/2 et 7/4). Il est clair que X est équimultiple de multiplicité 4 le long de $\mathbb{C}\times\{0\}$. Le lecteur est prié de vérifier que la courbe polaire $P_1(X,0)$ est vide. Par conséquent le couple $(X^o, \mathbb{C}\times\{0\})$ satisfait les conditions de Whitney, bien que la dimension de plongement des fibres saute pour $t = 0$.

1.2.1 Remarque : La théorie d'où est tiré cet exemple se trouve dans l'appendice à [Z 2]. On pourrait aussi, au vu de (Loc. cit.) vérifier que (X^o, Y) satisfait les conditions de Whitney en utilisant (Chap. III, Prop. 5.2.3). Le lecteur pourra vérifier qu'une résolution simultanée forte des singularités de X est décrite par : $z_1 = t^4$, $z_2 = t^6 - \frac{y}{2}t^7 + \frac{y^2}{16}t^{16} + \dots$, $z_3 = t^{13} - \frac{y}{4}t^{15} + \dots$

§ 2. Réponse à une question de Zariski.

Dans ([Z 3], § 3, Question C) Zariski demande si, étant donné deux sous-espaces X et Y d'un espace analytique non-singulier Z, l'ensemble B(X,Y) des points $y \in Y$ en lesquels le couple (X^{o},Y) ne satisfait pas les conditions de Whitney est un sous-ensemble analytique fermé de Y. (En fait Zariski étudie le cas où X est une hypersurface, et il avait prouvé que la réponse est affirmative si $\dim X = \dim Y + 1$.)

2.1 Proposition : Etant donné un sous-ensemble analytique fermé Y d'un espace analytique réduit équidimensionnel X, l'ensemble $B(X,Y)$ des points $y \in Y$ en lesquels le couple (X^{o},Y) ne satisfait pas les conditions de Whitney est un sous-ensemble analytique fermé rare de Y.

Démonstration : Décomposant Y en composantes irréductibles, on voit que l'on peut supposer Y équidimensionnel. La question est locale sur Y. On peut donc choisir un plongement local $X \subset \mathbb{C}^N$ et considérer le diagramme :

$$\begin{array}{ccc} E_Y C(X) & \xrightarrow{\tilde{e}_Y} & C(X) \\ {\scriptstyle \varkappa'_Y}\downarrow & \searrow^{\zeta} & \downarrow{\scriptstyle \varkappa} \\ E_Y X & \xrightarrow[e_Y]{} & X \subset \mathbb{C}^N \end{array}$$

où $\varkappa$ est le morphisme conormal, e_Y l'éclatement de Y dans X et $\varkappa'_Y$ le morphisme transformé strict de $\varkappa$ par e_Y.
D'après l'existence de stratifications par l'équidimensionnalité des fibres (cf. [Le-T 2]) l'ensemble $C(X,Y) = \{z \in \zeta^{-1}(Y)/\dim_z \zeta^{-1}(\zeta(z)) > N-2-\dim Y\}$ est un sous-ensemble analytique fermé de $E_Y C(X)$. Puisque le morphisme ζ est propre, son image $\zeta(C(X,Y))$ est un sous-ensemble analytique fermé de Y, rare dans Y puisque $\dim \zeta^{-1}(Y) = N-2$. D'après (Chap. V, 1.2) l'ensemble $B(X,Y)$ des points où (X^{o},Y) ne satisfait pas les conditions de Whitney est :

$$B(X,Y) = (\mathrm{Sing}\ Y) \cup \zeta(C(X,Y))$$

d'où le résultat. ■

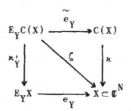

Remarque : Dans le cas où X est une hypersurface et Y son lieu singulier, supposé non singulier, une réponse affirmative avait été donnée dans [B-H-S].

§ 3. La stratification de Whitney canonique d'un espace analytique.

Rappelons la construction de [Lê-Te 1, 6.1.5] (voir aussi Chap. III, § 3). Notons comme en (Chap. IV, 3.2) $M^*_{X,x}$ la suite $(m_x(X), m_x(P_1(X,x)), \dots, m_x(P_{d-1}(X,x)))$ des multiplicités des variétés polaires associées au germe (X,x).
Etant donné un espace analytique complexe réduit localement équidimensionnel X définissons par récurrence une filtration de X par des sous-espaces analytiques fermés :

$$X = F_o \supset F_1 \supset \dots \supset F_k \supset \dots$$

comme ceci :

$F_o = X$; soit $X = \bigcup F_{o,j}$ la décomposition de X en composantes irréductibles
$F_1 = \bigcup_j \{x \in F_{o,j} / M^*_{X,x}$ ne prend pas la valeur qu'il prend en un point général de $F_{o,j}\}$.
Notons que cette valeur est $(1,0,0,\dots,0)$ avec $\dim F_{o,j} - 1$ zéros et que $F_1 = \mathrm{Sing}\, X$.
Supposons avoir défini $F_o, F_1, \dots, F_{k-1}$, et définissons F_k :
$F_k = \bigcup_{0 \le i \le k-2} B(F_i, F_{k-1})$ avec la notation introduite au § 2 ci-dessus, et l'on vérifie au moyen de (Chap. V, 1.2) que l'on peut aussi définir F_k comme ceci : soit $F_{k-1} = \bigcup F_{k-1,j_{k-1}}$ la décomposition de F_k en ses composantes irréductibles ; F_k est la réunion du lieu singulier de F_{k-1} et des sous-ensembles analytiques fermés en nombre fini définis comme suit: $\{x \in F_{k-1,j_{k-1}}$ / l'une des suites $M^*_{X,x}, M^*_{F_1,j_1,x}, \dots, M^*_{F_{k-2},j_{k-2},x}$ ne prend pas en x la valeur qu'elle prend en un point général de $F_{k-1,j_{k-1}}\}$.
D'après (Chap. V, 1.2) on a donc :

3.1 Proposition-Définition : La réunion des composantes connexes des $F_j - F_{j+1}$ $(0 \leq j)$ est une stratification de Whitney de X, appelée stratification de Whitney canonique de X.

Il suffit de remarquer que puisque F_{k+1} est rare dans F_k, la filtration stationne localement, et en fait globalement si X est de dimension bornée, et les composantes connexes des $F_j - F_{j+1}$ sont bien des sous-ensembles analytiques non-singuliers constructibles dans X.

3.2 Proposition : Etant donné une stratification de Whitney quelconque $X = \bigcup X_\alpha$ de X, chacun des fermés F_i définis plus haut est réunion de strates X_α.

Démonstration : On peut supposer les X_α connexes. Remarquons d'abord que si un fermé $F \subset X$ est réunion de strates d'uns stratification de Whitney (X_α), chaque composante irréductible de F est aussi réunion de strates. En effet chaque composante de F contient une strate X_α qui est dense dans cette composante, et donc si la strate X_β rencontre cette composante, elle rencontre l'adhérence de X_α, et donc est dedans, d'après la condition de frontière. L'espace entier $X = F_0$ est réunion de strates. Supposons, par récurrence, avoir montré que F_k est réunion de strates pour $k < i_0$ et montrons que F_{i_0} est réunion de strates Pour cela, il suffit de prouver que si $X_\alpha \cap F_{i_0} \neq \emptyset$, on a $X_\alpha \subset F_{i_0}$. Or si $X_\alpha \cap F_{i_0} \neq \emptyset$, d'après l'hypothèse de récurrence, X_α est une strate d'une stratification de Whitney de chaque F_k, $k < i_0$, et même de chaque composante irréductible F_{k,j_k} de F_k qui le contient, puisque X_α est connexe. Supposons d'abord que $X_\alpha \cap \mathrm{Sing}(F_{i_0-1}) \neq \emptyset$. Si X_α n'est pas tout entier contenu dans $\mathrm{Sing}(F_{i_0-1}) \subset F_{i_0}$, la fonction $M^*_{F_{i_0-1,j}}$ de la composante irréductible $F_{i_0-1,j}$ de F_{i_0-1} qui contient X_α ne saurait être constante le long de X_α, contrairement à l'hypothèse selon laquelle X_α est une strate de $F_{i_0-1,j}$. Nous pouvons donc supposer que $X_\alpha \cap \mathrm{Sing}(F_{i_0-1}) = \emptyset$. Par conséquent si $X_\alpha \cap F_{i_0} \neq \emptyset$, par définition de F_{i_0}, cela signifie que pour un $k < i_0$, la fonction $M^*_{F_{k,j_k}}$ d'une composante irréductible F_{k,j_k} de F_k ne prend pas en $x_0 \in X_\alpha \cap F_{i_0}$ la valeur qu'elle prend en un point général d'une composante $F_{i_0-1,j}$. En particulier

$x_o \in F_{k,j_k}$ par semi-continuité (cf. Chap. IV, 6.1) et donc $X_\alpha \subset F_{k,j_k}$ d'après la remarque. Si la valeur que prend la fonction $M^*_{F_{k},j_k}$ en un point général de X_α est égale à celle qu'elle prend en un point général de $F_{i_o-1,j}$, nous obtenons aussitôt une contradiction si $X_\alpha \not\subset F_{i_o}$ puisque $M^*_{F_{k},j_k}$ doit être constante le long de X_α. Sinon, on a l'inclusion $X_\alpha \subset F_{k+1}$, donc la fonction $M^*_{F_{k+1},j_{k+1}}$ doit être constante le long de X_α, où $F_{k+1,j_{k+1}}$ est la composante irréductible de F_{k+1} qui contient X_α. On répète l'argument ci-dessus, ce que l'on peut faire tant que $k \leq i_o - 1$, et l'on prouve ainsi finalement l'inclusion $X_\alpha \subset F_{i_o}$. ∎

Remarque : Cela signifie que la stratification canonique est minimale parmi toutes les stratifications de Whitney de X.

Remarque : Le concept de stratification de Whitney canonique a été introduit par J. Mather (cf. [Ma], § 4) : il construit une stratification canonique, par un procédé dont celui décrit ici est un avatar, et montre que toute stratification de Whitney moins fine que la stratification canonique coïncide avec elle. On vient de voir, dans le cas analytique complexe, un résultat beaucoup plus précis : toute stratification de Whitney est plus fine que la stratification canonique.

§ 4. Réciproque du théorème de Thom-Mather (Tout ce paragraphe présente un travail en collaboration avec Lê D.T.).

D'abord quelques rappels :

4.1 Proposition ([Lê-Te], Prop. 6.1.8) : Soient X un espace analytique réduit, purement de dimension d, et $X = \cup X_\alpha$ une stratification de Whitney de X. Soient $x \in X_\alpha$ et $(X,x) \subset (\mathbb{C}^N, x)$ un plongement local de X au voisinage de x. Pour chaque entier $i \geq d_\alpha$, où $d_\alpha = \dim X_\alpha$, il existe un ouvert de Zariski dense W de la grassmannienne des plans de codimension i de $\mathbb{C}^N$, et pour chaque $L_o \in W$ un nombre réel $\eta > 0$ et un ouvert dense U de la boule de centre x et de rayon η dans $\mathbb{C}^N$, un nombre réel $\varepsilon \ll \eta$ tels que, notant $\mathbb{B}_\varepsilon$ la boule de centre x et de rayon ε dans $\mathbb{C}^N$, on ait : pour $t \in U$, la caractéristique d'Euler-Poincaré $\chi(X \cap (L_o + t) \cap \mathbb{B}_\varepsilon)$ ne dépend que de (X, X_α). On la notera $\chi_{i-d_\alpha}(X, X_\alpha)$. (Ici $L_o + t$ représente un plan parallèle à la direction L_o, passant par le point t

4.2 Théorème ([Lê-Te], Th. 6.1.9) : Soit $X = \cup X_\alpha$ une stratification de Whitney de X, et soit X_α la strate contenant $x \in X$. On a l'égalité

$$\chi_1(X, X_\alpha) - \chi_2(X, X_\alpha) = \sum_{\beta \neq \alpha} (-1)^{d_\beta - d_\alpha - 1} m_x(P_{d_\beta - d_\alpha - 1}(\overline{X}_\beta, x))(1 - \chi_1(X, X_\beta)) .$$

Remarques : 1) Dans (Loc. cit.) ce théorème est énoncé pour la stratification canonique. La preuve vaut en fait pour n'importe quelle stratification de Whitney.

2) Le cas particulier $i = d_\alpha + 1$ de la construction de 4.1, c'est-à-dire l'intersection $X \cap (L_o + t) \cap \mathbb{B}_\varepsilon$ avec codim $L_o = \dim X_\alpha + 1$ est ce que Goresky et MacPherson appellent le "complex link" de (X, X_α), objet important dans leur théorie de Morse sur les espaces singuliers, (cf. [G-M]).

3) Le Théorème 4.2 est la généralisation des formules de ([Te 1], Chap. I et II et [Te 12]) exprimant la multiplicité des variétés polaires relatives d'une hypersurface à singularité isolée $(X,0) \subset (\mathbb{C}^{d+1}, 0)$ en fonction de la suite μ^* des nombres de Milnor (cf. [Mi]) des sections planes générales de X par des plans généraux passant par 0. ($\mu^{(i)} = \mu(X \cap H^i, 0)$ où $\dim H^i = i$). Soit $f : (\mathbb{C}^{d+1}, 0) \to (\mathbb{C}, 0)$ un morphisme dont (X,0) est la fibre. On a

$$P_k(X,0) = P_k(f,0) \cap X \quad \text{pour } 0 \leq k \leq d-1$$

et dans (Loc. cit.) on prouve :

$$m_o(P_k(X,0)) = \mu^{(k+1)}(X,0) + \mu^{(k)}(X,0) .$$

Par ailleurs, en adaptant ([Mi]) on prouve que

$$\chi_i(X, \{0\}) = 1 + (-1)^{d-i} \mu^{(d+1-i)}(X,0) \quad \text{(avec les notations de 4.1)}$$

ce qui vérifie bien le Théorème 4.2 dans ce cas, avec $X_\alpha = \{0\}$, $X_\beta = X - \{0\}$.

4.3 Rappelons maintenant le Théorème de Thom-Mather.

4.3.1 Théorème (Thom-Mather dans le cas particulier des espaces analytiques complexes (cf. [Th], [Ma]) : Soit $X = \cup X_\alpha$ un espace analytique muni d'une stratification de Whitney. Pour tout point $x \in X_\alpha$, tout plongement local $X \subset \mathbb{C}^N$ au voisinage de x, toute rétraction locale $\rho : \mathbb{C}^N \to X_\alpha$, il existe un voisinage ouvert U de x dans $\mathbb{C}^N$ tel que, en posant $V = U \cap X_\alpha$ on ait : il existe un homéomorphisme compatible avec ρ et

$$\begin{array}{ccc} X \cap U & \simeq & (\rho^{-1}(x) \cap X \cap U) \times V \\ & {\scriptstyle \rho | X\cap U} \searrow \quad \swarrow {\scriptstyle pr_2} & \\ & V & \end{array}$$

induisant pour chaque $\overline{X}_\beta$ contenant X_α un homéomorphisme analogue

$$\begin{array}{ccc} \overline{X}_\beta \cap U & \simeq & (\rho^{-1}(x) \cap \overline{X}_\beta \cap U) \times V \\ & {\scriptstyle \rho | \overline{X}_\beta \cap U} \searrow \quad \swarrow {\scriptstyle pr_2} & \\ & V & \end{array}$$

Remarquons maintenant que d'après (Chap. V, 1.2) on a, en posant $d_\alpha = \dim X_\alpha$

4.3.2 Proposition : Si (X_β, X_α) satisfait les conditions de Whitney, pour tout plongement local $(\overline{X}_\beta, x) \subset (\mathbb{C}^N, x)$ au voisinage de $x \in X_\alpha \subset \overline{X}_\beta$, il existe un ouvert de Zariski dense N de l'espace des drapeaux

$$\mathcal{D} : T_{X_\alpha, x} \subset D_{N-d_\alpha-1} \subset D_{N-d_\alpha-2} \subset \dots \subset D_o = \mathbb{C}^N$$

d'espaces vectoriels contenant $T_{X_\alpha, x}$ tel que pour $\mathcal{D} \in W$ et tout sous-espace non-singulier H_i de codimension i contenant X_α et tel que $T_{H,x} = D_i$, on ait :

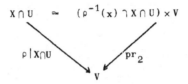

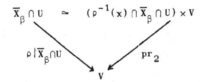

le couple de strates $([X_\beta \cap H_i]^o, Y)$ satisfait les conditions de Whitney.

En aboutant 4.3.2 à 4.3.1, on voit que l'on peut renforcer le théorème de Thom-Mather : Si (X_α) est une stratification de Whitney, non seulement chaque $\overline{X}_\beta$ est localement topologiquement trivial le long de X_α, au sens de 4.3.1, mais encore pour tout sous-espace non-singulier assez général H_i contenant X_α, l'intersection $\overline{X}_\beta \cap H_i$ est aussi localement topologiquement triviale le long de X_α. C'est cet énoncé renforcé qui admet une réciproque.

<u>4.4</u> <u>Théorème</u> (Lê-Teissier) : <u>Soit</u> X <u>un espace analytique complexe réduit purement de dimension</u> d, <u>et soit</u> $X = \bigcup X_\alpha$ <u>une partition de</u> X <u>en sous-espaces analytiques non-singuliers constructibles dans</u> X.
<u>Les conditions suivantes sont équivalentes</u> :

1) <u>Pour tout</u> $x \in X_\alpha$ <u>et tout plongement local</u> $(X,x) \subset (\mathbb{C}^N, x)$ <u>il existe un ouvert de Zariski dense</u> W <u>de l'espace des drapeaux</u>

$$\mathcal{D} : T_{X_\alpha, x} \subset D_{N-d_\alpha-1} \subset \dots \subset D_1 \subsetneq D_o = \mathbb{C}^N$$

<u>tel que pour</u> $\mathcal{D} \in W$ <u>et</u> H_i <u>non singulier de codimension</u> i <u>contenant</u> X_α <u>et avec</u> $T_{H_i, x} = D_i$, $\overline{X}_\beta \cap H_i$ <u>soit localement topologiquement trivial le long de</u> X_α <u>en</u> x, <u>pour tout</u> X_β <u>et</u> $0 \leq i \leq N-d_\alpha-1$.

2) <u>La décomposition</u> $X = \bigcup X_\alpha$ <u>est une stratification de Whitney de</u> X.

La démonstration détaillée sera donnée ailleurs. Nous venons de voir comment 2) ⇒ 1) résulte du Théorème de Thom-Mather. La preuve de 1) ⇒ 2) s'appuie essentiellement sur les résultats du Chapitre V et le Théorème 4.2 ci-dessus.

<u>Remarque</u> : Il serait plus simple de ne regarder que les plongements locaux $X \subset \mathbb{C}^N$ tels que l'image de X_α soit linéaire et de regarder alors les intersections $X \cap D_i$.

4.5 Réalisation de la partie B) du programme (cf. Introduction). Etant donné un espace analytique complexe X muni d'une stratification de Whitney $X = \cup X_\alpha$ il résulte de l'égalité de 4.2 par une récurrence facile que la collection des $\chi_i(X_\beta, X_\alpha)$ où $x \in X_\alpha \subset \overline{X}_\beta$ détermine la collection des multiplicités $m_x(P_k(\overline{X_\beta}, x))$ en x des variétés polaires des adhérences de strates $\overline{X_\beta}$ et est déterminée par elle. Ainsi la suite $M^*_{X,x}$ est bien déterminée par la topologie de l'ensemble stratifié X et de ses sections planes locales générales, au voisinage de x.

§ 5. Rapport avec le cas des hypersurfaces.

5.1 Proposition : Soient $X \subset \mathbb{C}^N$ un espace analytique réduit purement de dimension d, $Y \subset X$ un sous-espace analytique fermé réduit et $0 \in Y$ un point non-singulier de Y. Les conditions suivantes sont équivalentes :

i) Le couple (X^o, Y) satisfait les conditions de Whitney en 0.

ii) Il existe un ouvert de Zariski dense U de l'espace des projections linéaires $p : \mathbb{C}^N \to \mathbb{C}^{d+1}$ tel que pour $p \in U$, en posant $X_1 = p(X)$ on ait : X_1 est une hypersurface réduite de $\mathbb{C}^{d+1}$, $Y_1 = p(Y)$ est non-singulier en $0 = p(0)$ et le couple (X_1^o, Y_1) satisfait les conditions de Whitney en 0.

Démonstration : Cela résulte aussitôt de (Chap. V, 2.1 et Chap. IV, 5.5). On pourrait aussi utiliser les inégalités métriques de (Chap. III, 2.1).

Remarque : On prendra bien garde que l'inclusion $X_1^o \subset p(X^o)$ est en général stricte : la projection p "crée des singularités" et en particulier des points doubles évanescents sur X_1.

§ 6. Sur l'équisingularité à la Zariski (cf. [Z 1], [Z 4]).

Très sommairement, étant donné $X \subset \mathbb{C}^N$, un objet essentiel pour la théorie de Zariski est l'image Δ_1 par une projection générique $p : \mathbb{C}^N \to \mathbb{C}^d$ de la variété polaire $P_1\langle p\rangle$ (cf. Chap. IV, § 1). A priori, un avantage du point de vue des variétés polaires est que l'on n'a pas à prendre d'image, donc on n'a pas à

faire d'élimination, et elles sont plus calculables. Dans les cas très simples, par exemple lorsque Y est de codimension 1, la variété polaire $P_1<p>$ (resp. Δ_1) n'intervient que par le fait d'être vide ou non, et les résultats de transversalité du Chapitre IV font que les deux théories sont pratiquement interchangeables. Un excellent texte à consulter pour la généralisation du critère discriminant de Zariski au cas d'une famille de courbes quelconque est [Bu]. Dans le cas général, le rapport est encore bien mystérieux mais je crois qu'une conséquence des résultats de ce travail-ci est que dans le cas général le fait que $Y \subset X$ soit une strate d'une stratification de Whitney de X est <u>beaucoup plus faible</u> que le fait que X soit équisingulier "à la Zariski" le long de Y.
De ce point de vue, le problème de déterminer le rapport entre les conditions de Whitney et l'équisingularité à la Zariski , et le problème de caractériser "numériquement" l'équisingularité à la Zariski,(voir [B-H]) ou l'équisingularité au sens de la résolution simultanée forte (cf. Chap. III, 5.3.1) comme j'ai réussi à le faire ici pour les conditions de Whitney, restent largement ouverts . Il me semble que les variétés polaires locales joueront encore un rôle dans ces questions et je veux terminer par un problème dont la solution ferait, je crois, avancer l'étude du rapport entre variétés polaires et équisingularité à la Zariski.

<u>Définition</u> : Soit $Z \subset \mathbb{C}^N$ un sous-espace analytique réduit de dimension pure c dans $\mathbb{C}^N$. Soit $p: \mathbb{C}^N \to \mathbb{C}^{c+1}$ une projection linéaire. Nous dirons que p est une projection générale pour Z, relativement à une définition de l'équisingularité, si pour tout morphisme $h: \mathbb{D} \to \mathcal{P}$, où $\mathcal{P}$ désigne l'espace des projections linéaires de $\mathbb{C}^N$ dans $\mathbb{C}^{c+1}$, telle que $h(0) = p$, la famille des images $(h(t)(Z),t) \subset \mathbb{C}^{c+1} \times \mathbb{C}$ est équisingulière le long de $0 \times \mathbb{C}$ en 0×0.

<u>Problème</u> : Etant donné $X \subset \mathbb{C}^N$, montrer que pour tout k, $0 \leq k \leq d-1$, il existe un ouvert de Zariski dense W_k de l'espace des projections linéaires $p: \mathbb{C}^N \to \mathbb{C}^{d-k+1}$ tel que pour $p \in W_k$, la projection p soit générale pour la variété polaire $P_k<p>$, du point de vue des conditions de Whitney, ou mieux encore

de la résolution simultanée forte.

<u>Remarque</u> : Le cas $k = d-1$ est une partie essentielle du Lemme-clé du Chapitre V, et est dû à Briançon-Henry dans le cas $d = 2$, singularité isolée [B-H].

B I B L I O G R A P H I E

[Al] A.D. ALEKSANDROV : Theory of mixed volumes, 4 articles dans Mat. Sbornik 44 (N.S., t.2), p.947-972 et 1205-1238, et 45 (N.S., t. 2), p. 27-46 et 227-251. Une traduction par le Prof. J. Firey, en 1966-67, (Dept. of Math. University of Oregon) m'a aimablement été envoyée par le Prof. R. Schneider.

[Bbk 1] N. BOURBAKI : Algèbre commutative, Chap. V, § 1, Hermann, Paris.

[Bbk 2] N. BOURBAKI : Algèbre commutative, Chap. VI, § 1, Hermann, Paris.

[Bbk 3] N. BOURBAKI : Algèbre commutative, Chap. VIII, Masson, Paris (en préparation).

[B-C] F. BRUHAT et H. CARTAN : Sur la structure des sous-ensembles analytiques réels, Note aux C.R. Acad. Sc. Paris, t. 244 (1957) 988-990.

[B-G-G] J. BRIANÇON, A. GALLIGO et J.M. GRANGER : Déformations équisingulières des germes de courbes gauches réduites, Mémoire de la Société Mathématique de France, Nouvelle série, N° 1, 1980.

[B-H] J. BRIANÇON et J.P.G. HENRY : Equisingularité générique des familles de surfaces à singularités isolées, Bull. S.M.F., 108, 2 (1980) 259-281.

[B-H-S] J. BRIANÇON, J.P.G. HENRY et J.P. SPEDER : Les conditions de Whitney en un point sont analytiques, Note aux C.R. Acad. Sc. Paris, t. 282 (1976) 279.

[Bo] E. BÖGER : Zur theorie der saturation bei analytischen algebren, Math. Annalen, 221 (1974) 119-143.

[B-S] J.P. BRASSELET et M.H. SCHWARTZ : Sur les classes de Chern d'un ensemble analytique complexe, Astérisque N° 82-83 (1981) 93-147 (S.M.F.).

[B-Sk] J. BRIANÇON et H. SKODA : Sur la clôture intégrale d'un idéal de germes de fonctions holomorphes en un point de $\mathbb{C}^4$, Note aux C.R. Acad. Sc. Paris, t. 278 (1974) 949-951.

[Bu] R.O. BUCHWEITZ : On Zariski's criterion for equisingularity and non-smoothable monomial curves, Thèse d'Etat, Paris VII (1981), voir aussi Proc. A.M.S. Symp. on Singularities, Arcata 1981 (à paraître).

[Ca] Séminaire H. CARTAN 1960-1961, Publications de l'Institut Henri Poincaré.

[Ch] D. CHENIOT : : Sur les sections transversales d'un ensemble stratifié, Note aux C.R. Acad. Sc. Paris, t. 275 (1972), 915-916.

[G-H] P. GRIFFITHS et J. HARRIS : Principles of algebraic geometry, John Wiley, 1978.

[G-He] M. GIUSTI et J.P.G. HENRY : Minorations de nombres de Milnor, Bull. S.M.F., 108 (1980) 17-45.

[G-M] M. GORESKY et R. MacPHERSON : Morse theory on singular spaces, Preprint, Brown University 1980.

[Gr-R] L. GRUSON et M. RAYNAUD : Critères de platitude et de projectivité, Inventiones Math., 13 (1971) 1-89.

[H 1] H. HIRONAKA : Normal cones in analytic Whitney stratifications, Publ. Math. I.H.E.S. N° 36, P.U.F. 1970 (volume dédié à O. Zariski).

[H 2] H. HIRONAKA : Stratification and flatness, in Proc. Nordic Summer School "Real and complex singularities", Oslo 1976, Sijthoff and Noordhoff 1977.

[H 3] H. HIRONAKA : Introduction to real-analytic sets and real-analytic maps, Publ. Istituto Matematico "L. Tonelli" dell'Universita di Pisa, Pisa (1973).

[H-L-T] H. HIRONAKA, M. LEJEUNE-JALABERT et B. TEISSIER : Platificateur local et aplatissement local en géométrie analytique, in "Singularités à Cargèse", 1972, Astérisque N° 7-8 (1973).

[H-M] J.P.G. HENRY et M. MERLE : Limites d'espaces tangents et transversalité de variétés polaires, Actes de cette Conférence.

[Ke] G. KEMPF : On the geometry of a theorem of Riemann, Annals of Math., 98 (1973) 178-185.

[Kl 1] S. KLEIMAN : On the transversality of a general translate, Compositio Math. 28 (1974) 287-297.

[Kl 2] S. KLEIMAN : The enumerative theory of singularities, in Proc. Nordic Summer School "Real and complex singularities", Oslo 1976, Sijthoff and Noordhoff 1977.

[Le-T 1] M. LEJEUNE-JALABERT et B. TEISSIER : Dépendance intégrale sur les idéaux et équisingularité, Séminaire Ecole Polytechnique 1974, Publ. Inst. Fourier, St. Martin d'Hères F-38402 (1975).

[Le-T 2] M. LEJEUNE-JALABERT et B. TEISSIER : Normal cones and sheaves of relative jets, Compositio Math. 28, 3 (1974) 305-331.

[Lê-Te 1] LÊ Dũng Tráng et B. TEISSIER : Variétés polaires locales et classes de Chern des variétés singulières, Annals of Math., 114 (1981) 457-491.

[Li 1] J. LIPMAN : Reduction, blowing-up and multiplicities, Preprint Purdue University 1980, à paraître in Proceedings Conference on transcendental methods in Commutative algebra, George Mason University 1979.

[Li 2] J. LIPMAN : Relative Lipschitz saturation, American J. of Math., 97, 3 (1975) 791-813.

[Li-S] J. LIPMAN et A. SATHAYE : Jacobian ideals and a theorem of Briançon-Skoda, Michigan Math. J., 28 (1981) 199-222.

[Li-T] J. LIPMAN et B. TEISSIER : Pseudo-rational local rings and a theorem of Briançon-Skoda, Michigan Math. J., 28 (1981) 97-116.

[Ma] J. MATHER : Stratifications and mappings, in "Dynamical Systems", Academic Press 1973.

[Mi] J. MILNOR : Singular points of complex hypersurfaces, Princeton Univ. Press 1968.

[Na] M. NAGATA : Note on a paper of Samuel, Mem. Coll. Sc. Univ. of Kyoto, t. 30 (1956-57).

[Na 1] V. NAVARRO : Conditions de Whitney et sections planes, Inventiones Math., 61, 3 (1980), 199-266.

[Na 2] V. NAVARRO : Sur les multiplicités de Schubert locales des faisceaux algébriques cohérents, Preprint Univ. Politecnica de Barcelona

[No] A. NOBILE : Some properties of the Nash blowing-up, Pacific J. Math. 60 (1975), 297-305.

[P] F. PHAM : Fractions lipschitiziennes et saturation de Zariski, Actes du Congrès International des Mathématiciens, Nice 1970, tome 2, p. 649-654, Gauthier-Villars, Paris 1971.

[P-T] F. PHAM et B. TEISSIER : Fractions lipschitziennes d'une algèbre analytique complexe et saturation de Zariski, Preprint Centre de Mathématiques, Ecole Polytechnique, 1969.

[Pi] R. PIENE : Polar classes of singular varieties, Ann. Sc. E.N.S., 11 (1978).

[Ra] C.P. RAMANUJAM : On a geometric interpretation of multiplicity, Inventiones Math.,22 (1973) 63-67.

[Re] D. REES : A-transform of local rings and a theorem on multiplicities, Proc. Camb. Phil. Soc., 57 (1961) 8-17.

[R-S] D. REES et R.Y. SHARP : On a theorem of B. Teissier on multiplicities of ideals in local rings, J. London Math. Soc. 2nd Series, Vol. 18, part 3 (1978) 449-463.

[Sa] P. SAMUEL : Some asymptotic properties of powers of ideals, Annals of Math. Serie 2, t. 56 (1955).

[Se] J.P. SERRE : Algèbre locale et multiplicités, Springer Lecture Notes, N° 11 (1965).

[Te 1] B. TEISSIER : Cycles évanescents sections planes et conditions de Whitney in "Singularités à Cargèse", 1972, Astérisque N° 7-8 (1973).

[Te 2] B. TEISSIER : Cycles évanescents et résolution simultanée, I et II, in "Séminaire sur les singularités des surfaces 1976-77", Springer Lecture Notes N° 777 (1980).

[Te 3] B. TEISSIER : The hunting of invariants in the geometry of discriminants, Proc. Nordic Summer School "Real and complex singularities", Oslo 1976, Sijthoff and Noordhoff 1977.

[Te 4] B. TEISSIER : Jacobian Newton polyhedra and equisingularity, Proc. R.I.M.S. Conference on Singularities, April 1978, R.I.M.S Publ., Kyoto, Japon, 1978.

[Te 5] B. TEISSIER : Variétés polaires locales et conditions de Whitney, Note aux C.R. Acad. Sc., t. 290 (5 Mai 1980) 799.

[Te 6] B. TEISSIER : Variétés polaires locales : quelques résultats, in "Journées complexes", Nancy 1980, Publ. de l'Institut Elie Cartan, Nancy.

[Te 7] B. TEISSIER : Sur une inégalité à la Minkowski, Annals of Math., 106, 1 (1977) 38-44.

[Te 8] B. TEISSIER : On a Minkowski-type inequality for multiplicities, II, in : "C.P. Ramanujam, a tribute", Tata Institute, Bombay 1978, 347-361.

[Te 9] B. TEISSIER : Du théorème de l'index de Hodge aux inégalités isopérimétriques, Note aux C.R. Acad. Sc. Paris, t. 288 (17 Janvier 1979).

[Te 10] B. TEISSIER : Bonnesen-type inequalities in algebraic geometry, in "Seminar in Differential geometry (S.T. Yau)", Annals of Math. Studies 102, Princeton Univ. Press (1981) 85-105 (à paraître).

[Te 11] B. TEISSIER : Introduction to equisingularity problems, Proc. A.M.S.

Symp. in Pure Math., N° 29, Arcata 1974.

[Te 12] B. TEISSIER : Variétés polaires, I : Invariants polaires des singularités d'hypersurfaces, Inventiones Math., 40, 3 (1977) 267-292.

[Th] R. THOM : Ensembles et morphismes stratifiés, Bull. A.M.S., 75 (1969) 240-284.

[Ve] J.L. VERDIER : Stratifications de Whitney et théorème de Bertini-Sard, Inventiones Math., 36 (1976) 295-312.

[W] H. WHITNEY : Tangents to an analytic variety, Annals of Math., 81 (1964) 496-549.

[Z 1] O. ZARISKI : Foundations of a general theory of equisingularity ... Amer. J. of Math, 101, 2 (1979) 453-514.

[Z 2] O. ZARISKI : Modules de branches planes, Publications du Centre de Mathématiques, Ecole Polytechnique, F-91128 Palaiseau, 1973.

[Z 3] O. ZARISKI : Some open questions in the theory of singularities, Bull. A.M.S., 77, 4 (July 1971) 481-491.

[Z 4] O. ZARISKI : Collected papers, vol. IV : Equisingularity on algebraic varieties, MIT Press 1979.

[Z 5] O. ZARISKI et P. SAMUEL : Commutative algebra, vol. I, Van Nostrand, New York 1960.

***

Centre de Mathématiques de l'Ecole Polytechnique
F 91128 Palaiseau Cedex - France
"Laboratoire Associé au C. N. R. S. No 169"

REGULAR STRATIFICATIONS AND SUFFICIENCY OF JETS

DAVID TROTMAN

When in 1964 [20,21] Whitney defined his regularity conditions (a) and (b) and proved the existence of regular stratifications of analytic varieties, he explained how Thom required that a certain condition be generic so as to ensure stability of transversality. This condition I call (t)-regularity : a transversal to a stratum is locally transverse to neighbouring strata. In the analytic case it is equivalent to (a)-regularity [12] , and thus by [15] equivalent to stability of transversality to the stratification. In Thom's 1964 Bombay paper [11] he used this regularity condition to stratify jet spaces and sketched a proof of the topological stability theorem (an open dense set of smooth maps are topologically stable, now proved in [7]) , as well as the result that except for a set of infinite codimension in jet space, smooth maps between smooth manifolds are locally topologically finitely determined, hence topologically algebraic. In a recent Aarhus preprint [6] Andrew du Plessis proves these theorems using Whitney regular stratifications of spaces and mappings following closely Thom's indications. This is rather simpler than Varchenko's long proof of 1974 [17,18] , which used different arguments. A rather surprising proof that smooth maps are locally topologically finitely determined in general was found by T.-C. Kuo and Y.-C. Lu in 1979 [10] . Their theorem shows how various regularity conditions solve precisely the problem of when a jet is topologically sufficient.

<u>The Kuo-Lu Theorem</u>

Let $z = (z_1(x),\dots,z_p(x)) \in J^r(n,p)$.

Write $F_i(x, \lambda^{(i)}) = z_i(x) + \sum_{|\alpha| = r} \lambda_\alpha^{(i)} x^\alpha \qquad (1 \leq i \leq p)$, where $x^{\alpha} = x_1^{\alpha_1} \dots x_n^{\alpha_n}$, and $|\alpha| = \alpha_1 + \alpha_2 + \dots + \alpha_n$. Let $V_F = \{(x, \lambda) \in \mathbb{R}^n \times \Lambda \mid F_1(x, \lambda^{(1)}) = \dots = F_p(x, \lambda^{(p)}) = 0\}$, where Λ is isomorphic with $\mathbb{R}^N$, for some large N. Let $Y = 0 \times \Lambda$, and $X = V_F - (0 \times \Lambda)$.

Theorem (Kuo-Lu) : (1) z is V-sufficient in $\mathcal{E}^r$ if and only if (X,Y) is Whitney (b)-regular at 0,

(2) z is V-sufficient in $\mathcal{E}^{r+1}$ if and only if (X,Y) is Whitney (a)-regular at 0,

(3) every J^{r+s}-extension ($s \geq 1$) of z is V-sufficient in $\mathcal{E}^{r+s}$ if and only if (X,Y) is (t^s)-regular at 0, also if and only if z has at most a finite number of C^{r+s}-realisations $\{f^i\}$ such that the germs of the $(f^i)^{-1}(0)$ at 0 are topologically distinct.

Corollary : There is a proper subvariety $\Sigma \subset \Lambda$ depending on z such that for all $\lambda \notin \Sigma$, $F(x, \lambda)$ is V-sufficient in $\mathcal{E}^r$.

In the statement of the theorem there are some terms we should define : a jet z is V-sufficient in $\mathcal{E}^{r+s}$ ($s \geq 0$) if for all C^{r+s} realisations f_1, f_2 of z (i.e. $j^r(f_i) = z$, $i = 1,2$), $f_1^{-1}(0)$ and $f_2^{-1}(0)$ are homeomorphic germs. Recall that z is C^0-sufficient if there exists a local homeomorphism h such that $f_1 = f_2 \circ h$. If $p = 1$ the notions of V-sufficiency and C^0-sufficiency coincide [2]. Given a pair of disjoint smooth submanifolds X, Y in $\mathbb{R}^n$ we say (X,Y) is (t^s)-regular ($s \geq 1$) at 0 in Y if every submanifold S of class C^s transverse to Y at 0 is transverse to X near 0. Note again that (a)-regularity and (t^1)-regularity are equivalent here - as V_F is an algebraic variety - by Trotman [12].

Observe the relation of transversals to Y with realisations of z : a C^{r+s}-realisation f of a jet z in $J^r(n,p)$ can be written

$$f(x) = z + \sum_{|\alpha| = r} \lambda_{\alpha}(x)\, x^{\alpha}$$

where $\{\lambda_\alpha(x)\}$ are C^s functions of x, so giving a C^s map

$$\mathbb{R}^n \longrightarrow \Lambda \quad ,$$
$$x \longmapsto \{\lambda_\alpha(x)\}$$

whose graph in $\mathbb{R}^n \times \Lambda$ is a C^s submanifold, which for $s \geqslant 1$ is transverse to $Y = 0 \times \Lambda$.

At Dijon [14] in June 1978 I conjectured that (t^s) was equivalent with (h^s) : given C^s transversals S_1, S_2 to Y at 0, the germs at 0 of $S_1 \cap X$ and $S_2 \cap X$ are homeomorphic. This is true only when $s = 1$. That (t^s) is a consequence of (h^s) was proved in my thesis [13] for all $s \geqslant 1$ and the proof that (t^1) implies (h^1) will appear in [16]. These two implications are for arbitrary smooth stratified sets – no use is made of curve selection. For $s \geqslant 2$, (t^s) does not imply (h^s), as will be shown by the example below. It does still seem plausible though that (t^s) implies that the number of topological types of germs $X \cap S$ for S transverse to Y at 0 be finite ; the converse is quite easy to obtain by a slight addition to the proof [13] that (h^s) implies (t^s). We remark that the proof that (t^1) implies (h^1) for smooth stratified sets is rather subtle, but the following theorem for subanalytic sets is perhaps more immediately useful.

Theorem. *Let X, Y be subanalytic C^1 submanifolds of $\mathbb{R}^n$ with $0 \in Y = \overline{X} - X$. The following conditions are equivalent :*

(a) *the pair (X,Y) is Whitney (a)-regular at 0,*

(t^1) *every C^1 submanifold transverse to Y at 0 is transverse to X in some neighbourhood of 0,*

(h^1) *given C^1 submanifolds S_1, S_2 transverse to Y at 0, the germs at 0 of $S_1 \cap X$ and $S_2 \cap X$ are homeomorphic.*

Proof : That (a) and (t^1) are equivalent is proved by Trotman [12]; (h^1) implies (t^1) by Theorem 2.11 of my thesis [13]. Kuo [9] showed that (a) implies (h^∞), and via the smoothing lemma below we can deduce that (a) implies (h^1). Thus the theorem is proved.

Smoothing Lemma. Let X, Y be disjoint C^1 submanifolds of $\mathbb{R}^n$ with $0 \in Y \cap \bar{X}$. Let S_1, S_2 be C^1 submanifolds of $\mathbb{R}^n$ containing 0, such that for $i = 1, 2$, S_i is transverse to Y at 0, S_i is transverse to X near 0, and $\dim S_i = n - \dim Y$. Suppose that S_1 and S_2 are in general position, i.e. $T_0S_1 + T_0S_2 = \mathbb{R}^n$ if $2s \geq n$, and $T_0S_1 \cap T_0S_2 = \{0\}$ if $2s \leq n$.

Then there exists a C^1 diffeomorphism $\phi : (U,0) \longrightarrow (U',0)$, isotopic to the identity, where U, U' are neighbourhoods of 0 in $\mathbb{R}^n$, satisfying

1) $\phi(U \cap S_i) = U' \cap T_0S_i$, $(i = 1,2)$
2) $\phi(U \cap Y) = U' \cap T_0Y$,
3) $d\phi(0) = I_n$, the identity matrix,
4) $\phi(X - (S_1 \cap S_2 \cap X))$ is a C^∞ submanifold of $\mathbb{R}^n$.

If we further suppose that $(S_1 \cap S_2)$ is transverse to X near 0 we can make $\phi(X)$ a C^∞ submanifold.

The proof of the smoothing lemma will appear in [16], which contains a more general version, with many strata incident to Y. However the implication (a) $\Longrightarrow$ (h^1) is so far unproved when there is more than one incident stratum. The other implications of the theorem require no extra arguments when there are several incident strata.

An interesting consequence of the theorem above is that (a)-regularity is a necessary condition for the stability of the topological type of transversal intersections with a stratification. This is related to the question raised explicitly by Mather in 1970 [24], of finding the correct stratification of jet space for topological stability. Calculations by the Liverpool Group (Bedford, Bruce, Giblin, Gibson and Wall) show that the canonical (b)-regular stratification of [7] is finer than the stratification for topological stability. Even the first (lowest codimension) unimodal families are not canonical strata — this was shown by Bruce [3] for $\tilde{E}_6$, by Bruce and Giblin [5] for $\tilde{E}_7$, and by Wirthmüller [22] and Bruce [4] (using results of

Greuel and Hironaka) for $\tilde{E}_8$. In each case there is a stratification which is not (b)-regular although (a)-regularity and topological triviality hold. It will be useful to have a version of the above theorem with many incident strata as well as versions adapted to the jet space situation.

We shall conclude by describing a previously unpublished example which shows that (t^s)-regularity does not imply (h^s)-regularity when $s \geqslant 2$, and also that (t^2) and (t^1) are distinct conditions. This example, which arose out of a construction of Kuo and myself, together with observations of Koike and Kucharz [8] as described below, led to the Kuo-Lu theorem (of which we have given an extract at the beginning of this paper) characterising sufficiency in terms of regularity conditions on a pair of incident strata.

Example : Early in 1979 [8] S. Koike and W. Kucharz showed that although the 6-jet $x^3 - 3xy^5$ has uncountably many C^7 realisations (which follows from a theorem of Bochnak and Kuo [1] , since the jet is not V-sufficient in $\mathcal{E}^7$) it has precisely two C^8-realisations, and not one as had been conjectured by Thom.

In January 1979 , the author and Tzee-Char Kuo saw how to construct a semialgebraic example of a (t^2)-regular (h^2)-fault. Choose coordinates x , y , z in $\mathbb{R}^3$ and let Y be the y-axis. Let X be the union of the following five semialgebraic pieces :

$$X_1 = \{x = 0 , y^2 \leq z^3 , z > 0\} ,$$
$$X_2 = \{x < 0 , y^2 \leq z^3, x^2 = z^3\} \cup \{x^2 = z^3, x < 0, y < 0, y^2 > z^3\},$$
$$X_3 = \{x > 0 , y^2 \leq z^3, x^2 = z^3\} \cup \{x^2 = z^3, x > 0, y > 0, y^2 > z^3\},$$
$$X_4 = \{(x^2 + y^2 + z^3)^2 = 4z^3(x - y)^2 , y > 0 , z > 0 , y^2 > z^3\} ,$$
$$X_5 = \{(x^2 + y^2 + z^3)^2 = 4z^3(x - y)^2 , y < 0 , z > 0 , y^2 > z^3\} .$$

Then X is a C^1 submanifold of $\mathbb{R}^3$ and $X \cup Y$ is homeomorphic with a closed half-plane; see the figure overleaf.

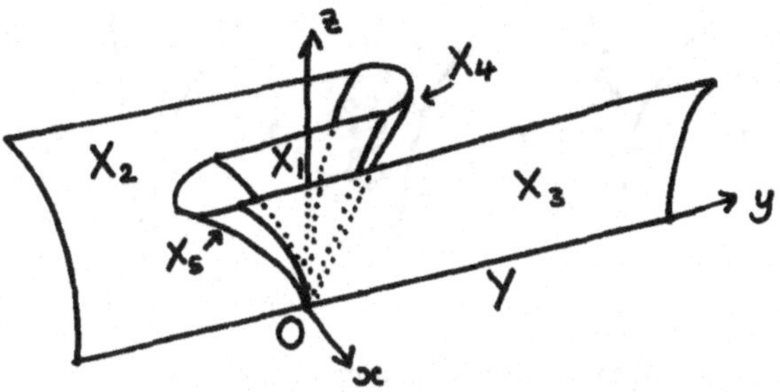

Inspection of the figure shows that (h^s) is not satisfied for any s, but (t^2) is satisfied : It is easy to check that a C^2 submanifold S transverse to Y at 0 will miss X_4 and X_5 in some neighbourhood U_1 of 0. Also because $\{x = 0\}$ is the only limiting tangent plane to X_1, X_2, X_3 at 0, there will be some neighbourhood U_2 of 0 on which S meets X_1, X_2 and X_3 transversely (if at all). Thus S will be transverse to X on $U = U_1 \cap U_2$. However (t^1) fails, because we may intersect X_5 with the surface $\{x^2 = z^3/2\}$ giving an arc α, such that $\overline{\alpha} = \alpha \cup \{0\}$ is a 1-dimensional C^1 submanifold-with-boundary of $\mathbb{R}^3$, and then we may choose a transversal S_α to Y containing $\overline{\alpha}$ and tangent to X_5 on α.

What are the possible topological types of $X \cap S$ for S a C^2 submanifold transverse to Y ?

Answer :  and  (two types).

What are the possible topological types of $X \cap S$ for C^1 transversals S ?

Answer : 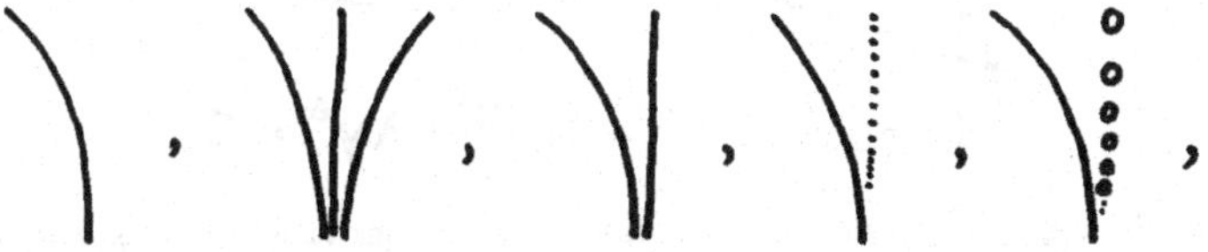

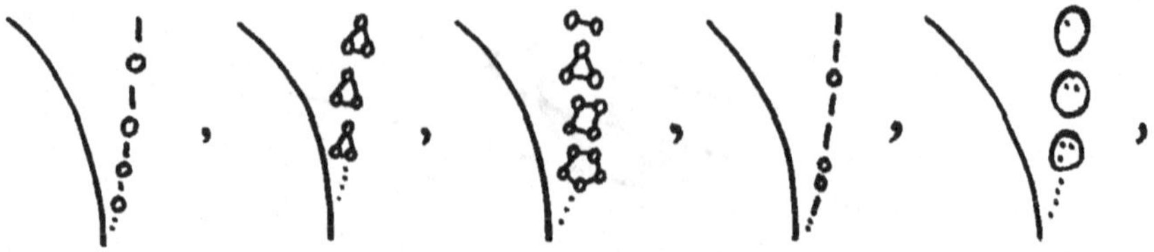

... et cetera. ("Comic Cuts")

In fact one can obtain, for the germ of $X \cap S$ at 0, all sequences $\{C_i\}_{i=1}^{\infty}$ where C_i is a compact (connected) subset of $\mathbb{R}^2$. This is easily seen to yield an uncountable number of topological types.

Strangely, it turned out that the figure representing the semialgebraic example given above very nearly represents the surface

$$V = \{x^3 - 3xy^5 + \lambda y^6 = 0\} \subset \mathbb{R}^2 \times \mathbb{R} ,$$

that is, the deformation of the jet $x^3 - 3xy^5$ (studied by Koike and Kucharz) along the y^6 direction in J^6. In fact this is the only important part of the whole deformation by monomials in H^6 used by Kuo and Lu in their theorem : $x^3 - 3xy^5 + \sum_{\alpha_1 + \alpha_2 = 6} \lambda_\alpha x^{\alpha_1} y^{\alpha_2}$. For justification of this assertion see Siersma's 1974 Amsterdam Thesis .

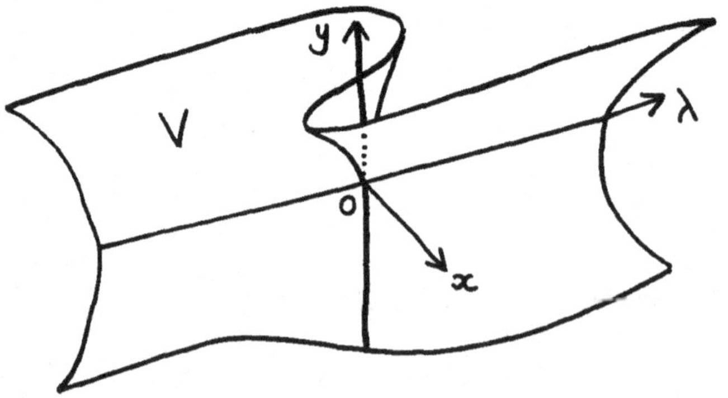

Figure: $x^3 - 3xy^5 + \lambda y^6 = 0$.

V may be shown to have the same properties as the semialgebraic example found jointly with Kuo: (t^2) holds, but (t^1) and (h^s) fail, for all s .

Thus we have a geometric explanation of the Koike-Kucharz phenomenon, via the Kuo-Lu theorem.

Similar examples show that (t^s) does not imply (h^s) for $s \geqslant 3$.

Addendum : Recently Satoshi Koike [22] has proved, for a jet z in $J^r(n,p)$, the equivalence of V-sufficiency in $\mathcal{E}_{r+s}$ $(s \geqslant 1)$ and (h^s)-regularity of $V_F - 0 \times \Lambda$ over $0 \times \Lambda$ (notation as in the Kuo-Lu theorem).

References

1. J. Bochnak and T.-C. Kuo, Different realisations of a non-sufficient jet, Indag. Math. 34 (1972), pp. 24-31.
2. J. Bochnak and S. Lojasiewicz, A converse of the Kuiper-Kuo theorem, Proceedings of Liverpool Singularities Sýmposium I, Lecture Notes in Mathematics 192, Springer, Berlin (1971), pp.254-261.
3. J. W. Bruce, A stratification of the space of cubic surfaces, Math. Proc. Camb. Phil. Soc. 87 (1980), pp.427-441.
4. J. W. Bruce, On the canonical stratification of complex analytic functions, Bull. London Math. Soc. 12 (1980), pp.111-114.
5. J. W. Bruce and P. J. Giblin, A stratification of the space of plane quartic curves, Proc. London Math. Soc., to appear.
6. A. A. du Plessis, On the genericity of topologically finitely-determined map-germs, Aarhus preprint (1980), to appear.
7. C. G. Gibson, K. Wirthmüller, A. A. du Plessis and E. J. N. Looijenga, Topological stability of smooth mappings, Lecture Notes in Mathematics 552, Springer, Berlin(1976).
8. S. Koike and W. Kucharz, Sur les réalisations de jets non-suffisants, C. R. Acad. Sci. Paris, Série A, 288 (1979), pp.457-459.
9. T.-C. Kuo, On Thom-Whitney stratification theory, Math. Annalen 234 (1978), pp.97-107.
10. T.-C. Kuo and Y.-C. Lu, Sufficiency of jets via stratification theory, Inventiones Math. 57 (1980), pp.219-226.

11. R. Thom, Local topological properties of differentiable mappings, Differential analysis, Oxford Univ. Press, London (1964), pp.191-202.

12. D. J. A. Trotman, A transversality property weaker than Whitney (a)-regularity, Bull. London Math. Soc. 8 (1976), pp.225-228.

13. D. J. A. Trotman, Whitney stratifications : faults and detectors, Warwick Thesis (1977), 93 pages.

14. D. J. A. Trotman, Interprétations topologiques des conditions de Whitney, Journées singulières de Dijon, Astérisque 59-60 (1979), pp.233-248.

15. D. J. A. Trotman, Stability of transversality to a stratification implies Whitney (a)-regularity, Inventiones Math. 50 (1979), pp.273-277.

16. D. J. A. Trotman, Transverse transversals and homeomorphic transversals, to appear.

17. A. N. Varchenko, Local topological properties of analytic mappings, Math. USSR-Izv. 7 (1973), pp. 883-917.

18. A. N. Varchenko, Local topological properties of differentiable mappings, Math. USSR-Izv. 8 (1974), pp. 1033-1082.

19. K. Wirthmüller, Universell Topologische Triviale Deformationen, Dissertation Regensburg.

20. H. Whitney, Local properties of analytic varieties, Diff. and Comb. Topology (ed. S. Cairns), Princeton (1965), pp.205-244.

21. H. Whitney, Tangents to an analytic variety, Ann. of Math. 81 (1965), pp.496-549.

22. S. Koike, On V-sufficiency and ($\overline{h}$)-regularity, Kyoto Univ. preprint (1980), 20 pages.

23. J. Mather, Problem section, Manifolds, Amsterdam 1970, Lecture Notes in Math. 197, Springer, Berlin (1971).

Faculté des Sciences,
Boulevard Lavoisier,
49045 ANGERS CEDEX,
France.